Virulence Mechanisms of Bacterial Pathogens

2nd EDITION

Virulence Mechanisms of Bacterial Pathogens

2nd EDITION

Edited by

James A. Roth
Center for Immunity Enhancement in Domestic Animals,
College of Veterinary Medicine,
Iowa State University, Ames, Iowa

Carole A. Bolin
U.S. Department of Agriculture,
Agricultural Research Service,
National Animal Disease Center,
Ames, Iowa

Kim A. Brogden
U.S. Department of Agriculture,
Agricultural Research Service,
National Animal Disease Center,
Ames, Iowa

F. Chris Minion
Veterinary Medical Research Institute,
Iowa State University, Ames, Iowa

Michael J. Wannemuehler
Veterinary Medical Research Institute,
Iowa State University, Ames, Iowa

ASM Press • *Washington, D.C.*

Copyright © 1995 American Society for Microbiology
1325 Massachusetts Ave., N.W.
Washington, DC 20005

Library of Congress Cataloging-in-Publication Data

Virulence mechanisms of bacterial pathogens / edited by James A. Roth
. . . [et al.. — 2nd ed.
    p.   cm.
"This volume resulted from the International Symposium on
Virulence Mechanisms of Bacterial Pathogens held in Ames, Iowa, June
6 to 8, 1994"—Acknowledgments.
    Includes index.
    ISBN 1-55581-085-3
    1. Virulence (Microbiology)—Congresses.   2. Pathogenic bacteria—
Congresses.   I. Roth, James A.   II. International Symposium on
Virulence Mechanisms of Bacterial Pathogens (1994 : Ames, Iowa)
    [DNLM: 1. Bacteria—pathogenicity—congresses.   2. Virulence—
congresses.   3. Bacterial Infections—physiopathology—congresses.
4. Bacterial Infections—prevention & control—congresses.
5. Bacterial Toxins—congresses.   QW 730 V821 1995]
QR175.V57   1995
616′.014—dc20
DNLM/DLC
for Library of Congress   95-2190
                                                            CIP

Cover photo: Enteropathogenic *Escherichia coli* adhering to the surface of human HeLa cells
and causing rearrangement of HeLa epithelial cell surfaces beneath adherent bacteria.
Photo is courtesy of B. Finlay and I. Rosenshine.

CONTENTS

V. STRATEGIES TO OVERCOME BACTERIAL VIRULENCE MECHANISMS
Section Editor: James A. Roth

VI. PAST, PRESENT, AND FUTURE STUDIES

ORGANIZING COMMITTEE

James A. Roth (Chair)
Center for Immunity Enhancement in Domestic Animals, Iowa State University,
Ames, IA 50011

Carole A. Bolin
U.S. Department of Agriculture, Agricultural Research Service, National Animal Disease
Center, P.O. Box 70, Ames, IA 50010

Kim A. Brogden
U.S. Department of Agriculture, Agricultural Research Service, National Animal Disease
Center, P.O. Box 70, Ames, IA 50010

Dawne Buhrow
Center for Immunity Enhancement in Domestic Animals, Iowa State University,
Ames, IA 50011

Thomas O. Bunn
U.S. Department of Agriculture, Animal and Plant Health Inspection Service,
National Veterinary Services Laboratories, P.O. Box 844, Ames, IA 50010

Nancy E. Clough
U.S. Department of Agriculture, Animal and Plant Health Inspection Service,
National Veterinary Services Laboratories, P.O. Box 844, Ames, IA 50010

Jane Galyon
Center for Immunity Enhancement in Domestic Animals, Iowa State University,
Ames, IA 50011

Shirley M. Halling
U.S. Department of Agriculture, Agricultural Research Service, National Animal Disease
Center, P.O. Box 70, Ames, IA 50010

Harley W. Moon
U.S. Department of Agriculture, Agricultural Research Service, National Animal Disease
Center, P.O. Box 70, Ames, IA 50010

F. Chris Minion
Veterinary Medical Research Institute, Iowa State University, Ames, IA 50011

Richard F. Ross
College of Veterinary Medicine, Iowa State University, Ames, IA 50011

Thaddeus B. Stanton
U.S. Department of Agriculture, Agricultural Research Service, National Animal Disease Center, P.O. Box 70, Ames, IA 50010

Kirsten L. Vadheim
U.S. Department of Agriculture, Agricultural Research Service, National Animal Disease Center, P.O. Box 70, Ames, IA 50010

Michael J. Wannemuehler
Veterinary Medical Research Institute, Iowa State University, Ames, IA 50011

AUTHORS

Anthony C. Allison
DAWA, Inc., Belmont, CA 94002

Joseph T. Barbieri
Department of Microbiology, Medical College of Wisconsin, Milwaukee, WI 53226

Andreas J. Bäumler
Department of Molecular Microbiology and Immunology, Oregon Health Sciences
University, 3181 S.W. Sam Jackson Park Road, Portland, OR 97201-3098

Paul J. Bertics
Department of Medical Microbiology and Immunology, 407 SMI, Madison, WI 53706

Nirupama Bhatnagar
Food and Drug Administration, Bethesda, MD 20892

Noel K. Childers
Department of Community and Public Health Dentistry, School of Dentistry,
University of Alabama at Birmingham, Birmingham, AL 35295

Barry C. Cole
Division of Rheumatology, Department of Internal Medicine, University of Utah Medical
Center, Salt Lake City, UT 84132

Diana R. Cundell
Laboratory of Molecular Infectious Diseases, Rockefeller University, 1230 York Avenue,
New York, NY 10021

Charles J. Czuprynski
Department of Pathobiological Sciences, School of Veterinary Medicine,
University of Wisconsin, Madison, WI 53706

Loren C. Denlinger
Department of Medical Microbiology and Immunology, 407 SMI, Madison, WI 53706

Mark T. Dertzbaugh
Applied Research Division, U.S. Army Medical Research Institute of Infectious Disease, Ft. Detrick, MD 21702-5011

Elsie M. Eugui
DAWA, Inc., Belmont, CA 94002

B. Brett Finlay
Biotechnology Laboratory and the Departments of Biochemistry and Molecular Biology and Microbiology and Immunology, University of British Columbia, Vancouver, British Columbia, Canada V6T 1Z3

Anne H. Fortier
EntreMed, Inc., Rockville, MD 20850

Jorge E. Galán
Department of Molecular Genetics and Microbiology, School of Medicine, State University of New York at Stony Brook, Stony Brook, NY 11794-5222

Fred Heffron
Department of Molecular Microbiology and Immunology, Oregon Health Sciences University, 3181 S.W. Sam Jackson Park Road, Portland, OR 97201-3098

Barbara Iglewski
Department of Microbiology and Immunology, University of Rochester Medical Center, Rochester, NY 14642

Kevin L. Knudtson
Division of Rheumatology, Department of Internal Medicine, University of Utah Medical Center, Salt Lake City, UT 84132

James L. Krahenbuhl
Laboratory Research Branch, G.W. Long Hansen's Disease Center, Louisiana State University, Baton Rouge, LA 25072

Kathleen Krueger
Department of Microbiology, Medical College of Wisconsin, Milwaukee, WI 53226

Suzanne M. Michalek
Department of Microbiology, University of Alabama at Birmingham, Birmingham, AL 35294-2170

Carol A. Nacy
EntreMed, Inc., Rockville, MD 20850

Luciano Passador
Department of Microbiology and Immunology, University of Rochester Medical Center, Box 672, Rochester, NY 14642

Tammy Polsinelli
Walter Reed Army Institute of Research, Washington, DC 20307-5100

Richard A. Proctor
Department of Medical Microbiology and Immunology, 407 SMI, Madison, WI 53706

Allen D. Sawitzke
Division of Rheumatology, Department of Internal Medicine, University of Utah Medical Center, Salt Lake City, UT 84132

Annette Siebers
Biotechnology Laboratory and the Departments of Biochemistry and Molecular Biology and Microbiology and Immunology, University of British Columbia, Vancouver, British Columbia, Canada V6T 1Z3

H. Smith
The Medical School, University of Birmingham, Edgebaston, Birmingham B15 2TT, United Kingdom

Peter W. Taylor
Ciba Pharmaceuticals, Horsham, West Sussex RH12 4AB, United Kingdom

Elaine I. Tuomanen
Laboratory of Molecular Infectious Diseases, Rockefeller University, New York, NY 10021

Rodney K. Tweten
Department of Microbiology and Immunology, University of Oklahoma Health Sciences Center, Oklahoma City, OK 73190

Eugene D. Weinberg
Department of Biology and Program in Medical Science, Indiana University, Bloomington, IN 47405

Rodney A. Welch
Department of Medical Microbiology and Immunology, University of Wisconsin—Madison Medical School, Madison, WI 53706

Kenneth H. Wilson
Division of Infectious Diseases, VA Medical Center and Duke University Medical Center, Durham, NC 27705

Michael K. Zierler
Department of Molecular Genetics and Microbiology, School of Medicine, State University of New York at Stony Brook, Stony Brook, NY 11794-5222

PREFACE

Remarkable advances have been made in understanding virulence mechanisms of bacterial pathogens since the first edition of this monograph appeared in 1988. The tools of molecular biology have been used to great advantage to further elucidate the molecular basis for specific aspects of bacterial virulence. This new information has also generated a whole new set of questions to be answered. It is a continual challenge to integrate the knowledge of the molecular basis of virulence mechanisms with the knowledge of clinical aspects of disease in order to understand the pathogenesis of infectious diseases.

The first edition of this monograph was well received and the Organizing Committee was very pleased with the original authors and their coverage of the topics. However, rather than ask those authors to update their chapters, we decided to identify, for the most part, a new set of authorities in the field to present their perspectives on the subjects covered. Therefore, this monograph should perhaps be considered volume 2 rather than a second edition of the original monograph. In selecting the authors for this monograph, the Organizing Committee sought out individuals that we perceived to be internationally recognized authorities on the topics they were asked to cover. We asked them to place an emphasis on conveying an understanding of the mechanisms of host-pathogen interactions rather than on presenting current research approaches and results.

Our goal was to produce a monograph that would be a useful source of information for molecular biologists wanting an understanding of how molecular mechanisms relate to the disease process, infectious disease specialists wanting a more thorough understanding of the cellular and molecular basis of pathogenesis, researchers attempting to elucidate pathogenic mechanisms in bacterial diseases that are not yet well characterized, industry scientists wishing to identify promising approaches to disease prevention and therapy, and faculty and graduate students wishing to gain an overview of the subject.

The Organizing Committee did decide to ask Harry Smith to update his summary chapter. For both meetings and for both monographs we asked him to provide his perspective on the state and future of studies on bacterial pathogenesis. His many years of experience and his important contributions to the understanding of bacterial virulence make him well qualified to provide insight and perspective on the current state and the future of studies on bacterial pathogenesis. His conclusion in 1994 was the same as that in 1987, "Our subject has made great progress and is in good heart. There is much to do in the future. What more could we wish?"

JAMES A. ROTH
December 1994

ACKNOWLEDGMENTS

This volume resulted from the International Symposium on Virulence Mechanisms of Bacterial Pathogens held in Ames, Iowa, June 6 to 8, 1994. The symposium was sponsored and supported by the following organizations.

Center for Immunity Enhancement in Domestic Animals
USDA ARS National Animal Disease Center
USDA APHIS National Veterinary Services Laboratories
Iowa State University Biotechnology Program
Iowa State University College of Veterinary Medicine

The Organizing Committee thanks the following for their generous financial support of the symposium.

Sponsors of specific events
Sanofi Animal Health, Overland Park, Kansas (Poster Session)
Hoffman-LaRoche, Inc., Nutley, New Jersey (Session I)

Supporters
Synbiotics Corporation Animal Health Division, San Diego, California
Boehringer Ingelheim Animal Health, St. Joseph, Missouri

Contributors
NOBL Laboratories, Sioux Center, Iowa
Ft. Dodge Laboratories, Ft. Dodge, Iowa
Schering-Plough Animal Health, Union, New Jersey
Pioneer Hi-Bred International, Inc., Microbial Genetics Division, Johnston, Iowa
Mallinckrodt Veterinary, Inc., Mundelein, Illinois

Other sponsors
USDA National Research Initiative Competitive Grants Program

BACTERIAL ADHERENCE, COLONIZATION, AND INVASION ON MUCOSAL SURFACES

ATTACHMENT AND INTERACTION OF BACTERIA AT RESPIRATORY MUCOSAL SURFACES

Diana R. Cundell and E. Tuomanen

There are many similarities between normal eukaryotic cell trafficking and the processes underlying bacterial pathogenesis. It is obvious that both require recognition of a destination by a cell surface-encoded address. However, it is equally true that both involve a series of actions and reactions between the adherent partners. The characteristics of a eukaryotic cell attached to a tissue matrix are different from those of the same cell unattached. This difference is dramatically demonstrated by tumor cells that acquire the ability to metastasize. Attachment therefore changes the cell. This concept is also true of prokaryotic organisms, such as bacteria, for which attachment is a necessary prerequisite for infection of host cells but is often followed by detachment and relocation to a new niche (43). The association of a bacterium with a eukaryotic cell or substratum changes both the bacterium and the cell to which it is attached in a dynamic and reciprocal fashion. This chapter compares and contrasts the mutual interactions of two pathogens with cells of the respiratory tract in order to illustrate the basic concept of the dynamic nature of reciprocal attachment.

Bacterial adhesion endows the pathogen with the ability to withstand normal host defense cleansing mechanisms on endothelial and mucosal surfaces (5, 43). Unattached organisms are eliminated by sneezing, mucociliary clearance, coughing, and blood flow, and those that are attached must constantly multiply on the tissue surface to avoid removal during normal cell surface turnover (desquamation). Adherence also confers a number of advantages on the bacterium, including enhanced toxicity to the host and increased resistance to deleterious agents (25).

Bacteria employ a number of mechanisms through which they attach to and/or penetrate host tissues. Bacteria adhere only to complementary surfaces, and adherence involves an interaction between structures on the surface of the bacterium (adhesins) and receptors on the substratum. Frequently, multiple ligands on the pathogen's surface serve to increase the strength and specificity of adherence when these ligands are engaged in concert. By targeting structures contained in matrix glycoproteins (53), integral membrane glycoproteins (54), or glycolipids (34), adhesins are proteins that engage in protein-carbohydrate or protein-protein interactions. It remains formally possible that adhesins are carbohydrates that engage cognate carbohydrates, as suggested in

Diana R. Cundell and E. Tuomanen Laboratory of Molecular Infectious Diseases, Rockefeller University, 1230 York Avenue, New York, New York 10021.

Virulence Mechanisms of Bacterial Pathogens, 2nd ed., Edited by J. A. Roth et al.
© 1995 American Society for Microbiology, Washington, D.C.

a few eukaryotic interactions (80), but this type of bacterial adherence has yet to be described.

Adhesins are normally exposed on the outer surface of the cell or borne on appendages such as fimbrae. Bacteria and most biological substrata are thought to be negatively charged (26). The arrangement of adhesins at some distance from the bacterial cell would therefore both assist in overcoming the repulsion forces and allow contact with receptors on the substratum at some distance from the bacterial cell (26).

The presence of a complementary receptor on the substratum does not, however, always equate with the ability of a bacterium to colonize that tissue. For example, *Escherichia coli,* which bears mannose-specific adhesins, does not colonize all mannose-containing substrata (25). This fact suggests that the adhesion process probably involves the correct presentation, orientation, and accessibility of both bacterial adhesins and host tissue receptors. There is a positive correlation between the ability of tissue cells to bind a bacterial pathogen and the susceptibility of the host to that pathogen. For example, *Bordetella pertussis* adheres well to human ciliated cells but not to the same cells from many other mammalian species that do not acquire pertussis (77). In addition to species specificity of infection, susceptibility of an individual within a species may be linked to adherence, presumably mediated by the presentation of specific receptors that are often in the form of blood group antigens (13). Adherence of *E. coli* to epithelial cells from patients with recurrent urinary tract infections may be up to five times greater than adherence to cells from infection-free individuals (6). Similarly, *Streptococcus pneumoniae* isolates from patients with otitis media show a greater propensity to adhere to nasopharyngeal cells than do those from patients with septicemia or meningitis, suggesting that these strains exhibit tissue tropism (3).

Pathogenic bacteria generally have the ability to differentiate among target hosts (species specificity) and among target tissues within the infected host (tissue tropism) (75). Bacteria that target the same tissue may not, however, display the same species specificity or be restricted to the same targets. For example, both the gram-negative organism *B. pertussis* and the gram-positive bacterium *S. pneumoniae* target the lung and produce infections of the lower respiratory tract (8, 69). Natural infection with *B. pertussis* results from adherence of the organism exclusively to human lung ciliated cells and is noninvasive (69). This degree of species and tissue tropism is highly unusual for a lung pathogen, and the lack of invasion during colonization distinguishes it from other pathogens such as *S. pneumoniae* for which mucosal attachment is followed by systemic spread of infection (8). Ciliary pathogens such as *B. pertussis* are few in number, have highly specialized adhesins, and can be specifically toxic for ciliated cells. In contrast, alveolar pathogens such as *S. pneumoniae* commonly cause systemic infections, and their requirements for localization are largely unknown. This chapter compares the very different mechanisms involved in the adherence of *B. pertussis* and *S. pneumoniae* to target host cells and identifies the role of adhesion in infections produced by these bacteria.

B. PERTUSSIS

Pertussis (whooping cough) is a noninvasive, acute respiratory disease of children and adults. Infection is associated with a protracted cough and a circulating leukocytosis. *B. pertussis* establishes infection by attaching itself to ciliated respiratory epithelial cells and remains attached to these cells during the course of the disease. *B. pertussis* also interacts with and can survive within alveolar macrophages, and this ability provides it with an opportunity to prolong the course of infection (60). *Bordetella* species exibit strict species and tissue tropisms. *B. pertussis* infection is limited to humans, whereas other *Bordetella* species produce respiratory disease in animals on a restricted species basis (77). Adherence is based on the highly restricted natural affinity of bordetellae for cilia (76). Electron micrographs have demonstrated that *B. pertussis* most frequently associates with the base of the ciliary shaft (76).

Although the bacterium is confined to the respiratory mucosa, several toxic components are released into the general circulation and account for the systemic manifestations of the disease (leukocytosis, histamine sensitivity, abnormal glucose metabolism, weight loss). Pertussis toxin (PT) is an ADP-ribosylating toxin with the ability to intoxicate a broad range of cells, leading to increased intracellular cyclic AMP and cellular dysfunction (35). An extracellular adenylate cyclase toxin similarly elevates eukaryotic cyclic AMP and distorts cellular functions (23).

The hallmark of pathology in the lung during whooping cough is destruction of the ciliated epithelium. Damage to ciliated cells occurs only when the bacteria are adherent, suggesting that adherence plays a role in directing local cellular toxicity (7). Infected ciliated cells develop a marked disarray and disruption of the ciliary necklace (42). As infection progresses, the ciliary necklace disappears, and the ciliated cells are killed and extruded from the epithelial surface (42). Several other bacteria produce substances that produce ciliary dyskinesia (irregular beat), ciliostasis, and ciliotoxicity (Table 1). The mechanism of *B. pertussis* cytotoxicity has been suggested to involve the generation of a low-molecular-weight (921-Da) tracheal cytotoxin (90) that is released from the bacterial peptidoglycan. This tracheal cytotoxin damages both human nasal (90) and hamster tracheal (22) organ cultures in vitro. The toxin is identical to the 1,6-anhydro disaccharide tetrapeptide (22) that forms

the ends of the interlocking chains of the bacterial cell wall (78). Studies using a hamster tracheal ring in vitro assay to define bioactivity have determined that the toxicity of this molecule is related to its ability to stimulate the generation of interleukin-1 (IL-1) and nitric oxide by epithelial cells (22).

Adherence of *B. pertussis* to Human Respiratory Cilia and Alveolar Macrophages

Most organisms interact with target host cells via fimbriae (43). *B. pertussis* is unusual in this respect. Although pertussis fimbriae bind to tissue-cultured cells, the role of these fimbriae in the pathogenesis of whooping cough is not yet clear (41). However, several possible adhesive molecules have been identified for *B. pertussis*. These include two nonfimbrial secreted proteins that are major virulence determinants: filamentous hemagglutinin (FHA) and PT (58, 74). The natural targets for FHA and PT are ciliated cells and macrophages (73, 74). The interaction between these two adhesins and target host cells is complex and shows unusual sophistication, involving both protein-carbohydrate and protein-protein interactions. In the following sections, the interactions between *B. pertussis* adhesins and target host cells are explored with a view to demonstrating how pathogens can mimic host cell recognition systems.

Identification of FHA and PT as *B. pertussis* Adhesins for Human Cilia

A series of mutants of *B. pertussis*, each deficient in one of several virulence factors, was created by transposon mutagenesis (87, 88). When in-

TABLE 1   Ciliotoxins released by respiratory tract pathogens

Bacterium	Ciliotoxin(s)	Reference
B. pertussis	Tracheal cytotoxin	23, 89
S. pneumoniae	Pneumolysin	65
P. aeruginosa	Pyocyanin, 1-hydroxyphenazine, and rhamnolipid	32, 33
H. influenzae	Lipooligosaccharide	28
	Peptidoglycan fragments	31

cubated with single human ciliated cells obtained by brushing of healthy human trachea during bronchoscopy, mutants unable to produce either FHA or PT were not able to adhere (87), suggesting that FHA and PT are important adhesins. Confirmation of this finding came from a series of reconstitution experiments in which purified FHA and PT were incubated with either the mutant strains or human ciliated cells (73). Two phenomena occurred in parallel: (i) the purified FHA and PT components could be visualized on the surface of the bacteria and cilia by immunofluorescence with anti-FHA or anti-PT antibodies, and (ii) importantly, nonadhering *B. pertussis* strains acquired normal adhesive capacity (87). The reconstitution of adherence by exogenous FHA and PT was dose dependent and specific and could be prevented by anti-FHA or anti-PT antibodies (73). These results together suggest that secreted FHA and PT can bind to the surfaces of both bacteria and cilia and form a stable bridging interaction.

The behavior of the soluble FHA and PT in vivo has been confirmed (74). In rabbits, virulent strains producing both adhesins localize to cilia (74). Mutants that lack both adhesins are cleared from the respiratory tract without inducing any lesions. Interestingly, mutants lacking only one adhesin also fail to attach to cilia but pass into the alveoli and produce different pathologies depending on their phenotype: FHA⁻ PT⁺ organisms produce pulmonary hemorrhage, and FHA⁺ PT⁻ organisms cause interstitial edema (74).

The ability of *B. pertussis* to secrete its adhesins into the environment and then recapture them is remarkable from a microbiologic standpoint and is not yet fully understood. Many other bacteria actively secrete surface components, but a secreted adhesin is normally considered lost to the function of adherence (37). Secreted *B. pertussis* adhesins remain functional when recaptured onto the bacterial surface. A model of the complex interaction among *B. pertussis,* its adhesins, and target human cells is represented in Fig. 1 and is discussed in detail in the following sections.

Structure and Functional Characteristics of *B. pertussis* Adhesin FHA

FHA is a 220-kDa secreted protein that contains several epitopes capable of recognizing receptors on host cells (Fig. 1). These include an N-terminal heparin-binding domain that binds sulfated polysaccharides and may be involved in hemagglutination (40), an N-terminal lectin domain that binds sialic acid and is involved in hemagglutination (38), a lectin domain for ciliated cells (46), a domain containing an RGD (arginine-glycine-asparagine) sequence that binds leukocyte integrin CR3 (48), and two regions that mimic binding sites on factor X of the coagulation cascade and bind to leukocyte CR3 (52, 57). Two regions of FHA also show approximately 30% sequence similarity to keratin and elastin (57).

The structural complexity of FHA is unusual for bacterial adhesins but is similar to features of eukaryotic adhesins (57) and extracellular binding proteins such as fibronectin (40). Because of its structural complexity, FHA can interact with a number of receptors on the host cell surface: ciliary glycoconjugates through the lectin domain (46), glycoconjugates in respiratory mucus by the heparin-binding domain (40), erythrocytes through the heparin-binding and N-terminal lectin domains (38, 40), and leukocyte integrins through the RGD and factor X-like domains (60).

Structure and Functional Characteristics of *B. pertussis* Adhesin PT

PT is a hexameric 105-kDa protein with an A-B architecture (57). The A protomer is composed of a single S1 subunit (26 kDa) containing the catalytic site for toxic ADP-ribosylation of cellular signal-transducing guanine nucleotide proteins. The B protomer possesses the cellular recognition domains; it is a pentamer consisting of four different subunits: S2 (22 kDa), S3 (22 kDa), S4 (two units, each 12 kDa), and S5 (11 kDa). By interacting with glycoproteins and glycolipids on different types of eukaryotic cells, the B oligomer is able to

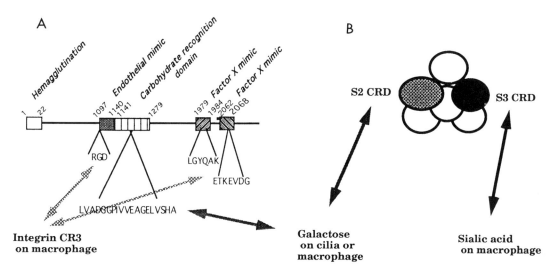

FIGURE 1 Central role played by FHA and PT in *B. pertussis* interactions with target host cells. (A) FHA is a large protein with multiple cell attachment domains. Two regions target the CR3 integrin (CD11b/CD18) on leukocytes. Two other regions target carbohydrates on mucus or ciliated cells. (B) PT presents carbohydrate recognition domains (CRD) in two subunits in its B oligomer for the purpose of cell recognition. S2 targets ciliated cells, and S3 targets macrophages.

produce its diverse effects: mitogenicity of T cells (90), attachment to eukaryotic cells (21, 57, 59, 73, 84), and delivery of the S1 unit to its target. Like FHA, PT possesses lectin activity (21). This activity appears to be located in the S2 and S3 subunits, which bind lactosamine-containing glycoconjugates and α2-6-linked sialic acid, respectively (Fig. 1) (21, 57, 59, 84). Despite >80% sequence homology between the S2 and S3 subunits, only S2 recognizes cilia (59), whereas S3 is involved in adherence to human macrophages (84). Interestingly, mutational exchange of amino acid residues 37 through 52 between S2 and S3 exchanges the lactosylceramide and ganglioside binding and, concomitantly, the recognition of macrophages versus ciliated cells (59). The S2 subunit affinity for lactose is lost by alterations in the amino acid sequence between residues 43 and 45 and residues 44 and 51 or by the introduction of proline at position 50 or 51 (59). As discussed below, this region of S2 and S3 has sequence similarity to mammalian C-type lectins.

A second lectin domain is found at the far amino-terminal regions (residues 18 through 23) of both S2 and S3. These regions have features of the *N*-acetylglucosamine (GlcNAc)-binding wheat germ agglutinin (WGA) lectin pocket at residues 62 through 67 (57). Mutation of this WGA-like region, the carboxy-terminal Asn-93 or Asn-105 of S2 or the Lys-105 of S3, results in abolition of all carbohydrate recognition (21, 57). The recent solution of the crystal structure of a dimeric form of PT supports the contention that the amino termini of S2 and S3 form composites of a plant (WGA-like) and an animal (selectin-like) lectin arranged in tandem (57).

Adherence of *B. pertussis* to Human Cilia: Role of Adhesin Lectin Domains

Although the precise structure and number of glycoconjugate receptors for *B. pertussis* on human cilia remain to be determined, the minimal receptor unit has been determined, on the basis of inhibition of adherence in vitro by use of haptens, lectins, and anticarbohydrate antibodies to be galactose-*N*-acetylglucosamine (Gal-GlcNAc) (75, 79). Interestingly, this same oligosaccharide sequence is also present in alveolar glycoprotein, which may therefore represent a natural competitive inhibitor of *B. pertussis* adherence in alveoli. Importantly, this sequence differs from that of *N*-acetylgalactosamine (GalNAc), an effective competitive inhibitor of adherence of *Bordetella bronchiseptica* to hamster trachea (45), suggesting a difference in structure between human and animal receptors. In addition, antibodies or lectins directed against Lewis blood group determinants are able to block the adherence of *B. pertussis* to human cilia (79). These reagents target Gal-GlcNAc residues substituted with fucose, an additional clue to receptor structure.

The interaction between *B. pertussis* and Gal-GlcNAc structures appears to be mediated through both FHA and PT, since, like the whole bacterium, both purified adhesins adhere to immobilized purified lactosylceramide and nonsialylated glycolipids prepared from ciliated cells on thin-layer chromatography plates (59, 79). Peptides derived from the lectin domains of FHA (residues 1141 through 1279) and PT subunit S2 (residues 40 through 54) also bind to lactosylceramide (46, 59). Importantly, comparison of the amino acid sequences of these domains revealed 20-amino-acid regions of significant similarity (57). This observation indicates that *B. pertussis* might use similar epitopes in two adhesins for recognition of a common carbohydrate specificity present in its two adhesins (57).

Piracy of *B. pertussis* FHA and PT Adhesins by Other Respiratory Tract Pathogens

B. pertussis infection is frequently associated with secondary infections caused by *Haemophilus influenzae, S. pneumoniae,* and *Staphylococcus aureus* (68). Adherence of these other respiratory tract pathogens to human cilia in vitro is normally low; e.g., *H. influenzae* demonstrates a low level of adherence, and *S. pneumoniae* and *S. aureus* are nonadherent (68). In contrast, these bacteria adhere remarkably well to human cilia pretreated with *B. pertussis* toxin FHA or PT (40). The presence of *B. pertussis*

adhesins also enhances the survival of unencapsulated pneumococci instilled into the lungs of rabbits (68). These results together indicate that heterologous species of bacteria can bind to *B. pertussis* adhesins and that such "piracy" may contribute to superinfection in mucosal diseases such as whooping cough.

Internalization of *B. pertussis* by Macrophages: Interactions with Integrins and Selectins

FHA and PT are also the adhesins that mediate *B. pertussis* attachment to alveolar macrophages (46, 57, 59, 84). However, while adherence to cilia appears to be a simple process of tissue tropism in which bacterial lectin domains interact with surface host cell glycoconjugate receptors (79), adherence of the bacteria to macrophages is more complex. The bacteria appear to use selectin-like components of PT to prime the macrophages for enhanced intracellular uptake mediated by binding between FHA and the integrin CD11b/CD18 (84).

FHA and the CD11b/CD18 Integrin

FHA mediates adherence of intact *B. pertussis* to macrophages by ligating two classes of surface molecules (48). One as yet unidentified receptor resembles that found on cilia in that *B. pertussis* binding is inhibitable by galactose (48) and is mediated by the lectin-binding domain of the molecule, which recognizes lactosylceramide (46). The second binding site functions in a carbohydrate-independent, protein-protein fashion and accounts for approximately half the binding of bacteria to leukocytes (48). This receptor has been identified by several criteria as the integrin CD11b/CD18 (CR3, $\alpha_M\beta_2$) (48, 60). The interaction of *B. pertussis* with this leukocyte-restricted integrin is discussed here in more detail to illustrate the different possible outcomes of a bacterial interaction with a eukaryotic cell adhesion molecule.

At least three strategies by which pathogenic bacteria ligate eukaryotic integrins have been described; these involve lectin binding, masking, and mimicry (75). In lectin binding, a bacterial protein may recognize a carbohydrate site decorating the integrin. This strategy is used by a number of bacterial species, including *E. coli, Histoplasma* spp., and group B streptococci. Masking involves adsorption of a serum protein, such as C3bi, onto the bacterial surface, which then serves as a bridge binding the bacteria to the C3bi–binding site on the integrin. This strategy is used by bacteria such as *Legionella* and *Mycobacteria* spp. The third strategy, mimicry, is the rarest and involves the production by the bacteria of a protein that contains a motif naturally bound directly by the CD11b/CD18 integrin. Mimicry appears to operate at three sites on FHA.

Several approaches have been taken to determine how FHA interacts with CD11b/CD18 (48). Interaction of FHA with the receptor as presented on cells is indicated by the ability of monoclonal antibodies directed against the alpha (CD11b, α_M) or beta (CD18, β_2) chain of the CD11b/CD18 receptor to decrease FHA-dependent bacterial adherence to leukocytes. In addition, phagocytes devoid of CD11b/CD18 integrin are unable to adhere to FHA-coated surfaces. Finally, purified FHA binds to the purified integrin (83). To permit identification of the amino acid positions corresponding to recognition sites on the FHA molecule, a number of synthetic peptides containing various residues of the molecule have been created. Of particular interest are peptides derived from regions of FHA that had sequence similarity to natural ligands of CD11b/CD18, particularly the RGD region and a factor X–like region. The biological activities of each of these peptides are discussed here as examples of how FHA mimics natural ligands for CD11b/CD18.

RGD triplets are recognized by several adhesin-promoting receptors of the integrin family (24). FHA contains a triplet RGD at residues 1097 through 1099 (48). The expected function of an RGD region is to serve as a receptor for the integrin-bearing cell, in this case as a receptor for leukocyte adhesion. To determine whether this region of FHA affected integrin function, overlapping 20-mer peptides

spanning amino acid residues 1077 through 1285 were tested for the ability to block adherence of leukocytes to endothelial cells (50). Only the 20-mer beginning with the RGD sequence blocked adherence. This peptide inhibited leukocyte recognition of endothelial cells as well as leukocyte transmigration across endothelial cell monolayers by >50%. Alteration of RGD to RAD inhibited adherence by ≈20%, suggesting that the RGD sequence was important to some but not all of the bioactivity of this region of FHA.

Several lines of evidence indicate that in addition to the RGD region, FHA interacts with the CD11b/CD18 integrin by factor X–like regions (52). Comparison of the amino acid sequences of FHA and the three CD11b/CD18 binding loops of factor X (amino acid residues 238 through 246, 366 through 373, and 422 through 430) revealed significant sequence similarity. Native FHA and four synthetic peptides representing the regions of FHA similar to these three factor X surface loops inhibited ^{125}I-factor X binding to CD11b/CD18. Three of the four peptides inhibited generation of procoagulant activity and prolonged the clotting time. The most effective synthetic FHA peptide, ETKEVDG, was a homolog of the factor X–CD11b/CD18 binding loop located between amino acid residues 366 through 373. This peptide is able to prevent adherence and transendothelial migration of leukocytes in vitro and to protect animals from blood-brain barrier injury during meningitis. These results suggest that FHA can mimic the natural CD11b/CD18 ligand factor X. This mimicry appears to disturb two functions of CD11b/CD18: procoagulation and endothelial adherence.

Given the multiplicity of binding interactions between *B. pertussis* and leukocytes, it is not surprising that these interactions lead to different fates. Comparison of the consequences of bacterial adherence mediated in an RGD-dependent versus a lectin domain–dependent fashion illustrate the diversity of fates the bacteria can engineer depending on the mechanisms of attachment. When an animal model of whooping cough was used, loss of the integrin-FHA interaction by mutation of the FHA RGD domain to RAD reduced the numbers of *B. pertussis* colonizing the lung 1,000-fold (60). Alveolar macrophages from bronchoalveolar lavages contained numerous wild-type bacteria but virtually no intracellular mutant bacteria. In contrast, elimination of the carbohydrate adherence mechanisms of *B. pertussis* by using either a competitive receptor analog or an anti-receptor antibody was sufficient to elute bacteria from the lung and thereby prevent pulmonary edema in the rabbit model (60). These findings suggest that FHA serves to enhance survival of *B. pertussis* and produce pulmonary pathology via two separate mechanisms. Persistence of bacteria in the lung is promoted through interaction between the RGD region of FHA and the macrophage integrin, which leads to uptake of the bacteria into the phagocyte without an oxidative burst, thereby promoting intracellular localization of bacteria. This state is not associated with pulmonary injury (60). In contrast, the presence of extracellular bacteria adherent to cilia and macrophages in carbohydrate-dependent interactions is associated with pulmonary pathology.

PT and the Selectins

Both S2 and S3 subunits of PT are involved in bacterial adherence to human macrophages (84). Recombinant S2 or S3 incubated with macrophages reduces bacterial adherence to a plateau value half that of control, whereas combination of S2 and S3 reduces adherence by >90%. The fact that S2 and S3 produce additive effects suggests that they are recognized by separate host cell receptors. S3-dependent adherence is reduced by sialic acid, and S2-dependent adherence is reduced by galactose. Antibodies to the Le a or Le x blood group determinants also decrease adherence, suggesting that the structures recognized by PT are borne by molecules that express fucosylated polylactosamines.

The ability of S2 and S3 subunits of PT to recognize the Lewis a and x determinants is a feature reminiscent of the selectin family of eukaryotic lectins (29, 81). Indeed, comparison of the amino acid sequences of S2 and S3 with the amino-terminal regions of the lectin domains of the selectins has revealed significant similarity between PT subunit residues 19 through 52 and selectin residues 15 through 46 (57). These regions have recently been shown to be superimposable in the three-dimensional crystal structures of PT S3 and E selectin (19, 57, 64). Thus, at the three-dimensional structure level, the cores of PT and the selectins are strikingly similar, and at the primary sequence level, the second alpha helices are very similar, and both lie in the basilar face of a lectin (59).

Implications of Selectin and Integrin Mimicry by *B. pertussis* Adhesins

Selectins are a triad of molecules that mediate reversible rolling of leukocytes on endothelia at sites of tissue inflammation (50, 51, 63). When carbohydrates on the neutrophil surface interact with selectins on inflamed endothelial cells, the leukocyte integrin CD11b/CD18 is upregulated and then promotes the transition from leukocyte rolling to transmigration (50, 51, 63) (Fig. 2A). This sequence also appears to be triggered by the sequential activities of PT and FHA (Fig. 2B). Molecular mimicry of selectin structure by PT (61) induces upregulation of the CD11b/CD18 receptor, which in turn serves as a specific ligand for FHA (60). Several studies indicate that PT and FHA resemble the natural ligands sufficiently to be able to behave as competitive inhibitors for selectins and integrins, resulting in impaired neutrophil adherence and margination (50, 51, 75). It is possible that these activities of PT and FHA contribute to the altered leukocyte trafficking that is a hallmark of whooping cough.

The ability of PT subunits to interfere with selectin functions has been demonstrated. The PT S2 and S3 subunits inhibit human neutrophil adherence to endothelial cells and E selectin-coated surfaces and upregulate the function of the leukocyte integrin CD11b/CD18 in vi-

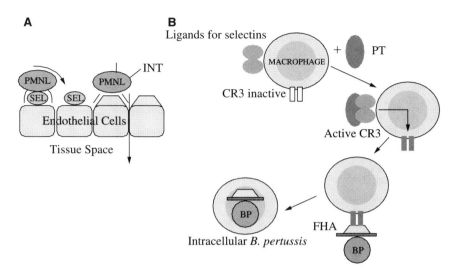

FIGURE 2 Mimicry between leukocyte trafficking systems and the interaction of *B. per-tussis* with leukocytes. (A) Leukocytes (PMNL) attach to selectins (SEL) on inflamed endothelial cells, resulting in upregulation of the CD11b/CD18 integrin (INT). CD11b/CD18 integrin upregulation promotes the transition from leukocyte rolling to transmigration. (B) Mimicry of selectins and integrins is also involved in *B. pertussis* (BP) internalization by macrophages. PT attaches to surface ligands for selectins (ovals) and activates the CR3 integrin (CD11b/CD18) (represented diagrammatically as a change from white to black rectangles). Bacterial FHA then ligates the activated integrin, leading to bacterial uptake. Combining the scenarios in panels A and B indicates that by molecular mimicry of selectins or integrins, PT or FHA can competitively impair neutrophil adherence and margination.

tro (50). Intravenous administration of the S2 and S3 subunits also reduces neutrophil accumulation in the cerebrospinal fluids of animals with meningitis (51). Similarly, the ability of FHA to interfere with integrin functions has been shown. Antibodies to FHA cross-react with human cerebral capillaries and prevent leukocyte accumulation in the brain in experimental models of meningitis (71). Peptides derived from the RGD and factor X regions also inhibit leukocytosis in this model. One of the important sequelae of the binding of anti-FHA antibody to cerebral capillaries in healthy animals is the induction of a reversible, enhanced permeability of the blood-brain barrier (71). Although the precise mechanism is unknown, the proposed similarity between FHA and a natural ligand for CD11b/CD18 on the endothelium suggests that this process may involve the loosening of endothelial tight junctions.

S. PNEUMONIAE

In contrast to the strict ciliary pathogenicity demonstrated by *B. pertussis,* pneumococcus causes a variety of localized and systemic infections that include otitis media, sepsis, meningitis, and pneumonia (8). Meningitis caused by *S. pneumoniae* is characterized by a higher mortality than is found with other meningeal pathogens, and neurological sequelae are frequent in survivors (62). *S. pneumoniae* is also the most frequent cause of bacterial pneumonia in children in developing countries and accounts for up to 76% of cases in adults (30).

Pneumococci generally enter the host via the nasopharynx, where they attach to epithelial cells and in some instances persist for several months (3, 30). Carriage of up to four different serotypes has been documented, and infections usually occur with the acquisition of a new serotype (3, 39). In experimental models, progression to pneumonia results from spread of the pneumococci by aerosol from the nasopharynx down into the lower respiratory tract. The early pneumonic lesion is characterized by fluid-filled alveoli containing pneumococci, which are frequently seen to line the alveolar walls, a distribution suggestive of a specific interaction that promotes retention in the alveolar space (91). Pneumococci readily gain access to the blood circulation from the alveolar space, suggesting an aggressive capacity to cross the vascular endothelial cells of the alveolar capillaries (47).

Invasive pneumococcal infection is characterized by particularly intense inflammation in which components of the pneumococcal cell wall play a central role (18, 39, 72). The clinical isolates of pneumococci that are unable to release cell wall show a significant attenuation in the production of pathophysiology (66, 70). In addition, mixtures of cell wall components can recreate the entire system complex of pneumonia, meningitis, and otitis media (18, 39, 67, 72), underlining the essential role of pneumococcal cell wall in the pathogenesis of inflammation. Pneumococcal cell wall binds to epithelia, endothelia, and macrophages. Cell wall fragments trigger the alternative pathway of complement activation in vitro (49) and induce leukocytosis and increased vascular permeability in models of meningitis and pneumonia (9, 72), the secretion of IL-1 from macrophages (49), and procoagulant activity on endothelial cells (16).

Pneumococcal Receptors on Host Cells

During the course of pneumococcal infections, the bacteria are presumed to attach to and interact primarily with nasopharyngeal cells, vascular endothelial cells, and lung cells. Although the precise structure and number of glycoconjugate receptors present on each of these cell types remain to be determined, the minimal receptor units have been defined on the basis of inhibition of adherence in vitro by simple and complex carbohydrates (2, 12) (Fig. 3).

Nasopharyngeal Receptors

Pneumococci bind to receptors on the nasopharynx that bear the neolactose series of glycocojugates containing GlcNAcβ1-3Gal (2). Structural analogs of the receptor can inhibit bacterial adherence and can elute bacteria already adherent to pharyngeal cells. In addition, erythrocytes coated with analogs of the receptor are agglutinated by pneumococci. The GlcNAcβ1-3Gal disaccharide has been identified as part of the saccharide chains of many glycolipids and glycoproteins and of the blood group antigens ABH, Lewis and Ii antigens (20). Pneumococcal adherence is impaired by human milk, which contains natural neolacto- and lactotetraose glycoconjugates (4). This impairment may play a role in preventing pneumococcal colonization in the newborn and could in part explain the suggested positive effect of breast feeding on otitis media (55).

Lung and Vascular Endothelial Cells

Studies of rabbit mixed and type II lung cell preparations have demonstrated that the type II pneumocyte is the preferred target for pneumococcal adherence (12). Several lines of evidence indicate that in contrast to cells from human upper airways, both type II pneumocytes and vascular endothelial cells interact with pneumococci via two separate classes of receptor defined by mannose linked to the disaccharides GalNAcβ1-4Gal and GalNAcβ1-3Gal (12). Interestingly, the receptors consist of the same saccharide substituents linked in distinctly different stereochemical configurations. Both receptor specificities are distinct from that in the nasopharynx. Carbohydrates that mimic the receptors (as indicated by the ability to block adherence when preincubated with pneumococcus or to elute adherent bacteria) include mannose, GalNAc, Gal (endothelial cells only), the GalNAcβ1-4Gal-

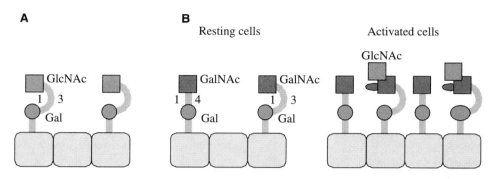

FIGURE 3 Pneumococcal receptor specificities on resting and activated cells. (A) Nasopharyngeal cells bear a single class of receptor for pneumococci: GlcNAcβ1-3Gal. (B) In contrast, in the resting state, both vascular endothelial cells and type II pneumocytes possess two separate classes of receptor for pneumococci: GalNAcβ1-3Gal and GalNAcβ1-4Gal. Cytokine stimulation of type II pneumocytes and vascular endothelial cells (right) results in receptor expansion and the appearance of a new specificity for GlcNAc within the GalNAcβ1-3Gal receptor population. Reciprocal changes in pneumococcal adhesive ligands match the changes in the eukaryotic cell surface in those strains that achieve successful colonization.

containing glycoconjugates asialo-GM1 and GM2, and the GalNAcβ1-3Gal-containing glycoconjugates forssman glycolipid and globoside. A combination of asialo-GM2 and globoside produces significantly greater antiadherence effects than either glycoconjugate used alone can produce. This suggests that these glycoconjugates define independent receptors. GalNAcβ1-4Gal-containing glycosphingolipids in the form of asialo-GM1 are present in significant amounts in lung tissue (36). Pneumococci and a number of other respiratory tract pathogens including *H. influenzae, Pseudomonas aeruginosa,* and *S. aureus,* can directly bind to purified immobilized analogs containing this disaccharide (36). These results suggest that glycoconjugate receptors for pneumococci differ between tissues and also that colonizing pathogens may compete for similar receptors on the same tissue.

Pneumococcal Adherence to Cytokine-Activated Epithelial and Endothelial Cells

Cytokines are thought to mediate many host responses to bacterial infection. Tumor necrosis factor and IL-1 play a central role in this

regard (61). The local generation of inflammatory cytokines can also dramatically alter the expression of potential receptors on cells. For example, activation of vascular endothelial cells by tumor necrosis factor and IL-1 increases the expression of a variety of cell surface receptors, including E selectin (50) and globotriosylceramide (82). These cytokines also increase the expression of glycoconjugate receptors for pneumococcus on cultures of type II pneumocytes and vascular endothelial cells (11) (Fig. 3). Adherence of pneumococci to pneumocytes increased ≈30% and adherence to endothelial cells increased ≈70% following cytokine stimulation. Enhanced adherence was associated with the appearance of a new receptor specificity for GlcNAc within the GalNAcβ1-3Gal receptor population. The change in adherence of pneumococci to activated cells illustrates an important concept in microbial pathogenesis. The initial attachment of pneumococci to resting cells involves one set of sugar specificities. Once engaged, however, the lung and vascular cells rapidly become activated and change the presentation of receptors on their surfaces. A virulent pneumococcus must match this change with the presentation of a new, appropriate cognate adhesin for the new receptor. This action–and–reaction scenario underlies the success of virulent pathogens and illustrates the dynamic nature of the response of both partners in an adherence interaction.

Biologically Active Components of the Pneumococcal Surface Involved in Adherence

Pneumococcal disease involves spread of the bacteria from nasopharyngeal cells to lung cells lining the alveolus and then invasion beyond vascular endothelial cells. This progression relies on the sequential recognition of GlcNAcβ1-3Gal receptors on nasopharyngeal cells (2) followed by GalNAcβ1-3Gal and GalNAcβ1-4Gal receptors on pneumocytes (12). Cytokine production from inflamed cells results in increased pneumococcal adherence associated with glycoconjugate receptor expansion and the appearance of a novel sugar specificity to GlcNAc (11). The following sections examine the multiple adhesive ligands on the pneumococcus that interact with these receptors.

Pneumococcal Cell Wall Components Involved in Adherence

Adherence of pneumococci to nasopharyngeal cells, type II pneumocytes, and vascular endothelium involves both cell wall and protein components of the pneumococcal surface (2, 12, 17). These interactions appear to be independent of capsular type. Both cell wall and protein components contribute approximately equally to the interaction with human vascular endothelial cells and type II pneumocytes (12, 17). Studies have shown that purified pneumococcal cell wall and its soluble subcomponents are able to compete with intact pneumococci for binding sites on endothelial cells and type II pneumocytes (12, 17). These results suggest that pneumococcal adherence to human vascular endothelial cells and type II pneumocytes is mediated by two mechanisms: one classic and one novel. The classic interaction involves a protein adhesin recognizing a carbohydrate. The novel mechanism involves

interactions between the bacterial cell wall (composed mostly of carbohydrates) and eukaryotic carbohydrates. The molecular details of this interaction remain unknown.

It appears that the expression of factors mediating colonization and adherence of pneumococci to epithelial cells undergoes phenotypic variation. Pneumococci display two different colonial morphologies when viewed with oblique, transmitted light: opaque and transparent (85, 86). The colonial phenotypes switch at frequencies of 10^{-3} to 10^{-6}. In an infant rat model of nasopharyngeal carriage, opaque variants are unable to colonize the nasopharynx, whereas transparent variants are stable and effective colonizers (85). Opaque and transparent strains of pneumococci also differ in their abilities to adhere to cultured human type II pneumocytes and vascular endothelial cells in vitro, most particularly to activated cells (86). The transparent forms of three separate pneumococcal strains show a slightly better capacity for adhering to lung and vascular cells than the opaque variants show (~25% difference). However, virulent pneumococcal adherence is further augmented by ≈70% in activated endothelial cells and ≈30% in lung cells (11). This enhanced adherence is observed only in transparent variants. Pathogens frequently genetically regulate expression of adhesive ligands. These results indicate the importance of this flexibility to outcome, defined here as colonization.

Genetic Strategies Used to Determine Cell Membrane Components Involved in Pneumococcal Adherence

The pneumococcal cell membrane contains an assembly of several hundred exported proteins that may represent many determinants of pneumococcal virulence. Genetic tools to survey mutants for altered phenotypes caused by recoverable genetic elements have recently been developed for the pneumococci (44). On the basis of translational fusions to the gene for alkaline phosphatase (*phoA*), several new exported proteins have been identified. These pneumococcal mutants with defects in exported proteins are now being used to identify bacterial elements involved in adherence to cultured human type II pneumocytes and vascular endothelial cells (10). From a bank of independent mutants, two mutants that demonstrate a >60% decrease in adherence to both cell types have been identified. Sequence analysis of the recovered loci shows that each encodes a distinct protein-dependent peptide permease, AmiA or PlpA (permease-like protein), suggesting that these proteins are involved in modulating pneumococcal adherence to epithelial and endothelial cells (10). Bacterial permeases form a family of membrane proteins (Table 2) that are responsible for the transport of a variety of substances across biological membranes. A number of streptococcal permeases have been implicated as adhesins involved in bacterial coaggregation and adherence to teeth (1, 14, 15, 27, 56). These structures all bear significant homology to one another and include FimA from *Streptococcus parasanguis* (14), SsaB from *Streptococcus sanguis* (15), PK488 from *Streptococcus gordonii* (1), and PsaA from *S. pneumoniae* (56). Distinct from this family, PlpA and AmiA from *S. pneumoniae* (44) demonstrate ≈60% sequence similarity to another permease, SarA from *S. gordonii,* which is responsible for the binding and transport of small peptides (27). These results suggest that a family of permeases involved in bacterial adherence exists in streptococcal species. The role of these permeases in pneumococcal adherence is currently under investigation.

SUMMARY

B. pertussis and *S. pneumoniae* are very distinct pathogens in terms of both tissue specificity and the strategies used in attachment and interaction with their human hosts. A comparison of these bacterial species is set out in Table 3. An understanding of these interactions suggests new strategies that can halt the course of the disease at various stages. Bacterial attachment is the primary prerequisite for infection to occur. Receptor analogs, including carbohydrates, antireceptor antibodies, and lectins, effectively block the adherence of *B. pertussis* and *S. pneumoniae* to target host cells and thus should be considered potential candidates for therapeutic intervention against disease. Complete elucidation of the target host cell glycoconjugates might allow the design of a highly competitive inhibitor of *B. pertussis* and *S. pneumoniae* adherence. According to current evidence, critical features of these analogs will be the interruption of bacterial interaction with the disaccharide Gal-GalNAc in the case of *B. pertussis* and with a combination of GalNAcβ1-3Gal, GalNAcβ1-4Gal, and GlcNAcβ1-3Gal for *S. pneumoniae*. The competitive inhibitor could be used directly, or alternatively, antibodies to the receptor could be designed. Many receptor analogs may be found naturally. For example, lactose or other sugars in breast milk or alveolar glycoprotein decrease bacterial interactions with lung cells. Sites of inflammation may also display an increased susceptibility to bacterial adherence, as evidenced by

TABLE 2 Bacterial permeases involved in adherence of streptococcal species to target surfaces of host

Streptococcal species	Permease	Target surface	Reference
S. parasanguis	FimA	Saliva and/or teeth	14
S. sanguis	SsaB	Saliva and/or teeth	15
S. gordonii	PK488	Saliva and/or teeth	1
S. pneumoniae	PsaA	Saliva and/or teeth	56
S. gordonii	SarA	Saliva and/or teeth	27
S. pneumoniae	PlpA	Endothelial cells and type II pneumocytes	10
S. pneumoniae	AmiA	Endothelial cells and type II pneumocytes	10

TABLE 3 Comparison of interactions between *B. pertussis* and *S. pneumoniae* and target human host cells[a]

Interaction	*B. pertussis*	*S. pneumoniae*
Tissue tropism	Ciliary pathogen	Alveolar pathogen
	Local infection	Systemic or local infections
Adhesion	Two adhesins of major importance: FHA and PT	Adhesins unknown. Cell wall teichoic acid and permeases may be involved.
	Adhesins released	Adhesins not released
Effect on leukocyte traffic	Decreased	Increased
	CD11b/CD18 interactions	CD11b/CD18 and PAF interactions
Toxicity	Ciliotoxicity via cell wall TCT	Endothelial cytotoxicity via cell wall pieces
	IL-1 mechanism	IL-1 mechanism
	PT	Pneumolysin
	Adenylate cyclase toxin	

[a]Abbreviations: PAF, platelet-activating factor; TCT, tracheal cytotoxin.

studies showing a dramatic increase in numbers of pneumococci adherent to cytokine-activated monolayer. These results confirm that cytokines are a double-edged sword during bacterial infection and suggest that some of the benefits derived from anticytokine therapies may relate to disturbances in microbial adhesion.

Characterization of bacterial components involved in adhesion provides clues as to how host cell recognition of the bacteria is mediated. In the case of *S. pneumoniae,* the adhesive ligands still remain to be determined. However, the ability of cell wall to mediate adherence raises the possibility that a new class of adherence interaction occurs. It is recognized that bacterial proteins can bind to either proteins or carbohydrates on eukaryotic cells. Cell wall-mediated adherence may indicate a carbohydrate-carbohydrate adherence motif. Current vaccination against *S. pneumoniae* is based on capsular polysaccharide antigens that are ineffective in the elderly and in very young children (39). Identification of virulence determinants, such as adhesins, that are proteins might eventually lead to a more broadly protective vaccine candidate.

Studies of the major adhesins of *B. pertussis,* FHA and PT, may also help elucidate the mechanisms by which leukocyte trafficking occurs. FHA and PT adhesins block leukocyte traffic through selectin and integrin mimicry (75). These structures mimic the natural ligands

to such a degree that they compete for binding sites on host cells. Biologically effective analogs of FHA and PT may provide a mechanism of reducing leukocyte trafficking in states of inflammation. In addition, anti-FHA antibodies open the blood-brain barrier. The mechanism(s) involved could provide clues as to how this occurs during meningitis. In addition, transient opening of the blood-brain barrier might also provide a method of delivering therapeutic agents to the brain.

REFERENCES

1. **Andersen, R. N., N. Ganeshkumar, and P. E. Kolenbrander.** 1993. Cloning of the *Streptococcus gordonii* PK488 gene, encoding an adhesin which mediates co-aggregation with *Actinomyces naeslundii* Pk606. *Infect. Immun.* **61:**981–987.
2. **Andersson, B., J. Dahmen, F. Torbjörn, H. Leffler, G. Magnusson, G. Noori, and C. Svanborg Eden.** 1983. Identification of an active disaccharide unit of a glycoconjugate receptor for pneumococci attaching to human pharyngeal epithelial cells. *J. Exp. Med.* **158:**559–570.
3. **Andersson, B., B. Eriksson, E. Falsen, A. Fogh, L. A. Hanson, O. Nylen, H. Petersen, and C. Svanborg Eden.** 1981. Adhesion of *Streptococcus pneumoniae* to human pharyngeal epithelial cells in vitro: differences in adhesive capacity among strains isolated from subjects with otitis media, septicemia, or meningitis or from healthy carriers. *Infect. Immun.* **32:**311–317.
4. **Andersson, B., O. Porras, L. A. Hanson, T. Lagergård, and C. Svanborg-Eden.** 1986. Inhibition of attachment of *Streptococcus pneumoniae* and *Haemophilus influenzae* by human milk and re-

ceptor oligosaccharides. *J. Infect. Dis.* **153:** 232–237.

5. **Beachey, E. H.** 1981. Bacterial adherence: adhesin receptor interactions mediating the attachment of bacteria to mucosal surfaces. *J. Infect. Dis.* **143:** 325–345.

6. **Beachey, E. H., B. Eisenstein, and I. Ofek.** 1982. *Adherence and Infectious Diseases. Current Concepts.* The Upjohn Co., Kalamazoo, Mich.

7. **Bemis, D. A., and S. C. Wilson.** 1985. Influence of potential virulence determinants on *Bordetella brontiseptica*-induced ciliostasis. *Infect. Immun.* **50:** 35–42.

8. **Burman, L. A., R. Norrby, and B. Trollfors.** 1985. Invasive pneumococcal infections: incidence, predisposing factors, and prognosis. *Rev. Infect. Dis.* **7:** 133–142.

9. **Cabellos, C., D. E. MacIntyre, M. Forrest, M. Burroughs, S. Prasad, and E. Tuomanen.** 1992. Differing roles for platelet activating factor during inflammation of the lung and subarachnoid space. The special case of *Streptococcus pneumoniae. J. Clin. Invest.* **90:** 612–618.

10. **Cundell, D. R., B. J. Pearce, A. Young, E. I. Tuomanen, and H. R. Masure.** 1994. Protein dependent peptide permeases from *Streptococcus pneumoniae* mediate cytoadherence to type II lung cells and to human endothelial cells. *J. Cell. Biochem.* **54**(Suppl. 18A):45.

11. **Cundell, D., and E. Tuomanen.** The molecular basis of pneumococcal infection: an hypothesis. *Clin. Infect. Dis.,* in press.

12. **Cundell, D. R., and E. I. Tuomanen.** Receptor specificity of adherence of *Streptococcus pneumoniae* to human type II pneumocytes and vascular endothelial cells *in vitro. Microbial Pathog.,* in press.

13. **Feizi, T., and R. A. Childs.** 1985. Carbohydrate structures of glycoproteins and glycolipids as differentiation antigens, tumour associated antigens and components of receptor systems. *Trends Biol. Sci.* **10:** 24–29.

14. **Fives-Taylor, P., J. C. Fenno, E. Holden, L. Linehan, L. Oligino, and M. Volansky.** 1991. Molecular structure of fimbria-associated genes of *Streptococcus sanguis,* p. 240–243. *In* G. M. Dunny, P. P. Cleary, and L. L. McKay (ed.), *Genetics and Molecular Biology of Streptococci, Lactococci and Enterococci.* American Society for Microbiology, Washington, D.C.

15. **Ganeshkumar, N., M. Song, and C. McBride.** 1988. Cloning of a *Streptococcus sanguis* adhesin which mediates binding to saliva-coated hydroxyapatite. *Infect. Immun.* **56:** 1150–1157.

16. **Geelen, S., C. Bhattacharyya, and E. Tuomanen.** 1992. Induction of procoagulant activity on human endothelial cells by *Streptococcus pneumoniae. Infect. Immun.* **60:** 4179–4183.

17. **Geelen, S., C. Bhattacharyya, and E. Tuo-**

manen. 1993. The cell wall mediates pneumococcal attachment to and cytopathology in human endothelial cells. *Infect. Immun.* **61:** 1538–1543.

18. **Giebink, G., P. Ripley-Petzoldt, S. Juhn, D. Aeppli, A. Tomasz, and E. Tuomanen.** 1988. Contribution of pneumococcal cell wall to experimental otitis media pathogenesis. *Ann. Otol. Rhinol. Laryngol.* **132**(Suppl.):28–30.

19. **Graves, B., R. Crowther, C. Chandran, J. Rumberger, S. Li, K.-S. Huang, D. Presky, P. Familletti, B. Wolitzky, and D. Burns.** 1994. Insight into E-selectin/ligand interaction from the crystal structure and mutagenesis of the lec/EGF domains. *Nature* (London) **367:** 532–538.

20. **Hakomori, S-I.** 1981. Blood group ABH and Ii antigens of human erythrocytes: chemistry polymorphism and their developmental change. *Semin. Hematol.* **18:** 39–62.

21. **Heerze L. D., P. C. S. Chong, and G. D. Armstrong.** 1992. Investigation of the lectin-like binding domains in pertussis toxin using synthetic peptide sequences. Identification of a sialic acid binding site in the S2 subunit of the toxin. *J. Biol. Chem.* **267:** 25810–25815.

22. **Heiss, L. N., S. A. Moser, E. R. Unanue, and W. E. Goldman.** 1993. Interleukin-1 is linked to the respiratory epithelial cytopathology of pertussis. *Infect. Immun.* **61:** 3123–3128.

23. **Hewlett, E. L.** 1984. Biological effects of pertussis toxin and *Bordetella* adenylate cyclase on intact cells and experimental animals, p. 168–171. *In* L. Lieve and D. Schlessinger, (ed.), *Microbiology—1984.* American Society for Microbiology, Washington, D.C.

24. **Hynes, R. O.** 1987. Integrins: a family of cell surface receptors. *Cell* **48:** 549–554.

25. **Ifek, I., and R. J. Doyle (ed.).** 1994. *Bacterial Adhesion to Cells and Tissues,* p. 513–546. Chapman & Hall, New York.

26. **Ifek, I., and R. J. Doyle (ed.).** 1994. *Bacterial Adhesion to Cells and Tissues,* p. 1–15. Chapman & Hall, New York.

27. **Jenkinson, H. F.** 1992. Adherence, coaggregation, and hydrophobicity of *Streptococcus gordonii* associated with expression of cell surface lipoproteins. *Infect. Immun.* **60:** 1225–1228.

28. **Johnson, A. P., and T. J. Inzana.** 1986. Loss of ciliary activity in organ cultures of rat trachea treated with lipo-oligosaccharide from *Haemophilus influenzae. J. Med. Microbiol.* **22:** 265–268.

29. **Johnston, G .I., R. Cook, and M. McEver.** 1989. Structural and biosynthetic studies of the granule membrane protein GMP-140, from human platelets and endothelial cells. *Cell* **56:** 1033–1044.

30. **Johnston, R. B.** 1991. Pathogenesis of pneumococcal pneumonia. *Rev. Infect. Dis.* **13**(Suppl. 6):S509–517.

31. **Kanthakumar, K., D. R. Cundell, G. Taylor, R. C. Read, K. Harrison, A. Rutman, E. R. Moxon, P. J. Cole, and R. Wilson.** 1995. Partial characterization of factors produced by *Haemophilus influenzae* which slow human ciliary beating in vitro. *Infect. Immun.*, in press.

32. **Kanthakumar, K., D. R. Cundell, G. W. Taylor, P. J. Wills, M. Johnson, P. J. Cole, and R. Wilson.** 1994. The effect of salmeterol and dibutyryl cyclic AMP on the cyclic AMP-mediated slowing of human ciliary beat *in vitro* produced by the *Pseudomonas aeruginosa* toxins 1-hydroxyphenazine and rhamnolipid. *Br. J. Pharmacol.* **112:**493–498.

33. **Kanthakumar, K., G. Taylor, K. W. T. Tsang, D. R. Cundell, A. Rutman, P. K. Jeffery, P. J. Cole, and R. Wilson.** 1993. Mechanism of action of *Pseudomonas aeruginosa* pyocyanin on human ciliary beat *in vitro*. *Infect. Immun.* **61:**2848–2853.

34. **Karlsson, K.-A.** 1989. Animal glycosphingolipids as membrane attachment sites for bacteria. *Annu. Rev. Biochem.* **58:**309–350.

35. **Katada, T., and M. Ui.** 1982. Direct modification of the membrane adenylate cyclase system by islet-activating protein due to ADP-ribosylation of a membrane protein. *Proc. Natl. Acad. Sci. USA* **79:**3129–3133.

36. **Krivan, H. C., D. D. Roberts, and V. Ginsburg.** 1988. Many pulmonary pathogenic bacteria bind specifically to the carbohydrate sequence GalNAcß1–4Gal found in some glycolipids. *Proc. Natl. Acad. Sci. USA* **85:**6157–6161.

37. **Lankford, C. E., and U. Legsomburana.** 1965. Virulence factors of choleragenic vibrios, p. 109–120. *In Proceedings: Cholera Research Symposium.* U.S. Government Printing Office, Washington, D.C.

38. **Leininger, E., J. G. Kenimer, and M. J. Brennan.** 1990. Surface proteins of *Bordetella pertussis*: role in adherence, p. 100–104. *In* C. R. Manclark (ed.), *Proceedings of the Sixth International Symposium on Pertussis.* U.S. Public Health Service, Department of Health and Human Services, Bethesda, Md.

39. **Masure, H. R., R. Austrian, and E. I. Tuomanen.** The pathophysiology of pneumococcal infection. *N. Engl. J. Med.*, in press.

40. **Mennozzi, F., R. Mutombo, G. Renauld, C. Gantiez, J. Hannah, E. Leininger, M. Brennan, and C. Locht.** 1994. Heparin-inhibitable lectin activity of the filamentous hemagglutinin of *Bordetella pertussis. Infect. Immun.* **62:**769–778.

41. **Mooi, F. R., W. H. Jansen, H. Brunings, H. Gielen, H. G. J. van der Heide, H. C. Walvoort, and P. A. M. Guinee.** 1992. Construction and analysis of *Bordetella pertussis* mutants defective in the production of fimbriae. *Microb. Pathog.* **12:**127–135.

42. **Muse, K. E.** 1980. Host cell membrane perturbations during experimental *Bordetella pertussis* infection. *Electron Microsc.* **2:**432–433.

43. **Ofek, I., and E. H. Beachey.** 1980. Bacterial adherence. *Adv. Intern. Med.* **25:**503–522.

44. **Pearce, B. J., Y. B. Yin, and H. R. Masure.** 1993. Genetic identification of exported proteins in *Streptococcus pneumoniae. Mol. Microbiol.* **9:**1037–1050.

45. **Plotkin, B. J., and D. A. Bemis.** 1984. Adherence of *Bordetella bronchiseptica* to hamster lung fibroblasts. *Infect. Immun.* **46:**697–702.

46. **Prasad, S. M., Y. Yin, E. Rozdzinski, E. Tuomanen, and H. R. Masure.** 1993. Identification of a carbohydrate recognition domain in filamentous hemagglutinin from *Bordetella pertussis. Infect. Immun.* **61:**2780–2785.

47. **Rake, G.** 1936. Pathogenesis of pneumococcus infection in mice. *J. Exp. Med.* **63:**17–31.

48. **Relman, D., E. Tuomanen, S. Falkow, D. T. Golenbock, K. Saukkonen, and S. D. Wright.** 1994. Recognition of a bacterial adhesin by an integrin: macrophage CR3 ($\alpha_M\beta_2$, CD11b/CD18) binds filamentous hemagglutinin of *Bordetella pertussis. Cell* **61:**1375–1382.

49. **Riesenfeld-Orn, I., S. Wolpe, J. F. Garcia-Bustos, M. K. Hoffmann, and E. Tuomanen.** 1989. Production of interleukin-1 but not tumor necrosis factor by human monocytes stimulated with pneumococcal cell surface components. *Infect. Immun.* **57:**1890–1893.

50. **Rozdzinski, E., W. N. Burnette, T. Jones, V. Mar, and E. Tuomanen.** 1993. Prokaryotic peptides that block leukocyte adherence to selectins. *J. Exp. Med.* **178:**917–924.

51. **Rozdzinski, E., T. James, W. N. Burnette, M. Burroughs, and E. Tuomanen.** 1993. Antiinflammatory effects in experimental meningitis of prokaryotic peptides that mimic selectins. *J. Infect. Dis.* **168:**1422–1428.

52. **Rozdzinski, E., J. Sandros, M. van der Flier, A. Young, B. Spellerberg, C. Bhattacharyya, and E. Tuomanen.** Inhibition of leukocyte-endothelial cell interactions by a bacterial adhesin which mimics coagulation factor X. *J. Clin. Invest.*, in press.

53. **Ruoslahti, E.** 1988. Fibronectin and its receptors. *Annu. Rev. Biochem.* **57:**375–413.

54. **Ruoslahti, E.** 1991. Integrins. *J. Clin. Invest.* **87:** 1–5.

55. **Saarinen, U.M.** 1982. Prolonged breast feeding as prophylaxis for recurrent otitis media. *Acta Paediatr. Scand.* **68:**691–694.

56. **Sampson, J. S., S. P. O'Connor, A. R. Stinson, J. A. Tharpe, and H. Russell.** 1994. Cloning and nucleotide sequence analysis of *psaA*, the

Streptococcus pneumoniae gene encoding a 37-kilo-dalton protein homologous to previously reported *Streptococcus* sp. adhesins. *Infect. Immun.* **62:** 319–324.

57. **Sandros, J., E. Rozdzinski, D. Cowburn, and E. Tuomanen.** 1994. Lectin domains in the adhesins of *Bordetella pertussis*: selectin mimicry linked to microbial pathogenesis. *Glycoconjugates* **11:**1–6.

58. **Sandros, J., and E. Tuomanen.** 1993. Attachment factors of *Bordetella pertussis*: mimicry of eukaryotic recognition molecules. *Trends Microbiol.* **1:**192–196.

59. **Saukkonen, K., W. N. Burnette, V. Mar, H. R. Masure, and E. Tuomanen.** 1992. Pertussis toxin has eukaryotic-like carbohydrate recognition domains. *Proc. Natl. Acad. Sci. USA* **89:**118–122.

60. **Saukkonen, K., C. Cabellos, M. Burroughs, S. Prasad, and E. Tuomanen.** 1991. Integrin-mediated localization of *Bordetella pertussis* within macrophages: role in pulmonary colonization. *J. Exp. Med.* **173:**1143–1149.

61. **Saukkonen, K., S. Sande, C. Cioffe, S. Wolpe, B. Sherry, A. Cerami, and E. Tuomanen.** 1990. The role of cytokines in the generation of inflammation and tissue damage in experimental gram-positive meningitis. *J. Exp. Med.* **171:**439–448.

62. **Schlech, W. F., III, J. I. Ward, J. D. Band, A. Hightower, D. W. Fraser, and C. V. Broome.** 1985. Bacterial meningitis in the United States, 1978 through 1981: the National Bacterial Meningitis Surveillance Study. *JAMA* **253:**1749–1754.

63. **Springer, T. A.** 1990. Adhesin receptors of the immune system. *Nature* (London) **346:**425–434.

64. **Stein, P., A. Boodhoo, G. Armstrong, S. Cockle, M. Klein, and R. Read.** 1994. Crystal structure of pertussis toxin. *Structure* **2:**45–57.

65. **Steinfort, C., R. Wilson, T. Mitchell, C. Feldman, A. Rutman, H. Todd, D. Sykes, J. Walker, K. Saunders, P. W. Andrew, G. J. Boulnois, and P. J. Cole.** 1989. Effect of *Streptococcus pneumoniae* on human respiratory epithelium in vitro. *Infect. Immun.* **57:**2006–2013.

66. **Täuber, M. G., M. Burroughs, U. M. Niemoller, H. Fuster, U. Borschberg, and E. Tuomanen.** 1991. Differences of pathophysiology in experimental meningitis caused by three strains of *Streptococcus pneumoniae*. *J. Infect. Dis.* **163:**806–811.

67. **Tomasz, A., and K. Saukkonen.** 1989. The nature of cell wall-derived inflammatory components of pneumococci. *Pediatr. Infect. Dis.* **8:**902–903.

68. **Tuomanen, E.** 1986. Piracy of adhesins: attachment of superinfecting pathogens to respiratory cilia by secreted adhesins of *Bordetella pertussis*. *Infect. Immun.* **54:**905–908.

69. **Tuomanen, E.** 1988. *Bordetella pertussis* adhesins, p. 75–93. *In* A. C. Wardlaw and R. Parton (ed.), *Pathogenesis and Immunity in Pertussis*. John Wiley & Sons, Inc., New York.

70. **Tuomanen, E., H. Pollack, A. Parkinson, M. Davidson, R. Facklam, R. Rich, and O. Zak.** 1988. Microbiological and clinical significance of a new property of defective lysis in clinical strains of pneumococci. *J. Infect. Dis.* **158:**36–43.

71. **Tuomanen, E., S. Prasad, J. George, A. Hoepelman, P. Ibsen, I. Heron, and R. Starzyk.** 1993. Reversible opening of the blood brain barrier by anti-bacterial antibodies. *Proc. Natl. Acad. Sci. USA* **90:**7824–7828.

72. **Tuomanen, E., R. Rich, and O. Zak.** 1987. Induction of pulmonary inflammation by components of the pneumococcal cell surface. *Am. Rev. Respir. Dis.* **135:**869–874.

73. **Tuomanen, E., and A. Weiss.** 1985. Characterization of two adhesins of *Bordetella pertussis* for human ciliated respiratory epithelial cells. *J. Infect. Dis.* **152:**118–125.

74. **Tuomanen, E., A. Weiss, R. Rich, F. Zak, and O. Zak.** 1986. Filamentous hemagglutinin and pertussis toxin promote adherence of *Bordetella pertussis* to cilia. *Dev. Biol. Stand.* **61:**197–204.

75. **Tuomanen, E. I.** 1992. Cell adhesion molecules in the development of bacterial infections, p. 297–306. *In* C. G. Gamberg, T. Mandrup-Poulsen, L. Wogensen Bach, and B. Hökfelt (ed.), *Leukocyte Adhesion. Basic and Clinical Aspects*. Elsevier Science Publishers, New York.

76. **Tuomanen, E. I., and O. Hendley.** 1983. Adherence of *Bordetella pertussis* to human respiratory epithelial cells. *J. Infect. Dis.* **148:**125–130.

77. **Tuomanen, E. I., J. Nedelman, J. O. Hendley, and E. L. Hewlett.** 1983. Species specificity of *Bordetella* adherence to human and animal ciliated respiratory epithelial cells. *Infect. Immun.* **42:**692–695.

78. **Tuomanen, E. I., J. Schwartz, S. Sande, K. Light, and D. Gage.** 1989. Unusual composition of peptidoglycan in *Bordetella pertussis*. *J. Biol. Chem.* **246:**11093–11098.

79. **Tuomanen, E. I., H. Towbin, G. Rosenfelder, D. Braun, G. Larson, G. C. Hanson, and R. Hill.** 1988. Receptor analogs and monoclonal antibodies that inhibit adherence of *Bordetella pertussis* to human ciliated respiratory epithelial cells. *J. Exp. Med.* **168:**267–277.

80. **Turley, E. A., and S. Roth.** 1980. Interactions between the carbohydrate chains of hyaluronate and chondroitin sulphate. *Nature* (London) **283:** 268–271.

81. **Tyrell, D., P. James, R. Narasinga, C. Foxall, S. Abbas, F. Dasgupta, and B. K. Bran-**

dley. 1991. Structural requirements for the carbohydrate ligand of E-selectin. *Proc. Natl. Acad. Sci. USA* **88**:10372–10376.

82. **van de Kar, N. C. A. J., L. A. H. Monnens, M. A. Karmali, and V. W. M. Hinsbergh.** 1992. Tumor necrosis factor and interleukin-1 induce expression of the verocytotoxin receptor globotriaosylceramide on human endothelial cells: implications for the pathogenesis of the hemolytic uremic syndrome. *Blood* **11**:2755–2764.

83. **Van Strijp, J. A. G., D. G. Russell, E. Tuomanen, E. J. Brown, and S. D. Wright.** 1993. Ligand specificity of purified complement receptor type three (CD11b/CD18, $\alpha_m\beta_2$, Mac-1). *J. Immunol.* **151**:3324–3336.

84. **van't Wout, J., W. N. Burnette, V. L. Mar, E. Rozdzinski, S. D. Wright, and E. I. Tuomanen.** 1992. Role of carbohydrate recognition domains of pertussis toxin in adherence of *Bordetella pertussis* to human macrophages. *Infect. Immun.* **60**:3303–3308.

85. **Weiser, J. N., R. Austrian, P. K. Sreenivasan, and H. R. Masure.** 1994. Phase variation in pneumococcal opacity: relationship between colonial morphology and nasopharyngeal colonization. *Infect. Immun.* **62**:2582–2589.

86. **Weiser, J. N., E. I. Tuomanen, D. R. Cundell, P. K. Sreenivasan, R. Austrian, and H. R. Masure.** 1994. The effect of colony opacity variation on pneumococcal colonization and adhesion. *J. Cell. Biochem.* **54**(Suppl. 18A):55.

87. **Weiss, A. A., E. L. Hewlett, G. A. Myers, and S. Falkow.** 1983. Tn5-induced mutations affecting virulence factors of *Bordetella pertussis. Infect. Immun.* **42**:33–41.

88. **Weiss, A. A., E. L. Hewlett, G. A. Myers, and S. Falkow.** 1984. Pertussis toxin and extracytoplasmic adenylate cyclase as virulence factors of *Bordetella pertussis. J. Infect. Dis.* **150**:219–222.

89. **Wilson, R., R. Read, M. Thomas, A. Rutman, K. Harrison, V. Lund, B. Cookson, W. Goldman, H. Lambert, and P. Cole.** 1991. Effects of *Bordetella pertussis* on human respiratory epithelium in vivo and in vitro. *Infect. Immun.* **59**:337–345.

90. **Witvliet, M. H., M. I. Vogel, E. H. J. Wiertz, and J. T. Poolman.** 1992. Interaction of pertussis toxin with human T lymphocytes. *Infect. Immun.* **60**:5085–5090.

91. **Wood, W. B., Jr.** 1941. Studies on mechanism of recovery in pneumococcal pneumonia; action of type-specific antibody on pulmonary lesion of experimental pneumonia. *J. Exp. Med.* **73**:201–222.

PARADIGMS IN BACTERIAL ENTRY INTO HOST CELLS

Michael K. Zierler and Jorge E. Galán

2

A number of microorganisms have developed the ability to invade (i.e., enter) nonphagocytic cells. This property is essential for their pathogenicity because, presumably, the intracellular environment offers these organisms protection against the host's own immunologic defenses, provides nutrients required for microbial growth, or simply allows the organism to reach more favorable destinations. In general, invasive organisms fall into two categories: those that have an obligate requirement for the intracellular environment in order to replicate (obligate intracellular parasites) and those that, even though intracellular in their natural hosts, have retained the ability to replicate in nonliving media (facultative intracellular parasites) (40). A number of enteric bacteria, such as *Salmonella, Yersinia,* and *Shigella* spp., and certain strains of *Escherichia coli,* are typical examples of the latter.

All mammalian cells can internalize small molecules and solutes, but only professional phagocytes (macrophages and polymorphonuclear leukocytes) are specially adapted to internalize large particles. Nonprofessional phagocytes (e.g., epithelial and endothelial cells,

fibroblasts, etc.) do not usually take up large particles; therefore, parasites able to gain entrance to these cells can do so because they have adapted in a way that promotes their own intake. Thus, bacterial entry is the result of a rather sophisticated manipulation of the host cell machinery by these pathogens. The mechanisms behind this process are beginning to emerge, and it is now clear that bacteria can gain access to cells in at least two different ways: by internalization as a result of high-affinity interactions between bacterial ligands and host receptors (exemplified by the *Yersinia* invasin paradigm) and by entry due to signaling and subsequent modulation of the host cell cytoskeleton (exemplified by the *Salmonella* paradigm) (reviewed in reference 12). In this chapter, we highlight the salient features of our current knowledge of these two paradigms.

THE *YERSINIA* SPP. INVASIN PARADIGM: ENTRY DUE TO HIGH-AFFINITY INTERACTIONS OF BACTERIAL LIGANDS WITH HOST RECEPTORS

The invasin protein of *Yersinia pseudotuberculosis* was originally identified by its ability to confer upon the normally noninvasive strain of *E. coli* HB101 the ability to enter cultured mammalian cells (27). Subsequent studies have established that this outer membrane protein confers

Michael K. Zierler and Jorge E. Galán   Department of Molecular Genetics and Microbiology, School of Medicine, State University of New York at Stony Brook, Stony Brook, New York 11794-5222.

Virulence Mechanisms of Bacterial Pathogens, 2nd ed., Edited by J. A. Roth et al.

invasive properties by its ability to bind to multiple β1 integrin receptors on the host cell surface (28). Integrins are heterodimeric transmembrane proteins found on the surfaces of many eukaryotic cells. They are composed of a variety of α chains that combine with specific β chains to form a receptor with distinct ligand specificity. In general, integrins promote cell adhesion to extracellular matrices as well as cell-to-cell adhesion (25).

Binding to integrins is a common theme among bacterial, viral, and parasitic pathogens (reviewed in reference 29). It appears that a large number of microbial pathogens have evolved mechanisms of binding to integrins as a way to associate with host cells. Besides *Yersinia* spp. a number of other microbial pathogens, such as *Bordetella pertussis, Leishmania mexicana, Histoplasma capsulatum, Borrelia burgdorferi,* echovirus 1, and adenovirus, directly bind host integrin receptors. Other pathogens, such as *Streptococcus* spp., *Staphylococcus aureus,* and *Pseudomonas aeruginosa,* bind a variety of ligands such as collagen, laminin, and fibronectin, which then associate with integrin receptors. The consequences of binding integrin receptors depend on the microorganism (26). In some cases, binding to an integrin receptor results in association of the microorganism with the cell surface without subsequent internalization, but in other cases, integrin binding is followed by uptake of the microbe.

Why, then, do some ligands (e.g., invasin) promote entry instead of attachment? It appears that internalization mediated by invasin is the result of a high-affinity interaction between this molecule and members of the β1-chain integrin receptor family (61). Bacteria coated with low-affinity ligands for the α5 β1 integrin receptor attach to the cell surface but are not internalized, while those coated with high-affinity ligands for the same receptor enter cells with high efficiency. Furthermore, overexpression of the α5 β1 integrin receptor increases the efficiency of bacterial internalization, indicating that receptor number is also important in determining whether a microorganism will be internalized or will remain on

the cell surface. Therefore, affinity of binding as well as receptor number determines the fate (i.e., internalization versus surface attachment) of a microorganism bound to an integrin receptor.

The importance of invasin in the pathogenesis of enteropathogenic *Yersinia* spp. has recently been investigated (45). A strain of *Yersinia enterocolitica* carrying a defined mutation in the *inv* gene was defective at colonizing Peyer's patches. However, this mutant and the wild-type strain had similar 50% lethal doses. These results indicate that invasin plays a crucial role in the initiation of *Y. enterocolitica* infection and early colonization in the mouse model of infection but is less important in the establishment of a lethal systemic infection. These data also underscore the shortcomings of the 50% lethal dose as a sole measure of virulence.

Besides invasin, other microbial determinants that promote entry into host cells by interacting with integrin receptors have been identified. The *Yersinia* protein YadA, for example, mediates an alternative entry pathway utilized by enteropathogenic *Yersinia* spp. YadA is a 60-kDa protein encoded in the virulence-associated plasmid of these microorganisms (5, 65). Monoclonal antibodies directed to the β1 chain of the integrin receptors inhibit YadA-mediated bacterial entry into cultured cells, indicating that this pathway may resemble that mediated by invasin. A number of microorganisms (e.g., *Leishmania* spp., *Legionella pneumophila,* and *H. capsulatum*) bind integrins that contain the β2 chain, in particular $\alpha_{mac}\beta2$ (4, 47, 53, 54, 57). This integrin is found in phagocytic cells, and many microorganisms gain access to these cells by interacting with this receptor. Presumably, gaining access to the phagocytic cell by interacting with the $\alpha_{mac}\beta2$ receptor allows these pathogens to avoid certain killing mechanisms such as the production of H_2O_2 (64). Uptake of this group of microorganisms via this pathway may occur by direct binding of microbial determinants to the $\alpha_{mac}\beta2$ integrin (e.g., *L. mexicana* and *Bordetella pertussis*) or by binding the complement component C3bi, which itself is a ligand of the $\alpha_{mac}\beta2$ integrin (e.g., *L. pneumophila*).

How the high-affinity interaction of a microbial ligand with an integrin receptor leads to microbial entry remains poorly understood. It is possible that this high-affinity interaction causes conformational changes in the receptor, triggering transmembrane signaling events leading to bacterial uptake. It has been shown that tyrosine kinase inhibitors block invasin-mediated uptake, suggesting that certain signaling events are indeed required for bacterial internalization (51).

THE *SALMONELLA* PARADIGM: ENTRY DUE TO SIGNALING AND SUBSEQUENT MODULATION OF HOST CELL CYTOSKELETON

Unlike the internalization mediated by ligands that bind integrins, *Salmonella* internalization involves cytoskeletal rearrangements that lead to dramatic changes in the plasma membrane at the point of direct contact between the bacterium and the host cell (60). *Salmonella* internalization appears to be the result of a complex interaction between the host cell and the bacterium. As a consequence of this interaction, surface appendages are assembled on the surfaces of salmonellae, a process that appears to be required for the subsequent triggering of signaling pathways in the host cell that leads to bacterial uptake (18). What follows is a brief review of the genetic basis of *Salmonella* entry as well as the host-cell signaling events that lead to bacterial uptake.

Molecular Genetic Basis of *Salmonella* Entry

The molecular genetic basis of *Salmonella* entry into host cells is very complex, a fact reflected by the large number of genetic loci required for bacterial entry into cultured mammalian cells. A number of these loci encode well-characterized determinants such as flagella or the surface lipopolysaccharide that have been shown to be important for efficient entry of some *Salmonella* serotypes (30, 32, 36, 41). However, it is not clear whether the role of these determinants is simply to facilitate "productive" contact between the bacterium and the host cell or whether these determinants play a more specific role in bacterial entry.

Genetic analyses have also identified a number of bona fide loci required for entry. Using Tn*phoA* mutagenesis, Stone et al. (58) screened *Salmonella enteritidis* for alkaline phosphatase-positive colonies that were defective in entry and/or attachment. Thirteen mutants were found and classified into six groups based on their relative abilities to enter HEp-2, CHO, and MDCK cells in culture. Southern hybridization and mapping phages positioned the 13 mutants to nine loci on the chromosome. The large number of loci having different effects in different cell lines suggests that *Salmonella* entry into host cells is a multifactorial process and/or that these organisms encode alternative entry pathways. The multifactorial nature of *Salmonella* entry is also supported by the work of Betts and Finlay (4a), who found four loci in *Salmonella typhimurium* that were important for entry into Caco-2 colonic epithelial cells. The mutants were not tested for their abilities to bind to host cells, so in some cases the noninvasive phenotype could be due to reduced adherence.

Elsinghorst and coworkers (9) successfully transferred invasiveness from *Salmonella* spp. to *E. coli* by cloning 33 kb of DNA from *Salmonella typhi* Ty2 into *E. coli* HB101. This region was mutagenized and shown to contain a minimum of four separate loci required for invasion. A homologous region from the *S. typhimurium* chromosome was unable to confer invasiveness on *E. coli* HB101, indicating that there are differences in the invasion mechanisms of these two *Salmonella* microorganisms.

Work by several laboratories has identified a number of genetic loci required for entry that are clustered at 59 min on the *Salmonella* chromosome (3, 13, 22, 35, 58). It is now estimated that perhaps as much as 50 kb of DNA within this region is required for entry of *Salmonella* spp. into mammalian cells.

The ability of *Salmonella* spp. to gain access to host cells is modulated by the bacterial growth state and a variety of environmental conditions (e.g., oxygen tension and osmolarity) that are known to alter levels of DNA superhelicity (10, 14, 34). On the basis of these observations, Lee et al. designed a strategy to isolate mutants of *S. typhimurium* that are able to gain access to cultured epithelial cells when the mutants are grown under nonpermissive conditions (35). This approach led to the identification of three classes of mutations. One class affected a number of *che* genes and conferred smooth swimming behavior, which presumably increases the frequency of productive contact between the bacterium and the host cell. The second class was made up of mutations in the promoter of *rho*, a transcriptional termination factor. Loss of *rho* alters gene expression and therefore may alter the expression of genes encoding invasion factors. The third class of mutations was in a locus termed *hil* (hyperinvasive locus), which is required for entry and is located at min 59 on the *Salmonella* chromosome. Presumably, *hil* encodes a regulatory factor or a product that is rate limiting for invasion of organisms grown under nonpermissive conditions. Behlau and Miller (3) identified another gene required for entry, *prgH*, which is closely linked to *hil*. Interestingly, the expression of *prgH* is regulated by the PhoP-PhoQ response regulator system, which is known to regulate genes required for survival within macrophages. These results suggest a link in the regulation of two different yet related events: entry and intracellular survival.

Galán and Curtiss identified an *S. typhimurium* locus (called *inv*) required for entry into cultured epithelial cells by complementing a noninvasive strain of *S. typhimurium* with DNA from a virulent *S. typhimurium* strain (13). The introduction of mutated *inv* genes into the chromosome of wild-type *S. typhimurium* rendered this organism much less virulent in orally challenged mice.

The *inv* locus is an array of at least 14 contiguous genes that have been mapped to min 59 of the *S. typhimurium* chromosome. Nucleotide sequence and functional analysis of this locus have identified at least 14 genes (*invA*

through *invO* but not *invB*) required for entry of *S. typhimurium* into cultured epithelial cells (1, 6, 8, 15, 19, 33). Mutations in these genes prevented entry of these organisms into cultured mammalian cells but did not affect their abilities to attach to the surfaces of host cells, indicating that attachment and entry are genetically separate events.

Nucleotide sequence analysis has revealed that a number of the Inv proteins share sequence similarity with proteins required for either flagellar biosynthesis or the export of determinants required for the virulence of a number of animal and plant pathogens (Table 1) (6, 8, 15, 19, 33). It is now clear that the *inv* locus encodes a *sec*-independent protein secretion system that presumably is required for the surface presentation of factors necessary for the entry process or for the biogenesis of a supramolecular structure involved in the delivery of entry determinants. This system, which is shared by *Yersinia*, *Shigella*, *Pseudomonas*, *Xanthomonas*, *Erwinia*, and *Salmonella* spp. and *E. coli*, involves mechanisms independent of the classic *sec*-dependent export pathway and different from other well-established *sec*-independent protein export apparatuses, such as the one represented by the hemolysin or pullulanase export systems (55, 62). The genes encoding these systems are clustered in either the chromosome or large virulence-associated plasmids, but despite the homology among the different gene products, the genetic organization of these loci varies significantly among different bacteria.

The similarity among these protein secretion systems goes beyond the protein sequence homology, since in some instances, complementation between homologous genes from different species has been observed. For example, mutations in the *S. typhimurium* genes *invA* and *invL* (*spaP*) can be complemented by the cognate *Shigella flexneri* genes *mxiA* and *spa-24*, respectively (17, 22). Mutations in the *S. typhimurium invA* gene, however, could not be complemented by the cognate *Y. enterocolitica* gene *lcrD*, although chimeric proteins consisting of the N-terminal half of LcrD and the C-

TABLE 1   *Salmonella* Inv protein homologs in other mammalian and plant pathogens[a]

Salmonella protein	Homolog			
	Shigella spp.	*Yersinia* spp.	Plant pathogens[b]	Flagellar assembly
InvA	MxiA	LcrD	HrpO	FlhA
InvB	Spa15	?	?	FliH
InvC	Spa47	YscN	HrpE	FliI
InvE	MxiC	LcrE	?	?
InvF	MxiE	VirF	HrpB	?
InvG	MxiD	YscC	HrpA	?
InvK	Spa33	YscQ	HrpQ	FliN
InvL	Spa24	YscR	HrpT	FliP
InvM	Spa9	YscS	MopD	FliQ
InvN	Spa29	YscT	MopE	FliR
InvO	Spa40	YscU	HrpN	FlhB

[a]Sequences were obtained from GenBank Release 79.
[b]The nomenclature used corresponds to genes in *P. solanacearum* (except MopD and MopE, which are from *Erwinia carotovora*), but homologous genes in *Xanthomonas campestris, Pseudomonas syringae,* and *E. amylovora* have been described.

terminal half of InvA successfully complemented an *invA* mutant (17). These results indicate that although there is functional conservation among these systems, the systems have been tailored to export specific determinants in each pathogen. In fact, there is little similarity among the primary sequences of the targets of these export systems thus far identified in the different microorganisms. These targets include the *Yersinia* outer proteins, or Yops (59), the Harpins of plant pathogenic bacteria such as *Erwinia amylovora* and *Pseudomonas solanacearum* (23, 63), and the invasion protein antigens of *Shigella* spp. (Ipa proteins) (56). All these proteins lack typical signal sequences and are found loosely associated with the bacterial cell envelope or in the culture supernatant of these organisms. Recently, a target of the protein secretion apparatus encoded in the *inv* locus was identified (7). This protein, termed InvJ, is found in the culture supernatant of invasion-competent *S. typhimurium* despite the fact that the protein lacks a typical signal sequence. Mutations in *invG* and *invC,* which encode components of the protein secretion system, prevented the export of InvJ. Another potential target of the *inv*-encoded protein secretion system is InvI, which shares significant homology with the *Shigella* IpaB protein.

The existence of a protein secretion apparatus involved in the export of determinants of invasion or the assembly of a supramolecular structure necessary for bacterial entry is consistent with observations made by Ginocchio et al. (18) with high-resolution, low-voltage scanning electron microscopy. Contact with MDCK epithelial cells induced the transient formation of appendages on the surface of *S. typhimurium.* The appendages were approximately 60 nm in diameter and 0.3 to 1.0 μm in length, which makes them significantly shorter and thicker than other well-characterized surface structures such as flagella, type I fimbriae, or bundle-forming pili. No invasion appendages were seen on bacteria grown to be competent for entry but not exposed to MDCK cells. The transient presence of the invasion appendages was correlated with the ability of *Salmonella* spp. to enter MDCK cells. Four invasion-defective mutants of *S. typhimurium,* each containing a nonpolar mutation in a single member of the *inv* locus (*invA, invC, invE,* and *invG*) were defective in the transient formation of the surface appendages. Studies have shown that de novo protein synthesis is not required for *Salmonella* entry into cultured cells when the bacteria are grown under conditions that render them competent for entry (37). Consistent with these findings, the addition of chloramphenicol at levels that immediately block protein synthesis, does not prevent formation of the invasion appendages. To establish a large structure on the surface of the cell would likely require a battery of proteins. This is the case with the P pilus and the flagella of gram-negative bacteria. Assembly of P pilus requires at least 11 genes encoding proteins that regulate gene expression, form the pilus structure, and serve as chaperones, ushers, and modulators during pilus formation, while assembly of flagella requires the concerted actions of at least 44 proteins (24, 38). The complexity of the genetic basis of *Salmonella* invasion may be due to the requirement for these organisms to assemble a surface organelle needed for efficient entry into host cells.

Assembly of such a complex structure may require a coordinated regulation of the expression of the genes encoding the structure itself as well as the assembly apparatus. Evidence indicating that several regulatory mechanisms modulate the expression of genes required for *Salmonella* entry is beginning to accumulate. The expression of the genes of the *inv* locus is regulated by changes in the levels of DNA superhelicity in response to environmental conditions such as osmolarity (14). Increased *invA* expression, resulting from varied levels of DNA superhelicity, correlate with an increased ability of wild-type salmonellae to enter cultured epithelial cells. Furthermore, mutations in *topA,* which drastically influence the levels of DNA superhelicity and severely affect *invA* expression, impede the ability of *S. typhimurium* to enter cultured epithelial cells (14). An additional level of regulation of *inv* gene expression may involve InvF, which has significant homology to the AraC family of positive transcriptional activators (33). However, mutations in *invF* do not significantly change the levels of expression of the *inv* genes tested, suggesting that InvF might regulate the expression of other genes located elsewhere in the chromosome or that its role as regulator of the *inv* genes is exerted under conditions different from the ones tested in vitro. Additional regulatory mechanisms are likely to be encoded in the *hil* locus, since certain mutations in this locus alter the regulation of the invasion phenotype (35).

Host-Cell Signal Transduction Pathways and *Salmonella* Entry

The interaction of *Salmonella* spp. with cultured epithelial cells leads to profound alterations of the host cell plasma membrane at the point of bacterium-host cell contact. This rapid restructuring of the epithelial-cell plasma membrane bears many of the hallmarks of membrane ruffling, a phenomenon that is induced by cellular growth factors, oncogenes, and a variety of mitogens (2, 50). These changes in the plasma membrane are also accompanied by fluxes in the levels of intracellular free Ca^{2+} ($[Ca^{2+}]_i$) and dramatic rearrangements of the cytoskeleton underlying the site of bacterial contact (11, 43). A number of cytoskeletal proteins, including actin, α-actinin, ezrin, talin, tubulin, and tropomyosin, accumulate beneath the membrane ruffle (11). The induction of membrane ruffles appears to be essential for *S. typhimurium* entry, because mutants unable to induce ruffling and associated rearrangements in the underlying cytoskeleton are deficient in their abilities to enter cultured mammalian cells (15, 19). It is unclear whether the membrane ruffles induced by *Salmonella* spp. are equivalent to those induced by other stimuli. Since the molecular bases of ruffle formation and the functions of membrane ruffles are incompletely understood, it is quite possible that ruffles differ depending on the cell type and the stimulus. Indeed, although epidermal growth factor (EGF) induction of membrane ruffles in Henle-407 cells could rescue entry-defective *S. typhimurium,* insulin-induced membrane ruffles in the same cells could not (16).

The host-cell signaling pathway leading to *S. typhimurium* internalization into the Henle-407 intestinal epithelial cell line has been studied in some detail (43). One early response of Henle-407 cells to *S. typhimurium* binding is activation of the EGF receptor (16). Phosphorylation of the EGF receptor correlates with *Salmonella* entry, as an isogenic mutant defective in entry is incapable of activating the EGF receptor (16). Furthermore, addition of EGF to Henle-407 cells infected with this invasion-defective mutant results in the internalization of these organisms. Activation of the EGF receptor by *Salmonella* spp. leads to the activation (via phosphorylation) of mitogen-activated protein (MAP) kinase (43). This likely activates phospholipase A_2 (PLA_2), which in turn produces arachidonic acid from membrane phospholipids. Arachidonic acid is metabolized via the lipoxygenase pathway to the end product leukotriene D_4 (LTD_4). Either directly or indirectly, LTD_4 opens Ca^{2+} channels in the plasma membrane, causing an influx of Ca^{2+}. An increase in $[Ca^{2+}]_i$ is necessary for *Salmonella* entry into Henle-407 cells, since calcium chelators or nonspecific calcium channel blockers prevent bacterial entry (43).

Recent studies on the mechanisms of EGF-induced actin remodeling in a number of cell lines have implicated arachidonic acid metabolites in this process (46). In particular, it was shown that cortical actin polymerization, induced by EGF, is mediated by lipoxygenase metabolites. These observations are consistent with the requirements of PLA_2 and 5-lipoxygenase activities and the involvement of an EGF receptor pathway in *S. typhimurium* entry into Henle-407 cells. In a number of studies, arachidonic acid and its derivatives were shown to modulate the activities of a variety of GTPase-activating proteins or GDP dissociation inhibitors, thereby regulating the activities of a variety of small GTP-binding proteins. Therefore, arachidonic acid metabolites may influence cytoskeletal rearrangements through their influence on small actin-organizing, GTP-binding proteins.

Some of the GTP-binding proteins that might play a role in the uptake of *Salmonella* spp. are Rac, Rho, and Ras, which have been found to regulate growth factor-mediated membrane ruffling and stress fiber formation (2, 48, 49). Recent work, however, has shown that inhibiting these three GTPases by microinjecting neutralizing molecules into the host cell has no effect on the abilities of *Sal-monella* spp. to induce membrane ruffles in infected MDCK, HEp-2, and Swiss 3T3 fibroblast cells (31). The actual levels of *Salmonella* entry into these microinjected cells were not determined.

Salmonella spp. can enter virtually any cell line, some of which do not express the EGF receptor. This ability indicates that these organisms can activate more than one signal transduction pathway. In HeLa and B82 cells, for example, *Salmonella* can induce an increase in the concentration of inositol phospholipids, presumably as a result of the activation of phospholipase Cγ (PLCγ) (42, 52). Inositol phospholipids are largely responsible for the induction of calcium release from intracellular stores and therefore are likely to be the mediators of the intracellular increase of this cation that follows *Salmonella* interaction with these cells (42). Consistent with this notion are the observations that in these cell lines, inhibitors of PLCγ and chelators of intracellular but not extracellular Ca^{2+} block bacterial entry.

It is not clear how salmonellae trigger different signaling pathways in different cells. It is unlikely that salmonellae encode multiple effector molecules, each one recognizing a specific host cell receptor. In fact, mutations in the *inv* locus that affect entry into one cell line equally affect entry into other cell lines in which different signaling pathways have been implicated. More likely, salmonellae may use an effector molecule(s) that binds to moieties, such as oligosaccharides, commonly found on the extracellular portions of many transmembrane receptors. This would be similar to the action of some lectins, like concanavalin A and wheat germ hemagglutinin, that bind a variety of glycosylated transmembrane receptors and induce mitogenic responses in the cells, presumably by causing clustering of the receptors. The signal transduction pathway that salmonellae activate might then depend on the relative receptor concentrations on the surface of the host cell. For instance, the EGF receptor may be the one engaged by salmonellae in Henle-407 cells, because a high concentration of this receiver ($\sim 2 \times 10^6$ receptors per cell) is present in this cell line.

Although salmonellae can activate different signal transduction pathways in different host cells, the end result of the signaling events (i.e., cytoskeletal rearrangements leading to membrane ruffling and bacterial internalization) appears to be similar in every case. Therefore, it is possible that different signaling pathways share common effector molecules downstream of the activated receptors (Fig. 1). One candidate for this effector molecule is Ca^{2+}, because a rise in $[Ca^{2+}]_i$ is a prerequisite for entry of

salmonellae into epithelial cells (42, 43). Fluxes in $[Ca^{2+}]_i$ may affect salmonellae-induced membrane ruffling by altering the properties of various cytoskeletal proteins. A number of actin-bundling proteins, like gelsolin, villin, and fimbrin, will sever actin filaments following an increase in $[Ca^{2+}]_i$ (39). This might result in depolymerization of the subcortical actin and the microfilaments that stabilize the microvilli, providing a reservoir of actin monomers that participate in membrane ruffling around the bound bacteria. The reorganization of actin may also be assisted by profilin, a molecule that

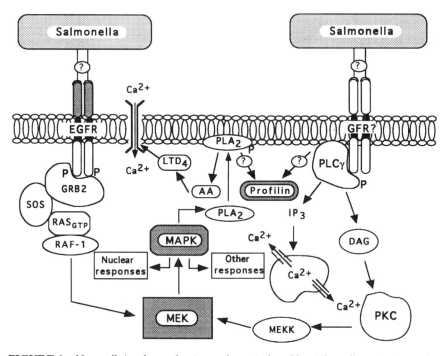

FIGURE 1 Host cell signal transduction pathway induced by *Salmonella* entry. In Henle-407 cells, salmonellae bind to and activate the EGF receptor (EGFR). Phosphorylation of the EGFR occurs, leading to activation of MAP kinase (MAPK). This likely activates PLA_2, which in turn produces arachidonic acid (AA) from membrane phospholipids. AA is metabolized via the lipoxygenase pathway to the end product LTD_4. Either directly or indirectly, LTD_4 opens Ca^{2+} channels in the plasma membrane, causing an influx of Ca^{2+}. An increase in $[Ca^{2+}]_i$ is necessary for *Salmonella* entry. In other cell types (e.g., HeLa and B82 cells), salmonellae bind to similar growth factor receptors (GFR?). This increases the concentration of inositol phospholipids (IP_3), presumably as a result of the activation of phospholipase Cγ (PLCγ). Inositol phospholipids then induce calcium release from intracellular stores. Receptor activation of PLCγ and possibly PLA_2 releases profilin from its association with the plasma membrane. Profilin then increases actin polymerization by elevating nucleotide exchange at the barbed end of actin filaments and by using the free energy of ATP hydrolysis associated with actin polymerization to facilitate actin assembly.

increases actin polymerization by elevating nucleotide exchange at the barbed ends of actin filaments and by using the free energy of ATP hydrolysis associated with actin polymerization to facilitate actin assembly (20, 44). Receptor activation of PLCγ and possibly other phospholipases, like PLA$_2$, releases profilin from its association with the plasma membrane. Cytoplasmic profilin is then available to interact with actin.

The nature of cellular responses induced by salmonellae suggests that signal transduction pathways triggered by different receptors intersect at one or more molecules. One point of convergence might be the MAP kinase (MEK) (Fig. 1). It has been suggested that MAP kinase plays a role in the induction of membrane ruffling, because it has been shown that upon stimulation by growth factors, one isoform of MAP kinase (p41^{mapk}) is localized to the membrane ruffle (21). MAP kinase also indirectly alters levels of gene transcription. *Salmonella* activation of MAP kinase might activate genes that lead to the production of cytokines, which are important in the pathogenesis of the inflammatory diarrhea induced by salmonellae.

CONCLUSIONS

We have briefly described two contrasting general mechanisms of bacterial entry into nonphagocytic cells. Despite the mechanistic differences, some common features may yet emerge, particularly considering that host-cell signal transduction pathways leading to cytoskeletal rearrangements are likely to be involved in both systems. It is also possible that some signaling events triggered at the cell surface, although independent of entry mechanisms, have a profound impact on the outcome of the infection. For example, signaling events leading to the production of host cell molecules capable of modulating inflammatory or immune responses may be important in the generation of pathology or in the establishment of bacterial colonization. As more mechanisms of bacterial entry are unraveled, we will likely find that other microorganisms utilize similar invasion strategies. In addition, increasing our understanding of bacterial entry will enhance our knowledge of eukaryotic cell biology.

REFERENCES

1. **Altmeyer, R. M., J. K. McNern, J. C. Bossio, I. Rosenshine, B. B. Finlay, and J. E. Galán.** 1993. Cloning and molecular characterization of a gene involved in *Salmonella* adherence and invasion of cultured epithelial cells. *Mol. Microbiol.* **7:**89–98.
2. **Bar-Sagi, D., and J. R. Ferasmico.** 1986. Induction of membrane ruffling and fluid-phase pinocytosis in quiescent fibroblasts by the ras proteins. *Science* **233:**1061–1068.
3. **Behlau, I., and S. J. Miller.** 1993. A PhoP-repressed gene promotes *Salmonella typhimurium* invasion of epithelial cells. *J. Bacteriol.* **175:** 4475–4484.
4. **Bellinger, K. C., and M. A. Horwitz.** 1990. Complement component C3 fixes selectively to the major outer membrane protein (MOMP) of *Legionella pneumophila* and mediates phagocytosis of liposome-MOMP complexes by human monocytes. *J. Exp. Med.* **172:**1201–1210.
4a. **Betts, J., and B. B. Finlay.** 1992. Identification of *Salmonella typhimurium* invasiveness loci. *Can. J. Microbiol.* **38:**852–857.
5. **Bliska, J. B., M. C. Copass, and S. Falkow.** 1993. The *Yersinia pseudotuberculosis* adhesin YadA mediates intimate bacterial attachment to and entry into HEp-2 cells. *Infect. Immun.* **61:** 3914–3921.
6. **Collazo, C., and J. E. Galán.** 1993. Mutagenesis analysis of the *inv* locus of *Salmonella typhimurium*, abstr. B-71, p. 38. *Abstr. Gen. Meet. Am. Soc. Microbiol.*
7. **Collazo, C. M., M. K. Zierler, and J. E. Galán.** Functional analysis of the *Salmonella typhimurium* invasion genes *invI* and *invJ* and identification of a target of the protein secretion apparatus encoded in the *inv* locus. *Mol. Microbiol.* **15:**25–38.
8. **Eichelberg, K., C. Ginocchio, and J. E. Galán.** 1994. Molecular and functional characterization of the *Salmonella typhimurium* invasion genes *invB* and *invC*: homology of InvC to the F$_0$F$_1$ ATPase family of proteins. *J. Bacteriol.* **176:** 4501–4510.
9. **Elsinghorst, E. A., L. S. Baron, and D. J. Kopecko.** 1989. Penetration of human intestinal epithelial cells by *Salmonella*: molecular cloning and expression of *Salmonella typhi* invasion determinants in *Escherichia coli*. *Proc. Natl. Acad. Sci. USA* **86:**5173–5177.
10. **Ernst, R. K., D. M. Domboski, and J. M. Merrick.** 1990. Anaerobiosis, type 1 fimbriae, and growth phase are factors that affect invasion of

Hep-2 cells by *Salmonella typhimurium*. *Infect. Immun.* **58**:2014–2016.

11. **Finlay, B. B., and S. Ruschkowski.** 1991. Cytoskeletal rearrangements accompanying *Salmonella* entry into epithelial cells. *J. Cell Sci.* **99**:283–296.

12. **Galán, J. E.** 1994. Interactions of bacteria with non-phagocytic cells. *Curr. Opin. Immunol.* **6**:590–595.

13. **Galán, J. E., and R. Curtiss III.** 1989. Cloning and molecular characterization of genes whose products allow *Salmonella typhimurium* to penetrate tissue culture cells. *Proc. Natl. Acad. Sci. USA* **86**:6383–6387.

14. **Galán, J. E., and R. Curtiss III.** 1990. Expression of *Salmonella typhimurium* genes required for invasion is regulated by changes in DNA supercoiling. *Infect. Immun.* **58**:1879–1885.

15. **Galán, J. E., C. Ginocchio, and P. Costeas.** 1992. Molecular and functional characterization of the *Salmonella typhimurium* invasion gene *invA*: homology of InvA to members of a new protein family. *J. Bacteriol.* **17**:4338–4349.

16. **Galán, J. E., J. Pace, and M. J. Hayman.** 1992. Involvement of the epidermal growth factor receptor in the invasion of cultured mammalian cells by *Salmonella typhimurium*. *Nature* (London) **357**:588–589.

17. **Ginocchio, C. C., and J. E. Galán.** 1995. Functional conservation among members of the *Salmonella typhimurium* InvA family of proteins. *Infect. Immun.* **63**:729–732.

18. **Ginocchio, C., S. B. Olmsted, C. L. Wells, and J. E. Galán.** 1994. Contact with epithelial cells induces the formation of surface appendages on *Salmonella typhimurium*. *Cell* **76**:717–724.

19. **Ginocchio, C., J. Pace, and J. E. Galán.** 1992. Identification and molecular characterization of a *Salmonella typhimurium* gene involved in triggering the internalization of salmonellae into cultured epithelial cells. *Proc. Natl. Acad. Sci. USA* **89**:5976–5980.

20. **Goldschmidt-Clermont, P. J., L. M. Machesky, S. K. Doberestein, and T. D. Pollard.** 1991. Mechanism of the interaction of human platelet profilin with actin. *J. Cell Biol.* **113**:1081–1089.

21. **Gonzalez, F. A., A. Seth, D. L. Raden, D. S. Bowman, F. S. Fay, and R. J. Davis.** 1993. Serum-induced translocation of mitogen-activated protein kinase to the cell surface ruffling membrane and the nucleus. *J. Cell Biol.* **122**:1089–1101.

22. **Groisman, E. A., and H. Ochman.** 1993. Cognate gene clusters govern invasion of host epithelial cells by *Salmonella typhimurium* and *Shigella flexneri*. *EMBO J.* **12**:3779–3787.

23. **He, S. Y., H.-C. Huang, and A. Collmer.** 1993. *Pseudomonas syringae* pv. *syringae* Harpin[Pss]: a protein that is secreted via the Hrp pathway and elicits the hypersensitive response in plants. *Cell* **73**:1255–1266.

24. **Hultgren, S. J., S. Abraham, M. Caparon, P. Falk, J. St. Geme III, and S. Normark.** 1993. Pilus and nonpilus bacterial adhesins: assembly and function in cell recognition. *Cell* **73**:887–901.

25. **Hynes, R. O.** 1992. Integrins: versatility, modulation, and signaling in cell adhesion. *Cell* **69**:11–25.

26. **Isberg, R. R.** 1991. Discrimination between intracellular uptake and surface adhesion of bacterial pathogens. *Science* **252**:934–938.

27. **Isberg, R. R., and S. Falkow.** 1985. A single genetic locus encoded by *Yersinia pseudotuberculosis* permits invasion of cultured animal cells by *Escherichia coli*. *Nature* (London) **317**:262–264.

28. **Isberg, R. R., and J. M. Leong.** 1990. Multiple beta 1 chain integrins are receptors for invasin, a protein that promotes bacterial penetration into mammalian cells. *Cell* **60**:861–871.

29. **Isberg, R. R., and G. Tran Van Nhieu.** 1994. Binding and internalization of microorganisms by integrin receptors. *Trends Microbiol.* **2**:10–14.

30. **Jones, B. D., C. A. Lee, and S. Falkow.** 1992. Invasion by *Salmonella typhimurium* is affected by the direction of flagellar rotation. *Infect. Immun.* **60**:2475–2480.

31. **Jones, B. D., H. F. Paterson, A. Hall, and S. Falkow.** 1993. *Salmonella typhimurium* induces membrane ruffling by a growth factor-receptor-independent mechanism. *Proc. Natl. Acad. Sci. USA* **90**:10390–10394.

32. **Jones, G. W., and L. A. Richardson.** 1981. The attachment to, and invasion of HeLa cells by *Salmonella*: the contribution of mannose-sensitive and mannose-resistant haemagglutinating activities. *J. Med. Microbiol.* **127**:361–370.

33. **Kaniga, K., J. C. Bossio, and J. E. Galán.** The *Salmonella typhimurium* invasion genes *invF* and *invG* encode homologues to the PulD and AraC family of proteins. *Mol. Microbiol.* **13**:555–568.

34. **Lee, C. A., and S. Falkow.** 1990. The ability of *Salmonella* to enter mammalian cells is affected by bacterial growth state. *Proc. Natl. Acad. Sci. USA* **87**:4304–4308.

35. **Lee, C. A., B. D. Jones, and S. Falkow.** 1992. Identification of a *Salmonella typhimurium* invasion locus by selection of hyperinvasive mutants. *Proc. Natl. Acad. Sci. USA* **89**:1847–1851.

36. **Liu, S. L., T. Ezaki, H. Miura, K. Matsui, and E. Yabuuchi.** 1988. Intact motility as a *Salmonella typhi* invasion-related factor. *Infect. Immun.* **56**:1967–1973.

37. **Macbeth, K. J., and C. A. Lee.** 1993. Prolonged inhibition of bacterial protein synthesis

abolishes *Salmonella* invasion. *Infect. Immun.* **61:** 1544–1546.

38. **Macnab, R. M.** 1992. Genetics and biogenesis of bacterial flagella. *Annu. Rev. Genet.* **26:** 131–158.

39. **Mooseker, M. S.** 1985. Organization, chemistry and assembly of the cytoskeletal apparatus of the intestinal brush border. *Annu. Rev. Cell. Biol.* **1:** 209–241.

40. **Moulder, J. W.** 1985. Comparative biology of intracellular parasitism. *Microbiol. Rev.* **49:** 298–337.

41. **Mroczenski-Wildey, M. J., J. L. Di Fabio, and F. C. Cabello.** 1989. Invasion and lysis of HeLa cell monolayers by *Salmonella typhi*: the role of lipopolysaccharide. *Microb. Pathog.* **6:** 143–152.

42. **Pace, J., L. M. Chen, and J. E. Galán.** Unpublished data.

43. **Pace, J., M. J. Hayman, and J. E. Galán.** 1993. Signal transduction and invasion of epithelial cells by *Salmonella typhimurium. Cell* **72:** 505–514.

44. **Pantaloni, D., and M.-F. Carlier.** 1993. How profilin promotes actin filament assembly in the presence of thymosin β4. *Cell* **75:** 1007–1014.

45. **Pepe, J. C., and V. L. Miller.** 1993. *Yersinia enterocolitica* invasin: a primary role in the initiation of infection. *Proc. Natl. Acad. Sci. USA* **90:** 6473–6477.

46. **Peppelenbosch, M. P., L. G. J. Tertoolen, W. J. Hage, and S. W. de Last.** 1993. Epidermal growth factor-induced actin remodeling is regulated by 5-lipoxygenase products. *Cell* **74:** 565–575.

47. **Relman, D., E. Tuomanen, S. Falkow, D. T. Golenbock, K. Saukkonen, and S. D. Wright.** 1990. Recognition of a bacterial adhesion by an integrin: macrophage CR3 (alpha M beta 2, CD11b/CD18) binds filamentous hemagglutinin of *Bordetella pertussis. Cell* **61:** 1375–1382.

48. **Ridley, A. J., and A. Hall.** 1992. The small GTP-binding protein rho regulates the assembly of focal adhesions and actin stress fibers in response to growth factors. *Cell* **70:** 389–399.

49. **Ridley, A. J., H. F. Paterson, C. L. Johnston, D. Diekmann, and A. Hall.** 1992. The small GTP-binding protein rac regulates growth factor-induced membrane ruffling. *Cell* **70:** 401–410.

50. **Rijken, P. J., W. J. Hage, P. M. Van Bergen En Henegouwen, A. J. Verkieij, and J. Boonstra.** 1991. Epidermal growth factor induces rapid reorganization of the actin microfilament system in human A431 cells. *J. Cell Sci.* **100:** 491–499.

51. **Rosenshine, I., V. Duronio, and B. B. Finlay.** 1992. Tyrosine protein kinase inhibitors block invasin-promoted bacterial uptake by epithelial cells. *Infect. Immun.* **60:** 2211–2217.

52. **Ruschkowski, S., I. Rosenshine, and B. B. Finlay.** 1992. *Salmonella typhimurium* induces an inositol phosphate flux in infected epithelial cells. *FEBS Lett.* **74:** 121–126.

53. **Russell, D. G.** 1987. The macrophage-attachment glycoprotein gp63 is the predominant C3-acceptor site on *Leishmania mexicana* promastigotes. *Eur. J. Biochem.* **164:** 213–221.

54. **Russell, D. G., and S. D. Wright.** 1988. Complement receptor type 3 (CR3) binds to an Arg-Gly-Asp-containing region of the major surface glycoprotein, gp63, of *Leishmania* promastigotes. *J. Exp. Med.* **168:** 279–292.

55. **Salmond, G. P. C., and P. J. Reeves.** 1993. Membrane traffic wardens and protein secretion in Gram-negative bacteria. *Trends Biochem. Sci.* **18:** 7–12.

56. **Sansonetti, P. J.** 1992. Molecular and cellular biology of *Shigella flexneri* invasiveness: from cell assay systems to shigellosis. *Curr. Top. Microbiol. Immunol.* **180:** 1–19.

57. **Schnur, R. A., and S. L. Newman.** 1990. The respiratory burst response to *Histoplasma capsulatum* by human neutrophils: evidence for intracellular trapping of superoxide anion. *J. Immunol.* **144:** 4765–4772.

58. **Stone, B. J., C. M. Garcia, J. L. Badger, T. Hassett, R. I. F. Smith, and V. Miller.** 1992. Identification of novel loci affecting entry of *Salmonella enteritidis* into eukaryotic cells. *J. Bacteriol.* **174:** 3945–3952.

59. **Straley, S. C., E. Skrzypek, G. V. Plano, and J. B. Bliska.** 1993. Yops of *Yersinia* spp. pathogenic for humans. *Infect. Immun.* **61:** 3105–3110.

60. **Takeuchi, A.** 1967. Electron microscopic studies of experimental Salmonella infection. 1. Penetration into the intestinal epithelium by *Salmonella typhimurium. Am. J. Pathol.* **50:** 109–136.

61. **Tran Van Nhieu, G., and R. R. Isberg.** 1993. Bacterial internalization mediated by beta 1 chain integrins is determined by ligand affinity and receptor density. *EMBO J.* **12:** 1887–1895.

62. **Van Gijsegem, F., S. Genin, and C. Boucher.** 1993. Conservation of secretion pathways for pathogenicity determinants of plant and animal bacteria. *Trends Microbiol.* **1:** 175–180.

63. **Wei, Z. M., R. J. Laby, C. H. Zumoff, D. W. Bauer, S. Y. He, A. Collmer, and S. V. Beer.** 1992. Harpin, elicitor of the hypersensitive response produced by the plant pathogen *Erwinia amylovora. Science* **257:** 85–88.

64. **Wright, S. D., and S. C. Silverstein.** 1983. Receptors for C3b and C3bi promote phagocytosis but not the release of toxic oxygen from human phagocytes. *J. Exp. Med.* **158:** 2016–2023.

65. **Yang, Y., and R. R. Isberg.** 1993. Cellular internalization in the absence of invasin expression is promoted by the *Yersinia pseudotuberculosis yadA* product. *Infect. Immun.* **61:** 3907–3913.

MECHANISMS OF MUCOSAL COLONIZATION AND PENETRATION BY BACTERIAL PATHOGENS

B. Brett Finlay and Annette Siebers

3

Many bacterial pathogens interact with surfaces on the body in ways that can result in disease. These interactions often occur on mucosal epithelial cells in various body locations. Because of their surface exposure, such cells are the initial contact sites for incoming pathogens. Once adherent on such mucosal cells, pathogens behave in a variety of ways. Some organisms remain surface localized without invasion and systemic spread. Alternatively, several pathogens are capable of breaching host barriers, especially epithelial and endothelial linings, which normally constitute a primary defense against bacterial invasion. Those that penetrate these barriers use two basic mechanisms of entry into the host: either they penetrate through the host cells (transcellular route), or they pass between these cells, often after disruption of the intercellular junctions or injury to the cells (pericellular route). Bacteria that penetrate mucosal barriers are adept at exploiting host cell trafficking systems, thereby facilitating their passage across these barriers. Organisms that penetrate through epithelial cells have special virulence factors, including invasion mechanisms, and the capacity to survive and replicate within host cells. Having these pathogenic attributes collectively allows these organisms to function as intracellular parasites. Such organisms usually also have other virulence factors that promote survival and replication in deeper tissues and phagocytic cells that are encountered after the initial barrier penetration.

HOST DEFENSES AGAINST MUCOSAL COLONIZATION

Mucosal surfaces have several mechanisms of defense that prevent bacterial colonization and penetration through them (31, 45, 46). Host mechanical defenses, including nonspecific mechanisms, serve as barriers that are effective against most bacteria. These mechanisms include the sweeping actions of the ciliated epithelium that remove bacteria trapped in mucus and the continual mechanical actions of peristalsis in the stomach and intestines that propel bacteria through and out of the gastrointestinal tract, thereby decreasing the interactions of the bacteria with host cell surfaces. Flushing of the bladder and blinking of the eyes are other effective inhibitors of bacterial colonization.

The resident flora, especially in the upper respiratory and gastrointestinal tracts, plays a large role in inhibiting colonization by incoming pathogens. Microbial antagonism by the

B. Brett Finlay and Annette Siebers   Biotechnology Laboratory, Department of Biochemistry and Molecular Biology, and Department of Microbiology and Immunology, University of British Columbia, Vancouver, British Columbia, Canada V6T-1Z3.

Virulence Mechanisms of Bacterial Pathogens, 2nd ed., Edited by J. A. Roth et al.
© 1995 American Society for Microbiology, Washington, D.C.

resident flora can work in several ways. For example, normal flora may prevent adherence by occupying the same receptor sites that pathogens adhere to. Fierce competition for nutrients and production of antimicrobial metabolites by indigenous microorganisms also prevent establishment of incoming pathogens. Pathogens have evolved strategies for circumventing this competition. It is probably not by chance that *Vibrio cholerae* secretes a toxin that removes virtually all of the normal flora in the intestine, giving the remaining cholera organisms a few days of "microbial bliss" without competition from other bacteria.

The host has a plethora of other defenses to further protect mucosal surfaces. These include cellular defenses such as resident macrophages and polymorphonuclear leukocytes. In addition, several antibacterial factors are secreted by the host onto mucosal surfaces. Such factors include secretory immunoglobulin A, lysozyme, lactoferrin, and other antimicrobial peptides. However, most pathogens have developed strategies for avoiding such factors. For example, many pathogens that reside extracellularly on mucosal surfaces produce capsules that resist antibody and complement deposition (opsonization) and, ultimately, clearance by phagocytic cells. In addition, some can produce molecules that bind lactoferrin, thereby acquiring essential iron. Finally, many pathogens that reside on host mucosal surfaces have developed strategies for neutralizing the host immune system, such as production of immunoglobulin A proteases (28). Alternatively, several pathogens can vary their surface protein profiles (antigenic variation), thereby avoiding host immune surveillance (39). Since most of these topics are covered in other chapters, we concentrate here on mechanisms used by mucosal pathogens to colonize and penetrate mucosal barriers.

MUCUS PENETRATION

Mucosal surfaces are coated with a thick covering of mucus, which is composed of many carbohydrates. This layer is the first barrier that pathogens encounter when they enter the host. However, little is known about how colonizing pathogens penetrate this barrier. Some organisms have the capacity to break down mucus by using secreted enzymes. Another factor contributing to penetration of the mucous layer is motility. For example, motility appears to be involved in *V. cholerae* colonization (16). It was originally thought that motility was needed for *Salmonella* virulence and that motility aided in penetrating the mucous barrier, although studies in this area continue (47). Motility does enhance *Salmonella* invasion and penetration of epithelial cells, although it is not essential. However, other pathogens that penetrate mucosal surfaces and interact intimately with mucosal epithelial cells are nonmotile. Some examples include *Shigella* species and yersiniae (at 37°C). The mechanisms by which this penetration occurs and the role mucus plays in this process remain uncharacterized.

M cells (specialized epithelial cells that are discussed below) have little mucus coating their apical surfaces, in contrast to columnar epithelial cells, which have a much deeper mucous coating. Since it appears that most organisms penetrate through M cells (see below), the lack of a mucous barrier on M cells suggests that mucus probably does not play a significant role in colonization of these specialized cells. Alternatively, some bacterial toxins that lead to diarrhea also lead to loss of mucus. This loss may facilitate access to mucosal epithelial cells, although the toxin-producing organisms would need to avoid clearance during this process.

ADHERENCE

Adherence to mucosal surfaces plays a large role in mucosal colonization for nearly all mucosal pathogens. The actual mechanisms used for adherence often involve binding of bacterial surface appendages such as pili (fimbriae) to host cell surface receptors. Questions on such matters as apical or basolateral receptor expression and the implications for adherence are currently being addressed. Much research has been done in this area, including the characterization of the bacterial genes involved in pili synthesis and identification of the host receptors (31, 45, 46). Alternatively, bacteria can make nonfimbrial adhesins that can mediate adherence. Such examples include the afimbrial adhesins of *Escherichia coli* and the filamentous hemagglutinin of *Bordetella pertussis*. Some of these topics are covered in other chapters of this book.

In addition to adhering to mucosal surface receptors, many bacterial adhesins mediate bacterium-bacterium contact, resulting in formation of apically bound microcolonies. Some pathogens that mediate this type of adherence include enteropathogenic *E. coli* (EPEC) and *V. cholerae* (24). The role that this interbacterial adherence plays in mucosal colonization remains to be determined, although it is tempting to speculate that once a pathogen is successfully bound to a host surface, it can spread its good fortune to help its neighbors. Alternatively, as bacteria divide on the host surface, they can remain localized by binding to their siblings rather than directly to the host cell surface, which may be limited in area. This interbacterial adherence suggests either that bacteria express specialized receptors that resemble the host cell receptor or that adhesins can recognize different receptors on the bacteria and host cells. Alternatively, bacteria express different types of adhesins for interspecies (bacterium-host cell) and intraspecies (bacterium-bacterium) contact.

COLUMNAR EPITHELIAL CELLS

The intestinal barrier is a mixed-cell epithelium linked by tight junctions and consisting of columnar absorptive cells, goblet secretory cells, M cells, and Paneth cells. Paneth cells are phagocytic cells that produce lysozyme and several other antimicrobial products and are localized in the bottom of crypts in the small intestine (27). The main function of goblet cells is to secrete mucus to coat the epithelial surface. Most epithelial cells are columnar absorptive cells, whose major functions include providing a functional impermeable epithelial barrier and transporting nutrients.

Columnar epithelial cells are closely linked to neighboring epithelial cells. In addition to the zona adherens, which is a specialized structure involved in cell-cell adherence, tight junctions (zona occludens) form an impermeable seal between epithelial cells (40). These structures are supported by actin and form a belt near the apical surface of the cell. Tight junctions also divide epithelial cells into two domains, apical and basolateral, and the cell thus has a polarized distribution of proteins and lipids that differs significantly on these two surfaces. The apical (luminal) surface of an intestinal epithelial cell contains well-defined structures called microvilli. These structures form fingerlike projections and are supported by actin filaments. They also form a unique structure that mucosal pathogens adhere to. Some pathogens adhere to microvilli at their tips and do not distort the host cell surface. However, many other pathogens alter the apical surface, often denuding the microvilli and thus perhaps providing the pathogens additional surface area with which to interact. For example, *Salmonella* species denude the epithelial microvilli, resulting in a bulging (ruffling) of the apical surface (9, 15). Infected cells exhibiting such deformations are more susceptible to additional *Salmonella* invasion.

Several mucosal pathogens are unable to adhere to the apical surfaces of polarized epithelial monolayers. For example, *Yersinia* and *Shigella* species interact specifically only with basolateral surfaces (32). This specificity presents a major dilemma, since basolateral surfaces are not exposed to incoming pathogens. There are several possible ways of circumventing such a problem. As discussed below, M cells indiscriminately take up most bacteria, including those, such as *V. cholerae*, that are noninvasive. Pathogens that utilize such a portal of entry can then gain access to basolateral surfaces. Alternatively, *Shigella* species may initially trigger chemotaxis of phagocytic cells such as polymorphonuclear leukocytes. As these cells are moving toward the bacteria, they span the mucosal epithelial barrier and open the tight junctions, at which time Shigellae are either internalized by these cells and penetrate intracellularly or slip through the intercellular route to the basal area of the mucosa (37).

THE ROLE OF M CELLS

In contradiction to the obvious need for a physical barrier provided by the columnar cells, host protection by immunologic processes requires that macromolecules and particulates from the gut lumen have a way to pass through the mucosal barrier in order to reach the lymphoid system. The gut-associated lymphoid tissue as a whole consists of Peyer's patches, isolated lymphoid follicles, the appendix, and mesenteric lymph nodes (27). Of special interest for the interaction with enteric bacteria are the Peyer's patches, which are aggregates of subepithelial lymphoid follicles that are present throughout the small intestines of mammals but are most prominent in the ileum. The luminal cover of the dome-shaped follicles consists of a specialized follicle-associated epithelium whose cell composition is distinct from that of absorptive villus epithelium. Absorptive cells are still the predominant cell type in the follicle-associated epithelium, whereas mucus-producing goblet cells are very rare. An estimated 10% of the epithelium in humans and mice (52) consists of a cell type with a unique morphology called M cells.

These specialized cells overlying Peyer's patches were first described in 1974 by Owen and Jones (34) and were later named membranous epithelial cells because of their morphology. M cells are interspersed between absorptive epithelial cells in the follicle-associated epithelium and are attached to adjacent cells by tight junctions, desmosomes, and interdigitations (34). Their luminal surfaces are characterized by a paucity of overlying mucus. M-cell microvilli are sparse, short, and irregular and lack an organized terminal web. The antiluminal M-cell surface is invaginated to form an intracellular central pocket into which lymphocytes and macrophages migrate. Low levels of typical brush border digestive enzymes (e.g., alkaline phosphatase) indicate that digestion and absorption of luminal content are not the main task of M cells. M-cell morphology (abundance of transport vesicles, deficiency of lysosomes) suggests a function in the transepithelial transport of undigested molecules and particulates. The transported material is very likely to be presented to lymphoid cells because of the close physical relation existing between M cells and various antigen-presenting and -processing cells contained in their central pockets (48).

A vast array of foreign material attaches to and is taken up by M cells (27, 33): macromolecules (horseradish peroxidase, ferritin, lectins), viruses (reovirus, poliovirus, human immunodeficiency virus), bacteria (see below), and even protozoa (*Giardia lamblia*). This well-designed antigen delivery system can be subverted by some enteric bacterial pathogens so that it becomes the Achilles' heel of the host organism instead of the protective shield. These bacteria use transepithelial movement through M cells as an invasion route.

For example, *E. coli* RDEC-1 is the rabbit equivalent of the human pathogen EPEC. In a rabbit model of orogastric infection (23), strain RDEC-1 adhered specifically to M cells rather than to the absorptive cells of the lymphoid follicle epithelium, causing local damage of the microvilli. Strain RDEC-1 has developed mechanisms that prevent uptake of the bacteria by M cells for antigen processing. In this way the bacteria circumvent the mucosal immune response. As a result, only mild to moderate inflammation is induced.

V. cholerae is a noninvasive enteropathogen that causes diarrheal disease by colonizing the mucosal surface of the small intestine. Studies with a ligated-loop model (35) revealed that *V. cholerae* colonizes the epithelial surfaces of both villi and follicles, although bacterial uptake across the epithelium is specifically performed by M cells and not by surrounding enterocytes. Exocytosis into the M-cell pocket results in dissemination of free or macrophage-engulfed bacteria into the follicle dome.

Invasive pathogens appear to use M cells as the main portal of entry across the epithelial barrier. Such pathogens include *Campylobacter jejuni* (50), *Yersinia* species (22), *Shigella flexneri* (51), and *Salmonella* species (26, 29), although it appears that *Salmonella* species can also penetrate through columnar epithelial cells. The different ways in which pathogens interact with M cells or enterocytes in order to promote adherence (*E. coli* RDEC-1) or invasion (*Yersinia, Shigella,* and *Salmonella* spp.) are schematically represented in Fig. 1.

INVASION

Many pathogenic bacteria possess the capacity to enter nonphagocytic eukaryotic cells, especially cultured epithelial cells (49). The mechanisms used by these bacteria vary between organisms, as does the role of invasion in pathogenesis. However, most invasive organisms appear to exploit existing host signal transduction systems to facilitate a cytoskeletal rearrangement that mediates bacterial uptake (36, 43). The capacity to enter and then penetrate through (transcytose) epithelial barriers facilitates bacterial breaching of epithelial barriers. However, given the role of M cells as facilitators of transepithelial transport in the Peyer's patches as discussed above, it is difficult to identify examples of bacteria that invade exclusively through columnar epithelial cells yet avoid M-cell invasion. At best, such as with *Salmonella* species, bacteria can penetrate both cell types.

Alternatively, invasion may play a specific role during later stages of penetration. For example, both *Shigella* and *Yersinia* species can invade cells, yet neither can invade the apical surface of a polarized cell. It is thought that they penetrate through M cells, thereby gaining access to the basolateral surfaces of epithelial cells that they then invade. Possibly *Shigella* species also penetrate through or next to polymorphonuclear leukocytes spanning the epithelial barrier in the process of transmigration (37).

TOXINS AS PATHOGENIC TOOLS FOR DISRUPTING EPITHELIAL BARRIERS

As attested to by several chapters in this book, pathogenic bacteria produce a variety of toxins that have many activities and target sites. It is not unreasonable to expect that some pathogens might disrupt a mucosal surface via the action of a toxin. For example, *Pseudomonas*

aeruginosa, which can colonize skin epithelial barriers, secretes a barrage of degradative toxins that can degrade this barrier, allowing the organism access to deeper tissues (1). Other examples can be found in several gram-positive pathogens.

Few bacterial toxins that function to specifically disrupt epithelial barriers (as opposed to degradative toxins) have been identified. Cholera toxin penetrates epithelial cells; it alters ion transport but does not physically disrupt intercellular junctions. However, *V. cholerae* also secretes a toxin (Zot) that specifically disrupts tight junctions, and this action appears to contribute to the production of diarrhea (7).

SALMONELLA TYPHIMURIUM AND EPEC INTERACTIONS WITH EPITHELIAL CELLS: TWO MODEL MUCOSAL PATHOGENS THAT USE DIFFERENT MECHANISMS FOR COLONIZATION AND PENETRATION

Both *S. typhimurium* and EPEC have several features, including their amenability to molecular genetic techniques and their relative safety in the laboratory, that make them useful models for studying interactions with epithelial cells. We have been characterizing the interactions of these two enteric pathogens with epithelial cells to gain a better understanding of the mechanisms involved in mucosal colonization.

EPEC is a leading cause of infantile diarrhea in developing countries. Initially, this organism attaches to epithelial cells in distinct microcolonies by a process termed localized adherence that is mediated by the bundle-forming pili. Intimate contacts, that are characterized by a shortening and loss of apical microvilli and the formation of a pedestallike structure beneath the adherent organism are then established between the bacterium and the epithelial cell. Although EPEC can invade cultured epithelial cells (6), invasive disease and systemic spread are not characteristic of EPEC infections.

Like EPEC, *S. typhimurium* also has marked effects on epithelial cells (8). *Salmonella* species cause many human diseases, such as gastroenteritis (including diarrhea), typhoid fever, and

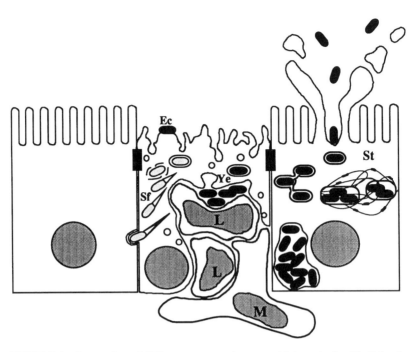

FIGURE 1 Interaction of different enteropathogens with intestinal epithelial cells. An M cell and two adjacent columnar absorptive cells are shown. Contained in the M-cell pocket are two lymphocytes (L) and one macrophage (M). The characteristic stages of the infection process of the different enteropathogens are as follows. (i) For *E. coli* RDEC-1 (Ec), enteroadherence, formation of a pedestal, and no transport by M cells; (ii) for *Yersinia enterocolitica* (Ye), transport through M cells in a vacuole, release into the M-cell pocket, extracellular replication, and resistance to uptake by macrophages; (iii) for *S. flexneri* (Sf), escape from M-cell transport vacuole into the cytoplasm, replication, intracellular movement by an actin tail, and lateral spread to adjacent enterocytes; and (iv) for *S. typhimurium* (St). Since *Salmonella* species destroy M cells upon contact, the invasion pathway through enterocytes is depicted: disruption of microvilli, membrane ruffling, formation and fusion of transport vacuoles, intracellular replication and formation of tubular lysosomes, and further multiplication, eventually leading to disruption of the enterocyte.

bacteremia. *S. typhimurium* interacts with various intestinal cells, including columnar epithelial cells and M cells, and causes marked surface rearrangements in infected cells (9). These organisms are capable of penetrating the intestinal barrier, presumably by passing through the cells, before entering the reticuloendothelial system and spreading to the liver or spleen.

INTERACTIONS WITH POLARIZED EPITHELIAL MONOLAYERS: A MODEL FOR STUDYING EPITHELIAL INTERACTIONS

The choice of the eukaryotic cell model is critical when bacterial interactions with host cells are being studied. The choices available for

model systems are numerous and include whole animal models (which are usually complex and expensive), primary cell cultures (which are difficult to use and inconsistent), and cultured cell lines (which often bear little resemblance to tissue cells). Most investigators use immortalized cell lines, since these are consistent and easy to use. Most epithelial cell lines are not polarized; that is, they do not form tight junctions or impermeable barriers, nor do they develop polarized apical and basolateral surfaces and defined microvilli. However, some cultured epithelial cell lines polarize when grown on permeable filter substrates (38, 40). Caco-2, derived from the human small intestine, is one such polarized epithelial cell line, and we have used this cell line to characterize the interactions that occur when EPEC or *S. typhimurium* infects polarized epithelial monolayers (Table 1) (4, 9).

The disruptions of polarized Caco-2 cells infected apically with EPEC that are observed by electron microscopy are nearly identical to those seen in vivo. There is a marked shortening of the microvilli in infected cells, and the characteristic attaching-effacing lesions and pedestals are formed on the apical surfaces below adherent organisms (4). Similarly, *S. typhimurium* infection of the apical surfaces of Caco-2 monolayers causes microvilli rearrangements resembling those seen in infected animals (9). The initial elongation and denuding of the host microvilli are followed by an extrusion on the apical surface of the epithelial cell (ruffling) that is localized to the area of the invading organism. *S. typhimurium* then invades the cell and takes up residence within a membrane-bounded vesicle inside the cell.

Since penetration through epithelial barriers is thought to contribute to pathogenesis for some pathogens, we also examined the abilities of EPEC and *S. typhimurium* to penetrate polarized Caco-2 cells. *S. typhimurium* penetrates these monolayers, appearing in the basolateral media between 1 and 2 h after apical infection

(9). In contrast, EPEC is not efficient at penetrating these monolayers, and only a small number of bacteria appear in the basolateral media at 12 h after infection (4). Mutants carrying defects in bacterial loci involved in signal generation in the host cell (see below) are also unable to penetrate Caco-2 monolayers (4).

The integrity of the polarized monolayer barrier is reflected by the transmonolayer electrical resistance, which is the sum of the effect of tight junctions and channels within the apical and basolateral cell membranes. *S. typhimurium* causes a complete loss in transepithelial electrical resistance within 2 to 3 h after apical addition. In contrast, EPEC causes a significant but incomplete drop in electrical resistance after 6 to 10 h, and this drop is reversible if the monolayers are treated with antibiotics. EPEC and *S. typhimurium* strains carrying mutations in loci involved in invasion and signal transduction are also unable to cause this decrease in resistance, indicating a correlation between these processes.

The mechanisms by which these two bacteria disrupt polarized monolayers are different. *S. typhimurium* causes depolarization of the epithelial cells and disruption of the tight junctions (9, 10). In contrast, EPEC does not disrupt the tight junctions and affects a transcellular component instead (4). It is possible that

TABLE 1 Comparison of EPEC and *S. typhimurium* interactions with epithelial cells[a]

Effect	EPEC	*S. typhimurium*
Cell surface disruption	Microvilli denuded, A/E lesion formed	Microvilli denuded transiently, membrane ruffling
Loss of transepithelial resistance	6–10 h, incomplete	2–3 h, complete
Monolayer disruption	Transcellular	Paracellular
Bacterial penetration	12 h, inefficient	1–2 h, efficient
Cytoskeletal rearrangements	>1.5 h, stable	30–60 min, transient
Cytoskeletal accumulation	Tightly focused beneath bacteria	Loosely arranged, ruffle-associated
Actin, α-actinin, talin, ezrin localization	+	+
Tropomyosin, tubulin	−	+
Keratin, vimentin, vinculin	−	−
Actin filaments required for invasion	+	+
Microtubules required for invasion	+	−
Tyrosine kinase activity	+ (Hp90)	−
Intracellular Ca^{2+} flux	+	+
IP_3 flux	+	+
MAP kinase activation	−	+

[a]Abbreviations: A/E, attaching-effacing; MAP, mitogen-activated protein.

similar disruptions in intestinal epithelial barriers contribute to diarrhea production for both of these pathogens.

CYTOSKELETAL REARRANGEMENTS

The disruption on the apical surfaces of epithelial cells caused by both EPEC and *S. typhimurium* indicates that these organisms affect the underlying host cell cytoskeleton. Both organisms cause marked rearrangements in cytoskeletal proteins in the vicinity of adherent or invading bacteria (Table 1) (11, 12). Besides actin, alpha-actinin, talin, and ezrin also colocalize to these bacterium-induced rearrangements. Other cytoskeletal components such as keratin, vinculin, and vimentin are not affected. Despite these similarities, the cytoskeletal rearrangements caused by EPEC and *S. typhimurium* are different. Tubulin and tropomyosin distribution are affected by *S. typhimurium* but not EPEC. The cytoskeletal rearrangements triggered by EPEC are tightly focused beneath the adherent bacteria, while those triggered by *S. typhimurium* are more loosely organized. The kinetics of formation also differ, as cytoskeletal rearrangements are seen soon after *S. typhimurium* addition and disappear after bacterial internalization, whereas EPEC-induced changes are not readily observed until approximately 1.5 h after bacterial addition, and the cytoskeleton remains localized beneath the bacteria for several hours. Both bacteria require functional actin filaments for invasion, since cytochalasin D blocks invasion of both bacteria. Interestingly, microtubule inhibitors also block uptake of EPEC but not of *S. typhimurium* into epithelial cells (6).

SIGNAL TRANSDUCTION

It is apparent that both *S. typhimurium* and EPEC need to trigger signals in epithelial cells to potentiate actin and other apical-surface rearrangements and bacterial uptake. Several groups have begun to identify the signals that are generated in host cells by invasive organisms in addition to the bacterial factors necessary for signal potentiation (3, 43). As observed

with the cytoskeletal rearrangements, some of the signals generated by EPEC and *S. typhimurium* in epithelial cells are similar, while others are different (summarized in Table 1).

Both EPEC and *S. typhimurium* trigger intracellular Ca^{2+} fluxes in epithelial cells (2), and chelating intracellular but not extracellular Ca^{2+} blocks *S. typhimurium* uptake into HeLa cells (44). Additionally, mutations in the invasion loci of *S. typhimurium* generate mutants that are unable to trigger Ca^{2+} release (21), indicating that release of intracellular Ca^{2+} is probably necessary for *Salmonella* invasion.

Both EPEC and *S. typhimurium* also trigger release of inositol phosphates (IPs), including inositol trisphosphate (IP_3), in infected epithelial cells (14, 44). This release is correlated with invasion (for *S. typhimurium*) and other signaling events (for EPEC). Additionally, avirulent mutants of both EPEC and *S. typhimurium* do not trigger release of IPs.

Another common signaling mechanism in eukaryotic cells is activation of tyrosine kinases, which results in tyrosine phosphorylation of specific proteins that then potentiate various signals. It is apparent that EPEC activates one or more host tyrosine kinases that are involved in propagating signal transduction events. Antibodies against phosphotyrosine stain EPEC-infected epithelial cells immediately beneath adherent bacteria in a pattern similar to that seen with cytoskeletal stains (41). Also, tyrosine kinase inhibitors are capable of blocking EPEC-induced signals, including IP fluxes and invasion, although they have no effect on *S. typhimurium* invasion (42). Additionally, EPEC triggers the tyrosine phosphorylation of a 90-kDa epithelial-cell protein, and this phosphorylation is inhibited by kinase inhibitors but not cytochalasins, indicating that tyrosine phosphorylation precedes actin rearrangement (41).

Another group has characterized many additional signals that are triggered by *S. typhimurium* in epithelial cells, including activation of epidermal growth factor receptor, stimulation of mitogen-activated protein kinase, and involvement of leukotriene D_4 (Fig. 2) (17,

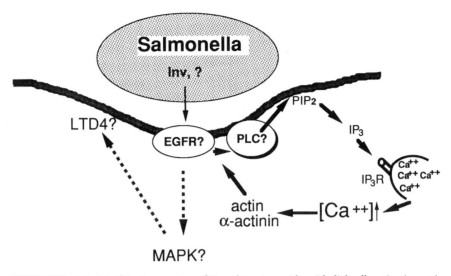

FIGURE 2 Model of the interaction of *S. typhimurium* with epithelial cells activating actin and actinin rearrangement. See text and reference 36 for details. EGFR, epidermal growth factor receptor; LTD$_4$, leukotriene D$_4$; MAPK, mitogen-activated protein kinase; PLC, phospholipase C; PIP$_2$, phosphatidylinositol 4,5-bisphosphate.

36). Additionally, they have characterized several bacterial loci (*inv*, for invasion) that are necessary for propagating signals in epithelial cells and mediating invasion. These events are described and reviewed elsewhere (8, 36).

Recently, several EPEC genes that are involved in transducing signals in epithelial cells have been identified. Originally isolated as Tn*phoA* mutants that do not invade epithelial cells, these mutants are grouped into five classes (5). One of these classes contained mutations in a gene, *eaeA*, that encodes intimin, a surface protein on EPEC (25). Mutations in *eaeA* are capable of triggering tyrosine phosphorylation and IP fluxes but are defective in focusing phosphotyrosine proteins and cytoskeletal components directly beneath the adherent bacteria (14, 41). At least two other genetic loci, *cfm* (class four mutant) and *eaeB*, are required for triggering signaling in epithelial cells (13, 41).

On the basis of these studies, a model of these events is summarized in Fig. 3. After initial attachment, which is mediated by the bundle-forming pilus and possibly the *eaeB* product, the products of *eaeB* and *cfm* activate

an epithelial-cell tyrosine kinase, which then tyrosine phosphorylates the 90-kDa protein (Hp90) and possibly other host proteins. An unidentified phospholipase C activity is presumably activated, and this results in cleavage of the membrane phospholipid phosphatidylinositol 4,5-bisphosphate (PIP$_2$) thereby releasing IP$_3$. This phospholipid then potentiates the release of sequestered calcium from intracellular stores. The localized calcium flux then activates molecules involved in actin rearrangement, which results in actin and related cytoskeletal proteins accumulating near the adherent bacteria. Finally, intimin (the *eaeA* product) and possibly the *eaeB* product sequester Hp90 and the actin-containing cytoskeleton immediately beneath the adherent organisms, forming a well-defined pedestal upon which EPEC resides.

CHARACTERIZATION OF THE INTRACELLULAR ENVIRONMENT OF *S. TYPHIMURIUM*

Although there is little doubt that the microenvironment found inside a membrane-bounded inclusion within a eukaryotic cell is

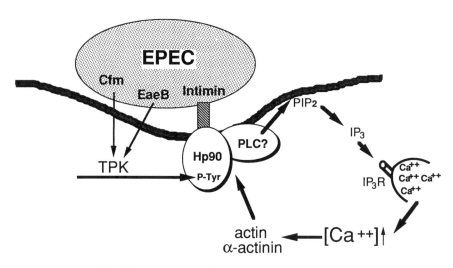

FIGURE 3 Model of the interaction of EPEC with epithelial cells activating actin and actinin rearrangement. See text for details. PLC, phospholipase C; TPK, tyrosine protein kinase.

different from culture medium, experimentally defining and characterizing this environment are difficult. One approach that has provided some clues about the intracellular environment is the measuring of gene fusion activity. For example, β-galactosidase activity was measured from *lacZ* transcriptional fusions made to genes regulated by Fe^{2+}, Mg^{2+}, and pH in *S. typhimurium* in MDCK epithelial cells (18). A fluorescent substrate was used to provide the sensitivity needed for measuring fusion activity of intracellular bacteria. Based on fusion activity, there appear to be low levels of free Fe^{2+} and Mg^{2+} in the vacuole. The vacuole also has a mild acidic pH and contains oxygen and lysine.

INTRACELLULAR TRAFFICKING AND REPLICATION OF *S. TYPHIMURIUM*

Little is known about the bacterial factors needed for intracellular survival and replication. We used an enrichment scheme to identify *S. typhimurium* transposon mutants that are unable to replicate within cultured epithelial cells (30). Ten auxotrophic and three prototrophic mutants were identified. The prototrophic replication-defective mutants were severely attenuated for virulence in mice,

indicating that intracellular growth may be necessary for *S. typhimurium* pathogenesis.

When *S. typhimurium* enters a cell, it triggers extensive capping of host surface proteins linked to the actin filament system (19). However, none of these cytoskeletally linked proteins, except major histocompatibility complex class I heavy chain, are internalized into the bacterium-containing vacuoles. Once inside, *S. typhimurium* bypasses the late endosome stage. Soon after internalization, the *S. typhimurium*-containing vacuole fuses with lysosomal glycoproteins (20). After approximately 4 h, *S. typhimurium* begins to multiply within the vacuole. At this time, infected cells contain long filamentous structures originating from the bacterial vacuole that are stained with antibodies against lysosomal glycoproteins (20). These tubular lysosomes are not seen in uninfected cells, and viable intracellular bacteria are also needed to trigger formation of these structures.

The formation of these structures appears to be linked to intracellular replication, since the intracellular-replication-defective mutants are unable to induce filament formation. Although the function of these structures remains undefined, one hypothesis is that by modifying a host vesicular transport pathway, these intracellular organisms can obtain a nutrient supply that enhances intracellular growth.

CONCLUSIONS

The interactions that occur between organisms that colonize and penetrate mucosal surfaces and mucosal epithelial cells are complex. The bacteria need several factors that allow them to avoid nonspecific defenses and the host immune system. Because of the selective pressures placed on the different pathogens, various mechanisms have evolved to circumvent potential hazards. Many pathogens appear to utilize M cells as a way of entering the host, since these cells are the weak link in the intestinal mucosal barrier.

As for *S. typhimurium* and EPEC, their interactions with epithelial cells represent a series of signals being passed between the bacteria and the host cells. It is clear that both bacteria exploit host signaling pathways for their own benefit. Although some of the signals triggered by each bacterium are shared, it is also apparent that different mechanisms are used by each organism. The bacterial effects on epithelial cells are closely linked to alterations in the cytoskeleton for both pathogens. In the case of *S. typhimurium,* the transient cytoskeletal rearrangement is used for facilitating bacterial uptake, while EPEC uses actin filaments to form a stable pedestal on the epithelial-cell surface. Both bacteria interact with the apical surfaces of polarized epithelial cells, causing alterations in microvilli distribution and disruption of epithelial integrity. However, EPEC affects the cell barrier function by a transcellular mechanism, while *S. typhimurium* disrupts tight junctions and eventually destroys polarized monolayers. Once internalized into a vacuole, *S. typhimurium* resides in an intracellular environment that is different from that found extracellularly. Fusions to genes that are known to be regulated by environmental parameters can be used to probe various characteristics of this environment. Closely linked to intracellular replication is the formation of novel host structures that

are filamentous in shape and contain lysosomal glycoproteins. Mutants that are unable to grow intracellularly also do not trigger formation of these structures.

As the specific mechanisms used by different pathogens to colonize and penetrate mucosal surfaces are elucidated, we will obtain a clearer idea of the basic requirements of mucosal pathogens for causing disease. It is the hope of workers in the field that once this information is gathered, rational attempts can be made to utilize the information in designing therapeutic agents to prevent prevalent mucosal infections.

ACKNOWLEDGMENTS

We thank the members of our laboratory who have contributed to this field.

Work in the laboratory of B.B.F. is supported by operating grants from the Medical Research Council of Canada and the Canadian Bacterial Disease Centre of Excellence and a Howard Hughes International Research Scholar award.

REFERENCES

1. **Azghani, A. O., L. D. Gray, and A. R. Johnson.** 1993. A bacterial protease perturbs the paracellular barrier function of transporting epithelial monolayers in culture. *Infect. Immun.* **61:**2681–2686.

2. **Baldwin, T. J., W. Ward, A. Aitken, S. Knutton, and P. H. Williams**. 1991. Elevation of intracellular free calcium levels in HEp-2 cells infected with enteropathogenic *Escherichia coli. Infect. Immun.* **59:**1599–1604.

3. **Bliska, J. B., J. E. Galan, and S. Falkow.** 1993. Signal transduction in the mammalian cell during bacterial attachment and entry. *Cell* **73:**903–920.

4. **Canil, C., I. Rosenshine, S. Ruschkowski, M. S. Donnenberg, J. B. Kaper, and B. B. Finlay.** 1993. Enteropathogenic *Escherichia coli* decreases the transepithelial electrical resistance of polarized epithelial monolayers. *Infect. Immun.* **61:**2755–2762.

5. **Donnenberg, M. S., S. B. Calderwood, R. A. Donohue, G. T. Keusch, and J. B. Kaper**. 1990. Construction and analysis of TnphoA mutants of enteropathogenic *Escherichia coli* unable to invade HEp-2 cells. *Infect. Immun.* **58:**1565–1571.

6. **Donnenberg, M. S., R. A. Donohue, and G. T. Keusch**. 1990. A comparison of HEp-2 cell invasion by enteropathogenic and enteroinvasive *Escherichia coli. FEMS Microbiol. Lett.* **57:**83–86.

7. **Fasano, A., B. Baudry, D. W. Pumplin, S. S. Wasserman, B. D. Tall, J. M. Ketley, and J.**

B. Kaper. 1991. *Vibrio cholerae* produces a second enterotoxin, which affects intestinal tight junctions. *Proc. Natl. Acad. Sci. USA* **88**:5242–5246.

8. **Finlay, B. B.** 1994. Molecular and cellular mechanisms of *Salmonella* pathogenesis. *Curr. Top. Microbiol.* **192**:163–185.

9. **Finlay, B. B., and S. Falkow**. 1990. *Salmonella* interactions with polarized human intestinal Caco-2 epithelial cells. *J. Infect. Dis.* **162**:1096–1106.

10. **Finlay, B. B., B. Gumbiner, and S. Falkow**. 1988. Penetration of *Salmonella* through a polarized Madin-Darby canine kidney epithelial cell monolayer. *J. Cell Biol.* **107**:221–230.

11. **Finlay, B. B., I. Rosenshine, M. S. Donnenberg, and J. B. Kaper**. 1992. Cytoskeletal composition of attaching and effacing lesions associated with enteropathogenic *Escherichia coli* adherence to HeLa cells. *Infect. Immun.* **60**:2541–2543.

12. **Finlay, B. B., S. Ruschkowski, and S. Dedhar**. 1991. Cytoskeletal rearrangements accompanying *Salmonella* entry into epithelial cells. *J. Cell Sci.* **99**:283–296.

13. **Foubister, V., I. Rosenshine, M. S. Donnenberg, and B. B. Finlay**. 1994. The *eaeB* gene of enteropathogenic *Escherichia coli* is necessary for signal transduction in epithelial cells. *Infect. Immun.* **62**:3038–3040.

14. **Foubister, V., I. Rosenshine, and B. B. Finlay**. 1994. A diarrheal pathogen, enteropathogenic *Escherichia coli* (EPEC), triggers a flux of inositol phosphates in infected epithelial cells. *J. Exp. Med.* **179**:993–998.

15. **Francis, C. L., M. N. Starnbach, and S. Falkow**. 1992. Morphological and cytoskeletal changes in epithelial cells occur immediately upon interaction with *Salmonella typhimurium* grown under low-oxygen conditions. *Mol. Microbiol.* **6**:3077–3087.

16. **Freter, R., P. C. OBrien, and M. S. Macsai**. 1981. Role of chemotaxis in the association of motile bacteria with intestinal mucosa: *in vivo* studies. *Infect. Immun.* **34**:234–240.

17. **Galan, J. E., J. Pace, and M. J. Hayman**. 1992. Involvement of the epidermal growth factor receptor in the invasion of cultured mammalian cells by *Salmonella typhimurium*. *Nature* (London) **357**:588–589.

18. **Garcia del Portillo, F., J. W. Foster, M. E. Maguire, and B. B. Finlay**. 1992. Characterization of the micro-environment of *Salmonella typhimurium*-containing vacuoles within MDCK epithelial cells. *Mol. Microbiol.* **6**:3289–3297.

19. **Garcia-del Portillo, F., M. G. Pucciarelli, W. A. Jefferies, and B. B. Finlay**. 1994. *Salmonella typhimurium* induces selective aggregation and internalization of host cell surface proteins during

invasion of epithelial cells. *J. Cell Sci.* **107**:2005–2020.

20. **Garcia del Portillo, F., M. B. Zwick, K. Y. Leung, and B. B. Finlay**. 1993. *Salmonella* induces the formation of filamentous structures containing lysosomal membrane glycoproteins in epithelial cells. *Proc. Natl. Acad. Sci. USA* **90**:10544–10548.

21. **Ginocchio, C., J. Pace, and J. E. Galan**. 1992. Identification and molecular characterization of a *Salmonella typhimurium* gene involved in triggering the internalization of salmonellae into cultured epithelial cells. *Proc. Natl. Acad. Sci. USA* **89**:5976–5980.

22. **Grützkau, A., C. Hanski, H. Hahn, and E. O. Riecken**. 1990. Involvement of M cells in the bacterial invasion of Peyer's patches: a common mechanism shared by *Yersinia enterocolitica* and other enteroinvasive bacteria. *Gut* **31**:1011–1015.

23. **Inman, L. R., and J. R. Cantey**. 1983. Specific adherence of *Escherichia coli* (strain RDEC-1) to membranous (M) cells of the Peyer's patch in *Escherichia coli* diarrhea in the rabbit. *J. Clin. Invest.* **71**:1–8.

24. **Iredell, J. R., and P. A. Manning**. 1994. The toxin-co-regulated pilus of *Vibrio cholerae* O1: a model for type 4 pilus biogenesis. *Trends Microbiol.* **2**:187–192.

25. **Jerse, A. E., J. Yu, B. D. Tall, and J. B. Kaper**. 1990. A genetic locus of enteropathogenic *Escherichia coli* necessary for the production of attaching and effacing lesions on tissue culture cells. *Proc. Natl. Acad. Sci. USA* **87**:7839–7843.

26. **Jones, B. D., N. Ghori, and S. Falkow**. 1994. *Salmonella typhimurium* initiates murine infection by penetrating and destroying the specialized epithelial M cells of the Peyer's patches. *J. Exp. Med.* **180**:15–23.

27. **Keren, D. F.** 1992. Antigen processing in the mucosal immune system. *Semin. Immunol.* **4**:217–226.

28. **Klauser, T., J. Pohlner, and T. F. Meyer**. 1993. The secretion pathway of IgA protease-type proteins in gram-negative bacteria. *Bioessays* **15**:799–805.

29. **Kohbata, S., H. Yokoyama, and E. Yabuuchi**. 1986. Cytopathogenic effect of *Salmonella typhi* GIFU 10007 on M cells of murine ileal Peyer's patches in ligated ileal loops: an ultrastructural study. *Microbiol. Immunol.* **30**:1225–1237.

30. **Leung, K. Y., and B. B. Finlay**. 1991. Intracellular replication is essential for the virulence of *Salmonella typhimurium*. *Proc. Natl. Acad. Sci. USA* **88**:11470–11474.

31. **Mims, C. A.** 1987. *The Pathogenesis of Infectious Disease*. Academic Press, London.

32. **Mounier, J., T. Vasselon, R. Hellio, M. Lesourd, and P. J. Sansonetti**. 1992. *Shigella flex-*

neri enters human colonic Caco-2 epithelial cells through the basolateral pole. *Infect. Immun.* **60:** 237–248.

33. **Neutra, M. R., and J.-P. Kraehenbuhl.** 1993. The role of transepithelial transport by M cells in microbial invasion and host defense. *J. Cell Sci.* **17:** 209–215.

34. **Owen, R. L., and A. L. Jones.** 1974. Epithelial cell specialization within human Peyer's patches: an ultrastructural study of intestinal lymphoid follicles. *Gastroenterology* **66:**189–203.

35. **Owen, R. L., N. F. Pierce, R. T. Apple, and W. C. Cray, Jr.** 1986. M cell transport of *Vibrio cholerae* from the intestinal lumen into Peyer's patches: a mechanism for antigen sampling and for microbial transepithelial migration. *J. Infect. Dis.* **153:**1108–1118.

36. **Pace, J., M. J. Hayman, and J. E. Galan.** 1993. Signal transduction and invasion of epithelial cells by *S. typhimurium. Cell* **72:**505–514.

37. **Perdomo, J. J., P. Gounon, and P. J. Sansonetti.** 1994. Polymorphonuclear leukocyte transmigration promotes invasion of colonic epithelial monolayer by *Shigella flexneri. J. Clin. Invest.* **93:**633–643.

38. **Pucciarelli, M. G., and B. B. Finlay.** 1994. Polarized epithelial monolayers: model systems to study bacterial interactions with host epithelial cells. *Methods Enzymol.* **236:**438–447.

39. **Robertson, B. D., and T. F. Meyer.** 1992. Genetic variation in pathogenic bacteria. *Trends Genet.* **8:**422–427.

40. **Rodriguez-Boulan, E., and W. J. Nelson.** 1989. Morphogenesis of the polarized epithelial cell phenotype. *Science* **245:**718–725.

41. **Rosenshine, I., M. S. Donnenberg, J. B. Kaper, and B. B. Finlay.** 1992. Signal transduction between enteropathogenic *Escherichia coli* (EPEC) and epithelial cells: EPEC induces tyrosine phosphorylation of host cell proteins to initiate cytoskeletal rearrangement and bacterial uptake. *EMBO J.* **11:**3551–3560.

42. **Rosenshine, I., V. Duronio, and B. B. Finlay.** 1992. Tyrosine protein kinase inhibitors block invasin-promoted bacterial uptake by epithelial cells. *Infect. Immun.* **60:**2211–2217.

43. **Rosenshine, I., and B. B. Finlay.** 1993. Exploitation of host signal transduction pathways and cytoskeletal functions by invasive bacteria. *Bioessays* **15:**17–24.

44. **Ruschkowski, S., I. Rosenshine, and B. B. Finlay.** 1992. *Salmonella typhimurium* induces an inositol phosphate flux in infected epithelial cells. *FEMS Microbiol. Lett.* **95:**121–126.

45. **Salyers, A. A., and D. D. Whitt.** 1994. *Bacterial Pathogenesis: a Molecular Approach.* American Society for Microbiology, Washington, D.C.

46. **Schaechter, M., G. Medoff, and B. I. Eisenstein.** 1989. *Mechanisms of Microbial Disease.* The Williams & Wilkins Co., Baltimore.

47. **Schmitt, C. K., S. C. Darnell, V. L. Tesh, B. A. Stocker, and A. D. OBrien.** 1994. Mutation of *flgM* attenuates virulence of *Salmonella typhimurium,* and mutation of *fliA* represses the attenuated phenotype. *J. Bacteriol.* **176:**368–377.

48. **Sneller, M. C., and W. Strober.** 1986. M cells and host defense. *J. Infect. Dis.* **154:**737–741.

49. **Tang, P., V. Foubister, M. G. Pucciarelli, and B. B. Finlay.** 1993. Methods to study bacterial invasion. *J. Microbiol. Methods* **18:**227–240.

50. **Walker, R. I., E. A. Schmauder-Chock, J. L. Parker, and D. Burr.** 1988. Selective association and transport of *Campylobacter jejuni* through M cells of rabbit Peyer's patches. *Can. J. Microbiol.* **34:**1142–1147.

51. **Wassef, J. S., D. F. Keren, and J. L. Mailloux.** 1989. Role of M cells in initial antigen uptake and in ulcer formation in the rabbit intestinal loop model of shigellosis. *Infect. Immun.* **57:**858–863.

52. **Wolf, J. L., and W. A. Bye.** 1984. The membranous epithelial (M) cell and the mucosal immune system. *Annu. Rev. Med.* **35:**95–112.

BACTERIAL ADAPTATION TO EXTRACELLULAR ENVIRONMENTS

RESISTANCE OF BACTERIA
TO COMPLEMENT

Peter W. Taylor

4

The complement system, first recognized over 100 years ago because of its ability to bring about dissolution of gram-negative bacteria (13, 66), plays a key role in the response of the host to microbial invasion and infection. A wide range of biological activities, including opsonization and phagocytosis of a variety of microorganisms, direct killing of many strains of gram-negative bacteria, neutralization of enveloped viruses, disposal of harmful immune complexes, and induction and modulation of the inflammatory response, may follow activation of complement. The importance of the complement system as a component of the overall defense of the host against infection is reflected both in the wide distribution of this system within the animal kingdom (3) and in the increased susceptibility to infection of individuals congenitally deficient in the biosynthesis of certain complement components (2).

Although gram-positive bacteria can act as efficient activators of the complement cascade either directly (97) or as a result of the binding of complement-activating antibodies to the cell surface (78), peptidoglycan acts as a barrier to attack by the late-acting, membrane-perturbing components of complement. As a re-

sult, complement activation on the surfaces of gram-positive bacteria leads not to direct killing but to opsonization. In contrast, a large number of gram-negative bacteria are susceptible to complement-mediated killing, and exposure of these cells to a suitable source of complement, such as serum or plasma, may lead to a remarkably rapid and efficient reduction in viability (98). Killing is sometimes accompanied by lysis of target bacteria due to the presence of the peptidoglycan-degrading enzyme lysozyme, but cell death can proceed at near-maximal rates in the absence of the enzyme (56, 98). Appropriate activation of either the classic or the alternative complement pathway may lead to destruction of the target cell population. Removal of any of the late-acting components results in complete loss of killing (38, 86), and it is clear that antibacterial activity is dependent on the assembly at the bacterial surface of the membrane attack complex of complement. This complex, also referred to as the C5b-9 complex, is assembled from the late-acting complement components, and its deposition onto the outer membrane initiates a series of incompletely understood events that culminate in irreversible loss of cell viability (96).

Some gram-negative strains appear to be refractory to the lethal effects of complement,

Peter W. Taylor   Ciba Pharmaceuticals, Horsham, West Sussex RH12 4AB, United Kingdom.

Virulence Mechanisms of Bacterial Pathogens, 2nd ed., Edited by J. A. Roth et al.
© 1995 American Society for Microbiology, Washington, D.C.

and isolates from sites of infection in the body that contain functionally effective concentrations of complement proteins are significantly more likely to be resistant to complement than are noninvasive organisms. For example, the majority of strains causing septicemia or bacteremia are complement resistant (68). However, other host defense mechanisms such as phagocytosis play a key role in determining the fate of potential pathogens, and it is clearly difficult to determine the precise contribution of any one mechanism during a concerted response to microbial attack. In addition, it is likely that structures at the bacterial surface are likely to confer resistance to more than one host defense mechanism. Thus, it has proven difficult to establish causal relationships between complement resistance and infectivity, although the bulk of evidence strongly suggests that complement-mediated humoral mechanisms participate in the control of some infections. Data from experimental infections in laboratory animal models also support this supposition.

Microorganisms have evolved a number of strategies for subverting complement attack; these mechanisms include failure to activate or bind early complement components, degradation of surface-bound proteins, and mechanisms for blocking the assembly of functional C5b-9 lesions on the surface. Gram-negative bacteria undoubtedly take advantage of a range of resistance mechanisms and most rely on expression at the cell surface of structures able to modulate complement binding and expression. Almost without exception, rough strains of gram-negative bacteria producing lipopolysaccharide devoid of O-specific side chains are highly susceptible to C5b-9-mediated killing, whereas smooth strains that synthesize a complete lipopolysaccharide are often complement resistant. In addition, capsular polysaccharides and outer membrane proteins increase complement resistance under certain circumstances (59). This chapter reviews current knowledge on the mechanisms of complement resistance and focuses on the strategies adopted by gram-negative bacteria. In addition, the role of complement resistance as a virulence factor in human and animal infections is assessed. As a clear understanding of the basis of complement resistance will emerge only from a full appreciation of the mechanism of complement-mediated bactericidal activity, I also examine the interactions of the key complement components with susceptible bacteria.

COMPLEMENT-MEDIATED KILLING OF TARGET BACTERIA

The complement system consists of about 20 glycoproteins that circulate in the extracellular fluid compartment and interact in a precise sequence of reactions that results in the production of biologically active cleavage fragments that promote opsonization and phagocytosis as well as direct cell damage (57). Two distinct pathways of complement activation, the classic and the alternative pathways, have been elucidated elsewhere in great detail with respect to their biochemical, physiologic, and molecular characteristics (89). A similar number of widely distributed cell surface proteins act as receptors for fragments of soluble complement proteins or as regulatory proteins controlling the activities of soluble complement proteins (Fig. 1).

Both the classic and the alternative pathways can be activated, often simultaneously, by susceptible gram-negative bacteria. The former is normally activated by interaction of an antibody of the appropriate class and subclass with an antigen on the bacterial surface, although under some circumstances, direct activation of the classic pathway without the intervention of antibody can occur (97). The initial step involves the binding of C1q to the activator; provided at least two of the six globular binding sites of this molecule interact with the activator, a conformational change occurs and is transmitted to the serine proteases C1r and C1s that are complexed with C1q. This movement allows autoactivation of C1r, which in turn activates C1s. C1s cleaves the next two components of the cascade, C4 and C2; the two larger fragments, C4b and C2a, constitute an enzymatically active complex that is covalently

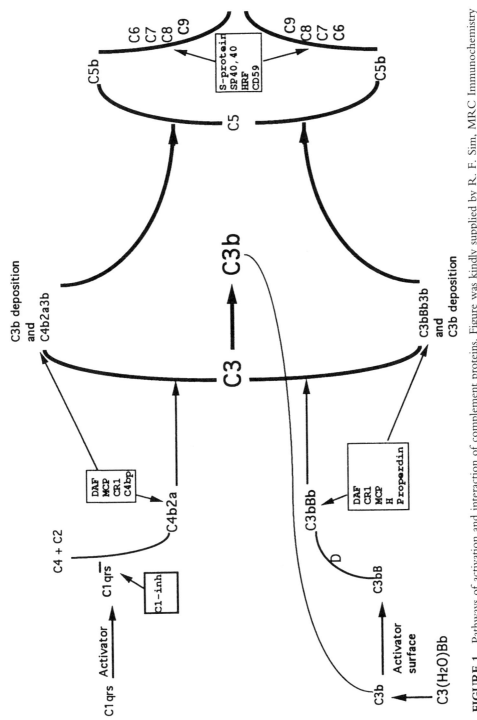

FIGURE 1 Pathways of activation and interaction of complement proteins. Figure was kindly supplied by R. F. Sim, MRC Immunochemistry Unit, University of Oxford. DAF, decay-accelerating factor; MCP, membrane cofactor protein; HRF, homologous restriction factor; MAC, membrane attack complex.

bound through C4b to the target surface. The C4bC2a complex represents the classic pathway C3 convertase and can cleave and activate C3. Like C4b, the larger of the two cleavage fragments, C3b, is able to bind covalently to the target surface by virtue of a thiolester group that is exposed on cleavage. It may also bind noncovalently to C4b to form a serine esterase complex (C4bC2aC3b) that can bind and cleave C5; binding is mediated through C3b, and this component of the complex also orients C5 for cleavage by C4bC2a.

The alternative pathway can be activated on the surfaces of certain bacteria in both an antibody-dependent and an antibody-independent fashion. Antibody-dependent activation is facilitated mainly by surface-bound immunoglobulin G. The nature of the bacterial surface is critical in determining whether or not activation occurs in the absence of antibody (97). Small amounts of C3b are continuously formed in vivo because of the activity of a C3 convertase of the alternative pathway. Initial convertase activity arises due to the slow, nonenzymatic hydrolysis of the internal thiolester bond within native C3b to form a C3b-like molecule, $C3(H_2O)$, that binds factor B. Factor D, a serine esterase present in an active form in blood, then cleaves $C3(H_2O)$-bound factor B to form a fluid-phase convertase that generates by proteolytic cleavage the initial metastable C3b that is capable of covalent attachment to the bacterial surface (79). Thus, the site of interaction of the alternative pathway proteins is transferred from the fluid phase to the target surface; bound C3b can then bind factor B to form, in the presence of factor P, a relatively stable C3-cleaving complex, C3bPBb, that can bind another molecule of C3b and acquire C5-cleaving activity. The formation and function of the alternative pathway C3 convertase are subject to control by the proteins factor I and factor H. Factor H competes with factors B and Bb for binding sites on C3b and $C3(H_2O)$; thus, H is able to prevent formation of the convertase by blocking the binding of factor B or by dissociating Bb from the convertase. Furthermore, C3b or $C3(H_2O)$, when complexed with H, is susceptible to cleavage and inactivation by the serine protease factor I. On some bacterial surfaces, particularly those rich in sialic acid, the apparent affinity of factor H for C3b is increased, with the consequence that the activation process does not proceed (49).

Following C5 cleavage by either of the C5-converting enzyme complexes, the larger cleavage fragment, C5b, spontaneously associates with native C6 and C7 to form a trimolecular C5b-7 complex that inserts into the hydrocarbon core of the target membrane (73). Binding of one C8 molecule (a trimer of α, β, and γ subunits) via its β subunit (61) to each C5b-7 complex gives rise to small transmembrane channels less than 1 nm in functional diameter (77). The primary role of membrane-inserted C5b-8 is, however, to act as an acceptor for multiple copies of C9, the major membrane-perturbing entity of the complement system, at the membrane attack site (90). The binding of one molecule of C9 to a C5b-8 complex initiates a process of C9 oligomerization, and if 16 to 18 C9 molecules are incorporated into the complex, the characteristic ringlike lesions first observed on complement-lysed erythrocytes are evident (9). The lesion is a hollow cylinder 15 to 16 nm in length with an internal diameter of 10 nm, and one end is rimmed by an annulus with an external diameter of 20 to 22 nm. The terminus distal to the annulus bears an apolar surface 4 nm in length that is embedded in the hydrophobic interior of the target membrane (6).

When globular monomeric C9 binds to the C8 α subunit (1) within C5b-8 on the target membrane, it undergoes a conformational change and doubles in length from 8 to 16 nm (72). There is evidence that C9 inserts into the membrane in its globular state prior to the conformational change (55), and it has been suggested that globular C9 inserts into the lipid domains perturbed by C5b-8 (91). Conformational change results in the appearance of a lipid-binding domain 4 nm in length that corresponds to the terminal hydrophobic domain on the hollow cylinder and allows binding and insertion of further C9 molecules. The molar concentration of C9 in serum is only about twice that of C8, so C5b-9 lesions generated on target membranes display a degree of heterogeneity with regard to C9 content (7). Although the incorporation of one C9 into the complex is sufficient to lyse target erythrocytes (8), this clearly does not happen with gram-negative bacteria. Efficient killing is correlated with the presence of multiple copies of C9 in individual C5b-9 complexes (5, 46, 54). For the killing of rough *Escherichia coli* cells, at least three copies of C9 in each complex are necessary, and for optimal killing, C9/C7 ratios of 6:1 have been demonstrated (46, 54). It is also clear that about 50 to 500 C5b-9 complexes, which is considerably more than is required for erythrocyte lysis, need to be deposited onto each target cell to effect loss of viability (12, 102). These observations strongly suggest that the classic C5b-9 lesion visualized as a ringlike structure buried in the target membrane is not per se the entity responsible for bacterial killing and that a precursor of the classic lesion is responsible for the damage. It has in fact been suggested that the classic lesion represents a noncytolytic or noncytocidal end-stage product of C9 insertion into the membrane (18, 22), and evidence for this has been obtained by using thrombin-cleaved C9 (C9^n) in serum bactericidal assays. C9^n can bind to C5b-8 but is unable to form the classic cylindrical lesion (18); it is, however, more effective than native C9 in killing susceptible *E. coli* (20) and *Salmonella minnesota* (103).

The gram-negative cell envelope represents a complex target for C5b-9 attack, and the mechanism of killing has not been precisely defined in molecular terms. It is clear, however, that loss of viability following exposure to lethal concentrations of complement is dependent on irreversible damage to the cytoplasmic membrane. Killing is strongly correlated to changes in the permeability and metabolic function of the cytoplasmic membrane. For example, the cytoplasmic membrane of an *E. coli* strain normally cryptic for β-galactosides becomes permeable to a β-galactosidase substrate following exposure to serum, and loss of

crypticity exactly parallels both killing and alkaline phosphatase release from the periplasmic space (24). Complement attack in the absence of lysozyme may also cause an efflux of small molecules such as proline and monovalent cations from the cytoplasm, but there appears to be no increase in permeability with respect to proteins (56, 108). Similarly, respiration and amino acid transport, both functions of an integral cytoplasmic membrane, are rapidly inhibited by C5b-9 deposition (54). Alterations in cytoplasmic membrane permeability are accompanied by limited degradation of the component phospholipids (99), but the bilayer retains its typical appearance when visualized by electron microscopy (107).

The observation that complement attack results in efflux of monovalent cations from the target bacterial cell (56, 108) suggests that C5b-9 deposition may facilitate the collapse of the potential across the cytoplasmic membrane. In an elegant series of experiments, Dankert and Esser studied the uptake of the lipophilic cation tetraphenylphosphonium and of proline and galactosides by rough *E. coli* cells following complement attack (19, 20). They detected a transient collapse of the membrane potential following assembly of nonlethal C5b-8 lesions on the bacterial surface; addition of C9 rapidly and irreversibly dissipated the ΔE_m and caused cell death. Because it has been established that inhibitors and uncouplers of oxidative phosphorylation, compounds that collapse the ΔE_m, can protect against complement attack (32, 98), Dankert and Esser suggested that an energized cytoplasmic membrane is essential for killing to take place once complement proteins have bound to the bacterial surface. It is likely that the cells attempt to reestablish a K$^+$ concentration gradient across the damaged ion-permeable cytoplasmic membrane; indirect evidence for this is provided by the observation that internal ATP pools decrease to undetectable levels during the complement killing reaction (51, 98) as the cell makes a vain attempt to restore ΔE_m.

The gram-negative cell envelope consists of three essential layers: the outer membrane, the peptidoglycan layer, and the cytoplasmic membrane. The two functionally distinct membranes, each with a thickness of 7.5 to 9.0 nm, are separated by a periplasmic gel of significant volume (36). It is clear from the dimensions of the lipid–binding domain of C5b-9 that individual complexes have the capacity to intercalate only into single-lipid bilayers. It seems reasonable to assume that C5-9 lesions form, at least initially, on the outer membrane, and there is evidence that deposition on the outer membrane is a prerequisite for the destabilization of the cytoplasmic membrane and for killing. For example, Wright and Levine (108) treated washed, viable *E. coli* cells carrying C5b-7 complexes with functionally pure C8 and C9 and observed rapid, C5b-9-mediated killing. Thus, since C5b-7 complex formation causes no loss of functional integrity of the outer membrane, subsequent lethal damage did not occur as a result of fresh C5b-9 formation on the cytoplasmic membrane following outer membrane disruption. Further evidence that the outer membrane represents the principal, and perhaps exclusive, locus of C5b-9 deposition was obtained by Taylor and Kroll (51, 99). They treated rough *E. coli* strains with lethal doses of complement and separated the outer and cytoplasmic membranes by density gradient centrifugation. Covalent binding of C3b to the outer but not to the cytoplasmic membrane was found early in the reaction sequence before the onset of viability loss. Binding of C5b-9 complexes exclusively to the outer membrane was first detected coincident with the onset of viability loss and increased rapidly during the active killing phase of the reaction. At no time during the reaction could either C3 products or C5b-9 complexes be found on the cytoplasmic membrane, but the situation was complicated by reduced recoverability of the cytoplasmic membrane due to limited phospholipid degradation. Electron microscopic studies confirmed that C5b-9 complexes were not deposited on the cytoplasmic membrane; in contrast, large numbers of cylindrical lesions were found in the outer

membranes during the active killing phase of the reaction (52). These observations, suggesting that C5b-9 deposition occurs exclusively on the outer membrane and that it mediates its lethal effect from this locus, are compatible with the earlier observation of Feingold and colleagues (24) that plasmolysis of complement-susceptible *E. coli* cells, which produces a spatial separation of outer and cytoplasmic membranes, protects the target bacteria from C5b-9 attack.

Complement attack on rough gram-negative bacteria results in the release of large amounts of phospholipid from the outer membrane by a detergentlike mechanism (37, 99). One consequence of this disruption is the release of marker proteins from the periplasmic space (24, 56, 98, 108), but this is not an essential feature of the bactericidal process, since complement-susceptible smooth *E. coli* strains do not release periplasmic markers following C5b-9 deposition (51). Phospholipid release and its sequelae may be related to activation of the outer membrane-located detergent-resistant phospholipase A, since mutants lacking this enzyme release no phospholipid when undergoing complement attack (99).

Many of the observations described in this section suggest that complement-mediated killing of gram-negative bacteria occurs as a result of physiologic processes pertaining to the assembly of C5b-9 lesions on the outer membrane rather than being simply the consequence of channel formation across a membrane bilayer. Further evidence for this comes from recent experiments performed by Tomlinson and coworkers (101, 102). These workers developed a system for fusing liposomes with rough gram-negative bacteria. They loaded liposomes with preformed C5b-9 complexes containing from 4 to 15 C9 molecules per complex and delivered an average of 1,900 complexes into the membranes of *E. coli* and *S. minnesota*. The liposomal phospholipid transferred to the outer membrane and subsequently equilibrated between the outer and cytoplasmic membranes, whereas the C5b-9 complexes remained exclusively in the outer membrane. Although the complexes functioned as water-filled channels across the outer membrane and appeared indistinguishable from complexes deposited on cells following exposure to lethal concentrations of serum, no loss of bacterial viability could be detected. The authors suggested that either there are critical sites in the outer membrane to which liposomally delivered C5b-9 complexes are unable to gain access or bacterial death is related to events occurring during polymerization of C9 on the cell surface (102). Clearly, simple insertion of tubular C5b-9 complexes into the outer membrane cannot account for the bactericidal phenomenon.

A hypothesis recently presented to account for these accumulated observations seems amenable to experimental investigation (96). It is reasonable to assume that perturbation requires interaction of C9 with the cytoplasmic membrane, and it is difficult to envisage that C9 bound to C5b-8 on the outer membrane would be able to penetrate sufficiently into the cytoplasmic membrane bilayer to cause dissipation of the membrane potential. Dankert and Esser (20) introduced native C9 into the periplasmic space and induced killing. This suggests that C9 undergoes a conformational change in the periplasmic space, possibly induced by the physical environment in this compartment, that facilitates insertion into the bilayer in the absence of C5b-8. If, as suggested by Stanley (90), the holes created in the lipid bilayer by C5b-8 are lined by the phospholipid polar head groups, then native, globular C9 molecules might evade capture by C5b-8 complexes in the outer membrane and escape into the periplasmic space to undergo conformational change and damage the cytoplasmic membrane (Fig. 2). Such a mechanism would be analogous to that of membrane-active colicins that may insert into the cytoplasmic membrane after undergoing a conformational change induced by the low pH in the periplasmic space (69).

DETERMINANTS OF COMPLEMENT RESISTANCE

Complement attack on susceptible target gram-negative bacteria appears to have its primary effect on cellular metabolic parameters; resistance to complement, however, is intimately linked to the capacity of structures at the bacterial surface to modulate either the activation of complement or the stable deposition of membrane attack complexes.

Because the alternative pathway can operate in the absence of an antigen-antibody reaction, it has been widely assumed that this pathway represents a phylogenetically more primitive defense mechanism than the classic pathway. As Frank has recently pointed out (27), this generalization may not always hold true, but it is clear that many pathogenic gram-negative organisms have developed surface structures that restrict activation of the alternative pathways as a means of survival in blood and tissue fluids. For example, *E. coli* K1 and *Neisseria meningitidis* B strains synthesize exopolysaccharides consisting of α-2,8-linked sialic acid that restrict activation of the alternative pathway by facilitating the binding of factor H to C3b (23). Therefore, in the absence of antibody, little complement deposition on the surface of these strains would be expected to occur. In practice, many sialylated strains are susceptible to complement attack (11, 21), although the mechanism of complement activation under these circumstances has not been defined.

Because of its key role in the regulation of complement activity at the cell surface, the extent of deposition of C3 cleavage products on the gram-negative surface in relation to killing has been determined by a number of groups. Reynolds and coworkers (81) found equivalent amounts of C3 products deposited on serum-resistant *S. typhimurium* cells and on the same strain that had been rendered susceptible to complement by treatment with Tris and EDTA. Fierer and Finley (25) also demonstrated, by immunofluorescence, equivalent deposition of C3 products on a serum-resistant

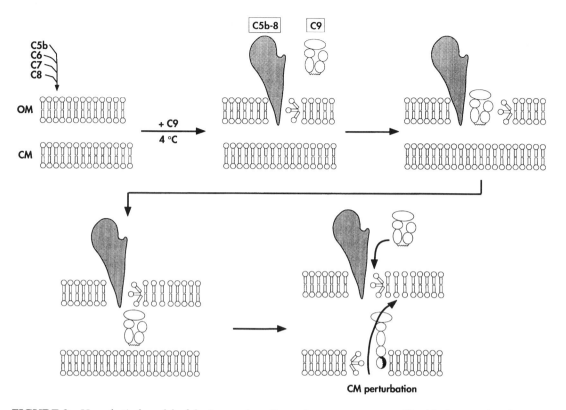

FIGURE 2 Hypothetical model of the interaction of complement component C9 with the cytoplasmic membrane (CM) in the absence of C5b-8 formation on the CM. OM, outer membrane. Refer to text for details.

strain before and after phenotypic conversion to susceptibility. However, these studies give no indication whether C3 was present as functionally active C3b or whether C3b had been cleaved by control proteins of the complement cascade to yield products, such as iC3b, that are inactive in bactericidal processes. The bulk of evidence suggests differences between susceptible and resistant strains with regard to the extent and precise surface location of C3b deposition. For example, there was less C3 deposition onto resistant strains of *Pseudomonas aeruginosa* expressing long and numerous lipopolysaccharide O-specific side chains than on O-side-chain-deficient susceptible strains from cystic fibrosis patients (84). Furthermore, C3b was present on the susceptible strains, whereas most of the C3 on the surfaces of resistant strains was in the form of iC3b.

In an elegant series of experiments, Leive, Joiner, and colleagues (33, 40, 43) showed that C3 binds almost exclusively to lipopolysaccharide after exposure of cells to serum. With a serum-resistant *Salmonella montevideo* strain bearing long and numerous O-specific side chains, C3 bound only to a small subset of the lipopolysaccharide molecules bearing the longest O-specific side chains. When the length of the O side chains was reduced by restricting the growth medium supply of mannose to this *pmi* mutant, C3 bound to lipopolysaccharide molecules with shorter O side chains. These studies imply that C3 binds to sites near the termini of long O side chains and that C3 is sterically constrained from interacting with short O side chains by the presence of even a small proportion of long O side chains. Thus, convertases may be formed at some distance from the bilayer of the outer membrane on O-side-chain-replete strains such that they are unable to ensure C5b-9 formation at the appropriate sites for killing. These remotely bound molecules may also be more accessible to complement control proteins and hence may be rapidly cleaved to inactive products. Differences in C3b binding between smooth, serum-resistant *Klebsiella pneumoniae* and rough mutants have also been observed (58). With the availability of monoclonal antibodies directed against well-defined neoepitopes present on activated, membrane-bound C3 fragments (50, 65), it may now be possible to define more precisely the molecular differences between C3 activation products on the surfaces of complement-susceptible and -resistant bacteria.

Since activated C3 is intimately involved in amplification of both complement pathways, it is presently unclear whether the failure of the activated complement pathways to kill some resistant strains is due to the presence of insufficient numbers of C3- and C5-convertase complexes on the surface. In any case, the critical event in determining whether or not killing takes place is almost certainly the stable deposition of C5b-8 complexes onto the outer membrane and the subsequent incorporation of multiple copies of C9 into the complex. For example, Taylor and Kroll (99) probed the outer membranes of serum-treated, complement-resistant *E. coli* cells with antibodies against a range of complement components and, in marked contrast to results with sensitive strains, found no evidence of C5, C7, or C9 deposition. Joiner et al. (44, 45) studied the interaction of complement with a smooth, serum-resistant *S. minnesota* strain and a deep rough susceptible mutant. On the mutant, the bulk of the C5b-9 generated was in stable association with the cell envelope and in all probability was intercalated into hydrophobic domains in the outer membrane. In contrast, the resistant parent activated complement efficiently, and C5b-9 complexes were formed on the cell surface, but they did not insert into the outer membrane and were subsequently released from the surface. Similar observations

have been made with other gram-negative species (42, 58).

Activation of the complement cascade leads to the efficient deposition of C5b-9 complexes on asymmetric model outer membranes containing lipopolysaccharide from deep rough strains, and the outer membrane appears to be an effective substrate for complement attack (87). O side chains may well protect the outer membrane from this deposition by steric hindrance alone, but alternatively, some components may inhibit C5b-9 deposition by reducing the fluidity of the target membrane. Galdiero and colleagues (29) have demonstrated an incomplete relationship between serum resistance and membrane fluidity determined by using fluorescence polarization techniques; the authors suggested that the incomplete nature of the correlation may be due to differences in the microheterogeneity of the membrane. C5b-9 failed to insert into *E. coli* membranes when the experiments were performed at temperatures below the phase transition point and with membrane phospholipids in a state of gel packing with their acyl chains in a restricted and ordered state (48). Above the transition point, killing occurred at a high rate, since C5b-9 inserted into the target membrane in which phospholipids were in a liquid crystal state and the acyl chains exhibited a high degree of molecular motion. It remains to be determined whether determinants of complement resistance have an effect on membrane fluidity sufficient to inhibit C5b-9-mediated damage and cell death.

The outer membrane is a highly asymmetric structure (64), with lipopolysaccharide molecules located exclusively on the outer surface and phospholipids occupying mainly the inner leaflet. Macromolecules on the outer surface are likely to be of critical importance in determining whether C5b-9 assembly and insertion occur, and there is now a very large body of evidence implicating lipopolysaccharide O side chains as a major structural determinant of complement resistance. Mutations from smooth to rough phenotype, usually but not invariably associated with loss of ability to synthesize O side chains, have long been known to lead to large increases in complement susceptibility (42). Growth in the presence of subinhibitory concentrations of some antibiotics leads to a reduction in the lengths of O side chains and a concomitant increase in complement susceptibility (63, 88). Modulation of the lengths of O side chains and the degree of substitution of core oligosaccharide by O side chains through manipulation of conditions of cell growth (43, 94) has emphasized the relationship between O-side-chain content and complement susceptibility. Evidence that increased coverage of lipid A-core with O antigen precludes access of C5b-9 to hydrophobic domains on the cell surface was provided by Goldman et al. (30) in a study of serum-resistant *E. coli* mutants. Additionally, lipopolysaccharide molecules from serum-resistant clinical isolates of *E. coli* contain more long O side chains than do those from susceptible isolates (75). Similar observations have been made for a number of other gram-negative species.

The presence of S lipopolysaccharide, replete with O side chains, is not sufficient, however, to confer complement resistance, and many smooth strains are susceptible to complement attack (28, 93). Hfr-mediated transfer of the *rfb* gene cluster, which determines the biosynthesis of O8-specific O side chains, to an extremely complement-susceptible rough *E. coli* strain yielded recombinants expressing the O8 mannan side chain to high levels. Recombinants were not completely resistant to complement but were killed at a much lower rate than the recipient was (100). Complement resistance could be further increased by inheritance of plasmid-borne genes determining synthesis of outer membrane proteins. Interestingly, inheritance of the *his*-linked genes for K27 exopolysaccharide production did not alter the serum response. Recombinants therefore behaved like many smooth, serum-susceptible, wild-type enterobacteria with respect to kinetics of complement killing. This study emphasized that complete resistance to complement is a multifactorial phenomenon, with expression of a lipopolysaccharide with a high degree of substitution of core with O side chains an essential prerequisite.

In contrast, the role of acidic exopolysaccharide capsules or microcapsules in the determination of complement resistance is unclear; many studies have examined the contribution of these structures, with highly variable results. The expression of an exopolysaccharide at the cell surface is not essential for resistance, and members of some groups of bacteria, such as salmonellae, are almost always acapsular and predominantly refractory to complement when isolated from infections. Conversely, some heavily capsulated organisms, such as many *Klebsiella* strains (4, 74), are frequently highly sensitive to complement. The most widely studied acidic exopolysaccharide in relation to complement resistance is the K1 antigen of *E. coli*. Many K1 isolates from cases of neonatal meningitis or urinary tract infection are susceptible to serum. In some cases, however, there is evidence that the expression of the polysialyl K1 antigen structure can modify the response to complement. The K1 antigen is structurally identical to an oncodevelopmental antigen in human kidney and brain (104) and as such may not be readily recognized by the immune system as nonself. There is evidence that the presence of this antigen on the surface of resistant strains restricts activation of both the alternative and the classic pathways (70, 71), and that reduction in expression of the antigen by either manipulation of the growth conditions (105) or mutation (53) may increase complement susceptibility. There is little evidence, however, that K1 is able to interfere with the insertion of the late-acting complement components into the outer membrane. In contrast, inactivation of genes involved in the synthesis of the *E. coli* K54 capsular polysaccharide by transposon mutagenesis results in a large increase in serum susceptibility that appears not to involve either relief of complement activation inhibition or differences in the surface binding of C3 (83).

A number of authors have shown that acquisition of certain plasmids, particularly of the FII incompatibility group, decreases the rate of killing of the recipient bacteria by human and animal serum (26, 76, 80). In general, the observed increases in resistance have been small and the effects have been manifest only when low concentrations of serum were used. In some studies, increased survival is apparent only when sera from certain animal species are utilized. For example, Ogata and Levine (67) compared the effect of acquisition of the R100 plasmid on the response of *E. coli* K-12 to various sera; they found virtually no effect with human serum but an increased survival with guinea pig and rabbit sera. These factors make comparison between studies difficult; much more pronounced effects can be demonstrated when smooth strains are used as hosts (10), emphasizing the multifactorial, interactive nature of complement resistance. In some instances, the plasmid-borne genes responsible for these effects have been cloned and sequenced. The *iss* gene on ColV plasmids, commonly present in clinical isolates of *E. coli,* is located outside the *tra* region of the plasmid and is closely linked to the genes for colicin V production (10). The gene on the IncF plasmid R6-5 that determines increased survival in serum is co-incident with the *traT* locus, one of two involved in surface exclusion in conjugation (60). Both *iss* and *traT* gene products are proteins located in the outer membrane. A *traT* gene was subsequently identified on ColV by DNA hybridization techniques (16), and it was demonstrated that both *iss* and *traT* contribute to increased survival in serum caused by ColV plasmids. However, the effect of *traT* could be demonstrated only when *iss* was deleted or when *traT* was present on a multicopy plasmid, indicating that ColV-mediated effects are due in the main to *iss*. There is no significant homology at the nucleotide level between the two genes (15). Recently, the *rck* gene, carried by the *Salmonella typhimurium* virulence plasmid, has been shown to decrease the susceptibility of strains of *E. coli* and *Salmonella* spp. to human serum (34); the gene encodes a 17-kDa outer membrane protein.

The mechanism by which these plasmid-encoded proteins decrease the serum suscepti-

bility of host strains is not known. The proteins that have been adequately characterized have functions additional to those associated with virulence and are generally present in high copy numbers in the outer membrane (92). They may therefore fortuitously reduce membrane fluidity and hence the capacity of C5b-9 complexes to intercalate into the outer membrane. The implication that they have not evolved specifically to increase survival in blood or tissue fluids is supported by the results of careful epidemiologic studies showing that the *traT* gene product is expressed as frequently in complement-susceptible as in complement-resistant populations of gram-negative bacteria (47, 62).

A number of non-plasmid-encoded proteins may also contribute to complement resistance. For example, the amount of an *E. coli* 46-kDa outer membrane protein that is expressed in larger quantities in a serum-resistant mutant correlates with phenotypically induced serum susceptibility (94). A protein modifying the response of *Neisseria gonorrhoea* to serum has been identified in strains producing disseminated gonococcal infection (35). An *ompA* mutant of a virulent *E. coli* K1 isolate was more sensitive to complement than the parent strain was; correction of the mutation in *ompA* with an *E. coli* K-12 *ompA* gene restored a level of resistance and virulence equivalent to those of the parent strain (106). It is likely that these proteins function as determinants of serum resistance only when a full complement of lipopolysaccharide O side chains are present at the cell surface.

ANTIBACTERIAL ACTIVITY OF COMPLEMENT IN IMMUNE DEFENSE

The host mounts a concerted immune attack against invading microorganisms, and it is consequently difficult to assess the precise contribution of humoral complement mechanisms to the control of gram-negative infections. Cellular defense plays such a prominent role in the eradication of potential pathogens that it is generally accepted that direct complement-mediated killing of gram-negative bacteria is of secondary importance in combatting invasion. Nevertheless, there are clear indications of correlation between resistance to complement and the capacity of strains to cause systemic infection. This correlation is particularly pronounced in the cases of bacteremia and septicemia, and an analysis of nine published studies indicates that 82% of isolates from such infections are resistant (95). In urinary infections, there is evidence that enterobacteria infecting renal tissue are more complement resistant than are those causing infections confined to the lower urinary tract (31). In that study, some patients with upper urinary tract infections were infected with strains that were susceptible to healthy human serum but were not killed by the patient's own serum owing to the presence of blocking antibody. As reviewed by Johnson (41), isolates from patients with pyelonephritis or cystitis are more commonly serum resistant than are those from feces or from patients with asymptomatic infections. Complement-susceptible strains from patients with asymptomatic bacteriuria are even more prevalent than complement-susceptible fecal strains, and successive isolates of the same organism show decreasing resistance (41). Thus, variants lacking surface structures associated with resistance are at an advantage in this environment, where there may be little or no contact with proteins of the humoral immune system.

There is a striking correlation between patients with homozygous deficiencies in the late-acting complement components and recurrent infections due to *N. meningitidis* and *N. gonorrhoeae* (82), suggesting a key role for humoral defense against these agents. Complementary to these observations are findings that gonococci isolated from patients with uncomplicated symptomatic local infections are complement susceptible, whereas those from patients with disseminated infections are almost always resistant (82, 85).

Epidemiologic studies have also demonstrated relationships between complement resistance and infections in animals. The large majority of enterobacteria causing bovine mas-

titis appear to be resistant to serum (14); the serum bactericidal system is likely to be an important defense against udder infection, as there is evidence that phagocytic cells in milk are less active than their blood counterparts (39). Direct complement activity also appears to play a role in the protection of cattle against brucellosis (17).

COMPLEMENT AND GRAM-POSITIVE BACTERIA

Gram-positive bacteria are not susceptible to direct killing by C5b-9 complexes, since the peptidoglycan layer, which underlies the outer membrane in gram-negative bacteria, is generally the outermost layer of the gram-positive cell. This thick layer acts as an impenetrable barrier to the components of the membrane attack pathway and thus protects the cytoplasmic membrane. However, the complement system plays a critical role in the control of gram-positive infections as a result of its ability to opsonize bacteria as a signal for their ingestion and destruction by phagocytes (59).

REFERENCES

1. **Abraha, A., and J. P. Luzio.** 1989. Inhibition of the formation of the complement membrane-attack complex by a monoclonal antibody to the complement component C8 α subunit. *Biochem. J.* **264**:933–936.
2. **Agnello, V.** 1978. Complement deficiency states. *Medicine* (Baltimore) **57**:1–23.
3. **Ballow, M.** 1977. Phylogenetics and ontogenetics of the complement systems, p. 183–204. *In* N. K. Day, and R. A. Good (ed.), *Biological Amplification Systems in Immunology*, vol. 2. *Comprehensive Immunology*. Plenum Publishing Corp., New York.
4. **Benge, G. R.** 1988. Bactericidal activity of human serum against strains of *Klebsiella* from different sources. *J. Med. Microbiol.* **27**:11–15.
5. **Bhakdi, S., G. Kuller, M. Muhly, S. Fromm, G. Siebert, and J. Parrisius.** 1987. Formation of transmural complement pores in serum-sensitive *Escherichia coli. Infect. Immun.* **55**:206–210.
6. **Bhakdi, S., and J. Tranum-Jensen.** 1978. Molecular nature of the complement lesion. *Proc. Natl. Acad. Sci. USA* **75**:5655–5659.
7. **Bhakdi, S., and J. Tranum-Jensen.** 1983. Membrane damage by complement. *Biochim. Biophys. Acta* **737**:343–372.

8. **Bhakdi, S., and J. Tranum-Jensen.** 1986. C5b-9 assembly: average binding of one C9 molecule to C5b-8 without poly-C9 formation generates a stable transmembrane pore. *J. Immunol.* **136:**2999–3005.

9. **Biesecker, G., P. Lachmann, and R. Henderson.** 1993. Structure of complement poly-C9 determined in projection by cryo-electron microscopy and single particle analysis. *Mol. Immunol.* **30:**1369–1382.

10. **Binns, M. M., D. L. Davis, and K. G. Hardy.** 1979. Cloned fragments of the plasmid ColV,I-K94 specifying virulence and serum resistance. *Nature* (London) **279:**778–781.

11. **Björkstein, B., R. Bortolussi, L. Gothefors, and P. G. Quie.** 1976. Interaction of E. coli strains with human serum: lack of relationship to K1 antigen. *J. Pediatr.* **89:**892–897.

12. **Born, J., and S. Bhakdi.** 1986. Does complement kill *E. coli* by producing transmural pores? *Immunology* **59:**139–145.

13. **Buchner, H.** 1889. Über die bakterientotende Wirkung des zellfreien Blutserums. *Zentralbl. Bakteriol. Parasitenkd. Infektionskr. Hyg. Abt. 1 Orig.* **5:**817–823.

14. **Carroll, E. J., and D. E. Jasper.** 1977. Bactericidal activity of standard bovine serum against coliform bacteria isolated from udders and the environment of dairy cows. *Am. J. Vet. Res.* **38:**2019–2022.

15. **Chuba, P. J., M. A. Leon, A. Banerjee, and S. Palchaudri.** 1989. Cloning and DNA sequence of plasmid determinant iss, coding for increased serum survival and surface exclusion, which has homology with lambda DNA. *Mol. Gen. Genet.* **216:**287–292.

16. **Chuba, P. J., S. Palchaudri, and M. A. Leon.** 1986. Contributions of *traT* and *iss* genes to the serum resistance phenotype of plasmid ColV2-K94. *FEMS Microbiol. Lett.* **37:**135–140.

17. **Corbeil, L. B., K. Blau, T. J. Inzana, K. H. Nielsen, R. H. Jacobson, R. B. Corbeil, and A. J. Winter.** 1988. Killing of *Brucella abortus* by bovine serum. *Infect. Immun.* **56:**3251–3261.

18. **Dankert, J. R., and A. F. Esser.** 1985. Proteolytic modification of human complement protein C9: loss of poly(C9) and circular lesion formation without impairment of function. *Proc. Natl. Acad. Sci. USA* **82:**2128–2132.

19. **Dankert, J. R., and A. F. Esser.** 1986. Complement-mediated killing of *Escherichia coli:* dissipation of membrane potential by a C9-derived peptide. *Biochemistry* **25:**1094–1100.

20. **Dankert, J. R., and A. F. Esser.** 1987. Bacterial killing by complement: C9-mediated killing in the absence of C5b-8. *Biochem. J.* **244:**393–399.

21. **Di Ninno, V. L., and V. K. Chenier.** 1981.

Activation of complement by *Neisseria meningitidis*. *FEMS Microbiol. Lett.* **12:**55–60.

22. **Esser, A. F.** 1991. Big MAC attack: complement proteins cause leaky patches. *Immunol. Today* **12:**316–318.

23. **Fearon, D. T.** 1978. Regulation by membrane sialic acid of β1H dependent decay-association of amplification C3 convertase of the alternative complement pathway. *Proc. Natl. Acad. Sci. USA* **75:**1971–1975.

24. **Feingold, D. S., J. N. Goldman, and H. M. Kuritz.** 1968. Locus of the lethal event in the serum bactericidal reaction. *J. Bacteriol.* **96:**2127–2131.

25. **Fierer, J., and F. Finley.** 1979. Lethal effect of complement and lysozyme on polymyxin-treated, serum-resistant, Gram-negative bacilli. *J. Infect. Dis.* **140:**581–588.

26. **Fietta, A., E. Romero, and A. G. Siccardi.** 1977. Effect of some R factors on the sensitivity of rough *Enterobacteriaceae* to human serum. *Infect. Immun.* **18:**273–282.

27. **Frank, M. M.** 1992. The mechanism by which microorganisms avoid complement attack. *Curr. Opin. Immunol.* **4:**14–19.

28. **Frank, M. M., and L. F. Fries.** 1988. The role of complement in defence against bacterial disease. *Baillière's Clin. Immunol. Allergy* **2:**335–361.

29. **Galdiero, F., L. Sommese, C. Capasso, M. Galdiero, and M. A. Tufano.** 1991. Serum-mediated killing of *Salmonella typhimurium* and *Escherichia coli* mutants which share a different content of major proteins. *Microbiologica* **14:**119–130.

30. **Goldman, R. C., K. Joiner, and L. Leive.** 1984. Serum-resistant mutants of *Escherichia coli* O111 contain increased lipopolysaccharide, lack an O-antigen-containing capsule, and cover more of their lipid A core with O antigen. *J. Bacteriol.* **159:**877–882.

31. **Gower, P. E., P. W. Taylor, K. G. Koutsaimanis, and A. P. Roberts.** 1972. Serum bactericidal activity in patients with upper and lower urinary tract infections. *Clin. Sci.* **43:**13–22.

32. **Griffiths, E.** 1974. Metabolically controlled killing of *Pasteurella septica* by antibody and complement. *Biochim. Biophys. Acta* **462:**598–602.

33. **Grossman, N., K. A. Joiner, M. M. Frank, and L. Leive.** 1986. C3b binding but not its breakdown is affected by the structure of the O-antigen polysaccharide in lipopolysaccharide from salmonellae. *J. Immunol.* **136:**2208–2215.

34. **Hefferman, E. J., S. Reed, J. Hackett, J. Fierer, C. Roudier, and D. Guiney.** 1992. Mechanism of resistance to complement-mediated killing of bacteria encoded by the *Salmonella typhimurium* virulence plasmid gene *rck*. *J. Clin. Invest.* **90:**953–964.

35. **Hildebrandt, J. F., L. W. Mayer, S. P. Wang,**

and T. M. Buchanan. 1978. *Neisseria gonorrhoeae* acquire a new principal outer membrane protein when transformed to resistance to serum bactericidal activity. *Infect. Immun.* **20:**267–273.

36. **Hobot, J. A., E. Carlemaln, W. Williger, and E. Kellenberger.** 1984. Periplasmic gel: a new concept resulting from the reinvestigation of bacterial cell envelope structure. *J. Bacteriol.* **160:** 143–152.

37. **Inoue, K., T. Kinoshita, M. Okada, and Y. Akiyama.** 1977. Release of phospholipids from complement-mediated lesions on the surface structure of *Escherichia coli. J. Immunol.* **119:**65–72.

38. **Inoue, K., K. Yonemasu, A. Takamizawa, and T. Amano.** 1968. Studies on the immune bacteriolysis. XIV. Requirement of all nine components of complement for immune bacteriolysis. *Biken J.* **11:**203–206.

39. **Jain, N. C., and J. Lasmanis.** 1978. Phagocytosis of serum-resistant and serum-sensitive coliform bacteria (*Klebsiella*) by bovine neutrophils from blood and mastitic milk. *Am. J. Vet. Res.* **39:** 425–427.

40. **Jimenez–Lucho, V. E., K. A. Joiner, J. Foulds, M. M. Frank, and L. Leive.** 1987. C3b generation is affected by the structure of the O-antigen polysaccharide in lipopolysaccharide from *Salmonellae. J. Immunol.* **139:**1253–1259.

41. **Johnson, J. R.** 1991. Virulence factors in *Escherichia coli* urinary tract infection. *Clin. Microbiol. Rev.* **4:**80–128.

42. **Joiner, K. A.** 1985. Studies on the mechanism of bacterial resistance to complement-mediated killing and on the mechanism of action of bactericidal antibody. *Curr. Top. Microbiol. Immunol.* **121:**135–158.

43. **Joiner, K. A., N. Grossman, M. Schmetz, and L. Leive.** 1986. C3 binds preferentially to long-chain lipopolysaccharide during alternative pathway activation by *Salmonella montevideo. J. Immunol.* **136:**710–715.

44. **Joiner, K. A., C. H. Hammer, E. J. Brown, R. J. Cole, and M. M. Frank.** 1982. Studies on the mechanism of bacterial resistance to complement-mediated killing. I. Terminal complement components are deposited and released from *Salmonella minnesota* S218 without causing bacterial death. *J. Exp. Med.* **155:**797–808.

45. **Joiner, K. A., C. H. Hammer, E. J. Brown, and M. M. Frank.** 1982. Studies on the mechanism of bacterial resistance to complement-mediated killing. II. C8 and C9 release C5b67 from the surface of *Salmonella minnesota* S218 because the terminal complex does not insert into the bacterial outer membrane. *J. Exp. Med.* **155:** 809–819.

46. **Joiner, K. A., M. A. Schmetz, M. E. Sanders, T. G. Murray, C. H. Hammer, R. Dour-**

mashkin, and M. M. Frank. 1985. Multimeric complement component C9 is necessary for killing of *Escherichia coli* J5 by terminal attack complex C5b-9. *Proc. Natl. Acad. Sci. USA* **82:**4808–4812.

47. **Kanukollu, V., S. Bieler, S. Hull, and R. Hull.** 1985. Contribution of the *traT* gene to serum resistance among clinical isolates of enterobacteriaceae. *J. Med. Microbiol.* **19:**61–67.

48. **Kato, K., and Y. Bito.** 1978. Relationship between bactericidal action of complement and fluidity of cellular membranes. *Infect. Immun.* **19:**12–17.

49. **Kazatchkine, M. D., D. T. Fearon, and K. F. Austen.** 1979. Human alternative complement pathway: membrane-associated sialic acid regulates the competition between B and ß1H for cell-bound C3b. *J. Immunol.* **122:**75–81.

50. **Kemp, P. A., J. H. Spragg, J. C. Brown, B. P. Morgan, C. A. Gunn, and P. W. Taylor.** 1992. Immunohistochemical determination of complement activation in joint tissues of patients with rheumatoid arthritis and osteoarthritis using neoantigen-specific monoclonal antibodies. *J. Lab. Clin. Immunol.* **37:**147–162.

51. **Kroll, H.-P., S. Bhakdi, and P. W. Taylor.** 1983. Membrane changes induced by exposure of *Escherichia coli* to human serum. *Infect. Immun.* **42:** 1055–1066.

52. **Kroll, H.-P., W.-H. Voigt, and P. W. Taylor.** 1984. Stable insertion of C5b-9 complement complexes into the outer membrane of serum treated, susceptible *Escherichia coli* cells as a prerequisite for killing. *Zentralbl. Bakteriol. Mikrobiol. Hyg. Abt. 1 Orig. Reihe A* **258:**316–326.

53. **Leying, H., S. Suerbaum, H.-P. Kroll, D. Stahl, and W. Opferkuch.** 1990. The capsular polysaccharide is a major determinant of serum resistance in K-1 positive blood culture isolates of *Escherichia coli. Infect. Immun.* **58:**222–227.

54. **MacKay, S. L. D., and J. R. Dankert.** 1990. Bacterial killing and inhibition of inner membrane activity as a function of the sequential addition of C9 to C5b-8 sites. *J. Immunol.* **145:**3367–3371.

55. **Marazziti, D., J. P. Luzio, and K. K. Stanley.** 1989. Complement C9 is inserted into membranes in a globular conformation. *FEBS Lett.* **243:**347–350.

56. **Martinez, R. J., and S. F. Carroll.** 1980. Sequential metabolic expressions of the lethal process in human serum-treated *Escherichia coli:* role of lysozyme. *Infect. Immun.* **28:**735–745.

57. **McAleer, M. A., and R. B. Sim.** 1993. The complement system, p. 1–15. *In* R. B. Sim (ed.), *Activators and Inhibitors of Complement.* Kluwer Academic Publishing, Dordrecht, The Netherlands.

58. **Merino, S., S. Camprubi, S. Alberti, V. J. Benedi, and J. M. Thomas.** 1992. Mechanisms

of *Klebsiella pneumoniae* resistance to complement-mediated killing. *Infect. Immun.* **60:**2529–2535.

59. **Moffitt, M. C., and M. M. Frank.** 1994. Complement resistance in microbes. *Springer Semin. Immunopathol.* **15:**327–344.

60. **Moll, A., P. A. Manning, and K. N. Timmis.** 1980. Plasmid-determined resistance to serum bactericidal activity: a major outer membrane protein, the *traT* gene product, is responsible for plasmid-specified serum resistance in *Escherichia coli*. *Infect. Immun.* **28:**359–367.

61. **Monahan, J. B., and J. M. Sodetz.** 1981. Role of the ß subunit in interaction of the eighth component of human complement with the membrane-bound cytolytic complex. *J. Biol. Chem.* **256:**3258–3262.

62. **Montenegro, M. A., D. Bitter-Suermann, J. K. Timmis, M. E. Aguero, F. C. Cabello, S. C. Sanyal, and K. N. Timmis.** 1987. *traT* gene sequences, serum resistance and pathogenicity-related factors in clinical isolates of *Escherichia coli* and other Gram-negative bacteria. *J. Gen. Microbiol.* **131:**1511–1521.

63. **Nelson, D., T. E. S. Delahooke, and I. R. Poxton.** 1993. Influence of subinhibitory levels of antibiotics on expression of *Escherichia coli* lipopolysaccharide and binding of anti-lipopolysaccharide monoclonal antibodies. *J. Med. Microbiol.* **39:**100–106.

64. **Nikaido, H., and M. Vaara.** 1985. Molecular basis of bacterial outer membrane permeability. Microbiol. Rev. **49:**1–32.

65. **Nilsson, B., K.-E. Svensson, P. Borwell, and U. R. Nilsson.** 1987. Production of mouse monoclonal antibodies that detect distinct neoantigenic epitopes on bound C3b and iC3b but not on the corresponding soluble fragments. *Mol. Immunol.* **24:**487–494.

66. **Nuttal, G.** 1888. Experimente über die bakterienfeindliche Einflüsse des tierischen Körpers. *Zentralbl. Hyg. Infektionskr.* **4:**353–394.

67. **Ogata, R. T., and R. P. Levine.** 1980. Characterisation of complement resistance in *Escherichia coli* conferred by the antibiotic resistance plasmid R100. *J. Immunol.* **125:**1494–1498.

68. **Opferkuch, W.** 1984. Die Serumbakterizidie, p. 19–27. *In* C. Krasemann (ed.), *Infektiologisches Kolloquium No. 2. Derabwehrgeschwächte Patient.* Walter de Gruyter, Berlin.

69. **Pattus, R., D. Massotte, H. U. Wilmsen, J. Lakey, D. Tsernoglou, A. Tucker, and M. W. Parker.** 1990. Colicins: prokaryotic killer-pores. *Experientia* **46:**180–192.

70. **Pluschke, G., and M. Achtman.** 1984. Degree of antibody-independent activation of the classical complement pathway by K1 *Escherichia coli* differs with O antigen type and correlates with virulence

of meningitis in newborns. *Infect. Immun.* **43:**684–692.

71. **Pluschke, G., J. Mayden, M. Achtman, and R. P. Levine.** 1983. Role of the capsule and the O antigen in resistance of O18:K1 *Escherichia coli* to complement-mediated killing. *Infect. Immun.* **42:**907–913.

72. **Podack, E. R.** 1986. Molecular mechanisms of cytolysis by complement and by cytolytic lymphocytes. *J. Cell. Biochem.* **30:**133–170.

73. **Podack, E. R., and J. Tschopp.** 1984. Membrane attack by complement. *Mol. Immunol.* **21:**589–603.

74. **Podschun, R., D. Sievers, A. Fischer, and U. Ullmann.** 1993. Serotypes, hemagglutinins, siderophore synthesis, and serum resistance of *Klebsiella* isolates causing human urinary tract infections. *J. Infect. Dis.* **168:**1415–1421.

75. **Porat, R., R. Mosseri, E. Kaplan, M. A. Johns, and S. Shibolet.** 1992. Distribution of polysaccharide side chains of lipopolysaccharide determine resistance of *Escherichia coli* to the bactericidal activity of serum. *J. Infect. Dis.* **165:**953–956.

76. **Pramoonjago, P., M. Kaneko, T. Kinoshita, E. Ohtsubo, J. Takeda, K. Hong, R. Inagi, and K. Inoue.** 1992. The role of TraT protein, an anticomplementary protein produced in *Escherichia coli* by R100 factor, in serum resistance. *J. Immunol.* **148:**827–836.

77. **Ramm, L. E., M. B. Whitlow, and M. M. Mayer.** 1982. Size of the transmembrane channels produced by complement proteins C5b-8. *J. Immunol.* **129:**1143–1146.

78. **Ratnoff, W. D., D. T. Fearon, and K. F. Austen.** 1983. The role of antibody in the activation of the alternative complement pathway. *Springer Semin. Immunopathol.* **6:**361–371.

79. **Reid, K. B. M.** 1986. Activation and control of the complement system. *Essays Biochem.* **22:**27–68.

80. **Reynard, A. M., and M. E. Beck.** 1976. Plasmid-mediated resistance to the bactericidal effects of normal rabbit serum. *Infect. Immun.* **14:**848–850.

81. **Reynolds, B. L., U. A. Rother, and K. O. Rother.** 1975. Interaction of complement components with a serum-resistant strain of *Salmonella typhimurium*. *Infect. Immun.* **11:**944–948.

82. **Ross, S. C., and P. Densen.** 1984. Complement deficiency states and infection: epidemiology, pathogenesis and consequences of neisserial and other infections in an immune deficiency. *Medicine* (Baltimore) **63:**243–273.

83. **Russo, T. A., M. C. Moffitt, C. H. Hammer, and M. M. Frank.** 1993. Tn*phoA*-mediated disruption of K54 capsular polysaccharide genes in

Escherichia coli confers serum sensitivity. *Infect. Immun.* **61:**3578–3582.

84. **Schiller, N. L., R. A. Hatch, and K. A. Joiner.** 1989. Complement activation and C3 binding by serum-sensitive and serum-resistant strains of *Pseudomonas aeruginosa. Infect. Immun.* **57:** 1707–1713.

85. **Schoolnik, G. K., T. M. Buchanan, and K. K. Holmes.** 1976. Gonococci causing disseminated gonococcal infection are resistant to the bactericidal action of normal human sera. *J. Clin. Invest.* **58:**1163–1173.

86. **Schreiber, R. D., D. C. Morrison, E. R. Podack, and H.-J. Müller-Eberhard.** 1979. Bactericidal activity of the alternative complement pathway generated from eleven isolated plasma proteins. *J. Exp. Med.* **149:**870–882.

87. **Schröder, G., K. Brandenburg, L. Brade, and U. Seydel.** 1990. Pore formation by complement in the outer membrane of Gram-negative bacteria studied with asymmetric planar lipopolysaccharide/phospholipid bilayers. *J. Membr. Biol.* **118:**161–170.

88. **Schweinle, J. O., and M. Nishiyasu.** 1992. Sub-minimal inhibitory concentrations of cefmetazole enhance serum bactericidal activity *in vitro* by amplifying poly-C9 deposition. *J. Clin. Invest.* **89:**1198–1207.

89. **Sim, R. B., and R. Malhotra.** 1994. Interactions of carbohydrates and lectins with complement. *Biochem. Soc. Trans.* **22:**106–111.

90. **Stanley, K. K.** 1988. The molecular mechanism of complement C9 insertion and polymerisation in biological membranes. *Curr. Top. Microbiol. Immunol.* **140:**49–65.

91. **Stanley, K. K., M. Page, A. K. Campbell, and J. P. Luzio.** 1986. A mechanism for the insertion of complement component C9 into target membranes. *Mol. Immunol.* **23:**451–458.

92. **Sukupolvi, S., and C. D. O'Connor.** 1990. TraT protein, a plasmid-specified mediator of interactions between Gram-negative bacteria and their environment. *Microbiol. Rev.* **54:**331–341.

93. **Taylor, P. W.** 1976. Immunochemical investigations on lipopolysaccharides and acidic polysaccharides from serum-sensitive and serum-resistant strains of *Escherichia coli* isolated from urinary tract infections. *J. Med. Microbiol.* **9:**405–421.

94. **Taylor, P. W.** 1984. Growth environment effects on pathogenicity of Gram-negative bacteria, p. 10–21. *In* A. C. R. Dean, D. C. Ellwood, and C. G. T. Evans (ed.), *Continuous Culture,* vol. 8. *Biotechnology, Medicine and the Environment.* Ellis Horwood, Chichester, England.

95. **Taylor, P. W.** 1988. Bacterial resistance to complement, p. 107–120. *In* J. A. Roth (ed.), *Virulence Mechanisms of Bacterial Pathogens.* American Society for Microbiology, Washington, D.C.

96. **Taylor, P. W.** 1992. Complement-mediated killing of susceptible Gram-negative bacteria: an elusive mechanism. *Exp. Clin. Immunogenet.* **9:** 48–56.

97. **Taylor, P. W.** 1993. Non-immunoglobulin activators of the complement system, p. 37–68. *In* R. B. Sim (ed.), *Activators and Inhibitors of Complement.* Kluwer Academic Publishing, Dordrecht, The Netherlands.

98. **Taylor, P. W., and H.-P. Kroll.** 1983. Killing of an encapsulated strain of *Escherichia coli* by human serum. *Infect. Immun.* **39:**121–131.

99. **Taylor, P. W., and H.-P. Kroll.** 1984. Interaction of human complement proteins with serum-sensitive and serum-resistant strains of *Escherichia coli. Mol. Immunol.* **21:**609–620.

100. **Taylor, P. W., and M. K. Robinson.** 1980. Determinants that increase the serum resistance of *Escherichia coli. Infect. Immun.* **29:**278–280.

101. **Tomlinson, S., P. W. Taylor, and J. P. Luzio.** 1989. Transfer of phospholipid and protein into the envelope of Gram-negative bacteria by liposome fusion. *Biochemistry* **28:**8303–8311.

102. **Tomlinson, S., P. W. Taylor, and J. P. Luzio.** 1990. Transfer of preformed terminal C5b-9 complement complexes into the outer membrane of viable Gram-negative bacteria: effect on viability and integrity. *Biochemistry* **29:**1852–1860.

103. **Tomlinson, S., P. W. Taylor, B. P. Morgan, and J. P. Luzio.** 1989. Killing of Gram-negative bacteria by complement: fractionation of cell membranes after complement C5b-9 deposition on to the surface *Salmonella minnesota* Re595. *Biochem. J.* **263:**505–511.

104. **Troy, F. A.** 1992. Polysialylation: from bacteria to brains. *Glycobiology* **2:**5–23.

105. **Vermuelen, C., A. Cross, W. R. Byrne, and W. Zollinger.** 1988. Quantitative relationship between capsular content and killing of K1-encapsulated *Escherichia coli. Infect. Immun.* **56:** 2723–2730.

106. **Weiser, J. N., and E. C. Gotschlich.** 1991. Outer membrane protein A (OmpA) contributes to serum resistance and pathogenicity of *Escherichia coli* K-1. *Infect. Immun.* **59:**2252–2258.

107. **Wilson, L. A., and J. K. Spitznagel.** 1968. Molecular and structural damage to *Escherichia coli* produced by antibody, complement, and lysozyme systems. *J. Bacteriol.* **96:**1339–1348.

108. **Wright, S. D., and R. P. Levine.** 1981. How complement kills *E. coli.* 1. Location of the lethal lesion. *J. Immunol.* **127:**1146–1151.

QUORUM SENSING AND VIRULENCE GENE REGULATION IN *PSEUDOMONAS AERUGINOSA*

Luciano Passador and Barbara H. Iglewski

5

The success of any pathogen depends on its ability to sense its local environment and, in response, modulate the expression of those genes necessary to establish itself in a newly found niche. The ability to survive in different environments highlights the importance of understanding the regulation of genes encoding an organism's virulence factors and of identifying the environmental signals involved. The ability to adapt to different conditions is extremely important for a pathogen such as *Pseudomonas aeruginosa,* which can be found in environments ranging from soil to burn tissue to the lungs of cystic fibrosis (CF) patients and non-CF patients. The last two environments result in very different *P. aeruginosa* infections in that CF patient infections are chronic and localized and non-CF patient lung infections are typically acute and capable of dissemination.

In this chapter, we discuss a regulatory mechanism known as autoinduction or quorum sensing that has recently been shown to be involved in the regulation of the expression of the virulence genes of *P. aeruginosa*. Using *P. aeruginosa* as a model organism, we have be-

gun to study how a pathogen might use this quorum-sensing mechanism to establish itself and cause disease. The importance of quorum sensing as a regulatory mechanism is borne out by recent reports that suggest that other organisms, including known plant and animal pathogens, also utilize the mechanism and that many gram-negative organisms contain at least some of the components of this process.

P. AERUGINOSA AS A PATHOGEN

P. aeruginosa is a ubiquitous gram-negative bacterium found in soil, fresh water, plants, and animals, including humans. In keeping with its ability to occupy a large number of habitats, *P. aeruginosa* is metabolically versatile. *P. aeruginosa* can utilize a large number of organic compounds for growth and, not surprisingly, plays an important role in the degradation of organic matter in nature (44). Furthermore, it can survive on minimal nutrients and grow at temperatures as high as 42°C (44). These minimal nutritional requirements and its ability to tolerate a wide variety of physical conditions contribute to the ecological success of *P. aeruginosa* and ultimately to its effectiveness as a pathogen.

Given the ability of *P. aeruginosa* to adapt to a range of environments, it is not surprising that it is capable of causing a wide variety of infections in humans. In addition to the respiratory

Luciano Passador and Barbara H. Iglewski Department of Microbiology and Immunology, University of Rochester Medical Center, Box 672, Rochester, New York 14642.

Virulence Mechanisms of Bacterial Pathogens, 2nd ed., Edited by J. A. Roth et al.
© 1995 American Society for Microbiology, Washington, D.C.

infections mentioned earlier, *P. aeruginosa* involvement in skin and soft tissue infections, endocarditis, bacteremia, gastrointestinal infections, central nervous system infections, otitis, corneal keratitis, urinary tract infections, and infections of bones and joints has been well documented (51).

Aiding *P. aeruginosa* in its role as a pathogen are its intrinsic resistance to many antibiotics and its ability to produce a large array of both cell-associated and secreted virulence factors (37, 41). For obvious reasons, only a few of the myriad virulence factors are mentioned and briefly described here. Readers interested in other virulence factors will find several reviews helpful (9, 25, 37, 41, 51, 59, 65).

Cell-associated structures include pili, which are proposed to act as adhesins and allow the binding of *P. aeruginosa* to epithelial cells. Supporting this view are data suggesting that nonpiliated mutants exhibit decreased virulence in animal models (29). Furthermore, purified pili and antipilus antibodies block attachment of bacteria to epithelial cells (52). Flagella also appear to play a role in pathogenesis, as evidenced by the fact that nonmotile mutants of *P. aeruginosa* are less virulent than motile strains (15, 39). Motility is often used by bacterial pathogens in avoiding host defenses and disseminating themselves from one site to another.

Under certain conditions, the organism is capable of producing a thick coat of exopolysaccharide, termed alginate, which covers its exterior surface. Strains that produce alginate are said to have a mucoid phenotype and appear to be important in chronic respiratory infections, as evidenced by their high frequency of isolation from the lungs and sputa of CF patients. Further evidence of the importance of the mucoid phenotype comes from the observation that nonmucoid isolates reintroduced into the lungs of a rat quickly convert to the mucoid phenotype. The production of alginate most likely serves as an adherence mechanism and as a method of preventing opsonization and phagocytosis of the bacterium. It has even been suggested that by its activity as a permeability barrier, this alginate matrix may explain the resistance of *P. aeruginosa* to some aminoglycoside antibiotics. Expression of the genes required for the synthesis of alginate is an energy-consuming process that, not surprisingly, is tightly regulated by a complex set of controls. A description of the genes involved and the mechanism of regulation is outside the scope of this chapter but is well documented elsewhere (11, 12).

In addition to the cell-associated factors already described, *P. aeruginosa* is capable of secreting a large number of other products into the surrounding environment (41, 51). While the role of the cell-associated factors may be primarily to aid in the evasion of host defenses, the ability of *P. aeruginosa* to invade tissue is probably due to the production of extracellular products that destroy host defense components and tissue. This tissue destruction leads to breaches of physical barriers (e.g., skin) and dissemination. Exoproducts may also be responsible for supplying nutrients and creating physical conditions that render the local environment more conducive to the survival of *P. aeruginosa*. Included within the category of exoproducts are extracellular enzymes and toxins, which will be briefly discussed below. Our research efforts have concentrated on several of these exoproducts, specifically the proteases and exotoxin A. As a result, discussion of the role of exoproducts and the genetic regulation of their expression with respect to the quorum-sensing mechanism is limited to these few factors.

Most clinical isolates of *P. aeruginosa* produce several extracellular proteases. Three main extracellular proteases (LasB elastase, LasA elastase, and alkaline protease) produced by *P. aeruginosa* have been identified (40). The genes for LasB elastase (*lasB*), LasA elastase (*lasA*), and alkaline protease (*aprA*) have been cloned and characterized (5, 10, 23, 43, 48, 54). Although LasB is considered to be the major source of *P. aeruginosa* elastase activity, it is believed that elastolysis by *P. aeruginosa* results from the concerted activity of both LasA and LasB through a mechanism by which elastin is made a more suitable LasB substrate after being nicked or altered by LasA (25). The contribution of the proteases to pathogenesis and tissue damage was initially demonstrated by Liu (36), who showed that the protease fractions of *P. aeruginosa* cause hemorraghic lesions in the skin of animals. Subsequently, it was shown that purified LasB elastase is capable of degrading many proteins that are important for host defense, including immunoglobulins G and A, complement components, and airway lysozyme and of inhibiting gamma interferon and natural killer cell activity. The facts that proteolytic fragments of antibodies are present in the sputa of CF patients and that antibodies to LasB elastase are produced (14, 22) indicate that LasB elastase is produced by *P. aeruginosa* in the human lung during an infection. In addition to compromising host defenses, LasB elastase may directly contribute to the pathogenesis of *P. aeruginosa* infections by causing tissue damage, as evidenced by the enzyme's ability to destroy type III and type IV collagens and elastin (30, 40). Given that elastin accounts for nearly one-third of the protein in lung tissue, elastase may play a major role in lung damage of CF patients as a result of *P. aeruginosa* infections.

In addition to LasA and LasB, *P. aeruginosa* also produces an alkaline protease. Although alkaline protease is able to degrade laminin and other substrates, suggesting a possible role for this enzyme in tissue invasion and dissemination, a direct role in pathogenesis has not been clearly established. Experiments with cultured rabbit corneas suggest that alkaline protease can degrade corneal proteins and thus may play a role in corneal keratitis (62). By contrast, alkaline protease does not appear to be involved in lung epithelial-cell damage (66). Alkaline protease is capable of degrading complement components C1q and C3, two major recognition molecules in the complement cascade (31).

Therefore, the proteases appear to serve as virulence factors in that although they are not cytotoxic, they may be involved in the destruction of various tissues and the ultimate dissemination of the organism. In addition, the proteases may destroy components of the host defense system such as antibodies and complement, thus allowing the bacteria to survive in the blood and at secondary sites of infection. Finally, the proteases may provide nutrients for *P. aeruginosa* cells within the microenvironment of the infection site.

Exotoxin A, encoded by the *toxA* gene, is the most toxic product produced by *P. aeruginosa,* and data suggest that most clinical isolates of *P. aeruginosa* are capable of producing it (65). The toxin has ADP-ribosyltransferase activity similar to that of diphtheria toxin and inhibits host cell protein synthesis. Furthermore, the exotoxin A and diphtheria toxin appear to have essentially the same mechanisms of toxin activation, internalization, reduction, and translocation. Despite this similarity, antibodies to exotoxin A do not react with diphtheria toxin.

Expression of *toxA* is sensitive to levels of iron, with maximal expression seen in low-iron conditions (65). This iron regulation of *toxA* does not appear to be mediated directly by the Fur protein. Rather, the evidence indicates that Fur regulates RegA, a transcriptional regulator of *toxA*. On the basis of its cytotoxic effects, a role for exotoxin A in pathogenesis appears quite obvious. However, unlike the proteases, exotoxin A appears to mediate both localized and systemic diseases. Its cytotoxicity most likely results in local tissue damage, which may allow dissemination of the organism from one site to another. The toxin may also aid in the destruction of host immune cells, the end result of which would be successful continuance of the infection. With regard to a systemic involvement, many studies have now shown the lethality of exotoxin A in a variety of animals (68). A high mortality is associated with septicemia involving toxin-producing strains. Similar studies with nontoxinogenic strains indicate that such strains are less lethal (67).

Even though *P. aeruginosa* is metabolically and physically versatile and produces a large number of virulence products, it remains categorized as an opportunistic pathogen, able to cause disease only in immunocompromised individuals. *P. aeruginosa* possesses two general characteristics that relegate it to the status of an opportunistic pathogen. First, *P. aeruginosa* is not very successful at overcoming initial physical barriers such as the skin and gaining access to host tissues. Second, when *P. aeruginosa* enters a host, it is not particularly effective at avoiding host defense mechanisms. In healthy individuals, the normal defenses of the human body are sufficient to prevent infection by *P. aeruginosa*. However, when these defenses are breached, as in the cases of burns, wounds, and suppressive immunotherapy, *P. aeruginosa* can take advantage of the opportunity and establish itself. Once the organism is established, its arsenal of virulence products and its metabolic diversity allow what was once a common saprophyte to become a serious and often fatal pathogen.

The expression of most if not all virulence factors is not constitutive. Rather, their production appears to be highly regulated. For example, maximum protease production appears to be cell density dependent in that it occurs during the slower growth rates of the nutrient-limited late logarithmic and early stationary phases of bacterial growth (63, 64), times at which cell density is also maximal.

Studies (53) that have made use of *lasB::lacZ* gene fusions have led to a series of observations that may be summarized as follows. First, the expression of *lasB* is initiated during the late logarithmic or early stationary phase of the growth cycle in a cell density-dependent manner. Second, *lasB* expression is approximately 100-fold higher in *P. aeruginosa* than in *Escherichia coli*, suggesting that a regulatory component(s) is missing in *E. coli*. Finally, the introduction of multiple copies of the *lasB* promoter on plasmids results in an overall decrease in elastase yields, suggesting that there is dilution of a factor(s) present in limiting quantities in *P. aeruginosa* that is required for the production of elastase. Taken together, these observations led to the hypothesis that the expression of *lasB* requires an activator(s).

In the course of attempting to complement

P. aeruginosa PA103, a clinical isolate that maintains a *lasB* structural gene but does not produce elastase, Gambello and Iglewski (26) isolated and characterized a gene (*lasR*) whose product (LasR) proved to be a transcriptional activator of *lasB*. *P. aeruginosa* PA103 has not been extensively studied with respect to its genetic background and hence would pose a problem in structure and function studies of LasR. To alleviate the problem, Gambello and Iglewski used allelic exchange to create an insertionally inactivated *lasR* mutant in the well-characterized PAO strain that they designated *P. aeruginosa* PAO-R1. Northern (RNA) analysis of *lasB* mRNA indicated that expression of *lasB* is regulated transcriptionally. As expected, these studies showed that the *lasR* mutant PAO-R1 does not produce detectable levels of *lasB* mRNA. Introduction of the *lasR* gene in *trans* on a multicopy plasmid is able to complement the *lasR* mutation and restore the expression of *lasB*.

LasR is a transcriptional activator for several other virulence genes in addition to *lasB* (27, 61). The first of these genes to be identified was *lasA*. As mentioned previously, the full elastolytic phenotype of *P. aeruginosa* requires the product of both the *lasB* and the *lasA* genes. The finding that *lasA* is regulated by *lasR* came from investigations of the contribution of LasA to elastolytic activity (61). In the course of those studies, a strain (*P. aeruginosa* PAO-E64) that carries a nitrosoguanidine-induced mutation complementable by *lasA* was further manipulated through allelic exchange to contain an insertionally inactivated *lasB* gene to create *P. aeruginosa* PAO-E64-B1. As expected, this strain produces no detectable elastolytic activity. The lack of elastolysis resembles the phenotype of the *lasR*-negative strain PAO-R1. Northern analysis of *lasA* mRNA indicated that strain PAO-R1 does not express *lasA* and that complementation of PAO-R1 with plasmid-borne *lasR* restores expression of *lasA*, indicating that *lasA* expression is regulated by *lasR*.

Alkaline protease encoded by *aprA* and exotoxin A encoded by *toxA* are two other virulence products of *P. aeruginosa.* Exotoxin A has an ADP-ribosyltransferase activity that transfers the ADP-ribose moiety of NAD to mammalian elongation factor 2, resulting in a cessation of protein synthesis and ultimately cell death (65). A *P. aeruginosa* PAO gene designated *regA* that is involved in the regulation of *toxA* expression has been identified (65). RegA is a transcriptional regulator of *toxA* expression and is in part responsible for the expression of *toxA* with respect to the phase of bacterial growth and iron availability. Both *aprA* and *toxA* are also regulated by LasR (27). Northern analysis of *aprA* mRNA indicated that the *lasR* gene is required for *aprA* transcription. As with *lasA* and *lasB,* the *lasR*-negative mutant PAO-R1 does not exhibit any detectable *aprA* message. Complementation with *lasR* in *trans* restores *aprA* expression.

Regulation of *toxA* expression does not appear to be as dramatic as for the *lasA, lasB,* and *aprA* genes (27). PAO-R1 still retains approximately two-thirds of the wild-type toxin activity as measured by toxin activity assays of culture supernatants. These data suggest that *lasR,* although not absolutely required for *toxA* expression, contributes significantly to the enhancement of toxin production. Further evidence for the requirement of *lasR* is the finding that multiple copies of *lasR* in *trans* can elevate toxin production in PAO-R1 to levels higher than those seen in the parent strain. The mechanism by which *lasR* regulates *toxA* expression is still unknown. Studies utilizing *toxA::lacZ* fusions essentially mimic and thus support the data obtained from toxin assays, suggesting that LasR does not exert its effect on *toxA* directly. Results from *regA::cat* gene fusions also indicate that LasR does not regulate *toxA* expression by regulating expression of *regA.*

The studies mentioned previously, although indicating a requirement for *lasR* in the expression of various genes, did not quantitate the efficiency of restoration of expression when *lasR* was supplied in *trans.* More recent observations utilizing *P. aeruginosa* PAO-R1 indicate that supplying the *lasR* gene in *trans* is not sufficient to completely complement *lasB* ex-

pression to wild-type levels (45). These findings eventually led to a search for further regulatory genes and resulted in the identification and characterization of *lasI,* which is discussed below.

LasR AND LasI ARE HOMOLOGOUS TO OTHER BACTERIAL TRANSCRIPTIONAL ACTIVATOR SYSTEMS

When the amino acid sequence of LasR was first deduced, searches of protein databases indicated that the protein was highly similar to another bacterial protein, LuxR, which is known to be a transcriptional activator of the bioluminescence genes of the marine organism *Vibrio fischeri* (38). The greatest areas of similarity between the two proteins fall within two regions (26) (Fig. 1). The first is in the carboxy-terminal portion of the LuxR protein, which contains a helix-turn-helix motif and to which a DNA-binding role has been ascribed (8, 57, 58). The second homologous region is located in the amino-terminal third of the protein, which, for LuxR, is suggested to be involved in the binding of a small, diffusible signal molecule known as the *Vibrio* autoinducer (VAI) (8, 57, 58).

V. fischeri is a specific symbiont of certain marine fishes and squids. It lives within specialized structures known as light organs, where it exists essentially in pure culture and can attain cell densities of 10^{10} to 10^{11} cells per

FIGURE 1 Homologies between LuxR and LasR. The LuxR and LasR proteins are illustrated as rectangles with the amino (NH$_2$) and carboxy (COOH) terminals indicated. The putative autoinducer-binding and DNA-binding regions of each are represented as shaded areas. The numbers below each shaded area represent the amino acid positions at the beginning and end of each region. The percentages of identical residues within the shaded regions are also given. Adapted from reference 26.

ml. *V. fischeri* is also found free-living in sea-water, where cell densities are typically less than 10^2 cells per ml (18).

During growth, the organism produces a diffusible autoinducer molecule (VAI) that may accumulate within the surrounding environment. VAI is *N*-3-(oxohexanoyl) homoserine lactone. It has also been demonstrated that the *V. fischeri* cell membrane is permeable to VAI (35). In conditions of low cell density (such as in seawater), the extracellular concentration of VAI will be low, and VAI thus diffuses out of cells and into the extracellular environment, where it becomes diluted. However, when *V. fischeri* attains high cell densities, as in the light organ, VAI accumulates in the environment, and when VAI has accumulated to sufficiently high concentrations (approximately 10 nM), the luminescence genes is activated (18, 35).

In *V. fischeri,* the bioluminescence genes are regulated via a process first described as autoin-duction and now known as quorum sensing (24). This process involves a series of genes that encode the transcriptional activator (LuxR), an autoinducer synthase (LuxI), and the genes required for production of substrate and lucifer-ase (*luxCDABEG*). Bioluminescence genes are organized in two divergently transcribed units: a leftward unit that contains only *luxR* and a rightward operon that contains *luxICDABEG*. As mentioned above, use of point mutational analysis (8, 57, 58) indicates that LuxR protein is divided into two main regions: the amino-terminal third, which is involved in the binding of VAI, and the carboxy-terminal half, which is involved in the binding of *lux* DNA. In the process of autoinduction, LuxR becomes activated upon binding of VAI. This activation of LuxR occurs only at high cell densities (as seen in the light organs) when sufficient VAI has accumulated. The activated LuxR is then capable of interacting with an operator region containing dyad symmetry and located approximately 40 bp upstream of the transcriptional start of *luxICDABEG*. Studies utilizing mutations within this operator region indicate that the region is required for the activation of transcription of the *luxICDABEG* operon by

LuxR (3, 13, 56). Note that the rightward operon may also be activated, albeit at a much lower level, by LuxR alone (18).

Since the *luxI* gene is part of the *lux* operon, the autoinducer is able to regulate its own synthesis through a positive feedback circuit. As a result of activation of the *luxICDABEG* operon, the VAI concentration is elevated and bioluminescence is increased. In addition to stimulating *luxICDABEG* expression, the LuxR-VAI complex also represses expression of *luxR,* which results in a decreased level of LuxR. This regulation appears to be complex, since evidence for both transcriptional (16, 17) and posttranscriptional (19) regulation has been reported. However, as a result of the regulatory circuit in place, as LuxR levels decrease, the repression is eventually relieved. This leads to maintained levels of LuxR that allow high-level expression of bioluminescence. Thus, the *luxR* and *luxI* genes provide a means for cell-to-cell chemical communication by which *V. fischeri* populations can monitor their own cell densities.

The similarities between LuxR and LasR and between LuxI and LasI suggest that LasI might function similarly to LuxI in *V. fischeri*. To address this possibility, an assay system that utilized *P. aeruginosa* PAO-R1 in which the *lasR* gene was inactivated was used to investigate the function of LasI with respect to LasR and LasB (45). Previous studies showed that although *P. aeruginosa* PAO-R1 demonstrates no detectable elastase activity or antigen, the elastolytic phenotype may be restored if *lasR* is supplied in *trans* (26). By using a *lasB::lacZ* gene fusion in *P. aeruginosa* PAO-R1, it is possible to demonstrate that the addition of *lasR* results only in the partial restoration of *lasB* expression compared to that in the wild-type strain PAO1. Only when both *lasR* and *lasI* are present is maximal expression of *lasB* obtained (45). The results from these experiments indicate that both *lasR* and *lasI* are required for maximal expression of *lasB* and that *lasI* may indeed function in a manner similar to that of *luxI*.

To determine whether *lasI* is involved in production of a diffusible molecule, we quantitated the expression of a *lasB::lacZ* fusion in the presence or absence of *lasR* in *E. coli* when the strains were grown in spent medium from a control strain or a strain that expressed *lasI* (45). Both *E. coli* and *P. aeruginosa* strains expressing *lasI* from a multicopy plasmid were used as sources of spent medium. Only when spent media from strains expressing *lasI* were used for growth did *lasB* expression increase. These results indicate that *lasI* is involved in the production of a diffusible VAI-like molecule, which we termed the *Pseudomonas* autoinducer (PAI). Furthermore, both LasR and PAI are required for maximum *lasB* expression, since in experiments in which *lasR* was not present, ß-galactosidase production did not increase. In addition, the experiments utilizing *E. coli* strains indicate that LasR and PAI are sufficient for the expression of *lasB*.

Production of an autoinducer by *P. aeruginosa* was confirmed by studies in which the autoinducer was purified and identified (47). Supernatants of *P. aeruginosa* PAO were subjected to solvent extraction, and the products retained in the organic phase were separated by high-pressure liquid chromatography. The product obtained was analyzed by mass spectrometry, fast atom bombardment, and nuclear magnetic resonance to verify its structure. PAI was also chemically synthesized and shown to be active at the same concentrations as purified PAI.

PAI was identified as *N*-3-oxododecanoyl homoserine lactone and thus is structurally similar to other identified autoinducers in that it is composed of a homoserine lactone ring to which is attached a fatty acid side chain. Various autoinducers differ in the lengths of the side chains and in the groups present on the side chains (6, 7, 21, 69) (Fig. 2). Current studies using chemically synthesized analogs of PAI indicate that the length of the acyl side chain and the structure of the ring component of PAI are critical for activity of the molecule. In addition, modifications of the carbonyl groups on the acyl side chain are tolerated as long as they are not extreme (46).

Given the large degree of homology between the *V. fischeri* and the *P. aeruginosa* sys-

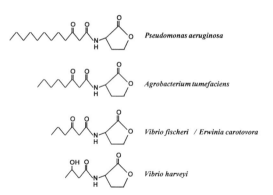

FIGURE 2 Structures of characterized autoinducer molecules. The data are from reference 47 for *P. aeruginosa*, reference 69 for *A. tumefaciens*, reference 21 for *V. fischeri*, reference 2 for *E. carotovora*, and reference 6 for *V. harveyi*.

tems, one might ask whether the two systems are interchangeable; i.e., can VAI or PAI activate the heterologous transcriptional activator to induce gene expression? Experiments addressing this question indicate that *lux* gene expression is stimulated by LasR and that *lasB* expression can be stimulated by LuxR but only in the presence of their cognate autoinducers (28). Neither LasR nor LuxR demonstrates appreciable activity in the presence of the heterologous autoinducer. These findings confirm the specificity of the interaction between autoinducer and activator protein and furthermore suggest that autoinduction-related elements may be conserved between the two systems. This is not to say that the interaction between the protein and its identified autoinducer is exclusive of any other autoinducerlike molecule binding. By way of example, it has been shown in both *V. fischeri* and *P. aeruginosa* that some analogs are capable of activation, albeit at lower levels than the identified cognate autoinducer. Furthermore, it is known that *P. aeruginosa* is capable of producing more than one autoinducerlike molecule (2, 47). Further complexity is suggested by the demonstration of the presence of another LuxR homolog in *P. aeruginosa*. RhlR is apparently involved in the synthesis of rhamnolipid biosurfactants (42). No autoinducer molecule has been identified for this protein. Thus, given the possibility of multiple activator proteins and multiple autoinducers, the findings suggest that the regulatory process used by *P. aeruginosa* may be more complex than that described for *V. fischeri*. Of course, the possibility also exists that other as yet unidentified *V. fischeri* genes are also regulated by LuxR-VAI or homologs of these components.

Nonetheless, with the identification of LasR and PAI and their role in regulating *lasB*, a process of autoinduction for *P. aeruginosa* elastase expression has been described (Fig. 3). As in the model for *P. aeruginosa* autoinduction, LasR interacts with PAI and becomes activated. The activated LasR then interacts with a region upstream of a gene (*lasB*), stimulating transcription of the gene.

As indicated above, LasR is involved in expression of several virulence genes. Through use of the *lasR*-negative mutant PAO-R1, we and our collaborators have been able to demonstrate that production of many *P. aeruginosa* exoproducts involves regulation by *lasR*. These products are listed in Table 1.

As mentioned earlier, a LuxR operator region upstream of the *luxR* and *luxI* genes has been defined. A region bearing extensive identity to the LuxR operator and containing dyad symmetry has been identified upstream of the transcriptional start of *lasB* (53). In addition, a sequence with partial identity to the LuxR operator and further upstream of *lasB* has been identified. Interestingly, point mutations in this partial site do not demonstrate a significant effect on expression of *lasB*. However, deletion of this region of partial identity as well as of regions upstream of this DNA tract appears to negatively affect *lasB* expression (53). These data suggest that sequences in and surrounding this region are important in the expression of *lasB*. In contrast to the results seen for the region of partial homology, point mutations in the region with high identity to the *lux* operator have indicated a requirement for this region in the LasR-regulated expression of *lasB*. These studies have also identified a number of critical residues within the region (53). The

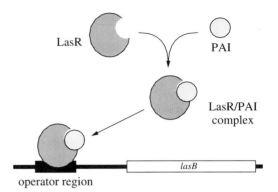

FIGURE 3 Model of the autoinduction mechanism for the *lasB* gene of *P. aeruginosa*. In the model, LasR interacts with PAI to form an LasR-PAI complex and to activate LasR. The complex binds specific LasR operator elements upstream of a gene (in this case *lasB*), resulting in expression of the target gene.

presence of a *lux* operator-homologous region upstream of *lasB* may also aid in explaining the interchangeability of the *lux* and *las* components described previously (28). The identification of such an operator region may allow the preliminary identification of those genes as possible candidates for regulation via a quorum-sensing mechanism. Note that no sequences with strong homology to the *lux* operator have yet been identified upstream of any other *P. aeruginosa* virulence genes, even those that are regulated by LasR-PAI. Those genes that do not have a *lux* operator region may in turn contain a *las* operator, which is as yet unidentified and which may be unique.

Aside from regulation via autoinduction, the *lux* genes are also subject to a requirement

TABLE 1 *P. aeruginosa* exoproducts regulated by LasR

Gene	Function
lasI	Autoinducer synthase
lasA	Elastase
lasB	Elastase
aprA	Alkaline protease
toxA	Exotoxin A
exoS	Exoenzyme S
?	Neuraminidase
?	Heat-stable hemolysin

for cyclic AMP (cAMP) and the cAMP receptor protein (CRP) (19). The binding of the cAMP-CRP complex to a region approximately 60 bp upstream of the transcriptional start of *luxR* activates transcription of the gene. The physiologic significance of this control is not clear, but the binding may play a role as a starvation signal.

The region upstream of the *lasR* transcriptional start also contains a stretch of DNA that perfectly matches a consensus *E. coli* cAMP-CRP-binding site (1). Studies have also suggested that expression of *lasR* may be modulated by cAMP-CRP, but the mechanism is not clear (1). Regardless, modulation of gene expression by cAMP-CRP may also prove to be conserved in quorum-sensing systems.

A HIERARCHY OF AUTOINDUCIBLE GENES

In order for *P. aeruginosa* to be a successful pathogen, it must be able to regulate the production of its virulence factors precisely with regard to when to express them and in what quantity. Recall that in the case of the *V. fischeri lux* genes, all genes required for the production of bioluminescence are present in an operon structure. The arrangement of the genes in such a manner allows for common control points for their regulation to ensure proper temporal and stoichiometric expression. The case of the *P. aeruginosa* virulence genes is more complex. The genes regulated by LasR are not organized into operon structures; rather, they are scattered around the chromosome of *P. aeruginosa*. As a result, there are no obvious common control points. This arrangement leads to the question of whether LasR regulates all the virulence genes equally. Or, to put it another way, do all the virulence genes depend on the presence of LasR to the same extent? Evidence to support the hypothesis of differential regulation comes from two main sources.

In the absence of LasR, *lasB*::*lacZ* expression is almost completely abolished (26, 45, 53). The same appears to be true for *lasI* according to Northern analysis of *lasI* mRNA in *P. aeru-*

ginosa PAO-R1 (45). Results from experiments monitoring the expression of both *lasI*::*lacZ* and *lasB*::*lacZ* gene fusions in the presence of LasR and increasing amounts of PAI indicate that the expression of *lasI* is much (approximately 10-fold) more sensitive to and thus more dependent on LasR-PAI than is *lasB* (55).

The second source of evidence for differential expression of virulence genes also comes from examining the expression of gene fusions of various virulence genes in the presence or absence of *lasR* (*P. aeruginosa* PAO or PAO-R1, respectively). These experiments show that the expression of *lasB* is much more dependent on the presence of *lasR* than is the expression of *lasA*, which is more dependent on *lasR* than is the expression of *toxA*. Taken together, the data from the above-mentioned experiments indicate that the expressions of various virulence genes differ in their dependencies on LasR.

As a result of these findings, we have proposed the existence of an autoinduction hierarchy in which some genes are more tightly regulated by LasR and PAI than are others (Fig. 4). The placement of the genes in this hierarchy is based on experimental determination of

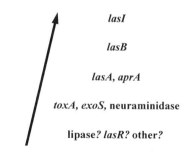

FIGURE 4 Hierarchy of *P. aeruginosa* gene expression by LasR-PAI. The genes known to be regulated by LasR are listed. The arrow indicates the hierarchy of the genes, from those least dependent on LasR (bottom) to those requiring LasR for their expression (top). For the sake of simplicity, not all known genes are listed. Question marks denote genes or gene products whose placements in the hierarchy are currently being investigated. Genes other than those presented here and in the text are possible.

the dependence of their expression on LasR. Some genes, such as the *lasR* gene itself, are currently being investigated. We do not mean that the genes listed in Table 1 and Fig. 4 are the only *lasR*-regulated genes in *P. aeruginosa*. There certainly may be others that are as yet unidentified.

The proposal of a hierarchy of gene expression may also explain the finding described above that not all genes contain a region with strong homology to the *lux* operator. Indeed, if LasR cannot interact with all the genes equally, one would expect this lack to be due to changes in the operator region, with some genes containing operators of higher affinity for LasR than others. In retrospect, the choice of *lasB* as the gene for the interchangeability studies mentioned previously (28) may have been fortuitous, in that, to date, it is the only gene that contains a strong *lux*-homologous operator sequence.

WHY QUORUM SENSING?

Given the various data presented here, the questions arise as to why an organism might evolve a process such as quorum sensing for gene regulation and what advantage such a system might afford a pathogen. There is no conclusive answer to either question, but a few comments might be made. First, as mentioned earlier, this type of system allows a bacterial population to monitor its cell density in order to allow cell density-dependent processes to be initiated. There are many instances in bacterial systems in which behavior is dependent on a multicellular effort. Classic examples include differentiation and fruiting-body development in myxobacteria, sporulation in *Bacillus* spp., and growth and differentiation in *Streptomyces* spp. (4, 20, 34). All these examples use cell–cell signaling to achieve cooperative behavior.

Second, quorum sensing may act to ensure the proper timing of production of various factors so as to guarantee the overwhelming of the host immune response. Perhaps if *P. aeruginosa* finds itself within its host at a low cell density and initiates the elaboration of virulence products, these products might elicit a host defense

response. Given the low cell density and the proportionately low concentrations of the virulence products, the infection might easily be cleared. By contrast, if production of the factors that elicit the host response is initiated at high cell densities, the infection might have a better probability of surviving these host challenges. The high degree of correlation between autoinduction and cell density has led to the adoption of a new terminology that replaces the term autoinduction with the term quorum sensing.

That quorum-sensing systems play an important role in the bacterial world is evidenced by the conservation of these systems among many organisms (2, 33, 50, 60). Furthermore, the conservation of autoinducer structure among various organisms strengthens the argument for a conserved and efficient gene regulatory system. Recent evidence (32) suggests that homoserine lactone, the ring component of autoinducer molecules, may actually play a role as a starvation signal for the organism. This finding suggests that the starvation signal molecule may also be used as a substrate for the pathway involved in autoinducer synthesis and may tie the nutritional status of the bacterial cell to the appearance of the quorum-sensing systems.

QUORUM SENSING IS AN ALMOST UBIQUITOUS PROCESS

As recently as 2 years ago, quorum sensing as it is now described was known only for *V. fischeri* and *Vibrio harveyi*, for which it was viewed as an interesting but esoteric method of gene regulation. Today it appears that many gram-negative organisms contain at least one of the components for such a regulatory system. Using a *lux*-based reporter system, Swift and coworkers have been able to screen large numbers of different organisms for the production of an autoinducerlike molecule (60). Given that not all autoinducerlike molecules may be able to stimulate a given LuxR homolog, the use of a *lux*-based reporter, although sufficient for screening for organisms that produce VAI and any other autoinducer molecule capable of interacting with LuxR, may be limited. Nonetheless, it provides a method for beginning to screen organisms for the possible presence of a quorum-sensing system. For some organisms in which components of the quorum-sensing mechanism have been identified, the system and the responses it regulates are well on their way to being understood. Aside from *Vibrio* spp. and *P. aeruginosa,* the most investigated systems include those of *Agrobacterium tumefaciens* (49, 69) and *Erwinia carotovora* (2, 33, 50).

In the plant pathogen *A. tumefaciens,* the Ti plasmid directs the plant to produce opines, which serve as nutrients for the bacterium. Also encoded by the plasmid are the *tra* genes, which are involved in conjugation, and the genes required for catabolism of the opines. Both of these sets of genes are induced in the presence of opines. In addition, these genes are also induced by a Lux-homologous TraR and TraI system, and a diffusible homoserine lactone derivative (*N*-3-oxooctanoyl homoserine lactone) is involved.

E. carotovora is also a plant pathogen. In this organism, a Lux-homologous system is used to regulate the synthesis of carbapenem antibiotics and cell wall-degrading enzymes. The autoinducer in this system has been identified as having the same structure as VAI. The LuxI and LuxR homologs in this organism have also been identified.

Quorum-sensing systems are turning out to be fairly common regulatory mechanisms for gram-negative organisms. A hierarchy of genes regulated by such a system might provide the bacterium, be it a pathogen or free-living in its preferred environment, with a powerful method of regulating the genes required for its survival.

ACKNOWLEDGMENT

Research in our laboratory was supported by grant AI33713 from the National Institutes of Health.

REFERENCES

1. **Albus, A., and B. H. Iglewski.** Unpublished data.
2. **Bainton, N. J., B. W. Bycroft, S. R. Chhabra, P. Stead, L. Gledhill, P. J. Hill, C. E. D. Rees, M. K. Winson, G. P. C. Salmond, G. S. A. B. Stewart, and P. Williams.** 1992. A general role for the *lux* autoinducer in bacterial cell signalling: control of antibiotic biosynthesis in *Erwinia*. *Gene* **116:**87–91.
3. **Baldwin, T. O., J. H. Devine, R. C. Heckel, J.-W. Lin, and G. S. Shadel.** 1989. The complete nucleotide sequence of the *lux* regulon of *Vibrio fischeri* and the *luxABN* region of *Photobacterium leiognathi* and the mechanism of control of bacterial bioluminescence. *J. Biolumin. Chemilumin.* **4:**326–341.
4. **Beppu, T.** 1992. Secondary metabolites as chemical signals for cellular differentiation. *Gene* **115:**159–165.
5. **Bever, R. A., and B. H. Iglewski.** 1988. Molecular characterization and nucleotide sequence of the *Pseudomonas aeruginosa* elastase structural gene. *J. Bacteriol.* **170:**4309–4313.
6. **Cao, J.-G., and E. A. Meighen.** 1989. Purification and structural identification of an autoinducer for the luminescence system of *Vibrio harveyi*. *J. Biol. Chem.* **264:**21670–21676.
7. **Cao, J.-G., and E. A. Meighen.** 1993. Biosynthesis and stereochemistry of the autoinducer controlling luminescence in *Vibrio harveyi*. *J. Bacteriol.* **175:**3856–3862.
8. **Choi, S.-H., and E. P. Greenberg.** 1992. Genetic dissection of DNA binding and luminescence gene activation by the *Vibrio fischeri* LuxR protein. *J. Bacteriol.* **174:**4064–4069.
9. **Coburn, J.** 1992. *Pseudomonas aeruginosa* exoenzyme S. *Curr. Top. Microbiol. Immunol.* **175:**133–143.
10. **Darzins, A., J. E. Peters, and D. R. Galloway.** 1990. Revised nucleotide sequence of the *lasA* gene from *Pseudomonas aeruginosa* PAO1. *Nucleic Acids Res.* **18:**6444.
11. **Deretic, V., C. D. Mohr, and D. W. Martin.** 1991. Mucoid *Pseudomonas aeruginosa* in cystic fibrosis: signal transduction and histone-like elements in the regulation of bacterial virulence. *Mol. Microbiol.* **5:**1577–1583.
12. **Deretic, V., M. J. Schurr, J. C. Boucher, and D. W. Martin.** 1994. Conversion of *Pseudomonas aeruginosa* to mucoidy in cystic fibrosis: environmental stress and regulation of bacterial virulence by alternative sigma factors. *J. Bacteriol.* **176:**2773–2780.
13. **Devine, J. H., G. S. Shadel, and T. O. Baldwin.** 1989. Identification of the operator of the *lux* regulon from the *Vibrio fischeri* strain ATCC7744. *Proc. Natl. Acad. Sci. USA* **86:**5688–5692.
14. **Doring, G., H. J. Obernesser, K. Botzenhart, B. Flehmig, N. Hoiby, and A. Hofmann.** 1983. Proteases of *Pseudomonas aeruginosa* in pa-

tients with cystic fibrosis. *J. Infect. Dis.* **147:**744–750.

15. **Drake, D., and T. C. Montie.** 1988. Flagella, motility and invasive virulence of *Pseudomonas aeruginosa. J. Gen. Microbiol.* **134:**43–52.

16. **Dunlap, P. V., and E. P. Greenberg.** 1985. Control of *Vibrio fischeri* luminescence gene expression in *Escherichia coli* by cyclic AMP and cyclic AMP receptor protein. *J. Bacteriol.* **164:**45–50.

17. **Dunlap, P. V., and E. P. Greenberg.** 1988. Analysis of the mechanism of *Vibrio fischeri* luminescence gene regulation by cyclic AMP and cyclic AMP receptor protein in *Escherichia coli. J. Bacteriol.* **170:**4040–4046.

18. **Dunlap, P. V., and E. P Greenberg.** 1991. Role of intercellular chemical communication in the *Vibrio fischeri*-Monocentrid fish symbiosis, p. 219–253. *In* M. Dworkin (ed.), *Microbial Cell-Cell Interactions.* American Society for Microbiology, Washington, D.C.

19. **Dunlap, P. V., and J. M. Ray.** 1989. Requirement for autoinducer in transcriptional negative regulation of the *Vibrio fischeri luxR* gene in *Escherichia coli. J. Bacteriol.* **171:**3549–3552.

20. **Dworkin, M.** 1991. Cell-cell interactions in myxobacteria, p. 179–216. *In* M. Dworkin (ed.), *Microbial Cell-Cell Interactions.* American Society for Microbiology, Washington, D.C.

21. **Eberhard, A., A. L. Burlingame, C. Eberhard, G. L. Kenyon, K. H. Nealson, and N. J. Oppenheimer.** 1981. Structural identification of autoinducer of *Photobacterium fischeri* luciferase. *Biochemistry* **20:**2444–2449.

22. **Fick, R. B., Jr., G. P. Nagel, S. U. Squier, R. E. Wood, J. B. Gee, and H. Y. Reynolds.** 1984. Proteins of the cystic fibrosis respiratory tract. Fragmented immunoglobulin G opsonic antibody causing defective opsonophagocytosis. *J. Clin. Invest.* **74:**236–248.

23. **Fukushima, J., S. Yamamoto, K. Morihara, Y. Atsumi, H. Takeuchi, S. Kawamoto, and K. Okuda.** 1989. Structural gene and complete amino acid sequence of *Pseudomonas aeruginosa* IFO 3455 elastase. *J. Bacteriol.* **171:**1698–1704.

24. **Fuqua, W. C., S. Winans, and E. P. Greenberg.** 1994. Quorum sensing in bacteria: the LuxR-LuxI family of cell density-responsive transcriptional regulators. *J. Bacteriol.* **176:**269–275.

25. **Galloway, D.** 1991. *Pseudomonas aeruginosa* elastase and elastolysis: recent developments. *Mol. Microbiol.* **5:**2315–2321.

26. **Gambello, M. J., and B. H. Iglewski.** 1991. Cloning and characterization of the *Pseudomonas aeruginosa lasR* gene, a transcriptional activator of elastase expression. *J. Bacteriol.* **173:**3000–3009.

27. **Gambello, M. J., S. Kaye, and B. H. Iglewski.** 1993. LasR of *Pseudomonas aeruginosa* is a transcriptional activator of the alkaline protease

gene (*apr*) and an enhancer of exotoxin A expression. *Infect. Immun.* **61:**1180–1184.

28. **Gray, K. M., L. Passador, B. H. Iglewski, and E. P. Greenberg.** 1994. Interchangeability and specificity of components from the quorum-sensing regulatory systems of *Vibrio fischeri* and *Pseudomonas aeruginosa. J. Bacteriol.* **176:**3076–3080.

29. **Hazlett, L. D., M. M. Moon, A. Singh, R. S. Berk, and X. L. Rudner.** 1991. Analysis of adhesion, piliation, protease production and ocular infectivity of several *P. aeruginosa* strains. *Curr. Eye Res.* **10:**351–362.

30. **Heck, L. W., K. Morihara, and D. R. Abrahamson.** 1986. Degradation of soluble laminin and depletion of tissue-associated basement membrane laminin by *Pseudomonas aeruginosa* elastase and alkaline protease. *Infect. Immun.* **54:**149–153.

31. **Hong, Y. Q., and B. Ghebrehiwet.** 1992. Effect of *Pseudomonas aeruginosa* elastase and alkaline protease on serum complement and isolated components C1q and C3. *Clin. Immunol. Immunopathol.* **62:**133–138.

32. **Huisman, G. W., and R. Kolter.** 1994. Sensing starvation: a homoserine lactone-dependent signalling pathway in *Escherichia coli. Science* **265:**537–539.

33. **Jones, S., B. Yu, N. J. Bainton, M. Birdsall, B. W. Bycroft, S. R. Chhabra, A. J. R. Cox, P. Golby, P. J. Reeves, S. Stephens, M. K. Winson, G. P. C. Salmond, G. S. A. B. Stewart, and P. Williams.** 1993. The *lux* autoinducer regulates the production of exoenzyme virulence determinants in *Erwinia carotovora* and *Pseudomonas aeruginosa. EMBO J.* **12:**2477–2482.

34. **Kaiser, D., and R. Losick.** 1993. How and why bacteria talk to each other. *Cell* **73:**873–885.

35. **Kaplan, H. B., and E. P. Greenberg.** 1985. Diffusion of autoinducer is involved in regulation of the *Vibrio fischeri* luminescence system. *J. Bacteriol.* **163:**1210–1214.

36. **Liu, P. V.** 1966. The roles of various fractions of *Pseudomonas aeruginosa* in its pathogenesis. 3. Identity of the lethal toxins produced in vitro and in vivo. *J. Infect. Dis.* **116:**481–489.

37. **Lory, S., and P. C. Tai.** 1985. Biochemical and genetic aspects of *Pseudomonas aeruginosa* virulence. *Curr. Top. Microbiol. Immunol.* **118:**53–69.

38. **Meighen, E. A.** 1991. Molecular biology of bacterial bioluminescence. *Microbiol. Rev.* **55:**123–142.

39. **Montie, T. C., D. Drake, H. Sellin, O. Slater, and S. Edmonds.** 1987. Motility, virulence and protection with a flagella vaccine against *Pseudomonas aeruginosa* infection. *Antibiot. Chemother.* **39:**233–248.

40. **Morihara, K., and Y. Homma.** 1985. *Pseudomonas* proteases, p. 49–79. *In* I. A. Holder (ed.),

Bacterial Enzymes and Virulence. CRC Press, Inc., Boca Raton, Fla.

41. **Nicas, T. I., and B. H. Iglewski.** 1985. The contribution of exoproducts to virulence of *Pseudomonas aeruginosa. Can. J. Microbiol.* **31**:387–392.

42. **Ochsner, U. A., A. K. Koch, A. Fiecther, and J. Reiser.** 1994. Isolation and characterization of a regulatory gene affecting rhamnolipid biosurfactant synthesis in *Pseudomonas aeruginosa. J. Bacteriol.* **176**:2044–2054.

43. **Okuda, K., K. Morihara, Y. Atsumi, H. Takeuchi, S. Kawamoto, H. Kawasaki, K. Suzuki, and J. Fukushima.** 1990. Complete nucleotide sequence of the structural gene for alkaline proteinase from *Pseudomonas aeruginosa* IFO 3455. *Infect. Immun.* **58**:4083–4088.

44. **Palleroni, N. J.** 1984. Family I. Pseudomonadaceae, p. 141–219. *In* N. R. Krieg and J. G. Holt (ed.), *Bergey's Manual of Systematic Bacteriology,* vol. 1. The Williams & Wilkins Co., Baltimore.

45. **Passador, L., J. M. Cook, M. J. Gambello, L. Rust, and B. H. Iglewski.** 1993. Expression of *Pseudomonas aeruginosa* virulence genes requires cell-to-cell communication. *Science* **260**:1127–1130.

46. **Passador, L., and B. H. Iglewski.** Unpublished data.

47. **Pearson, J. P., K. M. Gray, L. Passador, K. D. Tucker, A. Eberhard, B. H. Iglewski, and E. P. Greenberg.** 1994. Structure of the autoinducer required for expression of *Pseudomonas aeruginosa* virulence genes. *Proc. Natl. Acad. Sci. USA* **91**:197–201.

48. **Peters, J. E., and D. R. Galloway.** 1990. Purification and characterization of an active fragment of the LasA protein from *Pseudomonas aeruginosa:* enhancement of elastase activity. *J. Bacteriol.* **172**:2236–2240.

49. **Piper, K. R., S. B. von Bodman, and S. K. Farrand.** 1993. Conjugation factor of *Agrobacterium tumefaciens* regulates Ti plasmid transfer by autoinduction. *Nature* (London) **362**:448–450.

50. **Pirhonen, M., S. Flego, R. Heikinheimo, and E. T. Palva.** 1993. A small diffusible signal molecule is responsible for the global control of virulence and exoenzyme production in the plant pathogen *Erwinia carotovora. EMBO J.* **12**:2467–2476.

51. **Pollack, M.** 1990. *Pseudomonas aeruginosa,* p. 1673–1691. *In* G. L. Mandell, R. G. Douglas, and J. E. Bennett (ed.), *Principles and Practice of Infectious Disease.* Churchill Livingstone, Inc., New York.

52. **Rudner, X. L., R. S. Berk, and L. D. Hazlett.** 1993. Immunization with homologous *Pseudomonas aeruginosa* pili protects against ocular disease. *Regional Immunol.* **5**:245–252.

53. **Rust, L.** 1994. Ph.D. dissertation, University of Rochester, Rochester, N.Y.

54. **Schad, P. A., and B. H. Iglewski.** 1988. Nucleotide sequence and expression in *Escherichia coli* of the *Pseudomonas aeruginosa lasA* gene. *J. Bacteriol.* **170**:2784–2789.

55. **Seed, P. C., L. Passador, and B. H. Iglewski.** 1995. Activation of the *Pseudomonas aeruginosa lasI* gene by LasR and the *Pseudomonas* autoinducer PAI: an autoinduction regulatory hierarchy. *J. Bacteriol.* **177**:654–659.

56. **Shadel, G. S., J. H. Devine, and T. O. Baldwin.** 1990. Control of the *lux* regulon in *Vibrio fischeri. J. Biolumin. Chemilumin.* **5**:99–106.

57. **Shadel, G. S., R. Young, and T. O. Baldwin.** 1990. Use of regulated cell lysis in a lethal genetic selection in *Escherichia coli:* identification of the autoinducer-binding region of the LuxR protein from *Vibrio fischeri* ATCC 7744. *J. Bacteriol.* **172**:3980–3987.

58. **Slock, J., D. VanRiet, D. Kolibachuk, and E. P. Greenberg.** 1990. Critical regions of the *Vibrio fischeri* LuxR protein defined by mutational analysis. *J. Bacteriol.* **172**:3974–3979.

59. **Strom, M. S., and S. Lory.** 1993. Structure-function and biogenesis of the type IV pili. *Annu. Rev. Microbiol.* **47**:565–596.

60. **Swift, S., M. K. Winson, P. F. Chan, N. J. Bainton, M. Birdsall, P. J. Reeves, C. E. D. Rees, S. R. Chhabra, P. J. Hill, J. P. Throup, B. W. Bycroft, G. P. C. Salmond, P. Williams, and G. S. A. B. Stewart.** 1993. A novel strategy for the isolation of *luxI* homologues: evidence for the widespread distribution of a LuxR:LuxI superfamily in enteric bacteria. *Mol. Microbiol.* **10**:511–520.

61. **Toder, D. S., M. J. Gambello, and B. H. Iglewski.** 1991. *Pseudomonas aeruginosa* LasA: a second elastase under the transcriptional control of *lasR. Mol. Microbiol.* **5**:2003–2010.

62. **Twining, S. S., S. E. Kirschner, L. A. Mahnke, and D. W. Frank.** 1993. Effect of *Pseudomonas aeruginosa* elastase, alkaline protease and exotoxin A on corneal proteinases and proteins. *Invest. Ophthalmol. Vis. Sci.* **34**:2699–2712.

63. **Whooley, M. A., and A. McLoughlin.** 1983. The protonmotive force in *Pseudomonas aeruginosa* and its relationship to exoprotease production. *J. Gen. Microbiol.* **129**:989–996.

64. **Whooley, M. A., J. A. O'Callahan, and A. J. McLoughlin.** 1983. Effect of substrate on the regulation of exoprotease production by *Pseudomonas aeruginosa* ATCC 10145. *J. Gen. Microbiol.* **129**:981–988.

65. **Wick, M. J., D. W. Frank, D. G. Storey, and B. H. Iglewski.** 1990. Structure, function, and regulation of *Pseudomonas aeruginosa* exotoxin A. *Annu. Rev. Microbiol.* **44**:335–363.

66. **Wiener-Kronish, J. P., T. Sakuma, I. Kudoh, J. F. Pittet, D. Frank, L. Dobbs, M. L. Vasil, and M. A. Matthay.** 1993. Alveolar epithelial injury and plueral empyema in acute *P. aeruginosa* pneumonia in anesthetized rabbits. *J. Appl. Physiol.* **75**:1661–1669.

67. **Woods, D. E., S. J. Cryz, R. L. Friedman, and B. H. Iglewski.** 1982. Contribution of toxin A and elastase to virulence of *Pseudomonas aerugi-nosa* in chronic lung infections of rats. *Infect. Immun.* **36**:1223–1228.

68. **Woods, D. E., and B. H. Iglewski.** 1983. Toxins of *Pseudomonas aeruginosa:* new perspectives. *Rev. Infect. Dis.* **4**(Suppl.):S715–S722.

69. **Zhang, L., P. J. Murphy, A. Kerr, and M. Tate.** 1993. *Agrobacterium* conjugation and gene regulation by *N*-acyl-L-homoserine lactones. *Nature* (London) **362**:446–448.

ACQUISITION OF IRON AND OTHER NUTRIENTS IN VIVO[†]

Eugene D. Weinberg

6

During the past three decades, hundreds of laboratory studies and clinical observations concerning bacterial acquisition of host iron as a component of virulence have been published. Pathogenic cells transmitted either directly from the former to the new host or from non-host environments are generally unable to bring with them a sufficient quantity of stored iron for distribution to their offspring. Nearly all groups of potentially pathogenic bacteria thus far examined require the metal for activation of ribonucleotide reductase, aconitase, and various enzymes involved in electron transfer and oxygen metabolism.

Besides being essential for multiplication of the invading cells, iron also regulates the synthesis of many bacterial toxins. In one group to be discussed, however, manganese rather than iron is required for growth. In another group, manganese rather than iron regulates toxigenesis.

Not surprisingly, vertebrate hosts vigorously attempt to withhold iron from bacteria as well as from fungi, protozoa, helminths, and neo-plastic cells. In the contest for iron between vertebrate hosts and potential invaders, a considerable variety of constitutive and inducible mechanisms are available to the defenders (Table 1). Oddly, in invertebrate animals and in plants, the possible existence of comprehensive iron-withholding defense systems has not yet been ascertained.

In the past quarter century, numerous reviews of various aspects of iron procurement by plants, animals, and microorganisms have been published. Examples of reviews in which the association of microbial iron acquisition with virulence has been emphasized include those of Kochan (63), Weinberg (134, 136–138), Bullen et al. (21), Finkelstein et al. (43), Byers (22), Payne (99), Williams and Griffiths (147), Litwin and Calderwood (75), Woold-ridge and Williams (152), and Byers and Arceneaux (23).

This review focuses on current awareness of iron acquisition as a virulence mechanism of bacterial pathogens. Genetic regulation of bacterial iron acquisition is not emphasized, inasmuch as a recent comprehensive review is available (75). Moreover, this volume includes a discussion of regulation of gene expression for bacterial survival in the host (see chapter 5).

Potential invaders attempt to obtain host iron either (i) from such degraded products of

Eugene D. Weinberg   Department of Biology and Program in Medical Science, Indiana University, Bloomington, Indiana 47405.
[†]In recognition of Professors Ivan Kochan and Ashley A. Miles, whose pioneering visions of host-pathogen iron interactions continue to provide us with guidance and direction.

Virulence Mechanisms of Bacterial Pathogens, 2nd ed., Edited by J. A. Roth et al.
© 1995 American Society for Microbiology, Washington, D.C.

TABLE 1 The iron-withholding defense system[a]

Constitutive components
   Transferrin in plasma, cerebrospinal fluid, perspiration
   Lactoferrin in tears, nasal exudate, saliva, bronchial
      mucus, gastrointestinal fluid, hepatic bile, cervical
      mucus, seminal fluid, milk, neutrophil granules
   Ferritin within host cells

Processes induced at time of invasion by pathogens
   Suppression of assimilation of ≤80% of dietary iron[b]
   Suppression of iron efflux from macrophages that
      have digested effete erythrocytes to result in ≤70%
      reduction in plasma iron[b]
   Increased synthesis of ferritin to store sequestered
      iron[b]
   Release of neutrophils from bone marrow into
      peripheral circulation and then into site of invasion[b]
   Release of apolactoferrin from neutrophil granules
      followed by binding of iron in septic sites
   Macrophage scavenging of ferrated lactoferrin in areas
      of sepsis and of tumor cell clusters
   Hepatic release of haptoglobin and hemopexin (to
      bind extravasated hemoglobin and hemin,
      respectively)
   Synthesis of nitric oxide (from L-arginine) by
      macrophages to effect efflux of nonheme iron from
      invaders[c]
   Suppression of growth of microbial cells within
      macrophage phagosomes via downshift of
      expression of transferrin receptors of the host cells[c]
   Induction in B lymphocytes of synthesis of
      immunoglobulins to iron-repressible cell surface
      proteins that bind heme, ferrated transferrin, or
      ferrated siderophores

   [a]See reference 141.
   [b]Activated by interleukin-1 or -6 or by tumor necrosis factor
alpha.
   [c]Activated by gamma interferon.

hemoglobin as hemin or heme, (ii) directly from ferrated transferrin or ferrated lactoferrin, (iii) indirectly from ferrated transferrin or ferrated lactoferrin by means of siderophores, or (iv) from intracellular iron pools. Bacterial species often have strains that employ differing strategies, and the use of a specific method depends on the particular environment. Variation in strategy is especially important for strains that live inside a host at some times and outside the host at other times. Moreover, the differing biochemical environments in various tissues of a host might also necessitate shifts in iron acquisition strategy.

IRON CAN BE OBTAINED FROM HOST HEME

For over a century, medical microbiologists have recognized that the abilities of bacteria to lyse erythrocytes and digest hemoglobin contribute to virulence. It has long been apparent that with the exception of bartonellosis, the amount of erythrocyte destruction caused by hemolytic bacteria is generally insufficient to produce serious anemia. However, the liberation of even a small quantity of heme can provide the invader with sufficient iron to multiply extracellularly. The ability to bind hemin also might assist pathogens such as species of *Shigella* to coat themselves with the porphyrin and thus facilitate their endocytosis into potential host cells (100). Heme might also serve as one of the iron sources available within host cells for intracellular bacterial pathogens.

Examples of bacterial species that contain strains that can utilize heme iron are listed in Table 2. High virulence is possessed by such strains provided that, to obtain extracellular heme, they can lyse erythrocytes, digest hemoglobin, and bind and assimilate the porphyrin. Nevertheless, even heme-utilizing strains that are unable to lyse erythrocytes may be virulent in hosts who have such underlying hemolytic conditions as malaria, sicklemia, or autoantibodies to erythrocyte membranes.

Many of the species listed in Table 2 also have alternative iron aquisition strategies, as indicated in Tables 4, 6, and 8. For example, *Escherichia coli* is listed in Tables 2 and 6. Highly virulent strains of this species generally produce the invasion-promoting siderophore aerobactin. In one study, however, 21 human blood isolates were aerobactin negative; 18 of these strains were hemolytic (97). Similarly, in a set of 84 isolates of *Aeromonas* species (76), nearly all could use heme and hemoglobin, even when the latter was bound by haptoglobin. Additionally, 50 of the isolates produced a siderophore, ammonobactin, that enabled them to grow in the presence of transferrin that had a normal level of iron saturation. Thirty other isolates synthesized a siderophore, enterobactin, that enabled growth to occur in serum that

TABLE 2 Examples of bacterial species that contain strains that can obtain iron from heme compounds[a]

Acinetobacter calcoaceticus (144)
Actinobacillus pleuropneumoniae (77)
Aeromonas species (23, 51)
Bacteroides fragilis (98)
Campylobacter jejuni (102)
Escherichia coli (27)
Haemophilus ducreyi (68)
Haemophilus influenzae (148)
Neisseria gonorrhoeae (38, 82)
Neisseria meningitidis (9, 38, 71, 82)
Pasteurella haemolytica (48, 144)
Plesiomonas shigelloides (35)
Proteus species (65, 144)
Serratia marcescens (19, 104)
Shigella flexneri (100)
Staphylococcus aureus (45)
Streptococcus pneumoniae (125)
Streptococcus pyogenes (45)
Vibrio anguillarum (78)
Vibrio cholerae (123)
Vibrio parahaemolyticus (34)
Vibrio vulnificus (57)
Yersinia enterocolitica (124)
Yersinia pestis (101, 118)

[a]Numbers in parentheses are the relevant references.

contained transferrin that was fully iron saturated.

In a few species, heme compounds apparently are the sole source of host iron. *Streptococcus pneumoniae,* for instance, is unable to produce siderophores or utilize iron from transferrin or lactoferrin (125). Virulence is associated with formation of pneumolysin and utilization of hemin. A mutant strain defective in hemin utilization is markedly reduced in virulence.

In some cases, the ability to use hemoglobin iron can be demonstrated in laboratory animal models but is of much less importance in clinical situations. For example, *Neisseria meningitidis* is virulent in mice that have been iron loaded with either human or bovine hemoglobin or with human ferrated transferrin or lactoferrin (112). However, the pathogen is not virulent in mice that are iron loaded with bovine ferrated transferrin or lactoferrin. Furthermore, the pathogen causes meningitis in hu-

mans but not cattle. As noted below, *N. meningitidis* binds human but not bovine ferrated transferrin and lactoferrin. Possibly, the pathogen is unable to lyse bovine erythrocytes and thus cannot gain access to bovine hemoglobin.

Clinical and experimental observations that provide evidence that the ability to obtain heme iron from hosts can contribute to virulence are listed in Table 3. Note that attempts by the host to bind heme with hemopexin or albumin or to bind hemoglobin with haptoglobin often fail to prevent bacterial invaders from using the porphyrin iron. It may also be seen in Table 3 that strains that have lost hemolytic ability because of either the presence of antibody to hemolysin or the absence of the hemolysin gene can be virulent in the presence of exogenous iron.

IRON CAN BE OBTAINED DIRECTLY FROM HOST Fe-TRANSFERRIN OR Fe-LACTOFERRIN

Since 1980, strains of approximately 15 pathogenic bacterial species have been reported to bind the Fe-transferrin and/or the Fe-lactoferrin of one or two or, in a few cases, multiple mammalian host species (Table 4). It has not yet been demonstrated in every system that iron can be extracted from the host proteins. For example, although strains of *Aeromonas* can directly bind human and bovine lactoferrin (10, 28, 62), their iron is generally derived from heme or via the use of siderophores (23). However, if siderophore synthesis is suppressed, the ability of *Aeromonas* spp. to bind lactoferrin might be increased (10).

Strains of *Neisseria* that are pathogenic for humans can obtain iron from human transferrin, whereas commensal strains of this genus are unable to do so (81, 82, 119). Moreover, highly virulent (disseminating) strains of *Neisseria gonorrhoeae* can obtain the metal from human transferrin or lactoferrin whose iron saturation is as low as 5% (79).

In the absence of siderophores, the uptake of the metal from host iron-binding proteins requires direct contact of the latter with the

TABLE 3 Evidence that ability to obtain heme iron from hosts can be a factor of virulence

- Bacterial virulence is enhanced strongly in hosts who have underlying conditions that lead to erythrocyte rupture. Examples include presence of erythrocyte antibodies, chemicals such as phenylhydrazine, and hemolytic conditions such as sicklemia and malaria (135).
- In culture, many bacterial species (Table 2) contain strains that can use heme, hemin, or hemoglobin as the sole source of iron. Some but not all strains can also use heme-hemopexin, heme-albumin, or hemoglobin-haptoglobin as the sole source of iron (99, 103).
- Although each of three *Haemophilus* spp. can acquire iron from hemoglobin, only the major pathogen of the set, *H. influenza* type b, can obtain the metal from hemoglobin-haptoglobin (148).
- Mutant strains defective in acquisition (35, 101, 128) or utilization (125) of heme are markedly reduced in virulence.
- Synthesis of hemin receptor and heme-hemopexin receptor of *H. influenzae* type b (69, 150) and of hemin and hemoglobin receptors of *N. gonorrhoeae* (70, 72) is iron repressible.
- Hemolysin-positive strains are much more virulent than hemolysin-negative strains (145). Systemic isolates of *E. coli* tend to be hemolysin positive, whereas fecal, noninvasive strains tend to be hemolysin negative (36, 83).
- Virulence of hemolysin-positive strains is suppressed by antibody to hemolysin (74). The suppressive action of antibody to hemolysin is overcome by iron.
- Synthesis of hemolysin is often but not always repressible by iron (35, 77, 104, 107).
- Virulence of hemolysin-negative strains can be enhanced by injection of iron, hemoglobin, hemolysin, or a hemolysin-positive strain as well as by acquisition of a plasmid that contains a hemolysin-positive gene (74, 130).
- *Bacteroides fragilis* and *Streptococcus pyogenes* infections induce, respectively, immunoglobulins to bacterial heme receptors (98) and to streptolysin O.

bacterial cell surface and is energy dependent (79, 112). In contrast to host cells, the bacteria do not assimilate the iron-binding proteins (15, 79). The mechanism of iron extraction is not yet clear. Suggested mechanisms include a cell surface ferric reductase (for either transferrin or lactoferrin) or a localized fall in pH (for transferrin) with the subsequent release of the metal to an acceptor molecule (147). In *Neisseria* spp.,

the acceptor molecule has been proposed to be a 36-kDa ferric ion-binding protein that would function in periplasmic transport of the metal (29).

Generally, pathogens that obtain iron directly from host proteins have a much narrower host range than do those that utilize heme or produce siderophores. Strains of *Haemophilus somnus,* for instance, form receptors for bovine but not human transferrin; these bacteria are virulent for cattle but not humans (95). *Actinobacillus pleuropneumoniae* synthesizes only a swine-specific transferrin receptor and causes pneumonia only in pigs (49, 52).

Fortunately, the ferrated transferrin receptor proteins of bacterial invaders are quite distinct from those of the hosts. Accordingly, the bacterial proteins are antigenic. Humans recovering from neisserial meningitis (15) and pigs convalescing from actinobacillary pneumonia (49, 94) produce immunoglobulins to the respective bacterial transferrin receptors. Evidence that the ability to obtain iron directly from host transferrin or lactoferrin is a factor of virulence is contained in Table 5.

IRON CAN BE OBTAINED FROM HOST Fe-TRANSFERRIN OR Fe-LACTOFERRIN VIA USE OF SIDEROPHORES

As can be noted in Table 6, a wide variety of gram-negative and gram-positive bacteria can synthesize siderophores that are capable of extracting iron from host transferrin and, in some cases, lactoferrin. Siderophores are low–molecular-mass compounds (usually 0.5 to 1.0 kDa) with a high affinity for ferric iron. Their synthesis is generally derepressed in response to low iron. After the compounds are secreted, they bind the metal and facilitate its transport into the cell. Assimilation of the ferrated siderophores requires the assistance of specific receptor proteins at the cell surface. The ferrated siderophores may be transported into the cytoplasm, or the iron may be transferred to another iron-accepting molecule (22). Release of the metal from the siderophores probably involves reduction to the ferrous state.

TABLE 4 Examples of bacterial pathogens that possibly obtain iron by binding host ferrated transferrin or ferrated lactoferrin

Organism (reference)	Source[a]	
	Positive	Negative
Actinobacillus pleuropneumoniae (52, 94)	pTf	bTf, hTf
Aeromonas species (10, 28, 62)	bLf, hLf	
Bacteroides fragilis (129)	hTf	
Bordetella pertussis (80, 106)	aTf, hTf, hLf	
Helicobacter pylori (60)	hLf	aTf, hTf, bLf
Haemophilus influenzae type b (58, 87, 103, 111)	bTf, hTf, 1Tf, hLf	aTf, cTf, hLf, eTf, mTf
Haemophilus somnus (95)	bTf	aTf, hTf, pTf
Listeria monocytogenes (55)	bTf, eTf, hTf, mTf, oTf	
Mycoplasma pneumoniae (127)	hLf	hTf
Neisseria gonorrhoeae (72, 79, 81, 82, 146)	hTf, hLf	aTf, bTf, eTf, 1Tf
Neisseria meningitidis (59, 81, 82, 113, 119)	hTf, hLf	bTf, bLf
Pasteurella haemolytica (96)	bTf	aTf, eTf, hTf, pTf
Staphylococcus aureus (90)	bLf, hLf	
Staphylococcus epidermidis (89)	bLf	
Treponema pallidum (5)	hLf	

[a]Abbreviations: a, avian; b, bovine; c, canine; e, equine; h, human; l, lepine; p, porcine; o, ovine; Tf, transferrin; Lf, lactoferrin.

TABLE 5 Evidence that ability to obtain iron directly from host transferrin or lactoferrin can be a factor of virulence[a]

- In culture, positive strains can obtain iron by direct contact with at least one host source of either Fe-Tf or Fe-Lf. Strains cannot obtain iron if they are separated from the ferrated proteins by a membrane that excludes macromolecules (81, 82, 119).
- Positive strains contain iron-repressible or hemin-repressible surface protein receptors that bind Fe-Tf or Fe-Lf from specific hosts (86, 113, 119).
- Strains that lack receptors are avirulent (47, 81, 82, 119).
- Positive strains that are high in virulence can obtain iron from Tf that has low iron saturation (81, 82, 119).
- Positive strains are virulent for hosts that form Tf or Lf for which the pathogens have receptors. Normally, the strains are not virulent for hosts whose Tf or Lf cannot be bound (52, 95, 96, 112).
- The narrow host ranges of positive strains can be extended by introduction of Fe-Tf or Fe-Lf from susceptible hosts into normally resistant hosts (59, 112).
- Patients who are recovering from infection with positive strains produce immunoglobulins to the Tf or Lf receptor proteins of the invaders (15).

[a]Abbreviations: Tf, transferrin; Lf, lactoferrin.

Synthesis of the cell surface receptors is iron repressible. Some strains can express receptors for siderophores produced by other bacterial species or even by fungal species. Thus, in a mixed infection, such strains might have an advantage over bacteria that can utilize only their own ferrated siderophores.

The majority of bacterial siderophores are either hydroxamate or catechol compounds. Among enteric pathogens, synthesis of hydroxamates is more frequently associated with virulence than is formation of catechols (99). A hydroxamate, such as aerobactin, is more effective in removing iron from transferrin in serum than is a catechol, such as enterobactin, which tends to bind to albumin. Moreover, unlike aerobactin, enterobactin is degraded within the bacterial cell following transport and thus cannot be recycled. In *Pseudomonas aeruginosa* infections, elastase-induced cleavage of transferrin (20, 37) and lactoferrin (20) might facilitate siderophore capture of host iron. Evidence that the ability to use siderophores to obtain iron from host transferrin or lactoferrin can be a factor of virulence is contained in Table 7.

Some bacterial pathogens apparently employ siderophores for iron uptake in extracellular environments but not in intracellular niches. Mycobacterial species utilize the hydroxamate mycobactin to obtain iron from transferrin in in vitro culture systems at neutral pHs. Within host macrophages, however, the pH of phagocytic vacuoles is 4.5 to 6.0. Thus, mycobactin is not needed in the host intracellular niche, because transferrin cannot retain iron at acid pHs (66). *Mycobacterium paratuberculosis,* unable to form mycobactin, grows well in bovine ileal cells. Similarly, *Mycobacterium avium* and *Mycobacterium tuberculosis,* when growing in avian spleen and liver cells and murine spleen cells, respectively, require neither endogenous nor exogenous mycobactin (66).

Likewise, growth of aerobactin-minus strains of *Shigella flexneri* is impaired in extracellular low-iron cultures but not in HeLa cells (91, 100). Furthermore, enterobactin-minus strains of *Salmonella typhimurium* are unable to multiply in murine serum but retain their virulence in an intracellular environment (14).

IRON CAN BE OBTAINED FROM HOST INTRACELLULAR POOLS

Examples of bacterial pathogens that grow intracellularly are listed in Table 8. Evidence that their ability to obtain intracellular iron is a factor of virulence is contained in Table 9. Siderophores are produced neither by *Legionella pneumophila, Listeria monocytogenes, Rickettsia* species, nor *Yersinia pestis.* As cited above, strains of *Mycobacterium, Salmonella,* and *Shigella* spp. do not require siderophores for growth in intracellular habitats.

Strains of *Y. pestis* that absorb hemin in in vitro culture are virulent when injected into mice by an intraperitoneal or an intravenous route (118, 128). Their iron acquisition systems enable them to survive, at least for a time, in the low–iron extracellular regions of the host. Mutant nonpigmented strains unable to acquire extracellular iron are virulent if injected intravenously, because they are quickly endo-

TABLE 6 Examples of bacterial pathogens that can employ siderophores to obtain iron from transferrin or lactoferrin[a]

Acinetobacter baumannii (1)
Aeromonas hydrophila (13)
Aeromonas salmonicida (28)
Bordetella pertussis (3, 106)
Campylobacter jejuni (42)
Corynebacterium diphtheriae (108)
Escherichia coli (26, 85, 93)
Klebsiella pneumoniae (92)
Mycobacterium species (66)
Proteus mirabilis (40)
Pseudomonas aeruginosa (6, 37)
Pseudomonas pseudomallei (154)
Salmonella typhimurium (14)
Shigella flexneri (100)
Staphylococcus aureus (64)
Vibrio anguillarum (32)
Vibrio cholerae (117)
Vibrio parahaemolyticus (34)
Vibrio vulnificus (120)
Yersinia enterocolitica (56)

[a]Numbers in parentheses are the relevant references.

TABLE 7 Evidence that ability to obtain iron from host transferrin or lactoferrin via siderophores can be a factor of virulence

- Siderophores such as aerobactin can extract iron from ferrated transferrin (91) or ferrated lactoferrin (93).
- Siderophores are synthesized by extracellular bacteria in in vivo systems (53, 149).
- Siderophore receptor proteins are synthesized by extracellular bacteria in in vivo systems (54, 114, 115).
- Loss of ability to form siderophores is associated with loss of virulence in *Erwinia chrysanthemi* (39), *Pseudomonas aeruginosa* (151), *Vibrio anguillarum* (32), and *Yersinia enterocolitica* (56).
- Virulence associated with aerobactin-plus plasmid is greater than that associated with aerobactin-minus plasmid but is equivalent to the latter plus iron (26, 85, 93).
- Siderophore-minus strain of *Vibrio anguillarum* is avirulent but can grow in host if combined with siderophore-plus strain (149).
- Infected hosts produce immunoglobulins to siderophore receptor proteins (54, 114, 115, 126). These immunoglobulins can be used to confer passive immunity in potential hosts (17, 122).

TABLE 8 Examples of bacterial pathogens that can obtain iron from host intracellular pools[a]

Legionella pneumophila (24)
Listeria monocytogenes (55)
Mycobacterium avium (73)
Mycobacterium leprae (84)
Mycobacterium paratuberculosis (73)
Mycobacterium tuberculosis (73)
Rickettsia species (131)
Salmonella typhimurium (14)
Shigella flexneri (100)
Yersinia pestis (101, 118)

[a]Numbers in parentheses are the relevant references.

cytosed. However, when injected into peripheral tissues, the mutant strains are avirulent unless exogenous iron is provided.

Among possible sources of host intracellular iron are heme, iron released from transferrin at pH 4.5 to 6, and perhaps ferritin. The quantity of iron in epithelial-cell vacuoles has been estimated to be 1.0 μm (46). If the metal were unbound, this amount would be sufficient for growth of most bacterial pathogens. Little information on the quantity of iron in other host cell compartments or on the association of the

TABLE 9 Evidence that ability to obtain host iron intracellularly can be a factor of virulence

- Host cells provided with excessive iron support the growth of *Legionella* (24) and *Mycobacterium* (73) spp.
- Pathogens that use siderophores for extracellular growth do not need the compounds for intracellular growth (14, 66, 67, 91).
- Cells of *Yersinia pestis* inoculated at sites distant from host cells require the ability to obtain iron in extracellular tissues for survival. Cells inoculated close to host cells do not need this ability (101, 118).
- Ferritin can provide iron to *Yersinia pestis* (101, 118).
- Host cells attempt to defend themselves by lowering their intracellular iron burdens.
  1. Gamma interferon can induce reduction in expression of transferrin receptors (24).
  2. Gamma interferon can induce synthesis of nitric oxide to effect loss of nonheme iron (18, 41, 44, 140).
- Chloroquin can raise pH value in phagolysosomes, thus rendering transferrin iron less available to *Legionella pneumophila* (25) and *Mycobacterium tuberculosis* (33).

iron with specific high- or low-molecular-mass compounds is available (31).

The intracellular niches of pathogens are generally quite specific. For example, trypanasomes multiply in the cytoplasms of macrophages, whereas leishmaniae reside in phagolysosomes, and toxoplasmas, mycobacteria, chlamydiae, and legionellae live in phagosomes that do not fuse with lysosomes (24). Rickettsiae reside in the cytosols of endothelial cells (41). Different levels of availability of iron in the respective niches might well be a factor in the location of the pathogens.

The ability to reduce ferric iron would be necessary in obtaining the metal from ferritin and possibly from other intracellular sources. Virulent strains of *Legionella pneumophila* have been reported to possess an NADH-dependent iron reductase, whereas avirulent strains have an NADPH-dependent enzyme (61). In the intracellular macrophage environment in which these bacteria attempt to multiply, NADH remains at a constant level, whereas NADPH decreases markedly during phagocytosis. *Listeria monocytogenes* has been observed to employ a low-molecular-mass iron reductant that requires NADH (2).

LACTOBACILLUS REQUIRES NO IRON: WHY IS IT GENERALLY NONPATHOGENIC?

It has been apparent for many decades that lactobacilli predominate over other bacteria in environments in which availability of iron is very low. Examples include the guts of breast-fed infants, in which maternal lactoferrin provides an unbound-iron-binding capacity of 95% (109); the vaginas of adult women, in which unsaturated lactoferrin similarly functions in iron withholding; and nitrite-cured meats, in which the salt has combined with iron-sulfur centers to render the metal unavailable to nonlactobacillary contaminants (105, 116).

Prior to 1983, it was not clear whether lactobacilli possess an unusually efficient iron uptake system or, alternatively, whether they eschew iron. In 1983, Archibald reported that *Lactobacillus plantarum* has no mechanism for acquiring iron (7). The ribonucleotide reductase of this bacterium uses cobalt in place of iron. The various enzymes in lactobacilli (for which, in other bacteria, iron is the essential cofactor) employ cobalt or manganese. The V_{max} for manganese uptake in *Lactobacillus plantarum* is 5,000-fold greater than that in species of other bacterial genera (8).

Lactobacilli promptly lose their advantage over other bacteria when adequate iron becomes available. For instance, bovine milk has only 10% of the quantity of lactoferrin that human milk contains (thus permitting cellulose-digesting bacteria to grow in the calf rumen), and milk formula contains no lactoferrin. Moreover, milk formula is often supplemented with iron (143). Infants fed these breast milk substitutes quickly develop flora of toxigenic bacteria such as *Bacteroides, Clostridium, Escherichia, Pseudomonas,* and *Salmonella* spp. Thus, non-breast-fed infants, and especially those given milk formula with iron, are at greatly elevated risk of developing salmonellosis and botulism and of sudden death (143).

Despite their disdain for iron and therefore for host iron-withholding defense systems, lactobacilli rarely invade the systemic tissues of healthy vertebrate hosts (4). Fewer than 70 infections during the entire past half century have been reported in the literature. In each case, an underlying disease or immunosuppression was present. The comparatively low content of readily available manganese in various host tissues and the high demand of lactobacilli for this metal could explain at least in part the rarity of infections due to this organism. Apparently, lactobacilli are unable to synthesize manganophores that could extract the metal from host manganese-complexing compounds.

TRACE METAL CONTROL OF BACTERIAL TOXIGENESIS

The exotoxin of *Corynebacterium diphtheriae* was discovered in 1895, the control of its synthesis by iron was discovered in 1935, and the transcriptional site of action of the metal was discovered in 1975 (139). During the first third of the 20th century, the quantities of the pro-

tein that were synthesized in laboratory cultures remained unpredictable. By 1935, it was recognized that the variability was associated with the amount of available iron in the culture environment. One of the discoverers of this fact commented: "So narrow is the zone (of iron concentration) in which (diphtheria) toxin is obtained and so sharp the peak of maximal production that this simple uncontrolled factor must have played a greater role in any previous experiments than the specific conditions supposedly under investigation" (88).

In the late 1930s, the possible use of iron compounds to suppress toxigenesis in patients was considered. Microbial growth would, of course, be enhanced by this procedure. Fortunately, it was not put into practice because of the advent of sulfonamides and, in the 1940s, penicillin. During the next several decades, the synthesis of numerous other bacterial toxins as well as of diphtheria toxin was observed to be maximal at the iron concentration that is sufficient but not excessive for vegetative growth (133, 139). Examples include exotoxins of *Aeromonas hydrophila, Clostridium tetani, E. coli* (SLT-1), *Pasteurella haemolytica, P. aeruginosa, Shigella dysenteriae,* and *Vibrio cholerae* as well as such products of *P. aeruginosa* as hydrogen cyanide, alkaline protease, elastase, and lipase. Many features of the molecular basis of iron regulation of bacterial toxigenesis are shared with that of bacterial iron acquisition (75).

In some reports, such trace metals as manganese, copper, cobalt, and nickel appear to be active in place of iron in controlling toxigenesis (139). In those studies, supraphysiologic concentrations of the noniron metals were required for activity. The putative action of these extremely high quantities of reagent-grade noniron salts possibly was due to their overlooked contamination by trace amounts of iron.

In one system, however, manganese rather than iron is active at physiologic concentrations. This system comprises secondary metabolism (including toxigenesis) and differentiation of species of *Bacillus* (132, 133). The quantity of manganese required for growth of iron-dependent bacteria, including *Bacillus* species, is 0.01 to 0.1 μm; that required for secondary processes of *Bacillus* is approximately 50-fold higher. During the transition from vegetative growth to secondary metabolism, the V_{max} for manganese uptake in *Bacillus* species is increased 50-fold (8). It may be recalled that in *E. coli,* manganese can substitute for ferrous iron in the activation of the Fur protein (11). It might be useful to search for an accessory transcriptional factor in *Bacillus* spp. that requires manganese as a cofactor.

Inhibition of secondary metabolism but not of vegetative growth of *Bacillus* species at high concentrations of manganese has been observed in some but not all studies (133). In vitro synthesis of the protective antigenic component of *Bacillus anthracis* exotoxin, for instance, has been reported to require 5 μM and to be inhibited by 20 μM manganese (153). Thus, the content of manganese in animal skin (7 μM) and splenic tissue (5 μM) (30) would permit toxigenesis in those sites. Moreover, one of the components of anthrax toxin is a strong chelating agent that might function as a manganophore. Inasmuch as the manganese content of blood is only 0.5 μM, any toxin formed in the bloodstream is presumably derived from bacteria that had recently seeded the vascular system from lesions in the skin or spleen.

SUMMARY

Awareness of the need of microbial pathogens for host iron and of the vigorous iron-withholding defense mounted by hosts came slowly in the 1930s and 1940s. At that time, a few microbiologists considered developing iron-binding agents or siderophore antagonists that could serve as anti-infective agents. For example, a research group at Imperial Chemicals, Ltd., attempted to synthesize analogs of mycobactin for possible treatment of tuberculosis (121). By 1950, however, the profuse outpouring of natural and synthetic anti-infective agents that were relatively easy to obtain and that were, at that time, highly effective caused a decline in interest in medicinal attempts to deprive pathogens of iron.

Presently, some strains of commonly encountered pathogens are gaining resistance to nearly all available anti-infective agents. Thus, interest in attempts to strengthen the ability of hosts to withhold or even withdraw iron from pathogens as well as from neoplastic cells has revived. Eight groups of methods for such strengthening are listed in Table 10.

TABLE 10 Methods for suppression of microbial and neoplastic cell acquisition of host iron[a]

A. Reduction of excessive intake of ingested iron
  1. Decreased consumption of red meats (heme iron)
  2. Avoidance of processed foods that have been adulterated with inorganic iron or blood
  3. Decreased consumption of ethanol and ascorbic acid
  4. Elimination of iron tablets unless an iron deficiency has been correctly diagnosed
B. Reduction of excessive inhalation of iron (142)
  1. Elimination of use of tobacco (mainstream smoke of one pack of cigarettes contains 1.12 μg of iron)
  2. Employment of iron-free chrysotile asbestos in place of iron-loaded amosite, crocidolite, or tremolite asbestos
  3. Use of protective masks and clothing when mining or cutting ferriferous minerals
C. Reduction of excessive use of transfused blood
  1. On medical services, ≤35% of transfusions are given without unequivocal justification (110)
  2. Increased use of erythropoietin in place of blood when possible
D. Reduction of iron burden by regular donations of whole blood or erythrocytes
E. Increased use of iron chelators
  1. Use of human milk (high in lactoferrin) rather than milk formula (high in iron, lacking in lactoferrin) in nutrition of nurslings to reduce incidence of botulism, salmonellosis, and sudden death (143)
  2. Use of tea (iron-binding tannins) and bran (iron-binding phytic acid)
  3. Continued R&D of potential high- and low-molecular-mass iron chelator drugs (e.g., lactoferrin, biomimetic siderophores, α-ketohydroxy-pyridines)
F. Initiation of prompt therapy of chronic infectious and neoplastic diseases to forestall saturation of iron-withholding defense system (16)
G. Continued R&D of cytokines that induce iron withholding (interleukin-1, interleukin-6, tumor necrosis factor alpha, gamma interferon) and of enhancers of nitric oxide synthesis

Continued on following page

TABLE 10 *Continued*

H. Continued R&D of passive and active methods of immunization against surface receptor proteins that enable microbial and neoplastic cells to obtain host iron (12, 17, 50, 122)
  1. Bacteria: receptor proteins for ferrated transferrin, ferrated lactoferrin, ferrated siderophores, or heme
  2. Fungi: receptor proteins for ferrated siderophores
  3. Protozoa: receptor proteins for ferrated transferrin or ferrated lactoferrin
  4. Neoplastic cells: receptor proteins for ferrated transferrin

*See reference 141. R&D, research and development.

In conclusion, the outcome of nearly all encounters of bacterial pathogens with vertebrate hosts is determined in part by the ability of the host to withhold iron from the pathogen and the ability of the pathogen to extract the metal from the host. In most bacteria, iron is essential for DNA synthesis and for the activities of various enzymes involved in electron transfer and oxygen metabolism. In most infections, not enough iron can be carried into the host by cells of the inoculum.

Successful pathogens have developed one or more of several strategies for securing host iron. These strategies include (i) lysis of erythrocytes, digestion of hemoglobin, and binding and assimilation of heme; (ii) cell surface binding of ferrated transferrin and extraction and assimilation of the metal; (iii) production of siderophores that extract the metal from ferrated transferrin, with subsequent binding and assimilation of the ferrated siderophore or the metal; and (iv) assimilation of intracellular iron possibly derived from pools of low-molecular-mass host iron-binding compounds.

Toxigenesis of some bacterial pathogens is suppressed at the transcriptional level by concentrations of iron in excess of that required for bacterial growth. However, in species of *Bacillus,* although iron is required for growth, manganese controls toxigenesis.

ACKNOWLEDGMENT

This work was supported in part by a grant-in-aid from the Office of Research and the University Graduate School, Indiana University, Bloomington.

REFERENCES

1. **Actis, L. A., M. E. Tolmasky, L. M. Crosa, and J. H. Crosa**. 1993. Effect of iron-limiting conditions on growth of clinical isolates of *Acinetobacter baumannii. J. Clin. Microbiol.* **31:**2812–2815.
2. **Adams, T. J., S. Vartivarian, and R. E. Cowart**. 1990. Iron acquisition systems of *Listeria monocytogenes. Infect. Immun.* **58:**2715–2718.
3. **Agiato, L.-A., and D. W. Dyer**. 1992. Siderophore production and membrane alterations by *Bordetella pertussis* in response to iron starvation. *Infect. Immun.* **60:**117–123.
4. **Aguirre, M., and M. D. Collins**. 1993. Lactic acid bacteria and human clinical infection. *J. Appl. Bacteriol.* **75:**95–107.
5. **Alderete, J. F., K. M. Peterson, and J. B. Baseman**. 1988. Affinities of *Treponema pallidum* for human lactoferrin and transferrin. *Genitourin. Med.* **64:**359–363.
6. **Ankenbauer, R., S. Sriyosachati, and C. D. Cox**. 1985. Effects of siderophores on the growth of *Pseudomonas aeruginosa* in human serum and transferrin. *Infect. Immun.* **49:**132–140.
7. **Archibald, F.** 1983. *Lactobacillus plantarum,* an organism not requiring iron. *FEMS Microbiol. Lett.* **19:**29–32.
8. **Archibald, F., and M.-N. Duong**. 1984. Manganese acquisition by *Lactobacillus plantarum. J. Bacteriol.* **158:**1–9.
9. **Archibald, F. S., and I. W. DeVoe**. 1980. Iron acquisition by *Neisseria meningitidis* in vitro. *Infect. Immun.* **27:**322–334.
10. **Ascencio, F., A. Ljungh, and T. Wadstrom**. 1992. Characteristics of lactoferrin binding by *Aeromonas hydrophila. Appl. Environ. Microbiol.* **58:**42–47.
11. **Bagg, A., and J. B. Neilands**. 1987. Molecular mechanism of regulation of siderophore-mediated iron assimilation. *Microbiol. Rev.* **51:**509–518.
12. **Banerjee-Bhatnagar, N., and C. E. Frasch**. 1990. Expression of *Neisseria meningitidis* iron-regulated outer membrane proteins, including a 70-kilodalton transferrin receptor, and their potential use as vaccines. *Infect. Immun.* **58:**2875–2881.
13. **Barghouthi, S., R. Young, M. O. J. Olson, J. E. L. Arceneaux, L. W. Clem, and B. R. Byers**. 1989. Ammonobactin, a novel tryptophane- or phenylalanine-containing phenolate siderophore in *Aeromonas hydrophila. J. Bacteriol.* **171:**1811–1816.
14. **Benjamin, W. H., Jr., C. L. Turnbough, Jr., B. S. Posey, and D. F. Briles**. 1985. The ability of *Salmonella typhimurium* to produce the siderophore enterobactin is not a virulence factor in mouse typhoid. *Infect. Immun.* **50:**392–396.
15. **Black, J. R., D. W. Dyer, M. K. Thompson,**

and P. F. Sparling. 1986. Human immune response to iron-repressible outer membrane proteins of *Neisseria meningitidis*. *Infect. Immun.* **54:** 710–713.

16. Boelaert, J. R., G. A. Weinberg, and E. D. Weinberg. 1995. Is there a place for iron chelation in HIV-infection? Abstr., 6th Internat. Conf. on Oral Chelation, Nijmegin, The Netherlands, 1/26–1/30/95.

17. Bolin, C. A., and A. E. Jensen. 1987. Passive immunization with antibodies against iron-regulated outer membrane proteins protects turkeys from *Escherichia coli* septicemia. *Infect. Immun.* **55:** 1239–1242.

18. Boockvar, K. S., D. L. Granger, R. M. Poston, M. Maybodi, M. K. Washington, J. B. Hibbs, Jr., and R. L. Kurlander. 1994. Nitric oxide produced during murine listeriosis is protective. *Infect. Immun.* **62:**1089–1100.

19. Braun, V., H. Gunther, B. Neub, and C. Tantz. 1985. Hemolytic activity of *Serratia marcescens. Arch. Microbiol.* **141:**371–376.

20. Britigan, B. E., M. B. Hayek, B. N. Doebbeling, and R. B. Fick, Jr. 1993. Transferrin and lactoferrin undergo proteolytic cleavage in the *Pseudomonas aeruginosa*-infected lungs of patients with cystic fibrosis. *Infect. Immun.* **61:**5049–5055.

21. Bullen, J. J., H. J. Rogers, and E. Griffiths. 1978. Role of iron in bacterial infections. *Curr. Top. Microbiol. Immunol.* **80:**1–35.

22. Byers, B. R. 1987. Pathogenic iron acquisition. *Life Chem. Rept.* **4:**143–159.

23. Byers, B. R., and J. E. L. Arceneaux. 1994. Iron acquisition and virulence of the bacterial genus *Aeromonas*, p. 29–40. In J. A. Manthey, D. E. Crowley, and D. G. Luster (ed.), *Biochemistry of Metal Micronutrients in the Rhizosphere*. Lewis Publishers, Boca Raton, Fla.

24. Byrd, T. F., and M. A. Horwitz. 1989. Interferon gamma-activated human monocytes downregulate transferrin receptors and inhibit the intracellular multiplication of *Legionella pneumophila* by limiting the availability of iron. *J. Clin. Invest.* **83:** 1457–1465.

25. Byrd, T. F., and M. A. Horwitz. 1991. Chloroquin inhibits the intracellular multiplication of *Legionella pneumophila* by limiting the availability of iron. A potential new mechanism for the therapeutic effect of chloroquin against intracellular pathogens. *J. Clin. Invest.* **88:**351–357.

26. Carbonetti, N. H., S. Boonchai, S. H. Parry, V. Vaisanen-Rhen, T. K. Korhonen, and P. H. Williams. 1986. Aerobactin-mediated iron uptake by *Escherichia coli* isolates from human extraintestinal infections. *Infect. Immun.* **51:**966–968.

27. Cavalieri, S. J., G. A. Bohach, and I. S. Snyder. 1984. *Escherichia coli* α-hemolysin characteristics and probable role in pathogenicity. *Microbiol. Rev.* **48:**326–343.

28. Chart, H., and T. J. Trust. 1983. Acquisition of iron from *Aeromonas salmonicida*. *J. Bacteriol.* **156:**758–764.

29. Chen, C.-Y., S. A. Berish, S. A. Morse, and T. A. Mietzner. 1993. The ferric iron-binding protein of pathogenic *Neisseria spp.* functions as a periplasmic transport protein in iron acquisition from human transferrin. *Mol. Microbiol.* **10:**311–318.

30. Cotzias, G. C. 1962. Manganese, p. 404–442. In C. L. Comar and F. Bronner (ed.), *Mineral Metabolism*. Academic Press, Inc., New York.

31. Crichton, R. R., and R. J. Ward. 1992. Iron metabolism—new perspectives in view. *Biochemistry* **31:**11255–11264.

32. Crosa, J. H. 1989. Genetics and molecular biology of siderophore-mediated iron transport in bacteria. *Microbiol. Rev.* **53:**517–530.

33. Crowle, A. J., and M. H. May. 1990. Inhibition of tubercle bacilli in cultured human macrophages by chloroquin used alone and in combination with streptomycin, isoniazid, pyrazinamide, and two metabolites of vitamin D₃. *Antimicrob. Agents Chemother.* **34:**2217–2222.

34. Dai, J.-H., Y.-S. Lee, and H.-C. Wong. 1992. Effects of iron limitation on production of a siderophore, outer membrane proteins, and hemolysin and on hydrophobicity, cell adherence, and lethality for mice of *Vibrio parahaemoliticus*. *Infect. Immun.* **60:**2952–2956.

35. Daskaleros, P. A., J. A. Stoebner, and S. M. Payne. 1991. Iron uptake in *Plesiomonas shigelloides*: cloning of the gene for the heme-iron uptake system. *Infect. Immun.* **59:**2706–2711.

36. DeBoy, J. M., I. K. Wachsmuth, and B. R. Davis. 1980. Hemolytic activity in enterotoxigenic and nonenterotoxigenic strains of *Escherichia coli*. *J. Clin. Microbiol.* **12:**193–198.

37. Doring, G., M. Pfestorf, K. Botzenhart, and M. A. Abdallah. 1988. Impact of proteases on iron uptake of *Pseudomonas aeruginosa* pyoverdin from transferrin and lactoferrin. *Infect. Immun.* **56:** 291–293.

38. Dyer, D. W., E. P. West, and P. F. Sparling. 1987. Effects of serum carrier proteins on the growth of pathogenic *Neisseriae* with heme-bound iron. *Infect. Immun.* **55:**2171–2175.

39. Enard, C., A. Diolez, and D. Expert. 1988. Systemic virulence of *Erwinia chrysanthemi* 3937 requires a functional iron assimilation system. *J. Bacteriol.* **170:**2419–2425.

40. Evanylo, L. P., S. Kadis, and J. R. Maudsley. 1984. Siderophore production by *Proteus mirabilis*. *Can. J. Microbiol.* **30:**1046–1051.

41. Feng, H.-M., and D. H. Walker. 1993. Interferon-γ and tumor necrosis factor-α exert their

antirickettsial effect via induction of synthesis of nitric oxide. *Am. J. Pathol.* **143:**1016–1023.

42. **Field, L. H., V. L. Headley, S. M. Payne, and L. J. Berry.** 1986. Influence of iron on growth, morphology, outer membrane protein composition, and synthesis of siderophores in *Campylobacter jejuni. Infect. Immun.* **54:**126–132.

43. **Finkelstein, R. A., C. V. Sciortino, and M. A. McIntosh.** 1983. Role of iron in microbe-host interactions. *Rev. Infect. Dis.* **5:**S759–S777.

44. **Flynn, J. L., J. Chan, K. J. Triebold, D. K. Dalton, T. A. Stewart, and B. R. Bloom.** 1993. An essential role for interferon-γ in resistance to *Mycobacterium tuberculosis* infection. *J. Exp. Med.* **178:**2249–2254.

45. **Francis, R. T., Jr., J. W. Booth, and R. R. Becker.** 1985. Uptake of iron from hemoglobin and the haptoglobin-hemoglobin complex by hemolytic bacteria. *Int. J. Biochem.* **17:**767–772.

46. **Garcia-del Portillo, F., J. W. Foster, M. E. Maguire, and B. B. Finlay.** 1992. Characterization of the micro-environment of *Salmonella typhimurium*-containing vacuoles within MDCK epithelial cells. *Mol. Microbiol.* **6:**3289–3297.

47. **Genco, C. A., C.-Y. Chen, R. J. Arko, D. R. Kapczynski, and S. A. Morse.** 1991. Isolation and characterization of a mutant of *Neisseria gonorrhoeae* that is defective in the uptake of iron from transferrin and haemoglobin and is avirulent in mouse subcutaneous chambers. *J. Gen. Microbiol.* **137:**1313–1321.

48. **Gentry, M. J., A. W. Confer, E. D. Weinberg, and J. T. Homer.** 1986. Cytotoxin (leukotoxin) production by *Pasteurella haemolytica:* requirement for an iron-containing compound. *Am. J. Vet. Res.* **47:**1919–1923.

49. **Gerlach, G. F., C. Anderson, A. A. Potter, S. Klashinsky, and P. J. Willson.** 1992. Cloning and expressing of a transferrin-binding protein from *Actinobacillus pleuropneumoniae. Infect. Immun.* **60:**892–898.

50. **Gilmour, N. J. L., W. Donachie, A. D. Sutherland, J. S. Gilmour, G. E. Jones, and M. Quirie.** 1991. Vaccine containing iron-regulated proteins of *Pasteurella haemolytica* A2 enhances protection against experimental pasteurellosis in lambs. *Vaccine* **9:**137–140.

51. **Goebel, W., T. Chakraborty, and J. Kleft.** 1988. Bacterial hemolysins as virulence factors. *Antonie van Leeuwenhoek J. Microbiol.* **54:**453–463.

52. **Gonzalez, G. C., O. L. Caamano, and A. B. Schryvers.** 1990. Identification and characterization of a porcine-specific transferrin receptor in *Actinobacillus pleuropneumoniae. Mol. Microbiol.* **4:**1173–1179.

53. **Griffiths, E., and J. Humphreys.** 1980. Isolation of enterochelin from the peritoneal washings of guinea pigs lethally infected with *Escherichia coli. Infect. Immun.* **28:**286–289.

54. **Griffiths, E., P. Stevenson, and P. Joyce.** 1983. Pathogenic *Escherichia coli* express new outer membrane proteins when growing *in vivo. FEMS Microbiol. Lett.* **16:**95–99.

55. **Hartford, T., S. O'Brien, P. W. Andrew, D. Jones, and I. S. Roberts.** 1993. Utilization of transferrin-bound iron by *Listeria monocytogenes. FEMS Microbiol. Lett.* **108:**311–318.

56. **Heesemann, J., K. Hantke, T. Vocke, E. Saken, A. Rakin, J. Stojilikovic, and B. Berner.** 1993. Virulence of *Yersinia enterocolitica* is closely associated with siderophore production, expression of an iron-repressible outer membrane polypeptide of 65000 Da and pesticin sensitivity. *Mol. Microbiol.* **8:**397–408.

57. **Helms, S. D., J. D. Oliver, and J. C. Travis.** 1984. Role of heme compounds and haptoglobin in *Vibrio vulnificus* pathogenicity. *Infect. Immun.* **45:**345–349.

58. **Herrington, D. A., and P. F. Sparling.** 1985. *Haemophilus influenzae* can use human transferrin as a sole source for required iron. *Infect. Immun.* **48:**248–251.

59. **Holbein, B. E.** 1981. Enhancement of *Neisseria meningitidis* infection in mice by addition of iron bound to transferrin. *Infect. Immun.* **34:**120–125.

60. **Husson, M.-O., D. Legrande, G. Spik, and H. Leclerc.** 1993. Iron acquisition by *Helicobacter pylori:* importance of human lactoferrin. *Infect. Immun.* **61:**2694–2697.

61. **Johnson, W., L. Varner, and M. Poch.** 1991. Acquisition of iron by *Legionella pneumophila:* role of iron reductase. *Infect. Immun.* **59:**2376–2381.

62. **Kishore, A. R., J. Erdei, S. S. Naidu, E. Falsen, A. Forsgren, and A. S. Naidu.** 1991. Specific binding of lactoferrin to *Aeromonas hydrophila. FEMS Microbiol. Lett.* **83:**115–120.

63. **Kochan, I.** 1973. The role of iron in bacterial infections, with special consideration of host-tubercle bacillus interaction. *Curr. Top. Microbiol. Immunol.* **60:**1–30.

64. **Konetschny-Rapp, S., G. Jung, J. Meiwes, and H. Zahner.** 1990. Staphyloferrin A: a structurally new siderophore from staphylococci. *Eur. J. Biochem.* **191:**65–74.

65. **Koronakis, V., M. Crosa, B. Senior, E. Koronakis, and C. Hughes.** 1987. The secreted hemolysins of *Proteus mirabilis, Proteus vulgaris,* and *Morganella morganii* are genetically related to each other and to the alpha-hemolysin of *Escherichia coli. J. Bacteriol.* **169:**1509–1515.

66. **Lambrecht, R. S., and M. T. Collins.** 1993. Inability to detect mycobactin in *Mycobacteria*-infected tissues suggests an alternative iron acquisition mechanism by *Mycobacteria in vivo. Microb. Pathog.* **14:**229–238.

67. **Lawlor, K. M., P. A. Daskaleros, R. E. Robinson, and S. M. Payne.** 1987. Virulence of iron transport mutants of *Shigella flexneri* and utilization of host iron compounds. *Infect. Immun.* **55:** 594–599.

68. **Lee, B. C.** 1991. Iron sources for *Haemophilus ducreyi. J. Med. Microbiol.* **34:**317–322.

69. **Lee, B. C.** 1992. Isolation of an outer membrane hemin-binding protein of *Haemophilus influenzae* type b. *Infect. Immun.* **60:**810–816.

70. **Lee, B. C.** 1992. Isolation of haemin-binding proteins of *Neisseria gonorrhoeae. J. Med. Microbiol.* **36:**121–127.

71. **Lee, B. C., and P. Hill.** 1992. Identification of an outer-membrane haemoglobin-binding protein in *Neisseria meningitidis. J. Gen. Microbiol.* **138:** 2647–2656.

72. **Lee, B. C., and A. B. Schryvers.** 1988. Specificity of the lactoferrin and transferrin receptors in *Neisseria gonorrhoeae. Mol. Microbiol.* **2:**827–829.

73. **Lepper, A. W. D., and C. R. Wilks.** 1988. Intracellular iron storage and the pathogenesis of paratuberculosis. Comparative studies with other mycobacterial, parasitic or infectious conditions of veterinary importance. *J. Comp. Pathol.* **98:**31–53.

74. **Linggood, M. A., and P. L. Ingram.** 1982. The role of alpha hemolysin in the virulence of *Escherichia coli* for mice. *J. Med. Microbiol.* **15:**23–30.

75. **Litwin, C. M., and S. B. Calderwood.** 1993. Role of iron in regulation of virulence genes. *Clin. Microbiol. Rev.* **6:**137–149.

76. **Massad, G., J. E. L. Arceneaux, and B. R. Byers.** 1991. Acquisition of iron from host sources by mesophilic *Aeromonas* species. *J. Gen. Microbiol.* **137:**237–241.

77. **Maudsley, J. R., and S. Kadis.** 1986. Growth and hemolysin production by *Haemophilus pleuropneumoniae* cultivated in a chemically defined medium. *Can. J. Microbiol.* **32:**801–805.

78. **Mazoy, R., and M. L. Lemos.** 1991. Iron-binding proteins and heme compounds as iron sources for *Vibrio anguillarum. Curr. Microbiol.* **23:** 221–226.

79. **McKenna, W. R., P. A. Mickelsen, P. F. Sparling, and D. W. Dyer.** 1988. Iron uptake from lactoferrin and transferrin by *Neisseria gonorrhoeae. Infect. Immun.* **56:**785–791.

80. **Menozzi, F. D., C. Gantiez, and C. Locht.** 1991. Identification and purification of transferrin- and lactoferrin-binding proteins of *Bordetella pertussis* and *Bordetella bronchiseptica. Infect. Immun.* **59:**3982–3988.

81. **Mickelsen, P. A., E. Blackman, and F. P. Sparling.** 1982. Ability of *Neisseria gonorrhoeae, Neisseria meningitidis,* and commensal *Neisseria* species to obtain iron from lactoferrin. *Infect. Immun.* **35:**915–920.

82. **Mickelsen, P. A., and P. F. Sparling.** 1981. Ability of *Neisseria gonorrhoeae, Neisseria meningitidis,* and commensal *Neisseria* species to obtain iron from transferrin and iron compounds. *Infect. Immun.* **33:**555–564.

83. **Minshew, B. H., J. Jorgensen, G. W. Counts, and S. Falkow.** 1978. Association of hemolysin production, hemagglutination of human erythrocytes, and virulence for chicken embryos of extraintestinal *Escherichia coli* isolates. *Infect. Immun.* **20:**50–54.

84. **Momotani, E., N. Wuscher, P. Ravisse, and N. Rastogi.** 1992. Immunohistochemical identification of ferritin, lactoferrin and transferrin in leprosy lesions of human skin biopsies. *J. Comp. Pathol.* **106:**213–220.

85. **Montgomerie, J. Z., A. Bindereif, J. B. Neilands, G. M. Kalmanson, and L. B. Guze.** 1984. Association of hydroxamate siderophore (aerobactin) with *Escherichia coli* isolated from patients with bacteremia. *Infect. Immun.* **46:**835–838.

86. **Morton, D. J., J. M. Musser, and T. L. Stull.** 1993. Expression of the *Haemophilus influenzae* transferrin receptor is repressible by hemin but not elemental iron alone. *Infect. Immun.* **61:**4033–4037.

87. **Morton, D. J., and P. Williams.** 1990. Siderophore-independent acquisition of transferrin-bound iron by *Hemophilus influenzae* type b. *J. Gen. Microbiol.* **136:**927–933.

88. **Mueller, J. H.** 1941. The influence of iron on the production of diphtheria toxin. *J. Immunol.* **42:** 343–351.

89. **Naidu, A. S., J. Miedzobrodski, M. Andersson, L.-E. Nilsson, A. Forsgren, and J. L. Watts.** 1990. Bovine lactoferrin binding to six species of coagulase-negative staphylococci isolated from bovine intramammary infections. *J. Clin. Microbiol.* **28:**2312–2319.

90. **Naidu, A. S., J. Miedzobrodski, J. M. Musser, V. T. Rosdahl, S. A. Hedstrom, and A. Forsgren.** 1991. Human lactoferrin binding in clinical isolates of *Staphylococcus aureus. J. Med. Microbiol.* **34:**323–328.

91. **Nassif, X., M. C. Mazert, J. Mounier, and P. J. Sansonetti.** 1987. Evaluation with an *iuc*::Tn*10* mutant of the role of aerobactin production in the virulence of *Shigella flexneri. Infect. Immun.* **55:**1963–1969.

92. **Nassif, X., and P. J. Sansonetti.** 1986. Correlation of the virulence of *Klebsiella pneumoniae* K1 and K2 with the presence of a plasmid encoding aerobactin. *Infect. Immun.* **54:**603–608.

93. **Neilands, J. B., A. Bindereif, and J. Z. Montgomerie.** 1985. Genetic basic of iron assimilation in *Escherichia coli. Curr. Top. Microbiol. Immunol.* **118:**179–195.

94. **Niven, D. F., J. Donga, and F. S. Archibald.**

1989. Responses of *Haemophilus pleuropneumoniae* to iron restriction: changes in the outer membrane protein profile and the removal of iron from porcine transferrin. *Mol. Microbiol.* **3**:1083–1089.

95. **Ogunnariwo, J. A., C. Cheng, J. Ford, and A. B. Schryvers**. 1990. Response of *Haemophilus somnus* to iron limitation: expression and identification of a bovine-specific transferrin receptor. *Microb. Pathog.* **9**:397–406.

96. **Ogunnariwo, J. A., and A. B. Schryvers**. 1990. Iron acquisition in *Pasteurella haemolytica*: expression and identification of a bovine-specific transferrin receptor. *Infect. Immun.* **58**:2091–2097.

97. **Opal, S. M., A. S. Cross, P. Gemski, and L. W. Lyhte**. 1990. Aerobactin and alpha-hemolysin as virulence determinants in *Escherichia coli* isolated from human blood, urine and stool. *J. Infect. Dis.* **161**:794–796.

98. **Otto, B. R., W. R. Verweij, M. Sparrius, A. M. J. J. Verweij-van Vught, C. E. Nord, and D. M. MacLaren**. 1991. Human immune response to an iron-repressible outer membrane protein of *Bacteroides fragilis*. *Infect. Immun.* **59**:2999–3003.

99. **Payne, S. M.** 1988. Iron and virulence in the family Enterobacteriaceae. *Crit. Rev. Microbiol.* **16**:81–111.

100. **Payne, S. M.** 1989. Iron and virulence in *Shigella*. *Mol. Microbiol.* **3**:1301–1306.

101. **Perry, R. D., and R. B. Brubaker**. 1979. Accumulation of iron by yersiniae. *J. Bacteriol.* **137**:1290–1298.

102. **Pickett, C. L., T. Auffenberg, E. C. Pesci, V. L. Sheehn, and S. S. D. Jusuf**. 1992. Iron acquisition and hemolysin production by *Campylobacter jejuni*. *Infect. Immun.* **60**:3872–3877.

103. **Pidcock, K. A., J. A. Wooten, B. A. Daley, and T. L. Stull**. 1988. Iron acquisition by *Haemophilus influenzae*. *Infect. Immun.* **56**:721–725.

104. **Poole, K., and V. Braun**. 1988. Iron regulation of *Serratia marcescens* hemolysin gene expression. *Infect. Immun.* **56**:2967–2971.

105. **Reddy, D., J. R. Lancaster, and D. P. Cornforth**. 1983. Nitrite inhibition of *Clostridium botulinum*: electron spin resonance detection of nitric oxide complexes. *Science* **231**:769–772.

106. **Redhead, K., and T. Hill**. 1991. Acquisition of iron from transferrin by *Bordetella pertussis*. *FEMS Microbiol. Lett.* **77**:303–308.

107. **Riddle, L. M., T. E. Graham, and R. L. Amborski**. 1981. Medium for the accumulation of extracellular hemolysin and protease by *Aeromonas hydrophila*. *Infect. Immun.* **33**:728–733.

108. **Russell, L. M., S. J. Cryz, Jr., and R. K. Holmes**. 1984. Genetic and biochemical evidence for a siderophore-dependent iron transport system in *Corynebacterium diphtheriae*. *Infect. Immun.* **15**:143–149.

109. **Sanchez, L., M. Calvo, and J. H. Brock**. 1992. Biological role of lactoferrin. *Arch. Dis. Child.* **67**:657–661.

110. **Saxena, S., J. M. Weiner, A. Rabinowitz, J. Fridey, I. A. Shulman, and R. Carmel**. 1993. Transfusion practice in medical patients. *Arch. Intern. Med.* **153**:2575–2580.

111. **Schryvers, A. B.** 1989. Identification of the transferrin- and lactoferrin-binding proteins in *Haemophilus influenzae*. *J. Med. Microbiol.* **29**:121–130.

112. **Schryvers, A. B., and G. C. Gonzalez**. 1989. Comparison of the abilities of different protein sources of iron to enhance *Neisseria meningitidis* infections in mice. *Infect. Immun.* **57**:2425–2429.

113. **Schryvers, A. B., and L. J. Morris**. 1988. Identification and characterization of the transferrin receptor from *Neisseria meningitidis*. *Mol. Microbiol.* **2**:281–288.

114. **Sciortino, C. V., and R. A. Finkelstein**. 1983. *Vibrio cholerae* expresses iron-regulated outer membrane proteins in vivo. *Infect. Immun.* **42**:990–996.

115. **Shand, G. H., H. Anwar, J. Kaduragamuwa, M. R. W. Brown, S. H. Silverman, and J. Melling**. 1985. In vivo evidence that bacteria in urinary tract infection grow under iron-restricted conditions. *Infect. Immun.* **48**:35–39.

116. **Shank, J. L., R. H. Silliker, and R. H. Harper**. 1962. The effect of nitric oxide on bacteria. *Appl. Microbiol.* **10**:185–189.

117. **Sigel, S. P., J. A. Stoebner, and S. M. Payne**. 1985. Iron-vibriobactin transport system is not required for virulence of *Vibrio cholerae*. *Infect. Immun.* **47**:360–362.

118. **Sikkema, D. J., and R. R. Brubaker**. 1987. Resistance to pesticin, storage of iron and invasion of HeLa cells by yersiniae. *Infect. Immun.* **55**:572–578.

119. **Simonson, C., D. Brener, and I. W. DeVoe**. 1982. Expression of a high-affinity mechanism for acquisition of transferrin iron by *Neisseria meningitidis*. *Infect. Immun.* **36**:107–113.

120. **Simpson, L. M., and J. D. Oliver**. 1983. Siderophore production by *Vibrio vulnificus*. *Infect. Immun.* **41**:644–649.

121. **Snow, G. A.** 1970. Mycobactins: iron-chelating growth factors from mycobacteria. *Bacteriol. Rev.* **34**:99–125.

122. **Sokol, P. A., and D. E. Woods**. 1986. Characterization of antibody to the ferripyochelin-binding protein of *Pseudomonas aeruginosa*. *Infect. Immun.* **51**:896–900.

123. **Stoebner, J. A., and S. M. Payne**. 1988. Iron-regulated hemolysin production and utilization of heme and hemoglobin by *Vibrio cholerae*. *Infect. Immun.* **56**:2891–2895.

124. **Stojiljkovic, I., and K. Hantke**. 1992. Hemin uptake system of *Yersinia enterocolitica:* similarities with other TonB-dependent systems in Gram-negative bacteria. *EMBO J.* **11:**4359–4367.

125. **Tai, S. S., C.-J. Lee, and R. E. Winter**. 1993. Hemin utilization is related to virulence of *Streptococcus pneumoniae. Infect. Immun.* **61:**5401–5405.

126. **Todhunter, D. A., K. L. Smith, and J. S. Hogan.** 1991. Antibodies to iron-regulated outer membrane proteins of coliform bacteria isolated from bovine intramammary infections. *Vet. Immunol. Immunopathol.* **28:**107–115.

127. **Tryon, V. V., and J. B. Baseman**. 1987. The acquisition of human lactoferrin by *Mycoplasma pneumoniae. Microb. Pathog.* **3:**437–443.

128. **Une, T., and R. R. Brubaker**. 1984. In vivo comparison of avirulent Vwa⁻ and PGM⁻ or Pstʳ phenotypes of yersiniae. *Infect. Immun.* **43:**895–900.

129. **Verweij-Van Vught, A. M. J. J., B. R. Otto, F. Namavar, M. Sparrius, and D. M. MacLaren.** 1988. Ability of *Bacteroides* species to obtain iron from iron salts, haem compounds and transferrin. *FEMS Microbiol. Lett.* **49:**223–228.

130. **Waalwijk, C., D. M. MacLaren, and J. deGraaff.** 1983. In vivo function of hemolysin in the nephropathogenicity of *Escherichia coli. Infect. Immun.* **42:**245–249.

131. **Walker, D. H**. 1989. The rickettsia-host interaction, p. 79–92. *In* J. W. Moulder (ed.), *Intracellular Parasitism.* CRC Press, Inc., Boca Raton, Fla.

132. **Weinberg, E. D.** 1964. Manganese requirement for sporulation and other secondary biosynthetic processes of *Bacillus. Appl. Microbiol.* **12:**436–441.

133. **Weinberg, E. D**. 1970. Biosynthesis of secondary metabolites: roles of trace metals. *Adv. Microb. Physiol.* **4:**1–44.

134. **Weinberg, E. D**. 1971. Roles of iron in host-parasite interactions. *J. Infect. Dis.* **124:**401–410.

135. **Weinberg, E. D**. 1972. Systemic salmonellosis: a sequela of sideremia. *Tex. Rep. Biol. Med.* **30:**277–286.

136. **Weinberg, E. D**. 1974. Iron and susceptibility to infectious disease. *Science* **184:**952–956.

137. **Weinberg, E. D**. 1978. Iron and infection. *Microbiol. Rev.* **42:**45–66.

138. **Weinberg, E. D**. 1984. Iron withholding: a defense against infection and neoplasia. *Physiol. Rev.* **64:**65–102.

139. **Weinberg, E. D**. 1990. Roles of trace metals in transcriptional control of microbial secondary metabolism. *Biol. Metals* **2:**191–196.

140. **Weinberg, E. D**. 1992. Iron depletion: a defense against intracellular infection and neoplasia. *Life Sci.* **50:**1287–1297.

141. **Weinberg, E. D**. 1993. Iron withholding: the natural process and its modifications by pharmacologic agents, p. 97–109. *In* R. J. Bergeron and G. M. Brittenham (eds.), *The Development of Iron Chelators for Clinical Use.* CRC Press, Inc., Boca Raton, Fla.

142. **Weinberg, E. D.** 1993. Association of iron with respiratory tract neoplasia. *J. Trace Elem. Exp. Med.* **6:**117–123.

143. **Weinberg, E. D.** The role of iron in sudden infant death syndrome. *J. Trace Elem. Exp. Med.* **7:**47–51.

144. **Welch, R. A.** 1987. Identification of two different hemolysin determinants in uropathogenic *Proteus* isolates. *Infect. Immun.* **55:**2183–2190.

145. **Welch, R. A., E. P. Dellinger, B. Minshew, and S. Falkow.** 1981. Haemolysin contributes to virulence of extraintestinal *E. coli* infections. *Nature* (London) **294:**665–667.

146. **West, S. E. H., and P. F. Sparling.** 1985. Response of *Neisseria gonorrhoeae* to iron limitation: alterations in expression of membrane proteins without apparent siderophore production. *Infect. Immun.* **47:**388–394.

147. **Williams, P., and E. Griffiths**. 1992. Bacterial transferrin receptors—structure, function and contribution to virulence. *Med. Microbiol. Immunol.* **181:**301–322.

148. **Williams, P., D. J. Morton, K. J. Towner, P. Stevenson, and E. Griffiths.** 1990. Utilization of enterobactin and other exogenous iron sources by *Haemophilus influenzae, H. parainfluenzae* and *H. paraphrophilus. J. Gen. Microbiol.* **136:**2343–2350.

149. **Wolf, M. K., and J. H. Crosa.** 1986. Evidence for the role of a siderophore in promoting *Vibrio anguillarum* infections. *J. Gen. Microbiol.* **132:**2949–2952.

150. **Wong, J. C. Y., J. Holland, T. Parsons, A. Smith, and P. Williams.** 1994. Identification and characterization of an iron-regulated hemopexin receptor in *Haemophilus influenzae* type b. *Infect. Immun.* **62:**48–59.

151. **Woods, D. E., P. A. Sokol, and B. H. Iglewski.** 1982. Modulatory effect of iron on the pathogenesis of *Pseudomonas aeruginosa* mouse corneal infections. *Infect. Immun.* **35:**461–464.

152. **Wooldridge, K. G., and P. H. Williams.** 1993. Iron uptake mechanisms of pathogenic bacteria. *FEMS Microbiol. Rev.* **12:**325–348.

153. **Wright, G. G., M. A. Hedberg, and J. B. Slein.** 1954. Studies on immunity in anthrax. III. Elaboration of a protective antigen in a chemically defined, non-protein medium. *J. Immunol.* **72:**263–269.

154. **Yang, H., C. D. Kool, and P. A. Sokol**. 1993. Ability of *Pseudomonas pseudomallei* malleobactin to acquire transferrin-bound, lactoferrin-bound, and cell-derived iron. *Infect. Immun.* **61:**656–662.

BACTERIAL RESISTANCE TO CELLULAR DEFENSE MECHANISMS

ROLE OF MYCOBACTERIAL CONSTITUENTS IN REGULATION OF MACROPHAGE EFFECTOR FUNCTION

James L. Krahenbuhl

7

Identifying putative virulence factors of the mycobacteria is in general a most difficult task made only slightly easier by limiting discussion to the pathogenic mycobacteria important in human disease (i.e., *Mycobacterium tuberculosis, M. leprae, M. bovis, M. avium, M. intracellulare,* etc.). Excluding from consideration the mycobacteria that cause disease only in severely immunocompromised patients (i.e., organisms of the *M. avium-M. intracellulare* complex) does not necessarily make the identification of virulence factors any more straightforward. Discussion of only card-carrying human pathogens that cause tuberculosis (TB) and leprosy and attempts to identify unique or common virulence factors are stymied by the striking differences in pathogenesis of these diseases in humans and in the different animal models available to test virulence factor hypotheses.

TB AND LEPROSY

The most conspicuous difference between TB and leprosy in humans is the tissues involved in disease. However, resistance to both diseases has historically been linked to cell-mediated immunity (CMI), with the level of host resis-

tance hinging on complex interactions between lymphocytes of the helper and suppressor phenotypes, maturity of the macrophage in the lesions, and modulation of cytokines and other factors elaborated by the cells composing the lesion. In the vast majority of cases, it is safe to presume that CMI events follow infection with the tubercle or leprosy bacillus and that the macrophage plays a critical albeit undetected role in events that quietly culminate in the elimination of the organism without production of clinically detectable inflammation or disease. Thus, if a common denominator of host resistance and virulence for these mycobacteria that can lend itself to discussion and investigation exists, it is probably the preferred host cell for mycobacterial pathogens, the macrophage. The very brief discussion of TB and leprosy that follows focuses primarily on each disease from the perspective of the macrophage.

TB

In animal models of TB, adoptive transfer of different populations of T cells has shown that actual protection can be dissociated from delayed-type hypersensitivity (DTH [83]). TB is primarily a disease of the lungs, although blood-borne dissemination to other tissues can occur early in infection. Progressive disease fol-

James L. Krahenbuhl Laboratory Research Branch, G. W. Long Hansen's Disease Center, Louisiana State University, P.O. Box 25072, Baton Rouge, Louisiana 70894.

Virulence Mechanisms of Bacterial Pathogens, 2nd ed., Edited by J. A. Roth et al.
© 1995 American Society for Microbiology, Washington, D.C.

lows primary infection in 5 to 10% of cases, often in children. With the development of effective CMI, infection is contained and most of the organisms are killed, although viable bacilli may persist for years. This immunity may wane with age, and the individual may become susceptible to exogenous reinfection or may develop disease as a consequence of endogenous reinfection. To observe the importance of CMI and T cells in resistance to TB, one need only look at the markedly enhanced susceptibility of human immunodeficiency virus-positive patients to primary progressive TB or reactivated disease. Jones et al. (57) have shown that *M. tuberculosis* bacillemia is linked to the CD4 cell count in infected human immunodeficiency virus-positive patients (4% at >200 CD4$^+$ cells per μl; 49% at <100 CD4$^+$ cells per μl).

Dannenberg classifies human TB into four stages (27). The alveolar macrophages are the primary target of infection, which is initiated by inhalation of aerosols containing droplet nuclei bearing viable *M. tuberculosis*. Subsequent onset and course of clinical disease represent a complicated struggle between host and pathogen, with the disease alternating between phases of remission and exacerbation. In the first stage of TB, there can be destruction of bacilli by alveolar macrophages before growth and subsequent disease occur. In the second stage, however, the organism can grow within a developing pulmonary lesion in granulomas (tubercles) made up of blood-borne, monocyte-derived macrophages. In the third stage, the development of CMI and DTH influences events. Viable extracellular bacilli are found in the caseous necrotic tissue at the core of the tubercle, and the fates of intracellular bacilli in the walls of the granuloma are determined by an interplay between permissive healthy macrophages and microbicidal activated macrophages. Successful CMI might leave nonmultiplying bacilli encapsulated in a tubercle for decades, providing a source for reactivation of TB later in life or during an immunocompromising disease such as AIDS. In the fourth stage, the center of the pulmonary granuloma can liquefy, resulting in an enormous extracellular growth of bacilli. Cavitation and tissue destruction can release bacilli into the bronchial tree and serve as a source of transmissible aerosolized bacilli into the environment.

Leprosy

In leprosy, CMI or its absence plays a key role in resistance as well as pathogenesis. Leprosy predominantly affects the skin, peripheral nerves, and mucous membranes. Active infection with *M. leprae* is characterized by a broad spectrum of host responses, with great variability in histopathology and clinical course of infection (51). There is likewise a spectrum in the host humoral and CMI responses to infection. Individuals at the tuberculoid (TT) end of the spectrum manifest a strong DTH response to *M. leprae* antigens but produce relatively low levels of antibody. At the opposite end of the clinical spectrum, lepromatous leprosy (LL), there is a potent humoral antibody response but a progressive anergy in CMI specifically for antigens of the leprosy bacillus. In advanced LL, there is a characteristic pattern of sensory nerve loss due to uncontrolled growth of leprosy bacilli involving the dermal nerve fibers. Nerve damage in TT leprosy is due more to host response (CMI) to infection in and near nerve fibers than to bacillary multiplication. Borderline leprosy encompasses those types of disease between LL and TT and may upgrade or downgrade to either form.

Early indications that LL was associated with a general suppression of CMI have not held up. LL patients are not immunocompromised as defined by opportunistic infections and increased incidence of cancer. Pulmonary TB runs a normal course in LL patients, underscoring the specificity of the T-cell anergy to *M. leprae* antigens not shared by other mycobacteria. The CD4/CD8 ratio of T cells in the blood of TT and LL patients is normal (2:1). Across the leprosy spectrum, the local lesions are represented by collections of lymphocytes, macrophages, and other cell types organized within distinct three-dimensional structures (granulomas). However, in LL granulomas, there is a paucity of T cells, and the CD4/CD8 ratio is markedly lower (0.5:1) (72). In TT lesions, the ratio remains 2:1, and there is a distinct architectural arrangement of T cells, with CD4 cells in the center of the epithelioid-cell granuloma and CD8 cells limited to the periphery of the lesion.

The macrophage is clearly the predominant target cell for *M. leprae,* although it is not yet clear how the bacilli reach the involved sites. Although aerosol invasion of the nasal mucosa (26) and penetration of skin (56) are possibilities, the actual route of infection is not known. The presence of bacilli in the skin and mucous membranes may be more the result of suitable local growth conditions than of the proximity of these sites to the route of infection. Infected blood monocytes (35) or bone marrow promonocytes (96) would be ideal transport vehicles for hematogenous spread and deposition of bacilli in distant skin sites adjacent to peripheral nerve endings.

Regardless of the lack of a clear understanding of the details underlying the mechanisms of effective CMI in TB and TT leprosy or of T-cell anergy in LL, the failure of the macrophage to cope with *M. tuberculosis* or *M. leprae* is a conspicuous characteristic of clinical disease and is an issue central to understanding the mechanisms of host resistance to the tubercle and leprosy bacilli.

THE MACROPHAGE IN GENERAL

A general discussion of the origin and development of the macrophage; its phenotypic, functional, and secretory heterogeneity; and its afferent and efferent roles in CMI is beyond the scope of this chapter.

The mononuclear phagocyte system represents a cellular lineage in which cells that originate in the bone marrow enter the circulation and eventually transform into tissue macrophages (115). A few macrophages (<5%) divide locally in the tissues. Descendants of bone marrow-derived monoblasts and promonocytes include circulating and marginating blood monocytes, free macrophages in the peritoneal and pleural cavities, the macrophages in splenic sinusoids and lymph nodes, connective tissue histiocytes, Kupffer cells, alveolar macrophages, and the microglial cells of the brain.

The maturation of tissue macrophages is influenced by a number of stimuli and receptor-ligand interactions, giving the macrophages a constitutive and inducible diversity in functional activity in the steady state as well as in an inflammatory response. Macrophages are unique in their endocytic and phagocytic capacities for soluble substances and particulate matter and produce a vast array of secretory products to regulate the immune, hematopoietic, and inflammatory responses (77). In regulating the immune response to infection with intracellular pathogenic microorganisms (*M. leprae, M. tuberculosis*), the macrophage not only plays a key role in initiating and intensifying CMI but also is the principal cellular defensive arm that is stimulated and called upon to cope with the pathogen.

In their afferent immunoregulatory role, macrophages function as antigen-presenting cells (APC) in their interaction with T cells. All mammalian cells can endocytose and process soluble (protein) antigens, and some also express surface class II major histocompatibility complex (MHC) molecules. Many cells other than macrophages (i.e., dendritic cells, Langerhans cells, B cells, endothelial cells) express MHC class II antigens and can function as APC. However, macrophages are uniquely efficient APC because they not only can endocytose soluble antigens but also can phagocytose particulate antigens, including pathogenic microorganisms, and can degrade, process, and present antigens relevant for stimulation of CMI in an MHC class II-restricted context.

THE ROLE OF MACROPHAGES IN RESISTANCE TO INTRACELLULAR PATHOGENS

Primarily through the work of Mackaness and his colleagues in the 1960s (69) and of numerous workers since, we recognize the important role of the activated macrophage in host resistance to obligate and facultative intracellular pathogens. Activation modifies numerous biochemical, phenotypic, and functional properties of the macrophage, but chief among these changes is the induction of an enhanced microbicidal capacity. Whereas normal macrophages support the growth of intracellular microorganisms, activated macrophages kill or inhibit not only the pathogen that incited the CMI response but also a broad spectrum of phylogenetically unrelated intracellular microorganisms (65). Thus, macrophages activated by lymphokines released from sensitized T cells responding specifically to antigen (66) play a major role in resistance to a wide variety of obligate and facultative intracellular pathogens. Previous reports have demonstrated a role for the enhanced microbicidal activity of the activated mouse macrophage against a variety of mycobacterial pathogens, including *M. tuberculosis* (87), *M. bovis* (111), *M. microti* (40), *M. avium* (117), and *M. leprae* (1a). The key macrophage-activating lymphokine appears to be gamma interferon (IFN-γ) (78).

ANTIMICROBIAL MECHANISMS OF MACROPHAGES

Of the numerous morphologic and metabolic criteria that characterize activated macrophages, two enhanced metabolic pathways stand out: the enhanced production of reactive oxygen intermediates (ROI) (76, 79) and the increased production of L-arginine-derived reactive nitrogen intermediates (RNI) (33, 34). Mycobacterial constituents may modify these mechanisms that may underlie the enhanced microbicidal capacity of the activated macrophage for *M. tuberculosis* and *M. leprae* and are discussed in more detail below.

Briefly, depending on the maturity, location, and state of activation of the macrophage, appropriate stimulation of it is associated with a respiratory burst that leads to the production of ROI formed by the partial reduction of oxygen (61). The principal ROI are superoxide anion, hydrogen peroxide (H_2O_2), and hydroxyl radical (OH·). The molecular excitation of oxygen yields a fourth ROI, singlet oxygen. The baseline capacity of the blood monocyte to produce ROI is greater than that of more

mature tissue macrophages, in part because of the presence of cytoplasmic peroxidase granules in the former. However, the IFN-γ-activated macrophage has an enhanced ability to produce ROI (78) that underlies, in part, the enhanced nonspecific microbicidal activity of the activated macrophage. ROI have been implicated as a major (62, 98) antimicrobial effector mechanism. The abilities of *M. leprae* and *M. tuberculosis* or their constituents to neutralize the toxic effects of ROI may be an important virulence factor.

There are a number of reports (21, 26, 107) of oxygen-independent antimicrobial effects of activated macrophages on a wide variety of intracellular bacterial (including mycobacterial) (89, 97), fungal, and protozoal pathogens, suggesting that an alternative, nonoxidative mechanism may operate as well. A single amino acid, L-arginine, acts as the substrate for activated-macrophage-mediated cytotoxic activity against tumor cells (33, 34) and antimicrobial activity against a variety of intracellular pathogens, including fungi, bacteria (29), and protozoa (4, 50). A novel biochemical pathway that synthesizes nitrogen oxides from the terminal guanidino nitrogen atom of L-arginine produces RNI. Nitric oxide gas causes the same pattern of metabolic inhibition in tumor cells as that caused by activated macrophages and has a direct antifungal and antimycobacterial (20) function as well.

These data suggest that nitric oxide, a stable free radical with a relatively short half-life, is the actual effector molecule of this pathway. We have recently explored the role of the L-arginine-dependent pathway (1a) in the detrimental effects of activated macrophages on *M. leprae*. Others have linked the L-arginine-dependent pathway to the anti-*M. tuberculosis* effects of activated macrophages (20, 40).

STUDIES IN MICE AND MEN: A PARADOX

Two striking paradoxes are apparent when the antimycobacterial function of human macrophages is explored. First, unlike their murine counterparts, IFN-γ-activated human macrophages are not antimycobacterial. Second, the L-arginine-dependent RNI pathway has not been convincingly described in human macrophages, although it is a potent mechanism of antimycobacterial function in murine macrophages.

In well-controlled studies, human IFN-γ efficiently activates human macrophages to kill *Leishmania* spp., but in the case of *M. tuberculosis,* it decreases phagocytosis of the bacilli and appears to actually enhance intracellular replication (31). Similar findings with human macrophages were reported by Rook et al. (89), who also used *M. tuberculosis,* and by Toba et al. (112), who used *M. avium.* Our own studies clearly show that IFN-γ-activated human macrophages demonstrate a potent antimicrobial effect against the intracellular protozoan *Toxoplasma gondii* but fail to affect the metabolism (viability?) of *M. leprae* (64).

A supplemental pathway may be required for human macrophages. Rook's group (90) has shown that IFN-γ-treated human macrophages enzymatically convert the circulating inactive form of vitamin D (25-hydroxy-cholecalciferol) into its 1,25-dihydroxy form, calcitriol. A combination of IFN-γ and calcitriol induces detectable inhibition of *M. tuberculosis* in vitro, and there is evidence for this pathway in human TB lesions (9) and bronchoalveolar cells (17). We have explored this pathway in vitro with *M. leprae* and human macrophages and failed to show any effect, perhaps because of the sensitivity of the assay to viability of the uncultivable leprosy bacillus (1).

Cytokines other than IFN-γ have been employed with some success to activate human macrophages in in vitro models employing *M. avium.* Bermudez and Young (11) employed tumor necrosis factor alpha (TNF-α) alone or in combination with interleukin-2 (IL-2) to enhance *M. avium* killing by human macrophages, and Denis (28) reported that TNF-α and granulocyte-macrophage colony-stimulating factor stimulated human macrophages to inhibit virulent *M. avium* and kill avirulent *M. avium.* Denis and Gregg (30) and Shiratsuchi et al. (99) employed the cyclooxygenase inhibitor

indomethacin combined with IFN-γ to demonstrate activation of human macrophages to inhibit *M. avium.* Human macrophages treated with combinations of IFN-γ, TNF-α, lipopolysaccharide (LPS), and indomethacin do not have a deleterious effect on *M. leprae* (64).

The link between the L-arginine-dependent pathway and antimycobacterial macrophage effects in humans remains uncertain because of the difficulty in demonstrating an antimicrobial function for RNI in human macrophages by using cytokine induction procedures successful in mice (14, 20, 50). TNF-α and granulocyte-macrophage colony-stimulating factor activate human macrophages to inhibit *M. avium* (28), and the high-output production of RNI is implicated. Hibbs et al. (52) studied cancer patients receiving IL-2 immunotherapy and reported that a cytokine-inducible, high-output L-arginine–nitric oxide pathway exists in humans. Dumarey et al. (36) recently reported the induction of RNI (NO) in human-monocyte-derived macrophages with virulent strains of *M. avium* but not *M. tuberculosis* or *M. smegmatis.* The pathway for RNI synthesis was similar to that demonstrated in mouse macrophages, involving synthesis of NO from L-arginine by an inducible NO synthase. Only viable *M. avium* induced NO. Killed organisms or *M. avium* cell wall or cytosol fractions failed to induce NO.

CYTOKINE PRODUCTION BY MACROPHAGES AND IMPORTANCE OF THE LOCAL MICROENVIRONMENT

Studies of macrophage function largely employ murine peritoneal macrophages and human peripheral blood monocyte-derived macrophages because these cells are readily obtainable from easily accessible compartments and macrophage cell lines, which can be conveniently maintained in continuous culture. Comparatively little is known about the functions and properties of macrophages in other anatomical compartments, and even less is known about the function of macrophages isolated from granulomas such as those that characterize TB or leprosy.

As a secretory cell responding in vitro to mycobacteria and their soluble products, the macrophage is also the source of such potent immunoregulatory compounds as arachidonic acid cyclooxygenase products and cytokines such as IL-1, TNF-α, IL-6, IFN-α, IFN-β, IL-10, and transforming growth factor β (8, 114, 118). By attracting additional inflammatory cells and modulating the functions of these as well as of existing macrophages, macrophage secretory products play important roles in the initiation and maintenance of local TB and leprosy lesions. The interactions of cytokines in enhancing the antimycobacterial effects of macrophages are complex. Murine macrophages treated with IFN-γ and infected with *M. tuberculosis* produce enhanced amounts of both IL-10 and TNF-α. TNF-α acts as an endogenous cofactor to enhance IFN-γ-induced production of RNI and inhibition of mycobacteria. In contrast, IL-10 counteracts macrophage activation by downregulating TNF-α and RNI production and, subsequently, antimycobacterial activity (41).

Initially described as a cytokine that could induce the necrosis of specific tumors, TNF-α has now been shown to exert numerous and varied biological activities that are both protective and pathologic (7, 82, 116). TNF-α has been identified as one of the principal modulators of inflammation and is involved in triggering the release of RNI (34) and ROI (121), the regulation of numerous cytokines and arachidonic acid metabolites (59, 68, 120), and the induction of microbicidal activity and cell damage (37, 50, 100).

TNF-α plays a pivotal role in inflammatory phenomena that culminate in either pathogenesis or resistance in mycobacterial disease. The conversion of TNF-α from a protective to a pathologic role has been suggested by Rook and colleagues, who have found immunologic abnormalities resulting from *M. tuberculosis* infection, including elevated levels of agalactosyl immunoglobulin G (IgG) (88) and an increased sensitivity to the toxic effects of TNF-α (39). They suggested that this "TNF-α-enhancing activity" in *M. tuberculosis*-infected cells, which

could also be induced by crude *M. tuberculosis* supernatants, may be responsible for the tissue destruction seen in active tuberculosis.

In a landmark study, Kindler et al. (60) established the importance of TNF-α in the development of *M. bovis* BCG granulomas. Using repeated systemic administration of an anti-TNF-α monoclonal antibody in mice, they demonstrated striking histopathologic evidence of an inhibition of the formation and maintenance of BCG-induced granulomas. Inhibition of granuloma formation was associated with a decrease in antimycobacterial activity, as shown by overwhelming growth of BCG in the liver. Furthermore, in treated mice, these alterations were correlated with a marked decrease in TNF-α mRNA. These findings strongly suggest that TNF-α released by macrophages in the developing granuloma acts in an autocrine or paracrine manner to promote its own production and to contribute to bacterial elimination.

We recently (5) examined the regulatory role of TNF-α in murine TB by administering a recombinant adenovirus encoding a fusion protein consisting of the human 55-kDa TNF-α receptor extracellular domain and the mouse IgG heavy-chain domain (AdTNF-R). Mice that had been pretreated with AdTNF-R and subsequently infected with *M. tuberculosis* H37Ra possessed elevated bacterial burdens in all tissues examined. AdTNF-R pretreatment of mice with acute *M. tuberculosis* H37Rv infection resulted in extreme morbidity, and exorbitant bacterial growth was confirmed by CFU determination. Clearly, blocking TNF-α function in mice infected with *M. tuberculosis* exacerbates infection. Even more intriguing, however, are the effects of AdTNF-R treatment on mice with chronic *M. tuberculosis* H37Rv infection. Inhibition of TNF-α activity in chronically infected mice with numerous, well-formed, preexisting stable granulomas results in an intense and rapid exacerbation of the infection with augmented bacterial multiplication and development of tuberculous bronchopneumonia.

Finally, the marked ability of the virally expressed TNF-α receptor to exacerbate acute and chronic *M. tuberculosis* infections in mice supports the hypothesis suggested by Kindler et al. (60) that autocrine or paracrine release of TNF-α by granuloma macrophages perpetuates the synthesis and release of TNF-α, thus maintaining the granuloma and inhibiting bacterial growth. In our study, host control of *M. tuberculosis* growth is lost by even a temporary interruption of the TNF-α autoamplification process.

EFFECTS OF TUBERCLE BACILLI AND LEPROSY BACILLI ON MACROPHAGE FUNCTION

The ability of mycobacterial constituents to regulate TNF-α production by macrophages is discussed in detail below. There is also considerable evidence for regulation of macrophage function by the bacilli themselves.

Rook points out that the intracellular accumulation of relatively few tubercle bacilli results in the rounding up and death of monocytes or macrophages (87). This is especially true of virulent strains of *M. tuberculosis* such as H37Rv (38). *M. leprae,* in contrast, is relatively nontoxic for the macrophage; hundreds of viable organisms will exist in lepromatous macrophages with little or no effect on certain macrophage functions. Lepromatous macrophages from the footpads of *M. leprae*-infected *nu/nu* mice are engorged with leprosy bacilli yet retain their capacity to adhere and phagocytose via the Fc receptor (104, 105). These cells, however, are completely refractory to the ability of IFN-γ to enhance macrophage afferent and efferent effector functions.

Unresponsiveness to IFN-γ is evident in macrophages infected in vitro with viable *M. leprae* by the failure of IFN-γ (and LPS) to induce enhanced efferent (oxidative burst, killing of *T. gondii,* and cytotoxicity for tumor target cells) and afferent (enhanced expression of surface Ia antigen) effector function in *M. leprae*-infected macrophages (104–106). A high intracellular burden of *M. leprae* is required (>50:1), and only viable bacilli induce the defect. The development of defective activation

was closely correlated with an elevated prostaglandin E_2 (PGE$_2$) production by macrophages infected with live *M. leprae* that peaked at 48 to 72 h after infection. Others have shown that elevated levels of PGE$_2$ can dramatically downregulate a number of T-cell and macrophage functions, including T-cell proliferation, MHC class II expression, and macrophage cytotoxicity for tumor cells. In our model, the refractory response to IFN-γ was reversed by indomethacin.

TNF-α PRODUCTION IN RESPONSE TO BACILLI

Infection of macrophages in vitro with *M. tuberculosis* or *M. leprae* stimulates the release of TNF-α (42, 90). Falcone et al. (38) recently studied the abilities of various strains of virulent and avirulent mycobacteria to induce secretion of TNF-α in mouse macrophages; they found an inverse correlation related to the virulence of the inducing bacilli (H37Rv < BCG < H37Ra < *M. smegmatis*).

ROLE OF MYCOBACTERIAL CONSTITUENTS

Accumulating data suggest that *M. tuberculosis* and *M. leprae* and some of their constituents may be able to subvert macrophage function, play a role in the downregulation of macrophage effector function, and perhaps serve as virulence factors in the pathogenesis of TB and leprosy. Some of this evidence also applies to *M. avium-M. intracellulare* infections, although the role of putative virulence factors in this case appears to be more complicated, as these organisms are rarely pathogenic for the immunocompetent host. No ordered roster of macrophage antimicrobial functions is needed to establish mycobacterial infection in the host, but the following is a partial list of events that if overridden, blocked, or altered by mycobacterial constituents could tip the balance in favor of the pathogen:

- uptake
- phagosome-lysosome fusion
- phagosome acidification
- respiratory burst and ROI

- L-arginine production of RNI
- cytokine (especially TNF-α) secretion
- activation to an enhanced microbicidal state

Following uptake by the macrophage, mycobacterial pathogens are sequestered within phagosomes, where their subsequent survival is dependent on their abilities to evade killing and degradation by the host cell's digestive enzymes lying packaged in nearby lysosomes. Depending on the receptors employed by the macrophage to phagocytose the bacilli, subsequent events may favor intracellular survival. As discussed below, both *M. tuberculosis* and *M. leprae* are able to utilize novel receptor–ligand interactions to enter macrophages via complement or mannose receptors. If the macrophage has been activated, its lysosomal armamentarium and ROI and RNI pathways are enhanced. The fusion of lysosomes with phagosomes signals bad news for the organism, especially when this fusion occurs in conjunction with acidification of the phagosome, a situation that optimizes lysosomal enzyme activity. The mere ability of *M. tuberculosis* to produce ammonia would stymie acidification of the phagosome by raising intraphagosomal pH (44). The ROI and RNI antimicrobial systems afford additional barriers that a potential mycobacterial pathogen had better be ready to cope with if it is to survive and flourish.

The remainder of this chapter is concerned with the regulatory effects of mycobacterial constituents on macrophage function. By definition, mycobacterial virulence factors functioning at the level of the macrophage induce changes that foster the intracellular survival of the pathogen. Either the innate antimicrobial capacity of the macrophage is compromised or the role of the afferent or efferent participation of the infected macrophage in the CMI network is altered.

THE SULFATIDES

M. tuberculosis produces a family of five closely related sulfatides, the principal member of which is SL-I (12). A major component of the mycobacterial outer layer of lipids, sulfatides

are composed of the disaccharide trehalose esterified with one sulfate group and four fatty acids, three of which are long-chain methyl-branched fatty acids (46). Sulfatides were originally thought to be mycobacterial constituents that blocked the fusion of secondary lysosomes with the phagosome containing the mycobacteria (47), but Goren et al. subsequently concluded that the lysosomotropic nature of the sulfatides may alter this hypothesis (49). Sulfatide production by cultured *M. tuberculosis* was further linked by Goren et al. (48) to the rank order of virulence of various *M. tuberculosis* isolates in the guinea pig. Attenuated strains were virtually devoid of sulfatides, while large quantities of sulfatides were produced by the most virulent strains. Additional evidence indicated that sulfatide components of *M. tuberculosis* block the enhanced functions of IFN-γ- and LPS-primed human monocytes, including superoxide production and phagocytosis (84). More recently (16), these findings were extended to show that sulfatides block priming not only from IFN-γ and LPS but also from IL-1β, possibly by a mechanism that downregulates protein kinase C activity. Moreover, in the presence of LPS, sulfatides also markedly enhance monocyte secretion of IL-1β and TNF-α. The ability of these outer surface membrane components of *M. tuberculosis* to constrain macrophage bactericidal activity while augmenting granuloma formation through enhancement of cytokine production clearly suggests that the influence of mycobacterial constituents on macrophages may play a key role in the pathogenesis of TB.

CORD FACTOR

Cord factor has probably received more attention as a putative virulence factor than any other *M. tuberculosis* constituent, and the facts remain in doubt. The identification of a toxic surface factor on *M. tuberculosis* as cord factor by Bloch in 1950 (13) appeared to explain the peculiar serpentine aggregates or ropelike growth that occurred in cultured organisms as described by Middlebrook et al. in 1947 (71) and Robert Koch 63 years earlier (63). Subsequently, cord factor was identified as trehalose 6,6'-dimycolate (TDM) (81). Although the removal of TDM disrupted cord formation in *M. tuberculosis* (13), only recently did a report actually provide experimental evidence that TDM-coated beads aggregate to form organized structures that resemble the cords of cultured *M. tuberculosis* (10).

Bloch (13) also showed that *M. tuberculosis* H37Rv possesses high concentrations of cord factor, whereas the avirulent H37Ra strain has little. Removal of cord factor from *M. tuberculosis* decreases the capacity of this organism to cause progressive disease in mice. In *M. tuberculosis*-infected mice and guinea pigs, injections of TDM mixed with oil are toxic and enhance both acute and chronic infections (14). More recently, Silva et al. (108) removed TDM from BCG, reducing its ability to persist in the lungs; restoring TDM to this organism revived its ability to persist in vivo.

However, the evidence that TDM is a virulence factor is more complicated. The amount of TDM produced does not necessarily correlate with virulence (45); large amounts are found on some mycobacteria considered nonvirulent. Youmans (119) cautioned about the distinction of TDM as a virulence factor because of the need for admixture in mineral oil to obtain toxic manifestations. Intraperitoneal injection of mice with small amounts of TDM and oil results in cachexia, wasting, and, ultimately, death (110). High concentrations of TNF-α are found in plasma from the injected animals. In vitro, peritoneal macrophages treated with TDM in aqueous solution produce copious amounts of TNF-α (109).

PGL

Of the cell wall-associated constituents of the leprosy bacillus, phenolic glycolipid I (PGL-I) is the most notable. In a series of elegant studies, Brennan and his coworkers described the biochemically unique and immunologically specific structure of *M. leprae* PGL-I (55). The general structure of PGL-I consists of a trisaccharide moiety composed of a 3,6-dimethyl-β-D-glucose (1→4) 2,3-dimethyl-α-L-rhamnose (1→2) 3-methyl-α-L-rhamnose linked to a phthiocerol core through a phenolic group. The terminal disaccharide represents the major epitope of the molecule, and its characterization served as the basis for the development of synthetic neoglycolipids as serologic reagents.

PGL-I appears to be abundant in the outer surface of the bacillus and may form the electron transparent zone that surrounds *M. leprae* in transmission electron micrographs of infected tissues and macrophages. PGL-I appears to function as more than a protective "capsule" that serves as an interface between *M. leprae* and its host cell, the macrophage (32, 102). *M. leprae* is only a weak stimulus of the oxidative burst by macrophages (53, 67), possibly because of a specific downregulation of superoxide generation by a leprosy bacillus component, PGL-I (80, 113). As shown by Neill and Klebanoff (80), PGL-I itself may be an efficient scavenger of superoxide. These same workers contrived a clever model for coating *Staphylococcus aureus* with PGL-I and showed that it provides capable protection against the antimicrobial function of human macrophages. Chan et al. (19) employed sophisticated electron spin resonance techniques to confirm that PGL-I is a highly effective scavenger of hydroxyl radicals and superoxide anions. As *M. leprae* also possesses a superoxide dismutase (80), it appears that in an intracellular location, the leprosy bacillus is well equipped to deal with the macrophage's generation of ROI.

Surface PGL-I may also facilitate uptake of *M. leprae* by macrophages. The importance of heat-labile human serum components to ensuring maximum adherence and phagocytosis of *M. leprae* has been reported (92, 94, 95): the phagocytosis of *M. leprae* is mediated by complement receptors CR1 and CR3 on the surfaces of monocytes and by CR1, CR3, and CR4 on the surfaces of the more mature monocyte-derived macrophages. Phagocytosis is facilitated by the interaction of these receptors with complement component C3 fixed to the surface PGL-I antigen. The activation of C3b in nonimmune serum by PGL-I occurs by the classic complement pathway and is trig-

gered by low amounts of natural or cross-re-
acting IgG or IgM antibody in the serum. No
role for the mannose receptor in uptake of *M.
leprae* via a mannose-rich constituent has been
shown.

MYCOBACTERIAL LPS

Lipoarabinomannan (LAM) is the highly antigenic member of the family of major arabi-
nose- and mannose-containing phosphorylated
LPS from the cell walls of *M. tuberculosis, M.
leprae,* and other mycobacteria (54). Initially
identified by a diffuse band on a sodium dodecyl sulfate-polyacrylamide gel at approximately 35 kDa, LAM was thought to be a glycoprotein. Studies by Hunter and Brennan (54)
clarified the polysaccharide nature of LAM,
and additional studies from Brennan's group
have continued to elucidate the structure of
LAM, its heterogeneity, and its location in the
mycobacterial cell wall (70). A unique phosphatidylinositol lipid anchor at the reducing
end of LAM suggested a relationship with two
other mycobacterial LPS, lipomannan and
phosphatidylinositol mannoside (15). Chatterjee et al. (23–25) have shown lipomannan to
be distinct from phosphatidylinositol mannan
by virtue of its branched mannan core. Lipomannan and LAM may share a similar
phosphatidylinositol mannan core, but LAM
appears to possess an additional immunodominant arabinan extending from the mannan
core of lipomannan. Neither lipomannan nor
phosphatidylinositol mannan seem to be as biologically active as LAM. As shown by im-
munogold labeling (54), *M. tuberculosis* LAM is
located on the outer surface of the organism,
an ideal location from which to regulate macrophage function and contribute directly to intracellular survival of mycobacterial pathogens.

LAM is a potent downregulator of several
functions linked to host CMI. LAM inhibits
antigen responsiveness of human lymphocytes
(58) and antigen–induced proliferation of human CD4$^+$ T-cell clones (74). Pretreatment of
mouse macrophages with LAM from either *M.
leprae* or *M. tuberculosis* abrogates the ability of
IFN-γ to enhance ROI production (superox-

ide anion), microbicidal activity (*T. gondii*), and
cytotoxicity for tumor target cells (101–103).
Intact LAM is necessary to inhibit IFN-γ-me-
diated activation, as this capacity is lost when
the acyl side chains are removed by mild alkaline hydrolysis. Pretreatment with LAM is
essential; it results in the uptake of LAM-filled
cytoplasmic vesicles, as demonstrated with flu-
orescein-labeled anti-LAM monoclonal antibodies. LAM is also seen in cytoplasmic vesicles
of *M. leprae*-infected macrophages. The reports
of LAM inhibition of T-cell antigen recognition (58, 74) suggest that LAM inhibits antigen
presentation by macrophages and have been
confirmed by demonstrating that LAM treatment blocks IFN-γ–induced upregulation of Ia
antigen on mouse macrophages (101, 103).
INF-γ activation of human macrophages is also
blocked by LAM, and constitutively expressed
MHC class II antigen expression is markedly
downregulated by LAM treatment. LAM-
treated macrophages remain fully refractory to
IFN-γ activation for at least 5 days, and LAM-
induced unresponsiveness to IFN-γ is not due
to altered IFN-γ macrophage receptor binding
or concentration.

Clearly, LAM is able to interrupt transduction of the IFN-γ signal in macrophages. More
recent studies of the mechanisms of LAM effects on macrophage function by Chan and his
coworkers (18, 20) show that LAM acts at several levels to disable the oxidative antimicrobial
system by acting as a scavenger of ROI and
inhibiting protein kinase C activity. In human
macrophage cell lines, IFN-γ–inducible gene
transcription (γ.1 and HLA-DRβ genes) is
blocked by LAM treatment.

The chemical structure of LAM has been
intensively studied and suggests a structure-
function relationship that may underlie successful intracellular survival of certain mycobacteria. Although the phosphatidylinositol
mannan core of LAM appears to be conserved
in LAM preparations from different mycobacteria, Chatterjee and her colleagues have identified distinct differences in the nonreducing
end of the molecule (23). In LAM from the
virulent Erdman strain of *M. tuberculosis,* the

terminal linear and branched arabinose chains are "capped" with mannose oligosaccharides (ManLAM) (24). In contrast, LAM from an avirulent, rapidly growing strain of mycobacteria lacks the mannose caps on its arabinose side chains (AraLAM). As these termini are the portions of the LAM molecule that are presented on the surface of the organism, these structural differences can be linked to characteristics that may represent virulence factors expressed at the level of the host macrophage.

LAM may play a key role in receptor-mediated phagocytosis of mycobacteria. Schlesinger's group has recently described the importance of the mannose receptor (in addition to complement receptors) on human-monocyte-derived macrophages in mediating phagocytosis of two virulent strains of *M. tuberculosis* (Erdman and H37Rv) but not the avirulent H37Ra strain (91). These studies were extended in a recent report of a novel pathway in phagocytosis of *M. tuberculosis* in which the mannose-capped termini of ManLAM serve as the ligand for the mannose receptor on the macrophage (93). Polystyrene microspheres coated with ManLAM were selectively adherent to monocyte-derived macrophages. Confirmation of this novel receptor-ligand interaction was made by abrogating the effect with exomannosidase treatment of ManLAM, blocking ManLAM with antibody, and down-regulating the mannose receptor of monocyte-derived macrophages by capping the receptors on surfaces coated with mannose.

There is additional evidence that the nature of LAM on the mycobacterial surface may correlate with the relative virulence of the organism, especially as defined at the level of the macrophage. In a series of recent studies (3, 6, 25, 42), AraLAM from a nonvirulent strain of mycobacteria was shown to induce TNF-α production by murine and human macrophages. However, ManLAM from a virulent strain, *M. tuberculosis* Erdman, exhibited a decreased capability to stimulate TNF-α production by murine peritoneal macrophages (3, 6, 25). It was proposed (28, 86) that the presence of mannan residues in LAM from the virulent

strain may influence the induction of early gene responses, which in turn affects the magnitude of TNF-α production and the subsequent survival of *M. tuberculosis*. In an *M. avium* model, the abilities of various clinical isolates to replicate inside murine macrophages correlate not only with colony morphology but also with capacity to induce TNF-α production (43).

Our own laboratory (3) has confirmed that relatively low levels of TNF-α are produced upon stimulation of macrophages with ManLAM from virulent *M. tuberculosis* Erdman. In addition, LAM purified from *M. leprae,* another virulent mycobacterium, induces negligible TNF-α production. LAM from *M. leprae* also appears to contain mannose capping of the arabinan side chains, although not to the same extent as $ManLAM_{Erdman}$ (22). In addition, $LAM_{M. leprae}$ lacks the alkali-labile inositol phosphate units found in the AraLAM preparations (55). Whether this lack is of biological significance in the present context is unclear, since it is not known whether LAM_{Erdman} has similar features.

The finding that $LAM_{M. leprae}$ induces only very low levels of TNF-α provides ever more compelling evidence for the importance of the regulatory influence of this cell wall component in the pathogenesis of mycobacterial disease. It is important that Moreno et al. (75) utilized LAM isolated from the virulent H37Rv strain of *M. tuberculosis* to induce copious amounts of TNF-α in macrophages. In terms of virulence, H37Rv resembles more the Erdman strain of *M. tuberculosis* than an avirulent strain of mycobacteria. To our knowledge, structural studies of LAM_{H37Rv} have not been done, so it is not known if *M. tuberculosis* H37Rv LAM is ManLAM. If it is, TNF-α induction would not fit the pattern set for virulent *M. tuberculosis* Erdman. Of course, if LAM_{H37Rv} is AraLAM, the hypothesis is not served either. Of additional interest in this regard is the recent report by Prinzis et al. (85) that LAM_{BCG} is of the ManLAM type, showing no major structural differences from LAM_{Erdman}. As BCG is clearly attenuated in terms of virulence, the elementary terms of the ManLAM–TNF-α hypothesis are also not supported. However, a monoclonal antibody against LAM$_{Erdman}$ (CS-40) fails to react with LAM$_{BCG}$, suggesting that structural differences exist in spite of ManLAM similarities.

The facts that gram–negative LPS is such a potent inducer of TNF-α and that it is often a contaminant of tissue culture and microbiological systems have raised the question of whether the LAM molecule is solely responsible for its reported biological effects (73). In our own studies (3), the LAM employed was used only after it had been purified on polymyxin B Detoxigel columns (23). LPS was always included as a positive control, and all LPS and LAM preparations were tested by coculture with polymyxin B. The effects of LPS were almost totally abrogated by polymyxin B treatment, while the LAM-induced effects remained largely unchanged by polymyxin B treatment. Thus, we feel that any influence of LPS in our experimental system is highly remote.

Does the failure of virulent *M. tuberculosis* Erdman ManLAM to induce production of extracellular TNF-α by macrophages constitute a virulence factor? Data obtained with whole viable bacilli appear to indicate that it does (38). Virulent *M. tuberculosis* induced less TNF-α than avirulent strains or avirulent mycobacteria. Synthesis of TNF-α is controlled at the levels of both transcription and translation. In our own studies, macrophage RNA was analyzed by reverse transcriptase PCR procedures for the presence of TNF-α message to determine whether the discrepancies in the levels of TNF-α protein induced by the various LAM preparations correlated with differences in the expression of TNF-α mRNA. There was a direct correlation between the levels of TNF-α mRNA and the amount of TNF-α protein. Unlike our findings with the bioassay for TNF-α activity in culture supernatant medium, where the differences seemed absolute, the differences in TNF-α-specific mRNA induced by LAM$_{Erdman}$ and LAM$_{M. leprae}$ in comparison to AraLAM were relative. This suggests that the inabilities of LAM$_{Erdman}$ and LAM$_{M. leprae}$

to induce high concentrations of TNF-α protein are probably regulated both pre- and post-transcriptionally. It is precisely this difference in release of TNF-α that could underlie the ability of LAM to function as a virulence factor. TNF-α production in the microenvironment is a key event in granuloma formation, as determined by Kindler et al. (60), and granuloma maintenance, as shown by Adams et al. (5). Infection with virulent strains of *M. tuberculosis* (e.g., Erdman) could initiate progressive disease largely as a result of the failure of these strains to induce TNF-α production early in the course of infection, before the onset of specific immunity (25). In contrast, in advanced TB, TNF-α is clearly not restricted to the microenvironment and is linked to the pathologic manifestations of disease rather than to protection (87). LAM$_{M. leprae}$ fits the same pattern as LAM$_{Erdman}$ with regard to TNF-α production, but the pathogenicity of progressive LL (i.e., the slow growth of the organism and the organ systems involved) is quite distinct from the characteristics of TB.

Thus, TNF-α is an especially pleiotropic cytokine that may play a major role in inflammatory events that culminate in either pathogenesis or resistance in mycobacterial infections (39, 60, 87, 90). Production of TNF-α, however, is only one of a myriad of factors elicited by macrophages, many of which have been demonstrated in TB and across the spectrum of leprosy. As a secretory cell, the macrophage is also the source of such potent immunoregulatory compounds as arachidonic acid cyclooxygenase products, numerous cytokines, and toxic microbicidal and tumoricidal free radicals. Therefore, in addition to examining the production of TNF-α, we have examined the influence of the various LAM preparations on a number of other macrophage effector mechanisms.

We have previously shown that pretreatment of macrophages with AraLAM and LAM$_{M. leprae}$ blocked subsequent IFN-γ + LPS-induced macrophage activation (103). This inhibition was sustained for up to 5 days after LAM treatment (101). Our recent work shows that LAM_{Erdman} also blocks macrophage activation. Since PGE_2 and perhaps other cyclooxygenase metabolites of arachidonic acid metabolism in the macrophage membrane are potent downregulators of inflammation, we analyzed the relationship between prostanoids and the various LAM preparations in an effort to determine the mechanism whereby macrophage activation may be blocked. Indomethacin, an inhibitor of prostaglandin synthesis, greatly augmented TNF-α production by macrophages stimulated with AraLAM but failed to augment TNF-α induction by LAM_{Erdman} (3). In fact, in contrast to AraLAM, LAM_{Erdman} and $LAM_{M. leprae}$ were very poor simulators of PGE_2. These two lines of evidence suggest that prostanoids do not play key roles in the differential abilities of mycobacterial LAM from different sources to induce TNF-α. The inability of LAM_{Erdman} and $LAM_{M. leprae}$ to induce PGE_2 may be worthy of further investigation.

Both AraLAM and LAM_{Erdman} are potent inducers of RNI in IFN-γ-primed macrophages (3). This pathway has previously been shown to be important in macrophage-mediated antimicrobial effects on *T. gondii* (4) as well as on *M. leprae* (1) and *M. tuberculosis* (20, 29). Therefore, we questioned whether the LAM-induced RNI production in IFN-γ-primed macrophages correlated with their antimicrobial function. An assay for the killing of *M. leprae* was inappropriate, since *M. leprae* itself can serve as a second signal and trigger microbicidal events in IFN-γ-primed macrophages (2). Killing of *T. gondii* by IFN-γ-primed macrophages, however, requires a second signal such as LPS or TNF-α to trigger the microbicidal event in vitro (100). Sibley et al. (100) further showed that polymyxin B and not anti-TNF-α neutralizing antibody blocks LPS as a second signal, whereas anti-TNF-α antibody and not polymyxin B blocks TNF-α as a second signal. In our study, LAM from both the virulent and avirulent strains of *M. tuberculosis* could trigger IFN-γ-primed macrophages for inhibition of *T. gondii,* an effect that was not reversed in the presence of polymyxin B (3). The triggering capacity of LAM_{Erdman} was completely abro-

gated, however, by anti-TNF-α. This fact, supported by the demonstration of TNF-α mRNA, indicates that TNF-α is doubtlessly involved in the induction of macrophage activation by LAM_{Erdman}.

It is especially intriguing that both AraLAM and ManLAM can trigger antimicrobial activity and elevated levels of RNI, which are both manifestations of macrophage activation, but that ManLAM does so without inducing elevated levels of TNF-α in the supernatant medium. This could represent a novel mechanism by which LAM_{Erdman} can trigger primed macrophages. Alternatively, microbicidal activity and tumor cell cytotoxicity both require intimate contact with the activated macrophage. LAM_{Erdman} could induce a level of TNF-α that in the intracellular microenvironment of the parasitophorous vacuole or at the macrophage-tumor cell junction may be more than sufficient to initiate the series of events that culminate in microbicidal and cytocidal activities.

SUMMARY

An easily identifiable virulence factor for *M. tuberculosis* or *M. leprae* (or *M. avium-M. intracellulare* infection) remains elusive. Clearly, the interaction between the host and the mycobacterial pathogen is multifaceted. This discussion of the effects of mycobacterial constituents on the host macrophage confirms that there are multiple mycobacterial tactics for ensuring intracellular survival under the frankly hostile environment of the infected macrophage.

REFERENCES

1. **Adams, L. B.** Unpublished results.
1a. **Adams, L. B., S. G. Franzblau, Z. Vavrin, J. B. Hibbs, Jr., and J. L. Krahenbuhl.** 1991. L-arginine-dependent macrophage effector functions inhibit metabolic activity of *Mycobacterium leprae. J. Immunol.* **147:**1642–1646.
2. **Adams, L. B., Y. Fukutomi, and J. L. Krahenbuhl.** 1991. Mycobacterial cell wall constituents as second signals for activating macrophages, abstr. 135. *J. Leukocyte Biol.* **2**(Suppl.)**:**52.
3. **Adams, L. B., Y. Fukutomi, and J. L. Krahenbuhl.** 1993. Regulation of murine macrophage effector functions by lipoarabinomannan preparations from mycobacterial strains of varying virulence. *Infect. Immun.* **61:**4173–4181.

4. **Adams, L. B., J. B. Hibbs, Jr., R. R. Taintor, and J. L. Krahenbuhl.** 1990. Microbiostatic effect of murine-activated macrophages for *Toxoplasma gondii*. Role for synthesis of inorganic nitrogen oxides from L-arginine. *J. Immunol.* **144:** 2725–2729.

5. **Adams, L. B., C. M. Mason, J. K. Kolls, D. Scollard, J. L. Krahenbuhl, and S. Nelson.** 1995. Exacerbation of acute and chronic murine tuberculosis by administration of a TNF receptor-bearing adenovirus. *J. Infect. Dis.* **171(2):**400–405.

6. **Barnes, P. F., D. Chatterjee, J. S. Abrams, S. Lu, E. Wang, M. Yamamura, P. J. Brennan, and R. L. Modlin.** 1992. Cytokine production induced by *Mycobacterium tuberculosis* lipoarabinomannan. Relationship to chemical structure. *J. Immunol.* **149:**541–547.

7. **Barnes, P. F., D. Chatterjee, P. J. Brennan, T. H. Rea, and R. L. Modlin.** 1992. Tumor necrosis factor production in patients with leprosy. *Infect. Immun.* **60:**1441–1446.

8. **Barnes, P. F., S. Lu, J. S. Abrams, E. Wang, M. Yamamura, and R. Modlin.** 1993. Cytokine production at the site of disease in human tuberculosis. *Infect. Immun.* **61:**3482–3489.

9. **Barnes, P. F. G., R. L. Modlin, D. D. F. Bikle, and J. S. Adams.** 1989. Transpleural gradient of 1,25-dihydroxyvitamin D in tuberculosis pleuritis. *J. Clin. Invest.* **83:**1527–1532.

10. **Behling, C., B. Bennett, K. Takayama, and R. Hunter.** 1993. Development of a trehalose 6,6′-dimycolate model which explains cord formation by *Mycobacterium tuberculosis*. *Infect. Immun.* **61:**2296–2303.

11. **Bermudez, L. E. M., and L. S. Young.** 1988. Tumor necrosis factor, alone or in combination with IL-2, but not IFN-γ, is associated with macrophage killing of *Mycobacterium avium* complex. *J. Immunol.* **140:**3006–3013.

12. **Besra, G. S., and D. Chatterjee.** 1994. Lipids and carbohydrates of *Mycobacterium tuberculosis*, p. 285–306. *In* B. R. Bloom (ed.), *Tuberculosis: Pathogenesis, Protection, and Control*. American Society for Microbiology, Washington, D.C.

13. **Bloch, H.** 1950. Studies on the virulence of tubercle bacilli. Isolation and biological properties of a constituent of virulent organisms. *J. Exp. Med.* **91:**197–217.

14. **Bloch, H., and H. Noll.** 1955. Studies on the virulence of tubercle bacilli. The effect of cord factor on murine tuberculosis. *Br. J. Exp. Pathol.* **36:**8–17.

15. **Brennan, P. J., and C. E. Ballou.** 1967. Biosynthesis of mannophosphoinositides by *Mycobacterium phlei*. *J. Biol. Chem.* **242:**3046–3056.

16. **Brozna, J. P., M. Horan, J. M. Radenacher, K. M. Pabst, and M. J. Pabst.** 1991. Monocyte responses to sulfatide from *Mycobacterium tuberculosis*: inhibition of priming for enhanced release of superoxide associated with increased secretion of interleukin-1 and tumor necrosis factor alpha and altered protein phosphorylation. *Infect. Immun.* **59:** 2542–2548.

17. **Cadranel, J., A. J. Hance, B. Milleron, F. Paillard, G. M. Akoun, and M. Garabedian.** 1988. Vitamin D metabolism in tuberculosis. Production of 1,25(OH)2D3 by cells recovered by bronchoalveolar lavage and the role of this metabolite in calcium homeostasis. *Am. Rev. Respir. Dis.* **138:**984–989.

18. **Chan, J., X. Fan, S. W. Hunter, P. J. Brennan, and B. R. Bloom.** 1991. Lipoarabinomannan, a possible virulence factor involved in persistence of *Mycobacterium tuberculosis* within macrophages. *Infect. Immun.* **59:**1775–1761.

19. **Chan, J., T. Fujiwara, P. Brennan, M. McNeil, S. J. Turco, J.-C. Sibille, M. Snapper, P. Aisen, and B. R. Bloom.** 1989. Microbial glycolipids: possible virulence factors that scavenge oxygen radicals. *Proc. Natl. Acad. Sci. USA* **86:**2453–2457.

20. **Chan, J., Y. Xing, R. S. Magliozzo, and B. R. Bloom.** 1992. Killing of virulent *Mycobacterium tuberculosis* by reactive nitrogen intermediates produced by activated murine macrophages. *J. Exp. Med.* **175:**1111–1122.

21. **Chatterall, J. R., S. D. Sharma, and J. S. Remington.** 1986. Oxygen-independent killing by alveolar macrophages. *J. Exp. Med.* **163:**1113–1131.

22. **Chatterjee, D.** (Colorado State University). 1993. Personal communication.

23. **Chatterjee, D., S. W. Hunter, M. McNeil, and P. Brennan.** 1992. Lipoarabinomannan. Multiglycosylated form of the mycobacterial mannosylphosphatidylinositols. *J. Biol. Chem.* **267:** 6228–6233.

24. **Chatterjee, D., K. Lowell, B. Rivoire, M. R. McNeil, and P. J. Brennan.** 1992. Lipoarabinomannan of *Mycobacterium tuberculosis*. Capping with mannose residues in some strains. *J. Biol. Chem.* **267:**6234–6239.

25. **Chatterjee, D., A. D. Roberts, K. Lowell, P. J. Brennan, and I. M. Orme.** 1992. Structural basis of capacity of lipoarabinomannan to induce secretion of tumor necrosis factor. *Infect. Immun.* **60:**1249–1253.

26. **Chehl, S., C. K. Job, and R. C. Hastings.** 1985. Transmission of leprosy in nude mice. *Am. J. Trop. Med. Hyg.* **34:**1161–1166.

27. **Dannenberg, A.** 1994. Pathogenesis and immunology: basic facts, p. 17–41. *In* D. Schlossberg (ed.), *Tuberculosis*, 3rd ed. Spinger-Verlag, New York.

28. **Denis, M.** 1991. Tumor necrosis factor and granulocyte-macrophage colony stimulating factor

stimulate human macrophages to inhibit the growth of virulent *Mycobacterium avium* and to kill avirulent *M. avium:* killing effector mechanism depends on the generation of reactive nitrogen intermediates. *J. Leukocyte Biol.* **49:**380–387.

29. **Denis, M.** 1991. Interferon gamma-treated murine macrophages inhibit growth of tubercle bacilli via the generation of reactive nitrogen intermediates. *Cell. Immunol.* **132:**150–157.

30. **Denis, M., and E. O. Gregg.** 1991. Cytokine modulation of *Mycobacterium avium* in murine macrophages. Reversal of unresponsiveness to interferon gamma with indomethacin and interleukin-1. *J. Leukocyte Biol.* **65:**449–454.

31. **Douvas, G. S., D. L. Looker, A. E. Vatter, and A. F. Crowle.** 1986. Gamma interferon activates human macrophages to become tumoricidal and leishmanicidal but enhances replication of macrophage-associated mycobacteria. *Infect. Immun.* **50:**1–8.

32. **Draper, P., and R. F. W. Rees.** 1970. Electron transparent zone of mycobacteria may be a defense mechanism. *Nature* (London) **228:**860–861.

33. **Drapier, J.-C., and J. B. Hibbs, Jr.** 1988. Differentiation of murine macrophages to express nonspecific cytotoxicity for tumor cells results in L-arginine-dependent inhibition of mitochondrial iron-sulfur enzymes in the macrophage effector cells. *J. Immunol.* **140:**2829–2838.

34. **Drapier, J.-C., J. Wietzerbin, and J. B. Hibbs, Jr.** 1988. Interferon-γ and tumor necrosis factor induce the L-arginine-dependent cytotoxic effector mechanism in murine macrophages. *Eur. J. Immunol.* **18:**1587–1592.

35. **Drutz, D. J., S. M. O'Neill, and L. Levy.** 1974. Viability of blood-borne *Mycobacterium leprae. J. Infect. Dis.* **130:**288–292.

36. **Dumarey, C. H., V. Labrousse, N. Rastogi, B. B. Vargaftig, and M. Bachelet.** 1994. Selective *Mycobacterium avium*-induced production of nitric oxide by human monocyte-derived macrophages. *J. Leukocyte Biol.* **56:**36–40.

37. **Esparza, I., D. Mannel, A. Ruppel, W. Falk, and P. Krammer.** 1987. Interferon-γ and lymphotoxin or tumor necrosis factor act synergistically to induce macrophage killing of tumor cells and schistosomula of *Schistosoma mansoni. J. Exp. Med.* **166:**589–594.

38. **Falcone, V., E. B. Bassey, A. Toniolo, P. G. Conaldi, and F. M. Collins.** 1994. Differential release of tumor necrosis factor-α from murine peritoneal macrophages stimulated with virulent and avirulent species of mycobacteria. *FEMS Immunol. Med. Microbiol.* **8:**225–232.

39. **Filley, E., and G. A. W. Rook.** 1991. Effect of mycobacteria on sensitivity to the cytotoxic effects of tumor necrosis factor. *Infect. Immun.* **59:**2567–2572.

40. **Flesch, I., and S. Kaufman.** 1987. Mycobacterial growth inhibition by interferon gamma-activated bone-marrow macrophage and differential susceptibility among strains of *Mycobacterium tuberculosis. J. Immunol.* **138:**4400–4413.

41. **Flesch, I. E., J. H. Hess, I. P. Oswald, and S. H. E. Kaufman.** 1994. Growth inhibition of *Mycobacterium bovis* by IFN-γ stimulated macrophages: regulation by endogenous tumor necrosis factor-α and by IL-10. *Int. Immunol.* **6:**693–700.

42. **Fukutomi, Y., L. B. Adams, and J. L. Krahenbuhl.** 1991. Tumor necrosis factor in experimental leprosy. *Int. J. Lepr.* **59:**711–712.

43. **Furney, S. K., P. S. Skinner, A. D. Roberts, R. Appelberg, and I. M. Orme.** 1992. Capacity of *Mycobacterium avium* isolates to grow well or poorly in murine macrophages resides in their ability to induce secretion of tumor necrosis factor. *Infect. Immun.* **60:**4410–4413.

44. **Gordon, A. H., P. D'Arcy-Hart, and M. R. Young.** 1980. Ammonia inhibits phagosome-lysosome fusion in macrophages. *Nature* (London) **286:**79–81.

45. **Goren, M. B., and P. J. Brennan.** 1979. Mycobacterial lipids: chemistry and biologic activities, p. 63–193. *In* G. P. Youmans (ed.), *Tuberculosis,* The W. B. Saunders Co., Philadelphia.

46. **Goren, M. B., O. Brokl, P. Roller, H. M. Fales, and B. C. Das.** 1976. Sulfatides of *M. tuberculosis:* the structure of the principle sulfatide (SL-1). *Biochemistry* **15:**2728–2734.

47. **Goren, M. B., P. D'Arcy Hart, M. R. Young, and J. A. Armstrong.** 1976. Prevention of phagosome-lysosome fusion in cultured macrophages by sulfatides of *M. tuberculosis. Proc. Soc. Exp. Biol. Med.* **73:**2510–2513.

48. **Goren, M. B., J. M. Grange, V. R. Aber, B. W. Allen, and D. A. Mitchison.** 1982. Role of lipid content and hydrogen peroxide susceptibility in determining the guinea pig virulence of *M. tuberculosis. Br. J. Exp. Pathol.* **63:**693–700.

49. **Goren, M. B., A. E. Vatter, and J. Fiscus.** 1987. Polyanionic agents as inhibitors of phagosome-lysosome fusion in cultured macrophages: evolution of an alternative interpretation. *J. Leukocyte Biol.* **41:**111–121.

50. **Green, S. J., M. S. Meltzer, J. B. Hibbs, Jr., and C. A. Nacy.** 1990. Activated macrophages destroy intracellular *Leishmania major* amastigotes by an L-arginine-dependent killing mechanism. *J. Immunol.* **144:**278–284.

51. **Hastings, R. C., T. P. Gillis, J. L. Krahenbuhl, and S. G. Franzblau.** 1988. Leprosy. *Clin. Microbiol. Rev.* **1:**330–348.

52. **Hibbs, J. B., C. Westenelder, and R. Taintor.** 1992. Evidence for cytokine-inducible nitric oxide synthesis from L-arginine in patients receiving interleukin-2 therapy. *J. Clin. Invest.* **89:**867–877.

53. **Holzer, H., K. E. Nelson, V. Schauf, B. Crispen, and B. Anderson.** 1986. *Mycobacterium leprae* fails to stimulate phagocytic cell superoxide anion generation. *Infect. Immun.* **51:**514–520.

54. **Hunter, S., and P. Brennan.** 1990. Evidence for the presence of a phosphatidylinositol anchor on the lipoarabinomannan and lipomannan of *M. tuberculosis. J. Biol. Chem.* **265:**9272–9279.

55. **Hunter, S. W., H. Gaylord, and P. J. Brennan.** 1986. Structure and antigenicity of the phosphorylated lipopolysaccharide antigens from the leprosy and tubercle bacilli. *J. Biol. Chem.* **261:**12345–12351.

56. **Job, C. K., S. Chehl, and R. C. Hastings.** 1990. New findings on the mode of entry of *Mycobacterium leprae* in nude mice. *Int. J. Lepr.* **58:**726–729.

57. **Jones, B. E., S. M. Young, D. Antoniskis, P. T. Davidson, F. Kramer, and P. F. Barnes.** 1993. Relationship of the manifestations of tuberculosis to CD4 counts in patients with human immunodeficiency virus infection. *Am. Rev. Respir. Dis.* **148:**1292–1297.

58. **Kaplan, G., R. R. Gandhi, D. E. Weinstein, W. R. Levis, M. E. Patarroyo, P. J. Brennan, and Z. A. Cohn.** 1987. *Mycobacterium leprae* antigen-induced suppression of T cell proliferation *in vitro. J. Immunol.* **138:**3028–3024.

59. **Kettelhut, I. C., W. Fiers, and A. L. Goldberg.** 1987. The toxic effects of tumor necrosis factor *in vivo* and their prevention by cyclooxygenase inhibitors. *Proc. Natl. Acad. Sci. USA* **84:**4273–4277.

60. **Kindler, V., I.-P. Sappino, G. E. Grau, P.-F. Piguet, and P. Vassalli.** 1989. The inducing role of tumor necrosis factor in the development of bactericidal granulomas during BCG infection. *Cell* **56:**731–740.

61. **Klebanoff, S. J.** 1982. Oxygen-dependent cytotoxic mechanisms of phagocytes. *Adv. Host Def. Mech.* **1:**111–162.

62. **Klebanoff, S. J., and C. C. Shepard.** 1984. Toxic effect of the peroxidase-hydrogen peroxide-halide antimicrobial system on *Mycobacterium leprae. Infect. Immun.* **44:**534–536.

63. **Koch, R.** 1884. Die Aetiologie der Tuberkulose. *Mitt. Gesundheitsamt.* **2:**1–88.

64. **Krahenbuhl, J. L., and L. B. Adams.** 1994. The role of macrophages in resistance to the leprosy bacillus, p. 281–302. *In* B. S. Zwilling and T. K. Eisenstein (ed.), *Macrophage-Pathogen Interactions.* Marcel Dekker, Inc., New York,

65. **Krahenbuhl, J. L., J. S. Remington, and R. McLeod.** 1980. Cytotoxic and microbicidal properties of macrophages, p. 1631–1653. *In* R. van Furth (ed.), *Mononuclear Phagocytes—Functional Aspects. Proceedings of the Third Conference on Mononuclear Phagocytes.* Martinus Nijhoff, Amsterdam.

66. **Krahenbuhl, J. L., L. Rosenberg, and J. Remington.** 1973. The effects of anti-thymocyte serum on *in vitro* activation of macrophages to kill *Listeria monocytogenes. J. Immunol.* **11:**992–995.

67. **Launois, P. B., A. Maillere, and J. L. Dieye.** 1989. Human phagocyte oxidative burst activation by BCG, *M. leprae,* and atypical mycobacteria: defective activation by *M. leprae* is not reversed by interferon-gamma. *Cell. Immunol.* **124:**168–175.

68. **Lehmmann, V., B. Benninghoff, and W. Droge.** 1988. Tumor necrosis factor-induced activation of peritoneal macrophages is regulated by prostaglandin E_2 and cAMP. *J. Immunol.* **141:**587–591.

69. **Mackaness, G. B.** 1971. Resistance to intracellular infection. *J. Infect. Dis.* **123:**439–445.

70. **McNeil, M., and P. J. Brennan.** 1991. Structure, function and biogenesis of the cell envelope of mycobacteria in relation to bacterial physiology, pathogenesis and drug resistance; some thoughts and possibilities arising from recent structural information. *Res. Microbiol.* **142:**451–463.

71. **Middlebrook, G., R. J. Dubos, and C. H. Pierce.** 1947. Virulence and morphological characteristics of tubercle bacilli. *J. Exp. Med.* **86:**175–184.

72. **Modlin, R. L., J. Melancon-Kaplan, and S. M. Young.** 1988. Learning from lesions: patterns of tissue inflammation in leprosy. *Proc. Natl. Acad. Sci. USA* **85:**1213–1217.

73. **Molloy, A., G. Gaudernack, W. R. Levis, Z. A. Cohn, and G. Kaplan.** 1990. Suppression of T-cell proliferation by *Mycobacterium leprae* and its products. The role of lipopolysaccharides. *Proc. Natl. Acad. Sci. USA* **87:**973–977.

74. **Moreno, C., A. Mehlert, and J. Lamb.** 1988. The inhibitory effects of mycobacterial lipoarabinomannan and polysaccharides upon polyclonal and monoclonal human T cell proliferation. *Clin. Exp. Immunol.* **74:**206–210.

75. **Moreno, C., J. Taverne, A. Mehlert, C. A. W. Bate, R. J. Realey, A. Meager, G. A. W. Rook, and J. H. L. Playfair.** 1989. Lipoarabinomannan from *Mycobacterium tuberculosis* induces the production of tumor necrosis factor from human and murine macrophages. *Clin. Exp. Immunol.* **76:**240–245.

76. **Murray, H. W., B. Y. Rubin, S. M. Carriero, A. M. Harvis, and E. A. Jaffee.** 1985. Human mononuclear phagocyte antiprotozoal mechanisms: oxygen-dependent vs oxygen-independent activity against intracellular *Toxoplasma gondii. Infect. Immun.* **52:**151–158.

77. **Nathan, C. F.** 1987. Macrophages secretory products. *J. Clin. Invest.* **79:**319–326.

78. **Nathan, C. F., H. W. Murray, M. E. Wiebe, and B. Y. Rubin.** 1983. Identification of inter-

feron gamma as the lymphokine that activates human macrophage oxidative metabolism and antimicrobial activity. *J. Exp. Med.* **158**:670–674.

79. **Nathan, C. F., and R. K. Root.** 1977. Hydrogen peroxide release from mouse peritoneal macrophages. Dependence on sequential activation and triggering. *J. Exp. Med.* **146**:1648–1662.

80. **Neill, M., and S. J. Klebanoff.** 1988. The effect of phenolic glycolipid-1 from *Mycobacterium leprae* on the antimicrobial activity of human macrophages. *J. Exp. Med.* **167**:30–42.

81. **Noll, H., H. Bloch, J. Asselineau, and E. Lederer.** 1956. The chemical structure of the cord factor of *M. tuberculosis. Biochim. Biophys. Acta* **20**:299–318.

82. **Old, L. J.** 1990. Tumor necrosis factor, p. 1–30. *In* B. Bonavida and G. Granger (ed.), *Tumor Necrosis Factor: Structure, Mechanism of Action, Role in Disease and Therapy*. S. Karger, Basel.

83. **Orme, I. M.** 1987. The kinetics of emergence and loss of mediator T lymphocytes acquired in response to infection with *M. tuberculosis. J. Immunol.* **138**:293–298.

84. **Pabst, M. J., J. M. Gross, J. P. Brozna, and M. B. Goren.** 1988. Inhibition of macrophage priming by sulfatides from *M. tuberculosis. J. Immunol.* **140**:634–640.

85. **Prinzis, S., D. Chatterjee, and P. Brennan.** 1993. Structure and antigenicity of lipoarabinomannan from *Mycobacterium bovis* BCG. *J. Gen. Microbiol.* **139**:2649–2658.

86. **Roach, T. I. A., C. H. Barton, D. Chatterjee, and J. M. Blackwell.** 1993. Macrophage activation: lipoarabinomannan from avirulent and virulent strains of *Mycobacterium tuberculosis* differently induces the early genes c-fos, KC JE and tumor necrosis factor-α. *J. Immunol.* **150**:1886–1896.

87. **Rook, G. A. W.** 1988. Role of activated macrophages in the immunopathology of tuberculosis. *Br. Med. Bull.* **44**:611–625.

88. **Rook, G. A. W., and R. A. Attiyah.** 1991. Cytokines and the Koch phenomenon. *Tubercle* **72**:13–20.

89. **Rook, G. A. W., J. Steel, M. Ainsworth, and B. R. Champion.** 1986. Activation of macrophages to inhibit proliferation of *Mycobacterium tuberculosis*: comparison of the effects of recombinant gamma-interferon on human monocytes and murine peritoneal macrophages. *Immunology* **19**:333–338.

90. **Rook, G. A. W., J. Taverne, C. Leveton, and J. Steele.** 1987. The role of gamma-interferon, vitamin D$_3$ metabolites and tumor necrosis factor in the pathogenesis of tuberculosis. *Immunology* **62**:229–234.

91. **Schlesinger, L.** 1993. Macrophage phagocytosis of virulent but not attenuated strains of *M. tuberculosis* is mediated by mannose receptors in addi-

tion to complement receptors. *J. Immunol.* **150**:2920–2930.

92. **Schlesinger, L., and M. Horwitz.** 1990. Phagocytosis of leprosy bacilli is mediated by complement receptors CR1 and CR3 on human monocytes and complement component C3 in serum. *J. Clin. Invest.* **85**:1304–1314.

93. **Schlesinger, L., S. R. Hull, and T. M. Kaufman.** 1994. Binding of the terminal mannosyl units of lipoarabinomannan from a virulent strain of *M. tuberculosis* to human macrophages. *J. Immunol.* **152**:4070–4079.

94. **Schlesinger, L. S., and M. A. Horwitz.** 1991. Phenolic glycolipid-1 of *Mycobacterium leprae* binds complement component C3 in serum and mediates phagocytosis by human monocytes. *J. Exp. Med.* **174**:1031–1038.

95. **Schlesinger, L. S., and M. A. Horwitz.** 1991. Phagocytosis of *Mycobacterium leprae* by human monocyte derived macrophages is mediated by complement receptors CR1 (CD35), CR3 (CD11b/CD18), and CR4 (CD11C/CD18) and interferon gamma activation inhibits complement receptor function and phagocytosis of this bacterium. *J. Immunol.* **147**:1983–1984.

96. **Sen, R., P. K. Sehgal, U. Singh, M. S. Yadav, S. D. Chaudhary, and R. Sika.** 1989. Bacillaemia and bone marrow involvement in leprosy. *Ind. J. Lepr.* **61**:445–452.

97. **Sharp, A. K., and D. K. Banerjee.** 1985. Hydrogen peroxide and superoxide production by peripheral blood monocytes in leprosy. *Clin. Exp. Immunol.* **60**:203–206.

98. **Sharp, A. K., M. J. Colston, and D. K. Banerjee.** 1985. Susceptibility of *Mycobacterium leprae* to the bactericidal activity of mouse peritoneal macrophages and to hydrogen peroxide. *J. Med. Microbiol.* **19**:77–82.

99. **Shiratsuchi, H., J. L. Johnson, and J. J. Ellner.** 1991. Bidirectional effects of cytokines on growth of *Mycobacterium avium* within human monocytes. *J. Immunol.* **146**:3165–3170.

100. **Sibley, L. D., L. B. Adams, Y. Fukutomi, and J. L. Krahenbuhl.** 1991. Tumor necrosis factor-α triggers antitoxoplasmal activity of IFN-γ-primed macrophages. *J. Immunol.* **147**:2340–2345.

101. **Sibley, L. D., L. B. Adams, and J. L. Krahenbuhl.** 1990. Inhibition of interferon-gamma-mediated activation in mouse macrophages treated with lipoarabinomannan. *Clin. Exp. Immunol.* **80**:141–148.

102. **Sibley, L. D., S. G. Franzblau, and J. L. Krahenbuhl.** 1987. Intracellular fate of *Mycobacterium leprae* in normal and activated mouse macrophages. *Infect. Immun.* **55**:680–685.

103. **Sibley, L. D., S. W. Hunter, P. J. Brennan, and J. L. Krahenbuhl.** 1988. Mycobacterial li-

poarabinomannan inhibits gamma interferon-mediated activation of macrophages. *Infect. Immun.* **56**:1232–1236.

104. **Sibley, L. D., and J. L. Krahenbuhl.** 1987. *Mycobacterium leprae*-burdened macrophages are refractory to activation by gamma interferon. *Infect. Immun.* **55**:446–450.

105. **Sibley, L. D., and J. L. Krahenbuhl.** 1988. Defective activation of granuloma macrophages from *Mycobacterium leprae*-infected nude mice. *J. Leukocyte Biol.* **43**:60–66.

106. **Sibley, L. D., and J. L. Krahenbuhl.** 1988. Induction of unresponsiveness to gamma interferon in macrophages infected with *Mycobacterium leprae*. *Infect. Immun.* **56**:1912–1919.

107. **Sibley, L. D., J. L. Krahenbuhl, and E. Weidner.** 1985. Lymphokine activation of J774G8 cells and mouse peritoneal macrophages challenged with *Toxoplasma gondii*. *Infect. Immun.* **49**:760–764.

108. **Silva, C., S. Ekizlerian, and R. Fazioli.** 1985. Role of cord factor in the modulation of infection caused by mycobacteria. *Am. J. Pathol.* **118**:238–247.

109. **Silva, C., and L. Faccioli.** 1988. Tumor necrosis factor (cachectin) mediates induction of cachexia by cord factor from mycobacteria. *Infect. Immun.* **56**:3067–3071.

110. **Silva, C., I. Tincani, S. L. Brandao Fihlo, and L. Facciolo.** 1988. Mouse cachexia induced by trehalose dimycolate. *J. Gen. Microbiol.* **134**:1629–1633.

111. **Stach, J. L., G. Delgado, M. Strobel, J. Millan, and P. H. LaGrange.** 1984. Preliminary evidence of natural resistance to *Mycobacterium bovis* (BCG) in lepromatous leprosy. *Int. J. Lepr.* **52**:140–146.

112. **Toba, H., J. Crawford, and J. Ellner.** 1989. Pathogenicity of *Mycobacterium avium* for human monocytes: absence of macrophage-activating factor activity of human gamma interferon. *Infect. Immun.* **57**:239–244.

113. **Vachula, M., T. D. Holzer, and B. R. Andersen.** 1989. Suppression of monocyte oxidative response by phenolic glycolipid 1 of *Mycobacterium leprae*. *J. Immunol.* **142**:1696–1701.

114. **Valone, S. E., E. A. Rich, R. S. Wallis, and J. J. Ellner.** 1988. Expression of tumor necrosis factor α in vitro by human mononuclear phagocytes stimulated with whole *Mycobacterium bovis* BCG and mycobacterial antigens. *Infect. Immun.* **56**:3313–3315.

115. **van Furth, R.** 1988. Phagocytic cells: development and distribution of mononuclear phagocytes in normal steady state and inflammation, p. 281–295. *In* J. I. Gallin, J. I. Goldstein, and R. Snyderman (ed.), *Inflammation: Basic Principles and Clinical Correlates.* Raven Press, New York.

116. **Vassalli, P.** 1992. The pathophysiology of tumor necrosis factors. *Annu. Rev. Immunol.* **10**:411–452.

117. **Walker, L., and D. Lowrie.** 1981. Killing of *Mycobacterium microti* by immunologically activated macrophages. *Nature* (London) **293**:69–70.

118. **Wallis, R. S., H. Fujiwara, and J. J. Ellner.** 1986. Direct stimulation of monocyte release of IL-1 by mycobacterial protein antigens. *J. Immunol.* **136**:193–196.

119. **Youmans, G. P.** 1979. Biolic activities of mycobacterial cells and cell components, p. 53–54. *In* G. P. Youmans (ed.), *Tuberculosis.* The W. B. Saunders Co., Philadelphia.

120. **Zimmer, T., and P. P. Jones.** 1990. Combined effects of tumor necrosis factor-α, prostaglandin E$_2$, and corticosterone on induced Ia expression on murine macrophages. *J. Immunol.* **145**:1167–1175.

121. **Zimmerman, R., B. Marafino, A. Chan, P. Landre, and J. Winkelhake.** 1989. The role of oxidant injury in tumor cell sensitivity to recombinant tumor necrosis factor *in vivo*. *J. Immunol.* **142**:1405–1409.

MICROBIAL RESISTANCE TO MACROPHAGE EFFECTOR FUNCTIONS: STRATEGIES FOR EVADING MICROBICIDAL MECHANISMS AND SCAVENGING NUTRIENTS WITHIN MONONUCLEAR PHAGOCYTES

Andreas J. Bäumler and Fred Heffron

8

Pathogens have evolved various strategies to evade the immune response and multiply within the host. For invasive microorganisms, an intracellular location offers several advantages over an extracellular niche. The most obvious benefit is that host cells provide a place to hide from humoral defense mechanisms. Among different types of host cells that could be abused for colonization, phagocytes are frequently preferred by intracellular parasites, perhaps because the migration of these cells provides a mechanism for entry into or dissemination within the host. In addition, their highly phagocytic activity may render phagocytes easily accessible to microbes, and the similarity in bacterial growth requirements for multiplying within professional phagocytes and within protozoa might have facilitated the colonization of this mammalian habitat during evolution. In fact, the transition from a protozoan to a mammalian host is exemplified by *Legionella pneumophila,* a parasite of human alveolar macrophages, peripheral blood monocytes, and amebae. Among professional phagocytes, only mononuclear phagocytes seem to fulfill all requirements of an intracellular

niche. Their limited life span of approximately 1 day makes polymorphonuclear granulocytes (PMNs) unsuitable as an intracellular habitat. In contrast, mononuclear phagocytes have a life span of months, which renders them good targets for colonization by intracellular parasites. An apparent obstacle to colonizing professional phagocytes, however, is their antimicrobial activity.

PMNs possess a high antibacterial capacity that renders them potent killer cells for intracellular bacteria rather than a safe site for microbes in the host. The main goals of the microbe during contact with PMNs are most likely to survive and outlive the PMN, and the success of these efforts requires a high level of innate resistance. In contrast, colonization of mononuclear phagocytes by intracellular pathogens is characterized by multiplication within the host cell. In addition, a lower level of innate resistance might be required of the microbe, since mononuclear phagocytes change their antibacterial capacities depending on tissue localization and activation by exogenous stimuli. Resident tissue macrophages have a limited antibacterial capacity and are therefore better suited to support colonization by intracellular pathogens. The major effector functions of mononuclear phagocytes include generation of reactive oxygen intermediates, production

Andreas J. Bäumler and Fred Heffron   Department of Molecular Microbiology and Immunology, Oregon Health Sciences University, 3181 S.W. Sam Jackson Park Road, L220, Portland, Oregon 97201-3098.

Virulence Mechanisms of Bacterial Pathogens, 2nd ed., Edited by J. A. Roth et al.

of reactive nitrogen intermediates, limitation of intracellular iron availability, phagosome acidification and phagosome-lysosome fusion, and production of defensins (cationic antibacterial peptides). Most of these mechanisms are induced only by appropriate activation, which is optimally achieved following interaction with gamma interferon (IFN-γ). However, other interleukins may also activate macrophages (52, 54, 55).

The intracellular pathway used to colonize mononuclear phagocytes frequently influences the degree to which pathogens are exposed to the microbicidal mechanisms. The three major pathways are intraphagolysosomal, intraphagosomal, and extraphagosomal, the last two of which are reviewed in other chapters of this book. By using the intraphagosomal pathway, microbes avoid contact with lysosomal contents. The extraphagosomal pathway allows the parasite to escape from the hostile phagosomal or phagolysosomal environment into the cytoplasm of the host cell. These mechanisms therefore limit either the number of antimicrobial mechanisms faced by the bacterium or the time the pathogen is exposed to mononuclear phagocyte effector functions. However, none of these pathways eliminates contact between the microorganism and the host cell antimicrobial mechanisms. Thus, innate resistance to phagocytic killing mechanisms is required for colonization of mononuclear phagocytes no matter which intracellular pathway the pathogen follows. In addition, microbes residing in a vacuole of the host cell have to outmaneuver bacteriostatic host cell functions in order to multiply. This review focuses on effector functions utilized by mononuclear phagocytes and the microbial strategies employed to overcome or evade them.

ENTRY

The Mononuclear Lineage

Mononuclear phagocytes are a heterogeneous group of host cells that can undergo various phenotypic changes during their life cycles. Since developmental stages of mononuclear phagocytes differ in their microbicidal capacities, different developmental stages may not be equally suitable as intracellular habitats. In fact, different populations of macrophages differ in their microbicidal activities against *Salmonella typhimurium* (22). Therefore, targeting to a preferred mononuclear host cell type might be an important step for intracellular pathogens during infection.

Monocytes, which are produced in the bone marrow, represent the first stage of differentiation of mononuclear phagocytes. After circulating in the blood for about 24 h, monocytes are recruited by an unknown mechanism to enter the tissue and differentiate into resident macrophages. Depending on tissue localization, these mononuclear phagocytes are designated alveolar macrophages, peritoneal macrophages, microglia, Kupffer cells, or macrophages of the spleen. Resident macrophages possess low antibacterial capacity (89, 109). During inflammation, distinct signals induce the rapid migration of blood monocytes from the circulation to the site of inflammation. At this stage, mononuclear phagocytes are referred to as elicited macrophages and are phenotypically characterized by enhanced secretory and phagocytic activities. Finally, elicited macrophages can be further primed by exposure to cytokines derived from T cells, such as IFN-γ or interleukin-2 (IL-2). Cytokine-primed elicited macrophages have greatly enhanced microbicidal and microbiostatic capacities and are known as activated macrophages. This activation process can also occur in resident macrophages.

Phagocytic Receptors of Mononuclear Phagocytes

Numerous plasma membrane receptors that mediate phagocytosis can be found on mononuclear phagocytes; these include Fc receptor (FcR), complement receptor (CR), and carbohydrate receptors such as the macrophage mannose receptor (MMR). The level of expression of these receptors varies during the life cycles of mononuclear phagocytes. Ligation to receptors expressed by different developmental stages of mononuclear phagocytes often induces different signaling pathways, which has important consequences for the attached microorganism.

The FcR is expressed on all developmental stages of mononuclear phagocytes and mediates uptake of immunoglobulin-opsonized particles. Although phagocytosis via FcR is a constitutive property, it can be enhanced by various agents, including cytokines (60) and extracellular matrix proteins (19, 78, 117). Compared to monocytes, resident and activated macrophages show an enhancement of FcR-mediated phagocytosis. However, while internalization is not accompanied by increased generation of reactive oxygen intermediates in monocytes and resident macrophages (108), phagocytosis via FcR triggers a respiratory burst in elicited and activated macrophages.

Mononuclear phagocytes express CR type 1 (CR-1) specific for complement component C3b. Two other CRs, CR-3 and CR-4, mediate phagocytosis of particles coated with C3bi, a cleavage product of C3b. CR-3 also possesses a second binding site, which recognizes phospho-sugars and apparently contributes to the phagocytosis of carbohydrate-coated particles (128). Although CR-1, CR-3, and CR-4 are expressed on all mononuclear phagocytes, CR-3 is more abundant on the surfaces of resident macrophages (86).

Ligation to CRs does not automatically result in phagocytic uptake. Binding of C3b- or C3bi-coated erythrocytes to human monocytes is, for example, not followed by ingestion (107). However, after attachment of C3bi-opsonized bacteria, CR-3 is activated and is then capable of mediating phagocytosis (156). Furthermore, CR-mediated phagocytosis is developmentally regulated. Decreased phagocytosis via receptors for complement is observed upon activation of macrophages (45, 51, 134). Ligation of CR by particles coated with fragments of complement component C3 does not trigger an oxidative burst and the release of toxic oxygen metabolites (157, 158).

The MMR mediates uptake of unopsonized particles through specific recognition of man-nosyl- or fucosyl-glycoconjugates on the surfaces of bacteria. Like CR expression, the stage of macrophage differentiation has profound effects on MMR expression. For example, peripheral blood monocytes do not express MMR protein or activity when they are freshly isolated (137), whereas resident and elicited macrophages do express MMR activity (93, 142). Activation of macrophages by IFN-γ selectively downregulates MMR expression (118). Zymosan contains multiple mannose residues and is phagocytized via MMR (141), although in the presence of serum, there is cooperation with CR (43). Uptake of unopsonized zymosan does not trigger a significant respiratory burst (43). This is not the case with activated murine cells (17). However, the ability of the MMR to stimulate a respiratory burst during phagocytosis needs to be further assessed.

CR Provides a Safe Passage into Mononuclear Phagocytes

For several microbes, entry into mononuclear phagocytes seems to be mediated by more than one receptor. In fact, cooperativity has been demonstrated for several receptors and has been postulated to represent a common mechanism for phagocytosis. However, CRs play a role in internalization of many organisms and appear to provide a general pathway for entry of intracellular pathogens into mononuclear phagocytes (115). CRs are involved in the uptake of *Leishmania donovani* (18), *Leishmania major* (95), *Histoplasma capsulatum* (26), *Cryptococcus neoformans* (90), *Mycobacterium tuberculosis* (132), *Mycobacterium leprae* (133), *L. pneumophila* (115), *S. typhimurium* (71), *Listeria monocytogenes* (41), and *Rhodococcus equi* (66). Since mononuclear phagocytes secrete components of the alternative and classic complement pathways, complement may be generally available in tissue to mediate phagocytosis via CRs. Consistent with this idea, macrophages have been shown to deposit C3 on *Leishmania donovani* in the absence of serum (155). Since uptake of bacteria via CR does not trigger the release of reactive oxygen intermediates, entrance of intracellular pathogens by CR-mediated phagocytosis may allow the pathogens to avoid the toxic consequences of the oxidative burst. Consistent with this hypothesis, several intracellular pathogens, including *Toxoplasma gondii* (138, 153), *H. capsulatum* (42), *L. pneumophila* (73), *Salmonella typhi* (82, 94), and *S. typhimurium* (151), have been reported to elicit little or no oxidative burst upon entry into phagocytic cells.

Selective Targeting to Macrophages

In order for some pathogens to fulfill their virulence potential, they must ensure internalization by their preferred host cells. Parasites of mononuclear phagocytes seem not to depend on active mechanisms for invasion but instead take advantage of the phagocytic capacity of the host cell. Nevertheless, intracellular pathogens may play an active role in selecting a particular subset of professional phagocytes for colonization. This could be advantageous for the microbe, since phagocytes with high microbicidal capacities (e.g., IFN-γ-activated macrophages) do not support growth of a variety of intracellular pathogens in vitro (52, 53, 67, 74, 77, 102, 114, 120, 126). Pathogens may therefore use entry routes that allow them to select a particular phagocytic cell type. Consistent with this idea, activated macrophages exhibit reduced phagocytosis for a variety of intracellular microorganisms, including *Rickettsia* spp. (101), *Trypanosoma cruzi* (65), and *Leishmania major* (113). In each case, activated macrophages contain fewer ingested organisms than resident macrophages. This may be due to a downregulation of CR and MMR functions in activated macrophages (45, 51, 118). Targeting of microbes to their preferred host cell could be accomplished by attachment to MMR, which, in contrast to CR, is exclusively expressed on resident macrophages. In line with this hypothesis, binding of *Leishmania donovani* to IFN-γ-treated macrophages is mediated by CR-3, while increased attachment to resident macrophages is mediated by binding to MMR and CR-3 (96, 154). The MMR is also involved in mediating phagocytosis of *My-cobacterium avium* and *M. tuberculosis* (16, 123, 131). Interestingly, the MMR plays an important role in adherence of virulent but not of attenuated *M. tuberculosis* strains to macrophages (131). Thus, some pathogens seem to use the MMR as a recognition factor that selectively targets these organisms to resident or elicited but not activated macrophages.

MICROBICIDAL EFFECTOR FUNCTIONS

Reactive Oxygen Intermediates: Avoidance versus Resistance

The killing of most extracellular organisms by mononuclear phagocytes depends on the phagocyte's capacity to convert oxygen to microbicidal metabolites. This metabolic activity is significantly enhanced in mononuclear phagocytes obtained after chemically induced inflammation, bacillus Calmette-Guérin infection (75), or IFN-γ treatment (104). During phagocytosis via FcR, elicited or activated macrophages oxidize glucose through the hexose monophosphate shunt, thus producing NADPH, the substrate of NADPH oxidase. The NADPH oxidase of mononuclear phagocytes consists of cytosolic and membrane-bound components that form an electron transfer chain, allowing molecular oxygen to be converted into superoxide and other toxic oxygen intermediates such as hydrogen peroxide (139). This process is known as the phagocytosis-associated respiratory burst. The importance of oxidative cytocidal mechanisms to disease resistance is demonstrated by patients with chronic granulomatous disease. This genetic disorder can be caused by defects in different components of the NADPH oxidase and manifests itself in recurrent bacterial and fungal infections starting within the first years of life (10, 37).

Various microbial proteins mediating resistance to reactive oxygen species have been described and can be classified into enzymes that prevent and enzymes that repair damage caused by oxygen radicals (47, 62, 145). Many of these enzymes are present not only in most intracellular pathogens but also in most aerobic bacteria. Furthermore, there is a great deal of similarity in the overall defense strategies of bacteria and eukaryotic cells against oxidative stress. Nearly all aerobic cells, prokaryotic and eukaryotic, possess superoxide dismutase (SOD), catalase, and peroxidase to destroy active oxygen species. Cellular repair mechanisms also overlap extensively. For example, some bacterial and eukaryotic DNA repair enzymes specific for oxidative damage share extensive homology (119).

Enzymes that scavenge oxygen radicals can protect the cell from damage caused by reactive oxygen intermediates. The key defense against the superoxide radical is SOD. Nearly all bacteria that use oxygen as a terminal electron acceptor express one or more SODs. Similar enzymes are produced by eukaryotic cells. Indeed, SOD expressed from mammalian *sod* genes fully protects SOD–negative mutant bacteria from damage by superoxide radicals (106). Direct evidence for a function of SODs in virulence of pathogens of mononuclear phagocytes is missing. However, increased concentrations of a surface-located SOD of the facultative intracellular pathogen *Nocardia asteroides* have been correlated with virulence (14).

Several enzyme systems are available to detoxify hydrogen peroxide. Catalase is an enzyme that converts hydrogen peroxide into oxygen and water. The virulence of intracellular pathogens, including *N. asteroides* (13), *M. tuberculosis* (83), *Listeria monocytogenes* (152), and *Leishmania donovani* (99), has been related to their catalase content. However, catalase mutants of *Listeria monocytogenes* created by Tn*1545* mutagenesis show no decrease in virulence (62). In addition, more recent studies have shown that the resistance of *Mycobacterium intracellulare* to peroxide does not correlate with its catalase content and that the susceptibility of *M. tuberculosis* to killing by activated macrophages is not related to peroxide susceptibility (58, 110).

Peroxidases, present in several bacteria as well as in eukaryotic cells, are also capable of destroying hydrogen peroxide. Unlike catalase, peroxidases require NADPH or NADH as an electron source. Therefore, under conditions in which reducing power is limited, the protective role of peroxidases is likely to be small. Several bacteria synthesize glutathione, which can serve to maintain the reduced state of cellular proteins, as an antioxidant. A pool of reduced glutathione is maintained by glutathione synthetase and glutathione reductase activities. Like peroxidase, the glutathione system requires a source of reducing power in order to function. Under conditions of severe oxidative stress, these systems are inefficient, since reducing power is limited (47). There is also no experimental evidence for a role for peroxidases or the glutathione system in virulence of intracellular pathogens.

Mycobacteria are endowed with complex glycolipids that can scavenge toxic oxygen radicals (32, 33). There is, however, no direct evidence that these surface structures play a role in virulence by scavenging oxygen radicals.

In addition to the defense enzymes that destroy oxidants, several activities appear to repair damage caused by oxidants. Oxygen radicals can cause both DNA and membrane damage, either of which can be lethal. Damaged proteins are not likely to be the cause of oxygen radical-induced death, but the damaged molecules must be either repaired or degraded and replaced. Repair of membrane and protein damages are discussed elsewhere in this chapter. A large number of DNA repair systems implicated in defense against reactive oxygen intermediates have been summarized by Farr and Kogoma (47). The *recA* system, which mediates recombinational repair, has been explored most extensively. In *Escherichia coli, recA* mutants are very sensitive to killing by hydrogen peroxide (31), and it is the role of *recA* in the recombinational repair pathway that is important in cell survival of peroxide stress (70). Similarly, *recB* mutants are hypersensitive to hydrogen peroxide, and *sbcB* mutants that activate the *recF* pathway of recombination suppress the sensitivity of *recBC* mutants to hydrogen peroxide (69). In fact, Carlsson and Carpenter (31) reported that in *E. coli,* the *recA* gene product is more important than catalase in defense against hydrogen peroxide. *S. typhimurium* with mutations in the *recA* and *recBC* genes has a decreased ability to survive within macrophages in vitro as well as a decreased virulence for mice (24).

These examples indicate that the ability to tolerate a certain level of reactive oxygen species is important for at least some intracellular pathogens. However, as discussed above, the most prevalent strategy used by intracellular pathogens seems to be avoidance of reactive oxygen intermediates rather than resistance to them. Along these lines, it has been speculated that the ability of *L. pneumophila* to avoid toxic oxygen molecules may be important to its intracellular survival, because the bacterium is susceptible to relatively low concentrations of hydrogen peroxide (73, 91). A further example of avoidance or subversion of several macrophage functions is provided by members of the protozoan genus *Leishmania*. Initially, this avoidance involves adhering to and entering the macrophage without activating the macrophage's microbicidal responses, which can kill the parasite if they are triggered (3).

Resistance to Oxygen-Independent Killing Mechanisms

REACTIVE NITROGEN INTERMEDIATES

An important effector function of macrophages that has recently been implicated in combatting intracellular pathogens is the generation of reactive nitrogen intermediates. The enzyme nitric oxide synthetase generates nitric oxide (NO) from the terminal guanidino-nitrogen atoms of L-arginine. The generated NO is subsequently converted to NO_2^- and NO_3^- by a rapid oxidizing process. Under basal conditions, NO synthetase activity is negligible in resident macrophages, while stimulation with lipopolysaccharide (LPS) and IFN-γ produces massive enhancement of NO synthetase in a few hours (103). Reactive nitrogen intermediates have been implicated in the antimicrobial effect of activated macrophages against

Leishmania major, Toxoplasma gondii, and *M. leprae* (103). Furthermore, reactive nitrogen intermediates seem to be crucially involved in tuberculostatic activities of activated murine macrophages (34, 39, 56, 147). Microbial resistance mechanisms against reactive nitrogen intermediates have so far not been described.

ANTIBACTERIAL PEPTIDES

The electron-dense granules of some mammalian professional phagocytes, including rabbit, guinea pig, rat, and human PMNs and rabbit alveolar macrophages, contain small antibacterial peptides termed defensins. Defensins form channels in artificial membranes and may kill microbes by making their cytoplasmic membranes permeable (87, 88). There is evidence that defensins enter the phagosome during phagocytosis of bacteria, and the increase of the endosomal pH to basic levels for a brief period during phagocytosis provides optimal conditions for the activity of these antimicrobial peptides. Detailed studies of mechanisms of resistance to defensins have so far been conducted only with *S. typhimurium*. Resistance of *S. typhimurium* to antimicrobial peptides seems to be required for virulence, since defensin-susceptible mutants are attenuated in mice (48). In fact, defensins are the major constituents in phagosomes prepared from human PMNs ingesting *S. typhimurium* opsonized with complement or rabbit immunoglobulin G (76). In order to reach their primary target, the cytoplasmic membrane, defensins must cross the outer membranes of gram-negative bacteria. Owing to their positive charge, defensins are able to bind the anionic LPS (130). LPS structure is important for resistance to other cationic agents. *S. typhimurium* can synthesize LPS containing an increased amount of 4-aminoarabinose and larger amounts of ethanolamine (125, 149). This less acidic LPS confers resistance to polymyxin, a cationic lipopeptide, and bactericidal permeability-increasing protein, a 58-kDa strongly cationic protein present in azurophil granules of human and rabbit PMNs (136). *S. typhimurium* mutants that express this modified LPS constitutively are more resistant to intraphagocytic killing by PMNs (143). Furthermore, *S. typhimurium* mutants lacking the *N*-acetylglucosamine side chain of their LPS core are more sensitive to lysosomal extracts from PMNs (57). Two *Salmonella* genes, *dsbB* and *tolB*, which encode proteins that are not involved in LPS biosynthesis but are necessary for an intact outer membrane and for resistance to defensins, have recently been identified (100). Further, *S. typhimurium* mechanisms of resistance to defensins are controlled by a two-component regulatory system encoded by phoPQ (48, 61). Genes that are activated by PhoPQ are expressed after phagocytosis by macrophages (23). The PhoPQ-regulated gene(s) responsible for defensin resistance has not been identified yet.

LYSOSOMAL CONTENTS

There is evidence that some intracellular pathogens reside in a phagolysosome. Immunoelectron microscopic and immunofluorescence studies have demonstrated lysosome-associated membrane proteins in the membranes surrounding leishmaniae and a variety of lysosomal hydrolases in the vacuole lumen (7–9, 121). These data confirm earlier reports that leishmaniae reside in phagolysosomal compartments (2, 4). Lysosomes contain numerous degradative enzymes, including acid hydrolases, neutral proteinases, and lysozyme. Prior to lysosomal fusion, the pH of the phagosome decreases to acidic levels, thus supporting optimal activity of acidic lysosomal enzymes. Intracellular pathogens that reside in a phagolysosome of the host cell are therefore exposed to a hydrolytic milieu. Although often cited as a killing mechanism, lysosomal protease activity has not conclusively been shown to directly kill ingested microbes. The role of lysosomal proteases might be limited to degrading dead organisms.

The Macrophage-Induced Stress Response

Heat shock proteins are prominent antigens during infections with various intracellular pathogens, including *Trypanosoma cruzi*, *Leish-mania donovani*, *Plasmodium falciparum*, *M. tuberculosis*, *M. leprae*, *Chlamydia trachomatis*, *Coxiella burnetti*, *L. pneumophila*, and *Borrelia burgdorferi* (36). Heat shock proteins were first identified by their increased synthesis in response to elevated temperature, but it is now clear that these proteins are in fact inducible by almost any form of cellular stress in any type of cell, from prokaryotic to human. Many of the stressors known to induce heat shock protein synthesis (e.g., changes in pH and osmolarity and production of reactive oxygen and nitrogen species) also occur inside macrophages. These data therefore suggest that intracellular pathogens may produce elevated amounts of heat shock proteins inside macrophages. Indeed, the major heat shock proteins DnaK and GroEL are induced during growth of *S. typhimurium* within murine macrophages (23). The study by Buchmeier and Heffron (23) also showed that the synthesis of more than 30 *Salmonella* proteins is specifically induced by encountering the intracellular environment of a macrophage. The induction of a similar number of proteins is observed after growth of *L. pneumophila* within macrophages (84). Coordinate expression of stress proteins has been reported (e.g., heat shock induces reactive oxygen intermediate-detoxifying enzymes) in *S. typhimurium* (35). Therefore, two groups examined whether the macrophage-induced stress proteins (MIPs) can also be induced in the absence of phagocytes by exposure of *S. typhimurium* or *L. pneumophila* to various environmental conditions (1, 84). Several MIPs are induced by in vitro stress conditions, including heat shock, osmotic shock, phosphate starvation, carbon starvation, sulfur starvation, polymyxin B, and reactive oxygen intermediates. However, about one-fourth of the MIPs in *Salmonella* spp. and one-third of the MIPs in *Legionella* spp. are induced within macrophages but not by any of the stress conditions tested.

Several findings indicate that intracellular bacteria increase their MIP synthesis to protect themselves against intracellular killing and that mutants defective in MIP biosynthesis are impaired in survival within phagocytic host cells.

S. typhimurium mutants with a defect in *phoPQ* fail to express several MIPs (23). These mutants possess impaired abilities to survive within murine macrophages and are avirulent in mice (48, 49). Nutritional starvation is one of the stress factors that induces expression of several MIPs in vitro (1, 84). Starvation induces general resistance in *S. typhimurium* (140). In fact, a mutation in *katF,* which encodes an alternative sigma factor involved in starvation-induced protein synthesis, renders *S. typhimurium* avirulent for mice (46). As stated above, the MIPs include several heat shock proteins whose function has been receiving increasing attention. Elevated heat shock protein synthesis renders cells more resistant to subsequent, otherwise lethal insults. Some heat shock proteins function as chaperones that can not only protect cellular proteins from denaturation but also repair them (135). Other stress-induced gene products are proteases that degrade misfolded proteins, thus preventing the accumulation of these proteins to toxic levels. In addition, amino acids that result from this degradation can be utilized to form new proteins, a process that may be beneficial during nutrient limitation (122). The heat shock protein HtrA is a protease that functions in the periplasm, where it degrades abnormally folded proteins (146). *S. typhimurium* and *Brucella abortus htrA* mutants have decreased capacities for survival within macrophages (12, 49, 127). Similarly, a distinct protease, Prc, which also acts in the periplasm, reduces macrophage survival of *S. typhimurium* (12, 49). These proteases therefore seem to prevent the accumulation of misfolded proteins generated during growth within phagocytes. Along these lines, it has been speculated that the expression of flagellum subunits is responsible for the decrease in macrophage survival observed in some *S. typhimurium* mutants defective in flagellum assembly (12). The degradation of flagellum subunits (which are produced in several thousand copies per cell) by proteases like Prc and HtrA might impair the abilities of bacteria to degrade other abnormally folded periplasmic proteins generated after contact with macrophages. In conclusion, the stress response induced within macrophages seems to function in maintenance and repair of the bacterial cell.

MICROBIOSTATIC EFFECTOR FUNCTIONS

The Concept of Nutritional Immunity

The withholding of nutrients from the microbe by the host is a nonspecific defense mechanism described as nutritional immunity. Compounds that are limited in an intracellular environment have been identified by using auxotrophic mutants defective in intracellular growth. Interestingly, only a few auxotrophic mutations inhibit intracellular multiplication or virulence. In *S. typhimurium,* mutations in the biosynthetic pathways for purines and aromatic compounds cause a decrease in the ability to grow in murine macrophages in vitro, and strains carrying these mutations are avirulent in mice (49). From these studies it can be concluded that purines and suitable intermediates or end products of the aromatic pathway are not available for intracellular pathogens in vivo. Tryptophan deprivation has been investigated as a microbiostatic mechanism of IFN-γ-activated mononuclear phagocytes (30). IFN-γ induces the enzyme indoleamine 2,3-dioxygenase, which catalyzes the breakdown of tryptophan (30). In fibroblasts and uroepithelial cells, multiplication of *Toxoplasma gondii* and *Chlamydia psittaci* can be inhibited by tryptophan deprivation (29, 116). However, the role of tryptophan deprivation in the antimicrobial activities of mononuclear phagocytes is questionable, since induction of indoleamine 2,3-dioxygenase by IFN-α and IFN-β does not induce antitoxoplasma activity (98, 105). Moreover, the nutrients mentioned above can be synthesized by microbes, and limiting them does not therefore effectively limit multiplication of prototrophic pathogens. Thus, in order to develop an efficient nutritional immunity against infection, the host defense has focused on limiting the availability of a nutrient that cannot be synthesized by microbes, iron.

Macrophages and the Iron-Withholding Defense of the Host

Transferrin and lactoferrin are constitutive components of the iron-withholding system of the host and are present in all body fluids except urine. These iron-binding proteins have a normal level of iron saturation of about 30%, and the amount of free Fe^{3+} in equilibrium with the bound iron is about 10^{-18} M, which is far too low for bacterial growth. Therefore, the presence of transferrin and lactoferrin constitutes a very effective first line of defense against multiplication of microorganisms. Nevertheless, additional components of the iron-withholding defense are induced at the time of contact with a pathogen.

In response to infection, the serum iron content is reduced to about 30% of its normal level. At the same time, the iron contents of cells from the reticuloendothelial system, particularly in the liver and spleen, increase (150). These physiologic changes in the host are described as hypoferremia of infection. The principal mechanisms contributing to the hypoferremia are (i) a decrease in iron release from macrophages that have digested effete erythrocytes accompanied by increased intracellular synthesis of the iron storage protein ferritin and (ii) suppression of dietary iron assimilation in the intestine. The cytokines IL-1, IL-6, and tumor necrosis factor induce this host response. However, as discussed later, IFN-γ activates additional iron-withholding systems in macrophages.

Macrophages are main characters in the host's iron metabolism. Kupffer cells of the liver and macrophages of the red pulp of the spleen are most actively involved in iron recirculation. Approximately 360 billion senescent erythrocytes are removed daily by these phagocytes in humans (79). Part of the iron internalized during this process is released by macrophages by an unknown mechanism and is subsequently bound by the extracellular transferrin pool (44, 129). Secretion of iron bound to a secretory form of ferritin has also been reported (44, 81). Less iron is released after incubation of macrophages with tumor necrosis factor, a situation that may contribute to the development of hypoferremia during infection (21). However, part of the internalized iron is stored in hepatocytes and macrophages of the liver and spleen, the main organs for iron fixation, and this storage is increased during infection (150). Therefore, it is not surprising that the amount of the intracellular iron storage protein ferretin is greatest in macrophages and hepatocytes (20).

Intracellular Parasitism: A Strategy to Overcome the Iron-Withholding Defense?

While iron is very efficiently removed from body fluids, cells involved in iron storage might be the Achilles' heel of the iron-withholding defense. Intruders able to utilize iron hidden in macrophages could overcome the iron restriction and multiply in the host. The preference for growth in liver and spleen indicates that some intracellular pathogens like *Listeria monocytogenes, Leishmania donovani,* and *S. typhimurium* might actually use this strategy. In line with this theory is the observation that several other intracellular parasites, including *M. tuberculosis* and *L. pneumophila,* show a tropism for alveolar macrophages, which contain 100 times more ferritin than blood monocytes (6). In fact, pathogens that multiply either extracellularly or intracellularly seem to have developed distinct mechanisms for scavenging iron in the host. Either extracellular pathogens utilize host iron-binding compounds such as transferrin, lactoferrin, and heme, or they secrete low-molecular-weight iron chelators, termed siderophores, that are able to remove iron bound by host proteins. For example, in gram-negative bacteria, these iron uptake mechanisms are all mediated by outer membrane receptors that depend on the TonB protein for energizing this transport (11, 38, 59, 144). However, *tonB* mutants of *S. typhimurium* are fully virulent in mice, indicating that this intracellular pathogen acquires iron in vivo by using a TonB-independent pathway (15). Similarly, *L. pneumophila* does not produce a siderophore and cannot utilize iron from transferrin, lactoferrin, or heme. Nevertheless, the intracellular multiplication of *L. pneumophila* is iron dependent (27). Although siderophores stimulate growth of mycobacteria in vitro, iron acquisition in the host seems to be mediated by an alternative mechanism (85). Thus, while the mechanisms for iron acquisition by intracellular pathogens are poorly understood, it is clear that they are crucial in colonization of this host niche. In fact, intracellular parasitism might be a strategy for outmaneuvering the iron-withholding defense of the host.

The Iron Source within Macrophages

Iron internalized by macrophages enters the so-called labile intracellular pool. This pool consists of iron that is immediately available to the cell for metabolic processes (72). Iron not immediately required for metabolic activity is stored in ferritin or is released into the serum as discussed above. When required, iron from ferritin can again be released into the labile iron pool (50, 64). A part of the intracellular ferritin is denatured to form hemosiderin (68). Iron is removed more slowly from hemosiderin than from ferritin, and hemosiderin is generally considered to be a relatively inactive form of stored iron (111, 112).

Studies on the effect of desferrioxamine B and IFN-γ on intracellular multiplication of *Listeria monocytogenes, L. pneumophila*, and *Trypanosoma cruzi* provide evidence that these pathogens utilize the labile iron pool (5, 27, 92). The siderophore desferrioxamine B is a therapeutic substance for use in patients suffering from iron overload. It chelates iron in the labile iron pool, and treatment of macrophages with desferrioxamine B inhibits intracellular multiplication of *Listeria monocytogenes, L. pneumophila,* and *Trypanosoma cruzi* in vitro (5, 27, 92). Furthermore, treatment with desferrioxamine B increases the resistance of mice to infections with *Listeria monocytogenes* (148). IFN-γ-activated human monocytes and alveolar macrophages inhibit the intracellular multiplication of *L. pneumophila* by withholding iron from the intruders. This may in part be accomplished by downregulation of transferrin recep-

tors on IFN-γ-activated monocytes (27). In addition, IFN-γ-activated macrophages limit the available intracellular iron by decreasing intracellular ferritin levels (5, 28). The observed decrease in the concentration of ferritin may be due to enhanced conversion to hemosiderin, which is observed in activated macrophages, and has the net effect that less iron is available to the labile iron pool (63, 124). Taken together, these data indicate that intracellular pathogens residing in distinct compartments (phagosome or cytoplasm) of the mononuclear phagocyte have access to the labile iron pool. Upon activation with IFN-γ, macrophages are able to suppress intracellular multiplication by withholding iron from the intruders.

ACTIVATION OF MACROPHAGES AND RESISTANCE TO INFECTION

During infection, T cells and natural killer cells stimulated by LPS or antigens release cytokines, including IFN-γ. IFN-γ is capable of upregulating the pathways for several macrophage effector functions, i.e., the production of reactive oxygen and nitrogen intermediates (104, 147), the induction of indolamine 2,3-dioxygenase to catabolize tryptophan (30), and the withholding of intracellular iron (5, 27, 28). Treatment with IFN-γ reduces either intracellular growth or intracellular survival of various intracellular pathogens, including *M. bovis* (52, 53), *M. tuberculosis* (126), *Listeria monocytogenes* (120), *Leishmania major* (114), *L. pneumophila* (67, 102), *Brucella abortus* (74), and *S. typhimurium* (77), in macrophages in vitro. Administration of recombinant IFN-γ protects mice against *Listeria monocytogenes* and lethal *M. tuberculosis* infection, while neutralization of IFN-γ with antibody markedly exacerbates disease (25, 40, 80). Similarly, IFN-γ influences the rate of intracellular replication of salmonellae in mice, and depletion of IFN-γ greatly enhances the virulence of *S. typhimurium* (97). Therefore, the level of infection of various intracellular pathogens appears to be controlled by IFN-γ activation of macrophages. It is still unclear which macrophage effector functions are in each case responsible for the protective effect against a particular intracellular pathogen. However, these data indicate that during a successful infection, IFN-γ is released too late or at a suboptimal level. Intracellular pathogens can outmaneuver the resulting low level of macrophage effector functions of the nonimmune host and establish an infection. In contrast, during a secondary response, memory T cells respond to a lower dose of antigen and therefore release increased amounts of IFN-γ early during infection. Thus, T-cell immunity triggers the activation of macrophage effector functions more completely, and the host can successfully combat the intracellular infection.

ACKNOWLEDGMENTS

We thank R. Tsolis, P. Valentine, and H. Kusters for their critical reading of the manuscript. We greatly acknowledge I. Stoljiljkovic for discussing his ideas about the iron-withholding response of the host.

A. J. B. was supported by fellowship Ba1337/1-2 from the Deutsche Forschungsgemeinschaft.

REFERENCES

1. **Abshire, K. Z., and F. C. Neidhardt**. 1993. Analysis of proteins synthesized by *Salmonella typhimurium* during growth within a host macrophage. *J. Bacteriol.* **175**:3734–3743.

2. **Alexander, J.** 1981. *Leishmania mexiana*: inhibition and stimulation of phagosome-lysosome fusion in infected macrophages. *Exp. Parasitol.* **52**:261–270.

3. **Alexander, J., and D. G. Russel.** 1992. The interaction of *Leishmania* species with macrophages, p. 174–254. *In* J. R. Baker and R. Muller (ed.), *Advances in Parasitology*. Academic Press, Inc., New York.

4. **Alexander, J., and K. Vickermann.** 1975. Fusion of host cell secondary lysosomes with the parasitophorous vacuoles of *Leishmania mexicana*-infected macrophages. *J. Protozool.* **22**:502–508.

5. **Alford, C. E., T. E. King, and P. A. Campbell.** 1991. Role of transferrin, transferrin receptor and iron in macrophage listericidal activity. *J. Exp. Med.* **174**:459–466.

6. **Andreesen, R., J. Osterholz, H. Bodemann, K. J. Bross, U. Costabel, and G. W. Löhr.** 1984. Expression of transferrin receptors and intracellular ferritin during terminal differentiation of human monocytes. *Blut* **49**:195–202.

7. **Antoine, J. C., C. Jouanne, T. Lang, E. Prina, C. deCastellier, and C. Frehel.** 1991. Localization of major histocompatibility complex class II molecules in phagosomes of murine macrophages infected with *Leishmania amazonensis*. *Infect. Immun.* **59**:764–775.

8. **Antoine, J. C., C. Jouanne, A. Ryter, and V. Zilberfarb.** 1987. *Leishmania mexicana*: a cytochemical and quantitative study of lysosomal enzymes in infected rat bone marrow-derived macrophages. *Exp. Parasitol.* **64**:485–498.

9. **Antoine, J. C., E. Prina, C. Jouanne, and P. Bongrand.** 1990. Parasitophorous vacuoles of *Leishmania amazonensis*-infected macrophages maintain acidic pH. *Infect. Immun.* **58**:779–787.

10. **Baehner, R. L.** 1972. Disorders of leucocytes leading to recurrent infection. *Pediatr. Clin. N. Am.* **1**:935–956.

11. **Bäumler, A. J., R. Koebnik, I. Stojiljkovic, J. Heesemann, V. Braun, and K. Hantke.** 1993. Survey on newly characterized iron uptake systems of *Yersinia enterocolitica*. *Zentralbl. Bakteriol.* **278**:416–443.

12. **Bäumler, A. J., J. G. Kusters, I. Stojiljkovic, and F. Heffron.** 1994. *Salmonella typhimurium* loci involved in survival within macrophages. *Infect. Immun.* **62**:1623–1630.

13. **Beaman, B. L., S. M. Scates, S. E. Moring, R. Deem, and H. P. Misra.** 1983. Purification and properties of a unique superoxide dismutase from *Nocardia asteroides*. *J. Biol. Chem.* **258**:91–96.

14. **Beaman, L., and B. L. Beaman.** 1984. The role of oxygen and its derivatives in microbial pathogenesis and host defense. *Annu. Rev. Microbiol.* **38**:27–48.

15. **Benjamin, W. H., C. L. Turnbough, J. D. Goguen, B. S. Posey, and D. E. Briles.** 1986. Genetic mapping of novel virulence determinants of *Salmonella typhimurium* to the region between trpD and supD. *Microb. Pathog.* **1**:115–124.

16. **Bermudez, L. E., L. S. Young, and H. Enkel.** 1991. Interaction of *Mycobacterium avium* complex with human macrophages: roles of membrane receptors and serum proteins. *Infect. Immun.* **59**:1697–1702.

17. **Berton, G., and S. Gordon.** 1983. Modulation of macrophage mannosyl-specific receptors by cultivation on immobilized zymosan: effects on superoxide-anion release and phagocytosis. *Immunology* **49**:705–715.

18. **Blackwell, J. M., R. A. B. Ezekowitz, M. B. Roberts, J. Y. Channon, R. B. Sim, and S. Gordon.** 1985. Macrophage complement and lectin-like receptors bind *Leishmania* in the absence of serum. *J. Exp. Med.* **162**:324–331.

19. **Bohnsack, J. F., H. K. Kleinmann, T. Takahashi, J. J. O'Shea, and E. J. Brown.** 1985. Connective tissue proteins and phagocytic cell function. Laminin enhances complement and

Fc-mediated phagocytosis by cultured human macrophages. *J. Exp. Med.* **161:**912–923.

20. **Brock, J. H.** 1989. Iron and cells of the immune system, p. 81–108. *In* J. H. Brock (ed.), *Iron in Immunity, Cancer and Inflammation.* John Wiley & Sons, Inc., New York.

21. **Brock, J. H., and X. Alvarez-Hernández.** 1989. Modulation of macrophage iron metabolism by tumor necrosis factor and interleukin 1. *FEMS Microbiol. Immunol.* **47:**309.

22. **Buchmeier, N. A., and F. Heffron.** 1988. Intracellular survival of wild-type *Salmonella typhimurium* and macrophage-sensitive mutants in diverse populations of macrophages. *Infect. Immun.* **57:**1–7.

23. **Buchmeier, N. A., and F. Heffron.** 1990. Induction of *Salmonella* stress proteins upon infection of macrophages. *Science* **248:**730–732.

24. **Buchmeier, N. A., C. J. Lipps, M. Y. H. So, and F. Heffron.** 1993. Recombination-deficient mutants of *Salmonella typhimurium* are avirulent and sensitive to the oxidative burst of macrophages. *Mol. Microbiol.* **7:**933–936.

25. **Buchmeier, N. A., and R. D. Schreiber.** 1985. Requirement of endogenous interferon gamma production for resolution of *Listeria monocytogenes* infection. *Proc. Natl. Acad. Sci. USA* **82:**7404–7408.

26. **Bullock, W. E., and S. D. Wright.** 1987. Role of the adherence promoting receptors CR3, LFA-1, and p150,95 in the binding of *Histoplasma capsulatum* by human macrophages. *J. Exp. Med.* **165:**195–210.

27. **Byrd, T. F., and M. A. Horwitz.** 1989. Interferon gamma-activated human monocytes downregulate transferrin receptors and inhibit the intracellular multiplication of *Legionella pneumophila* by limiting the availability of iron. *J. Clin. Invest.* **83:**1457–1465.

28. **Byrd, T. F., and M. A. Horwitz.** 1990. Interferon gamma-activated human monocytes downregulate the intracellular concentration of ferritin: a potential new mechanism for limiting iron availability to *Legionella pneumophila* and subsequently inhibiting intracellular multiplication. *Clin. Res.* **38:**481A.

29. **Byrne, G. I., L. K. Lehmann, and G. J. Landry.** 1986. Induction of tryptophan catabolism is the mechanism for gamma-interferon-mediated inhibition of intracellular *Chlamydia psittaci* replication in T24 cells. *Infect. Immun.* **53:**347–351.

30. **Carlin, J. M., E. C. Borden, P. M. Sondel, and G. I. Byrne.** 1987. Biologic response modifier-induced indoleamin 2,3 dioxygenase activity in human peripheral blood mononuclear cell cultures. *J. Immunol.* **139:**2414–2418.

31. **Carlsson, J., and V. S. Carpenter.** 1980. The recA$^+$ gene product is more important than catalase and superoxide dismutase in protecting *Escherichia coli* against hydrogen peroxide toxicity. *J. Bacteriol.* **142:**319–321.

32. **Chan, J., X. Fan, S. W. Hunter, P. J. Brennan, and B. R. Bloom.** 1991. Lipoarabinan, a possible virulence factor involved in persistence of *Mycobacterium tuberculosis* within macrophages. *Infect. Immun.* **59:**1755–1761.

33. **Chan, J., T. Fujiwara, P. Brennan, M. McNeil, S. J. Turco, J.-C. Sibille, M. Snapper, P. Aisen, and B. R. Bloom.** 1989. Microbial glycolipids: possible virulence factors that scavenge oxygen radicals. *Proc. Natl. Acad. Sci. USA* **86:**2453.

34. **Chan, J., Y. Xing, R. S. Magliozzo, and B. R. Bloom.** 1992. Killing of virulent *Mycobacterium tuberculosis* by reactive nitrogen intermediates produced by activated murine macrophages. *J. Exp. Med.* **175:**1111–1122.

35. **Christman, M. F., W. Morgan, F. S. Jacobson, and B. N. Ames.** 1985. Positive control of a regulon for defenses against oxidative stress and some heat shock proteins in *Salmonella typhimurium. Cell* **41:**753–762.

36. **Cohen, I. R., and D. B. Young.** 1991. Autoimmunity, microbial immunity and the immunological homunculus. *Immunol. Today* **12:**105–110.

37. **Cohen, M. S., R. E. Isturiz, H. L. Malech, R. K. Root, C. M. Wilfert, L. Gutman, and R. H. Buckley.** 1981. Fungal infections in chronic granulomatous disease. The importance of the phagocyte in defense against fungi. *Am. J. Med.* **71:**56.

38. **Cornelissen, C. N., G. D. Biswas, J. Tsai, D. K. Paruchuri, S. A. Tompson, and P. F. Sparling.** 1992. Gonococcal transferrin-binding protein 1 is required for transferrin utilization and is homologous to TonB-dependent outer membrane receptors. *J. Bacteriol.* **174:**5788–5797.

39. **Denis, M.** 1991. Interferon-gamma-treated murine macrophages inhibit growth of tubercle bacilli via the generation of reactive nitrogen intermediates. *Cell Immunol.* **132:**150–157.

40. **Denis, M.** 1991. Involvement of cytokines in determining resistance and acquired immunity in murine tuberculosis. *J. Leukocyte Biol.* **50:** 495–501.

41. **Drevets, D. A., and P. A. Campell.** 1991. Roles of complement and complement receptor type 3 in phagocytosis of *Listeria monocytogenes* by inflammatory mouse peritoneal macrophages. *Infect. Immun.* **59:**2645–2652.

42. **Eissenberg, L. G., and W. E. Goldman.** 1986. *Histoplasma capsulatum* fails to trigger release of superoxide from macrophages. *Infect. Immun.* **55:** 29–34.

43. **Erzekowitz, R. A. B., R. B. Sim, G. G.**

MacPherson, and S. Gordon. 1985. Interaction of human monocytes, macrophages, and polymorphnuclear leukocytes with zymosan *in vitro*. *J. Clin. Invest.* **76:**2368–2376.

44. Esparza, I., and J. H. Brook. 1981. Release of iron by resident and stimulated mouse peritoneal macrophages following ingestion and digestion of transferrin-antitransferrin immune complexes. *Br. J. Haematol.* **49:**603–614.

45. Esparza, I., R. I. Fox, and R. D. Schreiber. 1986. Interferon-gamma-dependent modulation of C3b receptors (CR1) on human peripheral blood monocytes. *J. Immunol.* **136:**1360–1365.

46. Fang, F. C., S. J. Libby, N. A. Buchmeier, P. C. Loewen, J. Switala, J. Harwood, and D. G. Guiney. 1992. The alternate sigma factor KatF(RpoS) regulates *Salmonella* virulence. *Proc. Natl. Acad. Sci. USA* **89:**11978–11982.

47. Farr, S. B., and T. Kogoma. 1991. Oxidative stress rsponses in *Escherichia coli* and *Salmonella typhimurium*. *Microbiol. Rev.* **55:**561–585.

48. Fields, P. I., E. A. Groisman, and F. Heffron. 1989. A Salmonella locus that controls resistance to microbicidal proteins from phagocytic cells. *Science* **243:**1059–1062.

49. Fields, P. I., R. V. Swanson, C. G. Haidaris, and F. Heffron. 1986. Mutants of *Salmonella typhimurium* that cannot survive within the macrophage are avirulent. *Proc. Natl. Acad. Sci. USA* **83:**5189–5193.

50. Fillet, G., J. D. Cook, and C. A. Finch. 1974. Storage iron kinetics. VII. A biological model for reticuloendothelial iron transport. *J. Clin. Invest.* **53:**1527–1533.

51. Firestein, G. S., and N. J. Zvaifler. 1987. Down regulation of human monocytes differentiation antigens by interferon gamma. *Cell. Immunol.* **104:**343–354.

52. Flesch, I., and S. H. E. Kaufmann. 1987. Mycobacterial growth inhibition by interferon-gamma-activated bone marrow macrophages and differential susceptibility among strains of *Mycobacterium tuberculosis*. *J. Immunol.* **138:**4408–4413.

53. Flesch, I., and S. H. E. Kaufmann. 1988. Attempts to characterize the mechanisms involved in mycobacterial growth inhibition by gamma-interferon-activated bone marrow macrophages. *Infect. Immun.* **56:**1464–1469.

54. Flesch, I. E. A., and S. H. E. Kaufmann. 1990. Activation of tuberculostatic macrophage functions by interferon gamma, interleukin 4, and tumor necrosis factor. *Infect. Immun.* **58:**2675–2677.

55. Flesch, I. E. A., and S. H. E. Kaufmann. 1990. Stimulation of antibacterial macrophage activities by B cell stimulatory factor 2/interleukin 6. *Infect. Immun.* **58:**269–271.

56. Flesch, I. E. A., and S. H. E. Kaufmann. 1991. Mechanisms involved in mycobacterial growth inhibition by gamma-interferon-activated bone marrow macrophages: role of reactive nitrogen intermediates. *Infect. Immun.* **59:**3213–3218.

57. Friedberg, D., and M. Shilo. 1970. Interaction of gram-negative bacteria with the lysosomal fraction of polymorphnuclear leucocytes. Role of cell wall composition of *Salmonella typhimurium*. *Infect. Immun.* **1:**305–310.

58. Gangadhram, P. R. J., and P. K. Pratt. 1984. Susceptibility of *Mycobacterium intracellulare* to hydrogen peroxide. *Am. Rev. Respir. Dis.* **130:**309–311.

59. Goldberg, M. B., S. A. Boyko, J. R. Butterton, J. A. Stoebner, S. M. Payne, and S. B. Calderwood. 1992. Characterization of a *Vibrio cholerae* virulence factor homologous to the family of TonB-dependent proteins. *Mol. Microbiol.* **6:**2407–2418.

60. Griffin, F. M., and P. M. Mullinax. 1985. Effects of differentiation *in vivo* and of lymphokine treatment *in vitro* on the mobility of C3 receptors of human and mouse mononuclear phagocytes. *J. Immunol.* **135:**3394–3397.

61. Groisman, E. A., E. Chiao, C. L. Lipps, and F. Heffron. 1989. *Salmonella typhimurium phoP* is a transcriptional regulator. *Proc. Natl. Acad. Sci. USA* **86:**7077–7081.

62. Hassett, D. J., and M. S. Cohen. 1989. Bacterial adaption to oxidative stress: implications for pathogenesis and interaction with phagocytic cells. *FASEB J.* **3:**2574–2582.

63. Hershko, C. 1977. Storage iron regulation, p. 105–148. *In* E. B. Brown (ed.), *Progress in Haematology*. Grune and Stratton, New York.

64. Hershko, C., J. D. Cook, and C. A. Finch. 1974. Storage iron kinetics. VI. The effect of inflammation on iron exchange in the rat. *Br. J. Haematol.* **28:**67–75.

65. Hoff, R. 1975. Killing *in vitro* of *Trypanosoma cruzi* by macrophages from mice immunized with *T. cruzi* or BCG and absence of cross immunity on challenge *in vivo*. *J. Exp. Med.* **142:**299–311.

66. Hondalus, M. K., M. S. Diamond, L. A. Rosenthal, T. A. Springer, and D. M. Mosser. 1993. The intracellular bacterium *Rhodococcus equi* requires Mac-1 to bind to mammalian cells. *Infect. Immun.* **61:**2919–2929.

67. Horwitz, M. A., and S. C. Silverstein. 1981. Activated human monocytes inhibit the intracellular multiplication of the Legionnaires' disease bacterium. *J. Exp. Med.* **154:**1618–1635.

68. Hoy, T. G., and A. Jacobs. 1981. Ferritin polymers and the formation of haemosiderin. *Br. J. Haematol.* **49:**593–602.

69. Imlay, J. A., and S. Linn. 1986. Bimodal pattern of killing of DNA-repair-defective or anox-

ically grown *Escherichia coli* by hydrogen peroxide. *J. Bacteriol.* **166**:797–799.

70. **Imlay, J. A., and S. Linn.** 1987. Mutagenesis and stress responses induced in *Escherichia coli* by hydrogen peroxide. *J. Bacteriol.* **169**:2967–2976.

71. **Ishibashi, Y., and T. Arai.** 1990. Roles of the complement receptor type 1 (CR1) and type 3 (CR3) in phagocytosis and subsequent phagosome-lysosome fusion in *Salmonella*-infected murine macrophages. *FEMS Microbiol. Immunol.* **64**:89–96.

72. **Jacobs, A.** 1977. Low molecular weight intracellular iron transport compounds. *Blood* **50**:433–439.

73. **Jacobs, R. F., R. M. Locksley, C. B. Wilson, J. E. Haas, and S. J. Klebanoff.** 1984. Interaction of primate alveolar macrophages and *Legionella pneumophila*. *J. Clin. Invest.* **73**:1515–1523.

74. **Jiang, X., and C. L. Baldwin.** 1993. Effects of cytokines on intracellular growth of *Brucella abortus*. *Infect. Immun.* **61**:124–134.

75. **Johnston, R. B., C. A. Godzik, and Z. A. Cohn.** 1978. Increased superoxide anion production by immunologically activated and chemically elicited macrophages. *J. Exp. Med.* **148**:115–127.

76. **Joiner, K. A., T. Ganz, J. Albert, and D. Rotrosen.** 1989. The opsonizing ligand on *Salmonella typhimurium* influences incorporation of specific, but not azurophil granule constituents into neutrophil phagosomes. *J. Cell Biol.* **109**:2771–2782.

77. **Kagaya, K., K. Watanabe, and Y. Fukazawa.** 1989. Capacity of recombinant gamma interferon to activate macrophages for salmonella-killing activity. *Infect. Immun.* **57**:609–615.

78. **Kaplan, G., and G. Gaudernack.** 1982. *In vivo* differentiation of human monocytes. Differentiation in monocyte phenotypes induced by cultivation on glass or on collagen. *J. Exp. Med.* **156**:1011–1014.

79. **Kay, M. M. B.** 1989. Recognition and removal of senescent cells, p. 17–33. *In* M. D. Sousa and J. H. Brock (ed.), *Iron in Immunity, Cancer and Inflammation*. John Wiley & Sons, Inc., New York.

80. **Kinderlen, A. F., S. H. E. Kaufmann, and M.-L. Lohmann-Matthes.** 1984. Protection of mice against the intracellular bacterium *Listeria monocytogenes* by recombinant immune interferon. *Eur. J. Immunol.* **14**:964–967.

81. **Kleber, E. E., S. R. Lynch, B. Skikne, J. D. Torrance, T. H. Bothwell, and R. W. Charlton.** 1978. Erythrocyte catabolism in the inflammatory peritoneal monocyte. *Br. J. Haematol.* **39**:41–54.

82. **Kossack, R. E., R. L. Guerrant, P. Densen, J. Schadelin, and G. L. Mandell.** 1981. Diminished neutrophil oxidative metabolism after phag-

ocytosis of virulent *Salmonella typhi*. *Infect. Immun.* **31**:674–678.

83. **Kusunose, E., K. Ichihara, Y. Noda, and M. J. Kusunose.** 1976. Superoxide dismutase from *Mycobacterium tuberculosis*. *Biochem. J.* **80**:1343–1352.

84. **Kwaik, Y. A., B. I. Eisenstein, and N. C. Engelberg.** 1993. Phenotypic modulation by *Legionella pneumophila* upon infection of macrophages. *Infect. Immun.* **61**:1320–1329.

85. **Lambrecht, R. S., and M. T. Collins.** 1993. Inability to detect mycobactin in mycobacteria-infected tissues suggests an alternative iron acquisition mechanism by mycobacteria *in vivo*. *Microb. Pathog.* **14**:229–238.

86. **Law, S. K.** 1988. C3 receptors on macrophages. *J. Cell. Sci. Suppl.* **9**:67–97.

87. **Lehrer, R., T. Ganz, and M. E. Selsted.** 1991. Defensins: endogenous antibiotic peptides of animal cells. *Cell* **64**:229–230.

88. **Lehrer, R. I., and T. Ganz.** 1990. Antimicrobial polypeptides of human neutrophils. *Proc. Natl. Acad. Sci. USA* **76**:2169–2181.

89. **Lepay, D. A., C. F. Nathan, R. M. Steinman, H. W. Murray, and Z. A. Cohn.** 1985. Murine Kupffer cells. Mononuclear phagocytes deficient in the generation of reactive oxygen intermediates. *J. Exp. Med.* **161**:1079–1096.

90. **Levitz, S. M., and A. Tabuni.** 1991. Binding of *Cryptococcus neoformans* by human cultured macrophages: requirements for multiple complement receptors and actin. *J. Clin. Invest.* **87**:528–535.

91. **Locksley, R. M., R. F. Jacobs, C. B. Wilson, W. M. Weaver, and S. J. Klebanoff.** 1982. Susceptibility of *Legionella pneumophila* to oxygen-dependent microbicidal systems. *J. Immunol.* **129**:2192–2197.

92. **Loo, V. G., and R. G. Lalonde.** 1984. Role of iron in intracellular growth of *Trypanosoma cruzi*. *Infect. Immun.* **45**:726–730.

93. **Maynard, Y., and J. U. Baenzinger.** 1981. Oligosaccharide specific endocytosis by isolated rat hepatic reticuloendothelial cells. *J. Biol. Chem.* **256**:8063–8068.

94. **Miller, R. M., J. Garbus, and R. B. Hornick.** 1972. Lack of enhanced oxygen consumption by polymorphonuclear leucocytes on phagocytosis of virulent *Salmonella typhi*. *Science* **175**:1010–1011.

95. **Mosser, D. M., and P. J. Edelson.** 1985. The mouse macrophage receptor for iC3b (CR3) is a major mechanism in the phagocytosis of *Leishmania* promastigotes. *J. Immunol.* **135**:2785–2789.

96. **Mosser, D. M., and E. Handman.** 1992. Treatment of murine macrophages with interferon-gamma inhibits their ability to bind *Leishmania* promastigotes. *J. Leukocyte Biol.* **52**:369–376.

97. **Muotiala, A., and P. H. Mäkelä.** 1990. The role of IFN-gamma in murine *Salmonella typhimurium* infection. *Microb. Pathog.* **8:**135–141.

98. **Murray, H., A. Szuro-Sudol, D. Wellner, C. Rothermehl, M. Oca, and A. Granger.** 1988. Interferon induced tryptophan degradation and macrophage antimicrobial activity. *Clin. Res.* **36:**580A.

99. **Murray, H. W.** 1982. Cell-mediated immune response in experimental visceral Leishmaniasis. II. Oxygen dependent killing of intracellular *Leishmania donovani* amastigotes. *J. Immunol.* **129:** 351–356.

100. **Muy-Rivera, M., C. J. Lipps, Y. Cho, R. I. Lehrer, and F. Heffron.** A mutation in the tol import system (tolB) or in the disulfide bond formation (dsbB) genes of *Salmonella typhimurium* eliminates virulence in mice. Submitted for publication.

101. **Nacy, C. A., and M. S. Melzer.** 1984. Macrophages in resistance to rickettsial infections: protection against *Rickettsia tsutsugamushi* infections by treatment of mice with macrophage activating agents. *J. Leukocyte Biol.* **35:**385–396.

102. **Nash, T. W., D. M. Libby, and M. A. Horwitz.** 1988. IFN-gamma activated human alveolar macrophages inhibit the intracellular multiplication of *Legionella pneumophila*. *J. Immunol.* **140:**3978–3981.

103. **Nathan, C. F., and J. B. Hibbs.** 1991. Role of nitric oxide synthesis in macrophage antimicrobial activity. *Curr. Opin. Immunol.* **3:**65.

104. **Nathan, C. F., H. W. Murray, M. E. Wiebe, and B. Y. Rubin.** 1983. Identification of interferon-gamma as the lymphokine that activates human macrophage oxidative metabolism and antimicrobial activity. *J. Exp. Med.* **158:**670.

105. **Nathan, C. F., T. J. Prendergast, M. E. Wiebe, E. R. Stanley, E. Platzer, H. G. Remold, K. Welte, B. Y. Rubin, and H. W. Murray.** 1984. Activation of human macrophages. Comparison of other cytokines with interferon. *J. Exp. Med.* **160:**600–605.

106. **Natvig, D. O., K. Imlay, D. Touati, and R. A. Hallewell.** 1987. Human CuZn-superoxide dismutase complements superoxide dismutase deficient *Escherichia coli* mutants. *J. Biol. Chem.* **262:**14697–14701.

107. **Newman, S. L., R. A. Musson, and P. M. Henson.** 1980. Development of functional complement receptors during *in vitro* maturation of human monocytes into macrophages. *J. Immunol.* **125:**2236–2244.

108. **Newman, S. L., and M. A. Tucci.** 1990. Regulation of human monocyte/macrophage function by extracellular matrix. Adherence of monocytes to collagen matrices enhances phagocytosis of opsonized bacteria by activation of

complement receptors and enhancement of Fc receptors function. *J. Clin. Invest.* **86:**703–714.

109. **North, R. J.** 1970. The relative importance of blood monocytes and fixed macrophages to the expression of cell mediated immunity to infection. *J. Exp. Med.* **132:**521–534.

110. **O'Brian, S., P. S. Jackett, D. B. Lowrie, and P. W. Andrew.** 1991. Guinea pig alveolar macrophages kill *Mycobacterium tuberculosis in vitro*, but killing is independent of susceptibility to hydrogen peroxide or triggering of the respiratory burst. *Microb. Pathog.* **10:**199–207.

111. **O'Connell, M. J., B. Halliwell, C. P. Moorhouse, O. I. Aruoma, H. Baum, and T. J. Peters.** 1986. Formation of hydroxyl radicals in the presence of ferritin and haemosiderin. Is haemosiderin formation a biological protection mechanism? *Biochem. J.* **234:**727–731.

112. **O'Connell, M. J., R. J. Ward, H. Baum, and T. J. Peters.** 1986. *In vitro* and *in vivo* studies on the availability of iron from storage proteins to stimulate membrane lipid peroxidation, p. 29–37. *In* C. Rice-Evans (ed.), *Free Radicals. Cell Damage and Disease*. Richelieu Press, London.

113. **Panosian, C. B., and C. J. Wyler.** 1983. Acquired macrophage resistance to *in vitro* infection with leishmania. *J. Infect. Dis.* **148:**1049–1054.

114. **Passwell, J. H., R. Shor, and J. Shoham.** 1986. The enhancing effect of interferon-beta and -gamma on the killing of *Leishmania tropica* major in human mononuclear phagocytes *in vitro*. *J. Immunol.* **136:**3062–3066.

115. **Payne, N. R., and M. A. Horwitz.** 1987. Phagocytosis of *Legionella pneumophila* is mediated by human monocyte complement receptors. *J. Exp. Med.* **166:**1377–1389.

116. **Pfefferkorn, E. R., M. Eckel, and S. Rebhuhm.** 1986. Interferon-gamma supresses the growth of *Toxoplasma gondii* in human fibroblasts through starvation for tryptophan. *Mol. Biochem. Parasitol.* **20:**215–224.

117. **Pommier, C. G., S. Inada, L. F. Fries, T. Takahashi, M. M. Frank, and E. J. Brown.** 1983. Plasma fibronectin enhances phagocytosis of opsonized particles by human peripheral blood monocytes. *J. Exp. Med.* **157:**1844–1854.

118. **Pontow, S. E., V. Kery, and P. D. Stahl.** 1991. Mannose receptor. *Int. Rev. Cytol.* **137:** 221–244.

119. **Popoff, R. K., A. I. Spira, A. W. Johnson, and B. Demple.** 1990. Yeast structural gene (APN1) for the major apurinic endonuclease: homology to *Escherichia coli* endonuclease IV. *Proc. Natl. Acad. Sci. USA* **87:**4193–4197.

120. **Portnoy, D. A., R. D. Schreiber, P. Conelli, and L. G. Tilney.** 1989. Gamma interferon limits access of *Listeria monocytogenes* to the mac-

rophage cytoplasm. *J. Exp. Med.* **170:**2141–2146.

121. **Prina, E., J. C. Antoine, B. Wiederanders, and H. Kirschke.** 1990. Localization and activity of various lysosomal proteases in *Leishmania amazonensis*-infected macrophages. *Infect. Immun.* **58:**1730–1737.

122. **Reeve, C. A., and A. T. Bockman.** 1984. Role of protein degradation in survival of carbon-starved *Escherichia coli* and *Salmonella typhimurium*. *J. Bacteriol.* **157:**758–763.

123. **Roecklein, J. A., R. P. Schwarz, and H. Yeager.** 1992. Nonopsonic uptake of *Mycobacterium avium* complex by human monocytes and alveolar macrophages. *J. Lab. Clin. Med.* **119:**772–781.

124. **Roeser, H. P.** 1980. Iron metabolism in inflamation and malignant disease, p. 605–640. *In* A. J. A. M. Worwood (ed.), *Iron in Biochemistry and Medicine.* Academic Press, London.

125. **Roland, K. L., L. E. Martin, C. R. Esther, and J. K. Spitznagel.** 1993. Spontaneous pmrA mutants of *Salmonella typhimurium* LT2 define a new two-component regulatory system with a possible role in virulence. *J. Bacteriol.* **175:**4154–4164.

126. **Rook, G. A. W., J. Taverne, C. Leveton, and J. Steele.** 1987. The role of gamma-interferon, vitamin D3 metabolites and tumor necrosis factor in the pathogenesis of tuberculosis. *Immunology* **62:**229–234.

127. **Roop, R. M., T. W. Fletcher, N. M. Siranganathan, S. M. Boyle, and G. G. Schurig.** 1994. Identification of an immunoreactive *Brucella abortus* HtrA stress response protein homolog. *Infect. Immun.* **62:**1000–1007.

128. **Ross, G. D., J. A. Cain, and P. J. Lachmann.** 1985. Membrane complement receptor type three (CR3) has lectin-like properties analogous to bovine conglutinin and functions as a receptor for zymosan and rabbit erythrocytes as well as a receptor for iC3b. *J. Immunol.* **134:**3307–3315.

129. **Saito, K., T. Nishisato, G. A. Grasso, and P. Aisen.** 1986. Interaction of transferrin with iron-loaded rat peritoneal macrophages. *Br. J. Haematol.* **62:**275–286.

130. **Sawyer, J. G., N. L. Martin, and R. E. W. Hancock.** 1988. Interaction of macrophage cationic proteins with the outer membrane of *Pseudomonas aeruginosa. Infect. Immun.* **56:**693–698.

131. **Schlesinger, L. S.** 1993. Macrophage phagocytosis of virulent but not attenuated strains of *Mycobacterium tuberculosis* is mediated by mannose receptors in addition to complement receptors. *J. Immunol.* **150:**2920–2930.

132. **Schlesinger, L. S., C. G. Bellinger-Kawahara, N. R. Payne, and M. A.**

Horwitz. 1990. Phagocytosis of *Mycobacterium tuberculosis* is mediated by human monocyte complement receptors and complement component C3. *J. Immunol* **144:**2771–2780.

133. **Schlesinger, L. S., and M. A. Horwitz.** 1990. Phagocytosis of leprosy bacilli is mediated by complement receptors CR1 and CR3 on human monocytes and complement component C3 in serum. *J. Clin. Invest.* **85:**1304–1314.

134. **Schlesinger, L. S., and M. A. Horwitz.** 1991. Phagocytosis of *Mycobacterium leprae* by human monocyte-derived macrophages is mediated by complement receptors CR1 (CD35), CR3 (CD11b/CD18), and CR4 (CD11c/CD18) and interferon gamma activation of this bacterium. *J. Immunol.* **147:**1983–1994.

135. **Schröder, H., T. Langer, F.-U. Hertl, and B. Bukau.** 1993. DnaK, DnaJ and GrpE form a cellular chaperone machinery capable of repairing heat-induced protein damage. *EMBO J.* **12:**4137–4144.

136. **Shafer, W. M., S. G. Casey, and J. K. Spitznagel.** 1984. Lipid A and resistance of *Salmonella typhimurium* to antimicrobial granule proteins of human neutrophils. *Infect. Immun.* **43:**834–838.

137. **Shepherd, V. L., E. J. Campbell, R. M. Senior, and P. D. Stahl.** 1982. Characterization of mannose/fucose receptor on human mononuclear phagocytes. *J. Reticuloendothel. Soc.* **32:**423–431.

138. **Sibley, D. L., J. L. Krahlenbuhl, and E. Weidner.** 1985. Lymphokine activation of J774G8 cells and mouse peritoneal macrophages challenged with *Toxoplasma gondii. Infect. Immun.* **49:**760–764.

139. **Smith, R. M., and J. T. Curnette.** 1991. Molecular basis of chronic granulomatous disease. *Blood* **77:**673–686.

140. **Spector, M. P., Y. K. Park, S. Tirgari, T. Gonzales, and J. W. Foster.** 1988. Identification and characterization of starvation-regulated genetic loci in *Salmonella typhimurium* by using Mu d-directed *lacZ* operon fusions. *J. Bacteriol.* **170:**345–351.

141. **Speert, D. P., and S. C. Silverstein.** 1985. Phagocytosis of unopsonized zymosan by human monocyte derived macrophages: maturation and inhibition by mannan. *J. Leukocyte Biol.* **38:**655–658.

142. **Stahl, P., and S. Gordon.** 1982. Expression of a mannosyl-fucosyl receptor for endocytosis on cultured primary macrophages and their hybrids. *J. Cell Biol.* **93:**49–56.

143. **Stinavage, P., L. E. Martin, and J. K. Spitznagel.** 1989. O-antigen and lipid A phosphoryl groups in resistance of *Salmonella typhimurium* LT2 to nonoxidative killing in human polymorphonuclear cells. *Infect. Immun.* **57:**3894–3900.

144. **Stoljiljkovic, I., and K. Hantke.** 1992. Hemin uptake system of *Yersinia enterocolitica:* similarities with other TonB-dependent systems in Gram negative bacteria. *EMBO J.* **11:**4359–4367.

145. **Storz, G., L. A. Tartaglia, B. Farr, and B. N. Ames.** 1990. Bacterial defense against oxidative stress. *Trends Genet.* **6:**363–368.

146. **Strauch, K. L., and J. Beckwith.** 1988. An *Escherichia coli* mutant preventing degradation of abnormal periplasmic proteins. *Proc. Natl. Acad. Sci. USA* **85:**1576–1580.

147. **Stuehr, D. J., and M. A. Marletta.** 1987. Induction of nitrite/nitrate synthesis in murine macrophages by BCG infection, lymphokines, or interferon-gamma. *J. Immunol.* **139:**518–525.

148. **Sword, C. P.** 1966. Mechanism of pathogenesis in *Listeria monocytogenes* infection. I. Influence of iron. *J. Bacteriol.* **92:**536–542.

149. **Vaara, M., T. Vaara, M. Jenson, I. Helander, M. Nurminen, R. T. Rietschel, and P. H. Makela.** 1981. Characterization of the lipopolysaccharide from polymyxin-resistant pmrA mutants of *Salmonella typhimurium. FEBS Lett.* **129:**145–149.

150. **Vanotti, A.** 1957. The role of the reticuloendothelial system in iron metabolism, p. 172–187. *Physiopathology of the Reticulo-endothelial System.* Blackwell Scientific Publications, Oxford.

151. **Vladoianu, I.-R., H. R. Chang, and J.-C. Pechére.** 1990. Expression of host resistance to *Salmonella typhi* and *Salmonella typhimurium:* bacterial survival within macrophages of murine and human origin. *Microb. Pathog.* **8:**83–90.

152. **Welch, D. F., C. P. Sword, S. Brehm, and D. Duxanis.** 1979. Relationship between superoxide dismutase and pathogenic mechanisms of *Listeria monocytogenes. Infect. Immun.* **23:**863–872.

153. **Wilson, C. B., V. Tsai, and J. S. Remington.** 1980. Failure to trigger the oxidative burst by normal macrophages. Possible mechanisms for survival of intracellular pathogens. *J. Exp. Med.* **151:**328–346.

154. **Wilson, M. E., and R. D. Pearson.** 1988. Roles of CR3 and mannose receptors in the attachment and ingestion of *Leishmania donovani* by human mononuclear phagocytes. *Infect. Immun.* **56:**363–369.

155. **Wozencraft, A. O., G. Sayers, and J. M. Blackwell.** 1986. Macrophage type 3 complement receptors mediate serum-independent binding of Leishmania donovani. *J. Exp. Med.* **164:**1332–1337.

156. **Wright, S. D., and M. T. C. Jong.** 1986. Adhesion promoting receptors on human macrophages recognize *Escherichia coli* by binding to lipopolysaccharide. *J. Exp. Med.* **164:**1876–1888.

157. **Wright, S. D., and S. C. Silverstein.** 1983. Receptors for C3b and C3bi promote phagocytosis but not the release of toxic oxygen from human phagocytes. *J. Exp. Med.* **158:**2016–2023.

158. **Yamamoto, K., and R. Johnston.** 1984. Dissociation of phagocytosis from stimulation of the oxidative burst in macrophages. *J. Exp. Med.* **159:**405–416.

FRANCISELLA TULARENSIS, A MODEL PATHOGEN TO STUDY THE INTERACTION OF FACULTATIVE INTRACELLULAR BACTERIA WITH PHAGOCYTIC HOST CELLS

*Anne H. Fortier, Carol A. Nacy,
Tammy Polsinelli, and Nirupama Bhatnagar*

9

Successful intracellular pathogens develop unique strategies for survival in phagocytic cells, and these strategies are as diverse as the organisms themselves. The interactions between a particular facultative intracellular pathogen and its phagocytic host cell, however, are defined not only by pathogen factors but also by host factors, which are as diverse as those of the pathogen itself. Many phagocytic host cells with differing innate antimicrobial abilities interact with an invading pathogen during onset of infection and induction of immunity. The tremendous diversity in different host pathogen interactions makes a global description of such interactions nearly impossible. What is clear, however, is that there are two competing imperatives for each of these players: homeostasis, or maintenance of tissue integrity, is the priority of the host cell, and survival and replication are the priorities of the pathogen. The agenda for the infectious process is not set by either the host or the pathogen but is instead a dynamic process that changes as the pathogen persists. The host reacts and responds to pathogen persistence by escalating the intensity and diversity of its immunologic response, and the pathogen evades or subverts as many of these host defenses as it can.

In this chapter, we first describe host defensive strategies against infection, including barrier defenses, early inflammatory changes, and development of specific immunity. We then describe some of the more interesting bacterial mechanisms employed by selected intracellular organisms that ensure pathogen survival in the face of potent host resistance factors. Finally, we describe *Francisella tularensis* as a model intracellular pathogen. This bacterium interacts with host phagocytic cells that represent the entire gamut of host responses: resting tissue macrophages, inflammatory cells, and activated effector cells of immunity. While what we learn about each individual host-pathogen interaction may not be applicable to all pathogens or all interactions, we may gain an understanding of host capabilities for resistance to infectious diseases.

THE HOST IMPERATIVE: RESTORATION OF HOMEOSTASIS

All host responses during infection and development of immunity are driven by the host imperative to maintain or restore homeostasis. In an effort to conserve energy and preserve host tissue integrity, these protective host re-

Anne H. Fortier and Carol A. Nacy EntreMed, Inc., Rockville, Maryland 20850. *Tammy Polsinelli* Walter Reed Army Institute of Research, Washington, D.C. 20307-5100. *Nirupama Bhatnagar* Food and Drug Administration, Bethesda, Maryland 20892.

Virulence Mechanisms of Bacterial Pathogens, 2nd ed., Edited by J. A. Roth et al.
© 1995 American Society for Microbiology, Washington, D.C.

sponses occur in stages, each response building on the previous one. The first line of host defense to be overcome by the successful pathogen is that provided by the integrity of barrier defenses (physical and mechanical) seen in the gastrointestinal tract, on mucosal surfaces, and on the skin. *Pseudomonas* spp., *Bordetella* spp., and some enteropathogenic *Escherichia coli* strains do not traverse the mucosal linings of either the respiratory or the intestinal tract (19, 54, 63), and pathologic changes are localized to these mucosal surfaces. If the pathogen is inoculated into tissue that is not protected by a physical or mechanical barrier, the host has at its disposal a variety of effective innate and nonspecific early defense mechanisms that act quickly and locally to impart resistance to disease.

The first host response to invasion of tissue is inflammation: activation of complement, generation of chemotactic factors, recruitment of polymorphonuclear leukocytes (PMNs), and production of acute-phase reactants. During the initial inflammatory response to any invading microorganism, the first migratory cells to arrive at the site of infection are the PMNs, cells well equipped for the job of ingestion and elimination of bacterial invaders. Receptors for immunoglobulin G (FcR) and complement fragments (CR-3) that are present on PMNs function (alone or in concert) to promote phagocytosis of extracellular bacteria opsonized by antibody or complement components (44). Internalization exposes bacteria to PMN granules containing degradative enzymes, bacterial-permeability-increasing proteins, myeloperoxidase, and lactoferrin. Many bacterial pathogens are susceptible to the antibacterial effects of these early inflammatory cells (34, 41). PMNs, however, are short-lived cells that die in their effort to eliminate invading pathogens. In the process of PMN degranulation, intracellular components, which turn the extracellular tissue space into a truly noxious environment, are released into affected tissue. Many extracellular bacteria do not survive this toxic environment.

The PMN infiltrate is followed within hours by inflammatory macrophages newly immigrated into tissue from the circulation; these cells rapidly replace PMNs as the predominant cell type at the site of infection. If the PMNs successfully eliminate the viable pathogen from the site of infection, the function of inflammatory macrophages is to initiate healing of the affected area and restore tissue integrity. If the PMNs are not successful, inflammatory macrophages become the next line of tissue defense.

Compared with differentiated tissue macrophages, inflammatory macrophages have high levels of certain enzymes (14), increased oxidative metabolism (14), and a tremendous secretory capacity (4, 64). They arrive at the site of inflammation to restore tissue integrity and phagocytose cellular debris and any remaining infectious agents not eliminated by the PMNs. They also efficiently kill certain facultative intracellular pathogens, such as *Listeria monocytogenes,* without immune stimulation (15) but are less able to kill others (33).

If the host's natural barrier defenses and first-line inflammatory (i.e., innate) responses are unable to control infection, the host then unleashes its immune arsenal. As the pathogen resists, evades, or undermines the early defenses and colonizes tissue, the host responds to the pathogen by increasing the intensity of the previously activated defenses and activating new responses that address the pathogen specifically. Activation of antigen-specific T cells following challenge results in the secretion of a variety of cytokines that act on macrophages to change their physiology and function. The macrophages acquire new skills and become potent, nonspecific cytotoxic effector cells (52). Cytokine-activated macrophages have many potential antimicrobial effector activities not present in nonactivated macrophages, including downregulation of transferrin receptors for decreasing intracellular iron (9), degradation of tryptophan (10, 51), and production of such effector molecules as reactive oxygen and nitrogen intermediates (28, 58). Recent studies with inhibitors of nitric oxide (NO) synthesis suggest that NO is involved in the destruction of a number of intracellular and extracellular microbial agents by murine macrophages (18, 29, 35).

Tissue damage and systemic toxicity can result from production of toxic oxidative and nonoxidative metabolites by macrophages and dead or dying cells and from high circulating levels of cytokines with pathologic effects (especially tumor necrosis factor alpha and interleukin-1) (60, 62). If the pathogen is totally eliminated and the host recovers (sterile immunity), long-lived immunity results and the infected tissue returns to normal. If, however, the vigorous host response sets up a balance between pathogen persistence and tissue integrity (nonsterile immunity), reactivation of the pathogen and symptoms of disease can occur when the host immune system is compromised. Pathogens such as *Mycobacterium tuberculosis* and *Chlamydia* spp. (5) persist in tissues in the face of strong immunologic memory.

PATHOGEN QUALITIES THAT FAVOR SURVIVAL IN PHAGOCYTIC CELLS

Most pathogens that invade tissues and stimulate the inflammatory response are eliminated through ingestion and killing by PMNs. Production of a capsule is one mechanism that certain *E. coli* strains and other bacteria use to evade PMN phagocytosis (34). For others, such as *F. tularensis,* complement activation does not occur; thus, they avoid opsonic phagocytosis mediated by C3bi (56). If a pathogen cannot avoid phagocytosis by PMNs, then either it is killed or it finds other ways to prevent the lethal consequences of phagocytosis. *Bordetella pertussis,* for example, avoids intracellular destruction by preventing phagolysosome fusion (63). The successful pathogen must do more than simply avoid phagocytosis by PMNs, or even survive in these cells, since PMNs are so short-lived. For long-term survival, most intracellular pathogens must sequester themselves in a stable intracellular niche protected from the detrimental effects of extracellular complement and immunoglobulin.

Macrophages, whose life spans are measured in months rather than days, provide a mobile intracellular environment that satisfies the nutritional requirements of several interesting pathogens. Intracellular pathogens must enter cells in order to replicate, so they develop unique and effective mechanisms to ensure entry. Macrophages ingest most particulates and many bacteria by endocytosis or phagocytosis: both processes mobilize microfilaments to bring things into the cell. Cytochalasin B, a microfilament inhibitor (17), blocks uptake of certain intracellular bacteria (*Salmonella* spp., *Shigella* spp., *L. monocytogenes*) that are dependent on phagocytosis for entry into the cell (21, 38). *Chlamydia* spp. enter host cells by a mechanism insensitive to cytochalasin B (31, 69): elementary bodies, the infectious particles of *Chlamydia* spp., are apparently too small to be susceptible to cytochalasin B inhibition (17, 69). On the other hand, several *Rickettsia* spp. actively penetrate the host cell rather than utilize host-driven mechanisms of phagocytosis (68).

Once inside the cell, intracellular pathogens must subvert the hostile environment of a phagosome or phagolysosome in order to replicate. Most intracellular organisms have developed interesting and diverse strategies for survival and replication within macrophages. Some organisms fail to trigger a respiratory burst as they enter the cell (50), thereby avoiding toxic oxygen radicals; some neutralize the potent reactive oxygen intermediates released in the respiratory burst during phagocytosis (50). *Rickettsia* and *Listeria* spp. evade destruction in the phagolysosome, the digestive organelle of the macrophage, by replicating in the cytoplasm (66, 68); *Toxoplasma* and *Mycobacterium* spp. and *Bordetella pertussis* actively prevent lysosome fusion (3, 25, 36); and some intracellular pathogens, like the protozoan parasites *Leishmania* spp., replicate in the phagolysosome, using the low pH of this organelle to their enzymatic advantage (12, 13). The most successful of the intracellular pathogens (such as *Brucella, Yersinia, Francisella, Listeria, Legionella,* and *Salmonella* spp.) acquire multiple evasive strategies to ensure survival. Most of these pathogens, however, have not yet evolved mechanisms that allow them to survive within activated macrophages, a cell type that is qualitatively changed following immune activation (e.g., interaction with gamma interferon [IFN-γ]).

Cytokine-activated macrophages have at their disposal a number of effector mechanisms for the elimination of intracellular pathogens and the restoration of homeostasis. Among the potent extraordinary cytotoxic mechanisms stimulated during immune reactions are downregulation of transferrin receptors in order to limit the iron essential for certain pathogen survival, i.e., *Legionella* spp. (9), in human monocytes; degradation of tryptophan, a compound essential for survival of *Chlamydia* spp. (10); induction of nitric oxide to kill *M. tuberculosis* (18); and sequestration of iron from *Francisella* (22, 23) and *Leishmania* (28) spp.

F. TULARENSIS: A MODEL INTRACELLULAR PATHOGEN

There is no single paradigm of host-pathogen interactions that covers all bacteria and all host cells. The interaction between a particular pathogen and its host cell reflects an intimate and dynamic relationship peculiar to the agendas of the specific players. Understanding the unique characteristics of each host-pathogen relationship, however, adds to our appreciation of the depth of host capabilities for protection against threats to its delicate homeostatic balance and of pathogen capabilities for survival within the host.

F. tularensis, the etiologic agent of tularemia, is a tiny, pleomorphic, gram-negative bacterium that can infect mammals through either the dermal or the respiratory route. The usual method of transmission from infected animals to humans is through an accidental laceration or the creation of an aerosol during the skinning of infected game (trapper's fever). Depending on the route of infection, failure to treat tularemia with antibiotics results in up to 60% mortality in humans (65). *F. tularensis* is a fastidious organism that requires a complex

medium for isolation and growth in vitro (11). Both cysteine and a source of iron are nutritionally essential for extracellular replication of this pathogen (11). Although *F. tularensis* can be cultured on synthetic medium without cells, there are no data to support its extracellular growth in vivo in experimental animals or humans (72). In fact, PCR analysis of blood collected from *F. tularensis*-infected mice suggests that there is little to no bacteremia during the course of lethal disease with this pathogen. Organisms are not found in serum but only within circulating leukocytes (42). Further, electron microscopy studies of peritoneal cells from mice infected intraperitoneally show *F. tularensis* only within macrophages. In vivo, then, *F. tularensis* is an intracellular pathogen, and its principal host cell for intracellular replication is the macrophage (30, 45).

Although the mechanism(s) involved in the ingestion of *F. tularensis* by phagocytes is unknown, there are differences between macrophage populations even at this early phase of host-pathogen interaction. Given the same bacterial inoculum, alveolar macrophages ingest 10-fold more *F. tularensis* than do either resident or inflammatory peritoneal cells. Since both alveolar and peritoneal macrophages have equivalent nonspecific phagocytic abilities (53) (Table 1), these data suggest that different mechanisms may be involved in *F. tularensis* ingestion by different macrophage subpopulations.

Following entry, *F. tularensis* replicates within an endocytic vacuole in macrophages

TABLE 1 *Francisella* growth in murine macrophages[a]

Macrophage population	CFU/ml (10^{-3}) at:	
	2 h	72 h
Resident peritoneal	3	80,000
Inflammatory peritoneal	4	20,000
Resident alveolar	60	10,000

[a]Cells were adjusted to 10^6 macrophages per ml and cultured for 2 h as nonadherent cell pellets in polypropylene tubes with 10^6 *Francisella* CFU per ml. Following initial infection, cells were washed and analyzed for cell-associated bacteria by serial dilution of lysed cells and plate counts immediately (2 h) and at 72 h.

infected in vitro (1) or in vivo (22). *F. tularensis*-containing vacuoles rarely fuse with pre-labeled lysosomes (1). Whether these bacteria, like *Mycobacteria* spp. (3) and *Toxoplasma gondii* (36), actively prevent lysosomal fusion is not yet clear. The important characteristics of this intracellular compartment for pathogen survival were evaluated with lysosomotropic agents (NH₄Cl, chloroquine, and ouabain) that raise the intracellular pH (22). Each of these reagents blocks acidification by a different mechanism (47): regardless of the mechanism, however, *F. tularensis* cannot replicate in treated cells (Table 2). The requirement for a low intracellular pH may reflect *F. tularensis*'s strict requirement for iron for growth (67). Transferrin receptors at the macrophage cell surface transport transferrin-bound iron into the cell and release iron when endocytic vesicles containing internalized transferrin receptors become acidified. Low pH is the catalyst for this event (7, 16). That the acidophilic nature of *F. tularensis* is related to its iron requirements is supported indirectly by the fact that deferroxamine, a soluble iron chelator added to *F. tularensis*-infected macrophage cultures, suppresses intracellular bacterial growth (22) and is supported directly by reversal of the effects of

lysosomotropic agents with iron chelators that have a low affinity for iron at neutral pH (PP$_i$) but not by agents that have a high affinity for iron at a neutral pH (transferrin) (Table 2). Similar physiology characterizes the interaction between *Salmonella typhimurium* and murine macrophages (i.e., extracellular iron-loaded transferrin traffics to the bacterium-containing vacuole [55]). Infective *F. tularensis,* then, localizes to a macrophage-acidified vesicle but not necessarily a phagolysosome, and the importance of vacuolar acidification is its capacity to enhance the availability of iron for the intracellular pathogen.

All macrophage populations are susceptible to infection with *F. tularensis* in vitro (Table 1), and all are capable of sustaining intracellular replication of this pathogen (Fig. 1). Unlike *Listeria* and *Salmonella* spp., whose numbers are substantially depleted in the initial 30-min interaction with macrophages (15, 21), *F. tularensis* is resistant to this natural macrophage antimicrobial arsenal. Among other protective defenses, *F. tularensis* has catalase, which eliminates H₂O₂ produced by macrophages as an effector molecule. *F. tularensis* is, however, susceptible to hypochlorous acid produced by PMNs (41).

Compared with differentiated tissue macrophages, inflammatory macrophages have high levels of certain enzymes and increased oxidative capacity, and they are excellent cytotoxic cells for extracellular pathogens or tumor targets (14, 52). They also efficiently kill ingested *Listeria* spp. without further immune stimulation (15). *F. tularensis,* however, infects and replicates in inflammatory macrophages (Table 1). In fact, its generation time is actually shorter in inflammatory macrophages (2 to 4 h) than in differentiated tissue macrophages (6 to 8 h) (Fig. 1). The enhanced growth of *F. tularensis* in these cells is perhaps a reflection of the higher metabolic activity of inflammatory macrophages. The implication of these in vitro studies (yet to be confirmed by in vivo experiments) is that the inflammatory process, which is normally beneficial to the host, actually sup-

TABLE 2 Inhibition of intracellular macrophage growth with lysosomotropic agents and reversal with iron[a]

Cell treatment	CFU/ml (10^{-3}) at 72 h
Medium	20,000
+ Holotransferrin	16,000
+ Apotransferrin	23,000
+ Ferric pyrophosphate	19,000
NH₄Cl	50
+ Holotransferrin	3
+ Apotransferrin	6
+ Ferric pyrophosphate	2,200

[a]Cells were adjusted to 10⁶ macrophages per ml and cultured for 2 h as nonadherent cell pellets in polypropylene tubes with 10⁶ *Francisella* CFU/ml. Following initial infection, cells were washed and cultured with 20 mM NH₄Cl and either 2 mg of holotransferrin, 2 mg of apotransferrin, or 200 mg of ferric pyrophosphate per ml and then analyzed for cell-associated bacteria by serial dilution of lysed cells and plate counts at 72 h.

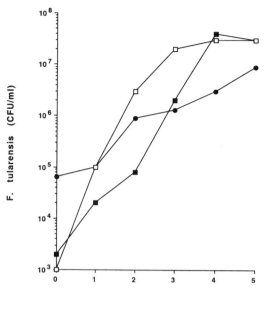

FIGURE 1 *F. tularensis* growth in macrophages. Alveolar (●) (53), inflammatory (induced by intraperitoneal inoculation of 1 ml of 3% Proteose Peptone broth 3 days prior to peritoneal cell harvest) (□) (23), and resident (■) peritoneal cells were harvested by lavage from groups of 5 to 10 mice (23). Cells were adjusted to a density of 10^6 macrophages per ml in Dulbecco modified Eagle medium (GIBCO, Grand Island, N.Y.) with 5% heat-inactivated fetal bovine serum (Sterile Systems, Inc., Logan, Utah) and cultured in 0.5-ml volumes in polypropylene tubes (no. 2063; Falcon, Oxnard, Calif.). Cells were infected with *F. tularensis* LVS at a multiplicity of infection of 1 for 2 h at 37°C with 5% CO_2, washed three times with medium by centrifugation at 700 x *g* at 4°C, and resuspended in 0.5 ml of fresh medium. *F. tularensis* LVS growth in macrophages was assessed by lysis of infected cells with 0.02% sodium dodecyl sulfate at the indicated times. Cell lysates were serially diluted in medium and plated on Mueller-Hinton modified plates (11) for an estimation of CFU.

plies the host cell in which *F. tularensis* replicates with greatest efficiency.

Generally speaking, pathogenic mechanisms used to overcome host defenses and establish infection are accompanied by the expression of one or more virulence factors that meet specific needs of the pathogen as it encounters these host defenses (46). Gram-negative organisms

usually have an endotoxin (lipopolysaccharide) as a component of their outer membrane. Some pathogenic organisms (i.e., *Shigella* and *Yersinia* spp.) also secrete cytotoxic exotoxins (37, 48). When *F. tularensis* is analyzed for classic exotoxins or endotoxins, none are found. *F. tularensis* has a unique lipopolysaccharide with reduced cell toxicity, reduced bioactivity in macrophage cultures, and reduced activity in the *Limulus* assay (57). *F. tularensis* does not have any characterized exotoxins, and cell-free supernatant fluids from *Francisella* cultures are not toxic for mice (24). Thus, all the evidence suggests that the major virulence attribute for *F. tularensis* is the ability to grow in cells, primarily macrophages.

We employed rifampin, which affects transcription by binding to the β subunit of RNA polymerase (71), for selecting mutant colonies resistant to this antibiotic and evaluated the resistant clones for infectivity in our lethal mouse model of *Francisella* infections (24). Organisms resistant to rifampin are frequently less virulent than parent rifampin-susceptible strains. For example, rifampin-resistant *Brucella* spp. with rough colony morphology show reduced virulence in mice (59). Rifampin-resistant mutants of *S. typhimurium* (70) and *Staphylococcus aureus* (49) are also less toxic in mice than their rifampin-susceptible counterparts. Manten and VanWijnggaarden (43) isolated rifampin-resistant strains of tubercle bacilli with low infectivity in guinea pigs.

We also isolated several rifampin-resistant *F. tularensis* clones with reduced virulence for mice (6). In addition, we observed differences between parent and rifampin mutants in response to stresses: rifampin-resistant mutants were unable to grow when the temperature was raised to 42°C and unable to survive the hostile intracellular environment of the macrophage. The inability of these mutants to grow in macrophages was observed both in vitro and in vivo (6). The intracellular environment, low pH, low iron, and toxic oxidative metabolites all invoke a stress, or heat shock, response in several different intracellular pathogens (8, 39, 61). The ability to respond to stressful environments like those found in the intracellular compartment occupied by *F. tularensis* is essential for expression of the virulent *Francisella* phenotype and is accompanied by the expression of certain heat shock proteins (6, 20).

Virulent *F. tularensis* is poorly controlled by the host's natural barrier defenses and first-line inflammatory responses. Macrophages exposed in vitro to cytokines from antigen-specific T lymphocytes, however, develop the capacity to kill intracellular *F. tularensis,* and the predominant cytokine responsible for this antimicrobial activity is IFN-γ (23). All inflammatory or differentiated macrophages acquire this capacity under the right set of circumstances, but the susceptibility of macrophages to activation factors differs. Thus, as little as 5 U of IFN-γ per ml induces 50% maximal intracellular killing activity in inflammatory macrophages, whereas 20 U of IFN-γ per ml induces an equivalent amount of intracellular killing in differentiated peritoneal cells. Differentiated alveolar macrophages are also exquisitely sensitive to IFN-γ, with half-maximal activation achieved at approximately 3 U/ml (23, 53).

Cytokine-activated macrophages have many potential antimicrobial effector activities not present in unstimulated macrophages. Recent studies with inhibitors of nitric oxide (NO) synthesis suggest that this molecule is involved in murine macrophage destruction of a number of intracellular and extracellular microbial targets (18, 29, 35). Nitric oxide, a short-lived reactive nitrogen intermediate, is rapidly oxidized to NO_2^- (26), which can be measured by a simple colorimetric test (27). NO_2^- in culture fluids, then, is a quantitative index of macrophage activation and killing capacity. Production of NO_2^- correlates with anti-*Francisella* activity in murine macrophages activated in vitro with IFN-γ (2, 23) (Table 3). A competitive inhibitor of L-arginine-derived NO, monomethyl-L-arginine blocks IFN-γ-induced anti-*Francisella* activity and production of NO_2^- in both resident and inflammatory peritoneal macrophage populations but not in alveolar macrophages (Table 3). The conclusion from these data is that production of NO is a common response of all murine macrophage populations to IFN-γ stimulation and that NO is the effector molecule used by activated peritoneal macrophages but not alveolar macrophages to eliminate intracellular *F. tularensis.*

L-Arginine-derived NO from activated macrophages exerts its effects by reacting with free iron to form iron-nitrosyl complexes and inactivating the enzymes requiring iron for activity in both the target cell and the macrophage itself (32, 40). Paralysis of iron-dependent enzymes involved in such essential processes as mitochondrial respiration, DNA synthesis, and the Krebs citric acid cycle enables macrophage-derived NO to affect a number of eukaryotic and prokaryotic targets (18, 29, 35). The absolute nutritional requirement of *F. tularensis* for iron suggests that the iron-nitrosyl complexes prevent the bacterium from acquiring this essential nutrient. Thus, IFN-γ induction of NO synthase in activated macrophages results in production of NO, complexing of free iron, and cytostasis of the bacterium. With continual deprivation of iron, *F. tularensis* eventually dies (23). Supporting this concept is the fact that the cytostatic and cytocidal activities of IFN-γ-activated peritoneal macrophages against *F. tularensis* can be completely reversed by supplying an unlimited source of exogenous iron.

In summary, murine macrophages are infected by and support the replication of *F. tularensis* in vitro. Macrophage populations can be activated by IFN-γ to eliminate this intracellular pathogen. Specifically, peritoneal macrophages eliminate the pathogen by generating NO and limiting the iron required for intracellular replication. To date, the bactericidal mechanism employed by IFN-γ-treated alveolar macrophages for anti-*Francisella* activity is unknown. What is known is that macrophage killing mechanisms are as diverse as pathogen evasion mechanisms. Like most successful pathogens that develop multiple evasive strategies for host defenses, the potent effector cells of host defense, the macrophages, have like-

TABLE 3 Correlation between killing and nitrite production in IFN-γ-stimulated murine macrophages[a]

Macrophage population	Cell treatment	Production of NO_2^- (μmol)	CFU/ml (10^{-3}) at 72 h
Resident peritoneal	Medium	0	2,000
	IFN-γ	8	5
	+ MMLA	1	800
	+ Holotransferrin	10	2,300
Inflammatory peritoneal	Medium	0	80,000
	IFN-γ		
	Alone	16	3
	+ MMLA	1	80,000
Alveolar	Medium	0	4,000
	IFN-γ	34	0
	+ MMLA	3	100
	+ Holotransferrin	26	10

[a]Cells were adjusted to 10^6 macrophages per ml and cultured for 2 h as nonadherent cell pellets in polypropylene tubes with 10^6 Francisella CFU/ml. Following initial infection, cells were washed and cultured with medium or 20 U of IFN-γ per ml and either 1 mM MMLA (a competitive inhibitor of L-arginine-derived NO) or 2 mg of holotransferrin per ml. After 72 h, cultures were analyzed for cell-associated bacteria by serial dilution of lysed cells and plate counts.

wise developed the necessary diversification required to combat infectious agents.

REFERENCES

1. **Anthony, L. S. D., R. D. Burke, and F. E. Nano.** 1991. Growth of *Francisella* spp. in rodent macrophages. *Infect. Immun.* **59**:3291–3296.
2. **Anthony, L. S. D., P. J. Morrissey, and F. E. Nano.** 1992. Growth inhibition of *Francisella tularensis* live vaccine strain by IFN-γ-activated macrophages is mediated by reactive nitrogen intermediates derived from L-arginine metabolism. *J. Immunol.* **148**:1829–1834.
3. **Armstrong, J. A., and P. D. Hart.** 1971. Response of cultured macrophages to *Mycobacterium tuberculosis*, with observations on fusion of lysosomes with phagosomes. *J. Exp. Med.* **134**:713–740.
4. **Auger, M. J., and J. A. Ross.** 1992. The biology of the macrophage, p. 1–74. *In* C. E. Lewis and J. O. McGee (ed.), *The Macrophage.* Oxford University Press, Oxford.
5. **Beatty, W. L., G. I. Byrne, and R. P. Morrison.** 1993. Morphologic and antigenic characterization of interferon γ-mediated persistent *Chlamydia trachomatis* infection *in vitro*. *Proc. Natl. Acad. Sci. USA* **90**:3998–4002.
6. **Bhatnagar, N., E. Getachew, S. Straley, J. Williams, M. Meltzer, and A. Fortier.** 1994. Reduced virulence of rifampicin-resistant mutants of *Francisella tularensis*. *J. Infect. Dis.* **170**:841–847.
7. **Brezkorovainy, A., and R. H. Zschocke.** 1974. Structure and function of transferrins. *Arzneimittelforschung* **24**:476–485.
8. **Buchmeier, N. A., and F. Heffron.** 1990. Induction of *Salmonella* stress proteins upon infection of macrophages. *Science* **248**:730–732.
9. **Byrd, T., and M. Horwitz.** 1989. Interferon gamma-activated human monocytes down-regulate transferrin receptors and inhibit the intracellular multiplication of *Legionella pneumophila* by limiting the availability of iron. *J. Clin. Invest.* **83**:1457–1465.
10. **Carlin, J. M., E. C. Borden, and G. I. Byrne.** 1989. Interferon-induced indoleamine 2,3-dioxygenase activity inhibits *Chlamydia psittaci* replication in human macrophages. *J. Interferon Res.* **9**:329–337.
11. **Chamberlain, R. E.** 1965. Evaluation of live tularemia vaccine prepared in a chemically defined medium. *Appl. Microbiol.* **13**:232–235.
12. **Chang, K. P., and D. M. Dwyer.** 1976. Multiplication of a human parasite (*Leishmania donovani*) in phagolysosomes of hamster macrophages *in vitro*. *Science* **193**:678–680.
13. **Channon, J. Y., and J. M. Blackwell.** 1985. A study of the sensitivity of *Leishmania donovani* promastigotes and amastigotes to hydrogen peroxide.

II. Possible mechanisms involved in protective H_2O_2 scavenging. *Parasitology* **91**:207–213.

14. **Cohn, Z. A.** 1978. The maturation and activation of mononuclear phagocytes: fact, fancy, and future. *J. Immunol.* **121**:813–814.

15. **Czuprynski, C., P. M. Henson, and P. A. Campbell.** 1984. Killing of *Listeria monocytogenes* by inflammatory neutrophils and mononuclear phagocytes from immune and nonimmune mice. *J. Leukocyte Biol.* **35**:193–208.

16. **Dautry-Varsat, A., and H. F. Lodish.** 1984. How receptors bring proteins and particles into cells. *Sci. Am.* **250**:52–58.

17. **Davis, M. W., and A. C. Allison.** 1978. Effect of cytochalasin B on endocytosis and exocytosis, p. 143–160. *In* S. W. Tannenbaum (ed.), *Cytochalasins: Biochemical and Cell-Biological Aspects.* Elsevier/North-Holland Biomedical Press, Amsterdam.

18. **Denis, M.** 1991. Interferon-gamma-treated murine macrophages inhibit growth of tubercle bacilli via the generation of reactive nitrogen intermediates. *Cell. Immunol.* **132**:150–157.

19. **Donnenberg, M. S., J. Yu, and J. B. Kaper.** 1993. A second chromosomal gene necessary for intimate attachment of enteropathogenic *Escherichia coli* to epithelial cells. *J. Bacteriol.* **175**:4670–4680.

20. **Ericsson, M., A. Tarnvik, K. Kuoppa, G. Sandstrom, and A. Sjostedt.** 1994. Increased synthesis of DnaK, GroEL, and GroES homologs by *Francisella tularensis* LVS in response to heat and hydrogen peroxide. *Infect. Immun.* **62**:178–183.

21. **Finlay, B. B., and S. Falkow.** 1988. Comparison of the invasion strategies used by *Salmonella choleraesuis, Shigella flexneri,* and *Yersinia enterocolitica* to enter cultured animal cells: endosome acidification is not required for bacterial invasion or intracellular replication. *Biochimie* **70**: 1089–1099.

22. **Fortier, A. H., D. A. Leiby, R. B. Narayanan, E. Asafoadjei, R. M. Crawford, C. A. Nacy, and M. S. Meltzer.** Growth of *Francisella tularensis,* LVS in macrophages: the acidic intracellular compartment provides essential iron required for growth. *Infect. Immun.,* in press.

23. **Fortier, A. H., T. Polsinelli, S. J. Green, and C. A. Nacy.** 1992. Activation of macrophages for destruction of *Francisella tularensis:* identification of cytokines, effector cells and effector molecules. *Infect. Immun.* **60**:817–825.

24. **Fortier, A. H., M. V. Slayter, R. Ziemba, M. S. Meltzer, and C. A. Nacy.** 1991. Live vaccine strain of *Francisella tularensis:* infection and immunity in mice. *Infect. Immun.* **59**:2922–2928.

25. **Freidman, R. L., K. Nordensson, L. Wilson, E. T. Akporiaye, and D. E. Yocum.** 1992. Uptake and intracellular survival of *Bordetella pertussis* in human macrophages. *Infect. Immun.* **60**: 4578–4585.

26. **Granger, D. L., J. B. Hibbs, J. R. Prefect, and D. T. Durack.** 1990. Metabolic fate of L-arginine in relation to microbistatic capability of murine macrophages. *J. Clin. Invest.* **85**:264–268.

27. **Green, L. C., D. A. Wagner, J. Glogowski, P. L. Skipper, J. S. Wishnok, and S. R. Tannenbaum.** 1982. Analysis of nitrate, nitrite, and ^{15}nitrate in biological fluids. *Anal. Biochem.* **126**: 131–134.

28. **Green, S. J., M. S. Meltzer, J. B. Hibbs, and C. A. Nacy.** 1990. Activated macrophages destroy intracellular *Leishmania major* amastigotes by an L-arginine-dependent killing mechanism. *J. Immunol.* **144**:278–283.

29. **Green, S. J., C. A. Nacy, and M. S. Meltzer.** 1991. Cytokine-induced synthesis of nitrogen oxides in macrophages: a protective host response to Leishmania and other intracellular pathogens. *J. Leukocyte Biol.* **50**:93–103.

30. **Green, S. J., C. A. Nacy, R. D. Schreiber, D. L. Granger, R. M. Crawford, M. S. Meltzer, and A. H. Fortier.** 1993. Neutralization of gamma interferon and tumor necrosis factor alpha blocks in vivo synthesis of nitrogen oxides from L-arginine and protection against *Francisella tularensis* infection in *Mycobacterium bovis* BCG-treated mice. *Infect. Immun.* **61**:689–698.

31. **Gregory, W. W., G. L. Byrne, M. Gardner, and J. W. Moulder.** 1979. Cytochalasin B does not inhibit ingestion of *Chlamydia psittaci* by mouse fibroblasts (L cells) and mouse peritoneal macrophages. *Infect. Immun.* **25**:463–469.

32. **Hibbs, J. B., Z. Vavrin, and R. R. Taintor.** 1987. L-Arginine is required for expression of the activated macrophage effector mechanism causing selective metabolic inhibition in target cells. *J. Immunol.* **138**:550–559.

33. **Hoover, D. L., and C. A. Nacy.** 1984. Macrophage activation to kill *Leishmania tropica:* defective intracellular destruction of amastigotes by macrophages elicited by sterile inflammatory agents. *J. Immunol.* **132**:1487–1493.

34. **Horwitz, M. A., and S. C. Silverstein.** 1980. Influence of the *Escherichia coli* capsule on complement fixation and on phagocytosis and killing by human phagocytes. *J. Clin. Invest.* **65**:82–94.

35. **James, S. L., and J. Glaven.** 1989. Macrophage cytotoxicity against schistosomula of *Schistosoma mansoni* involves arginine-dependent production of reactive nitrogen intermediates. *J. Immunol.* **143**:4208–4214.

36. **Jones, T. C., and J. G. Hirsch.** 1972. The interaction between *Toxoplasma gondii* and mammalian cells. II. The absence of lysosomal fusion with phagocytic vacuoles containing living parasites. *J. Exp. Med.* **136**:1173–1194.

37. **Keusch, G. T., and G. F. Grady.** 1972. The pathogenesis of *Shigella* diarrhea. 1. Enterotoxin production by *Shigella dysenteriae. J. Clin. Invest.* **51:**1212–1220.

38. **Kuhn, M., S. Kathariou, and W. Goebbel.** 1988. Hemolysin supports survival but not entry of the intracellular bacterium *Listeria monocytogenes. Infect. Immun.* **56:**79–82.

39. **Kwaik, Y. A., B. I. Eisenstein, and N. C. Engleberg.** 1993. Phenotypic modulation of *Legionella pneumophila* upon infection of macrophages. *Infect. Immun.* **61:**1320–1329.

40. **Lancaster, J. R., and J. B. Hibbs.** 1990. EPR demonstration of iron–nitrosyl complex formation by cytotoxic activated macrophages. *Proc. Natl. Acad. Sci. USA* **87:**1223–1228.

41. **Lofgren, S., A. Tarnvik, M. Thore, and J. Carlsson.** 1984. A wild and an attenuated strain of *Francisella tularensis* differ in susceptibility to hypochlorous acid: a possible explanation of their different handling by polymorphonuclear leukocytes. *Infect. Immun.* **43:**730–734.

42. **Long, G. W., J. J. Oprandy, R. B. Narayanan, A. H. Fortier, K. R. Poster, and C. A. Nacy.** 1993. Detection of *Francisella tularensis* in blood by polymerase chain reaction. *J. Clin. Microbiol.* **31:**152–154.

43. **Manten, A., and L. J. VanWijnggaarden.** 1969. Development of drug resistance to rifampicin. *Chemotherapy* **14:**93–100.

44. **Mantovani, B.** 1975. Different roles of IgG and complement receptors in phagocytosis by polymorphonuclear leukocytes. *J. Immunol.* **115:**15–17.

45. **McCrumb, F. R.** 1961. Aerosol infection of man with *Pasteurella tularensis. Bacteriol. Rev.* **25:**262–267.

46. **Mekalanos, J. J.** 1992. Environmental signals controlling expression of virulence determinants in bacteria. *J. Bacteriol.* **174:**1–7.

47. **Melman, I., R. Fuchs, and A. Helenius.** 1986. Acidification of the endocytic and exocytic pathways. *Annu. Rev. Biochem.* **55:**663–700.

48. **Montie, T. C., and S. J. Ajl.** 1970. Nature and synthesis of murine toxins of *Pasteurella pestis,* p. 1–34. *In* T. V. Montie, S. Kadis, and S. J. Ajl (ed.), *Microbial Toxins,* vol. III. Academic Press, Inc., New York.

49. **Moorman, D. R., and G. Mandell.** 1981. Characteristics of rifampicin resistant variants obtained from clinical isolates of *Staphylococcus aureus. Antimicrob. Agents Chemother.* **20:**709–713.

50. **Murray, H. W.** 1982. Pretreatment with phorbol myristate acetate inhibits macrophage activity against intracellular protozoa. *J. Reticuloendothel. Soc.* **31:**479–487.

51. **Murray, H. W., A. Szuro-Sudol, D. Wellner, J. Oca, A. Granger, D. Libby, C. Rothermel, and B. Rubin.** 1989. Role of tryptophan degradation in respiratory burst-independent antimicrobial activity of gamma interferon-stimulated human macrophages. *Infect. Immun.* **57:**845–849.

52. **Nacy, C. A., C. N. Oster, S. L. James, and M. S. Meltzer.** 1984. Activation of macrophages to kill *Rickettsiae* and *Leishmania:* dissociation of intracellular microbicidal activities and extracellular destruction of neoplastic and helminth targets. *Cont. Top. Immunobiol.* **13:**147–170.

53. **Polsinelli, T., M. S. Meltzer, and A. H. Fortier.** 1994. Nitric oxide independent killing of *Francisella tularensis* by interferon-γ stimulated murine alveolar macrophages. *J. Immunol.* **153:**1238–1245.

54. **Prince, A.** 1992. Adhesins and receptors of *Pseudomonas aeruginosa* associated with infection of the respiratory tract. *Microb. Pathog.* **13:**251–260.

55. **Rathman, M., D. Russell, and S. Falkow.** 1994. The intracellular fate of *Salmonella typhimurium,* abstr. D-50, p. 104. *Abstr. 94th Gen. Meet. Am. Soc. Microbiol. 1994.*

56. **Rhinehart, T., A. H. Fortier, and K. L. Elkins.** 1992. Characterization of the protective antibody response to *Francisella tularensis* strain LVS in mice, abstr. E-36, p. 150. *Abstr. 92nd Gen. Meet. Am. Soc. Microbiol. 1992.*

57. **Sandstrom, G., A. Sjostedt, T. Johansson, K. Kuoppa, and J. C. Williams.** 1992. Immunogenicity and toxicity of lipopolysaccharide from *Francisella tularensis* LVS. *FEMS Microbiol. Immunol.* **5:**201–210.

58. **Sasada, M., and R. B. Johnston.** 1980. Macrophage microbicidal activity: correlation between phagocytosis-associated oxidative metabolism and the killing of candida by macrophages. *J. Exp. Med.* **152:**85–91.

59. **Schurig, G. G., R. M. Roop, T. Bagchi, S. Boyle, D. Buhrman, and N. Sriranganathan.** 1991. Biological properties of RB 51, a stable rough strain of *Brucella abortus. Vet. Microbiol.* **28:**171–188.

60. **Smith, J. W., W. J. Urba, and B. D. Curti.** 1992. The toxic and hematologic effects of interleukin-1 alpha administered in a phase I trial to patients with advanced malignancies. *J. Clin. Oncol.* **10:**1141–1152.

61. **Sokolovic, Z., and W. Goebel.** 1989. Synthesis of listeriolysin in *Listeria monocytogenes* under heat shock conditions. *Infect. Immun.* **57:**295–298.

62. **Spriggs, D. R., and H. S. Jaffe.** 1993. Pathophysiological effects of TNF therapy on cancer patients, p. 165–170. *In* J. J. Oppenheim, J. Rossio, and A. Gearing (ed.), *The Clinical Applications of Cytokines.* Oxford University Press, New York.

63. **Steed, L. L., M. Setareh, and R. L. Friedman.** 1991. Intracellular survival of virulent *Bordetella pertussis* in human polymorphonuclear leukocytes. *J. Leukocyte Biol.* **50:**321–330.

64. **Takemura, R., and Z. Werb.** 1984. Secretory products of macrophages and their physiological function. *Am. J. Physiol.* **246:**C1–C9.

65. **Tarnvik, A.** 1989. Nature of protective immunity to *Francisella tularensis*. *Rev. Infect. Dis.* **11:** 440–451.

66. **Tilney, L. G., and D. A. Portnoy.** 1989. Actin filaments and the growth, movement and spread of the intracellular bacterial parasite, *Listeria monocytogenes*. *J. Cell Biol.* **109:**1597–1608.

67. **Traub, A., J. Mager, and N. Grossowicz.** 1955. Studies on the nutrition of *Pasteurella tularensis*. *J. Bacteriol.* **70:**60–69.

68. **Walker, T. S**. 1984. Rickettsial interactions with human endothelial cells in vitro: adherence and entry. *Infect. Immun.* **44:**205–210.

69. **Ward, M. E., and A. Murray.** 1984. Control mechanisms governing the infectivity of *Chlamydia trachomatis* for HeLa cells: mechanisms of endocytosis. *J. Gen. Microbiol.* **130:**1765–1780.

70. **Watanabe, T., K. Sugawara, and M. Watanabe.** 1971. Virulence of rifampicin resistant mutants of bacteria. *Jpn. J. Bacteriol.* **26:**192–199.

71. **Wehrli, W.** 1983. Rifampicin: Mechanism of action and resistance. *Rev. Infect. Dis.* **5**(Suppl.)407–410.

72. **White, J. D., J. R. Rooney, P. A. Prickett, E. B. Derrenbacher, C. W. Beard, and W. R. Griffith.** 1964. Pathogenesis of experimental respiratory tularemia in monkeys. *J. Infect. Dis.* **114:** 277–283.

BACTERIAL SUPERANTIGENS IN DISEASE

Allen D. Sawitzke, Kevin L. Knudtson, and Barry C. Cole

10

Infectious agents have long been recognized for their ability to overcome host defense mechanisms. Perhaps the newest mechanism makes use of the molecules known as superantigens (SAgs), which are capable of activating multiple compartments of the host immune system. Unlike conventional antigens, SAgs are capable of activating all T-cell clones expressing particular β-chain segments of the variable region (Vβ) of the α/β T-cell receptor for antigen (TCR) (156). As a consequence, SAgs are particularly potent T-cell mitogens. They also have an effect on B-cell function (149) that results in polyclonal-antibody production.

Infectious agents thus far known to possess SAgs include viruses, bacteria, and mycoplasmas (Table 1). SAgs in the form of soluble toxins have been identified in staphylococci (27, 103), streptococci (77), pseudomonads (99), clostridia (20), and mycoplasmas (36). Cell-associated SAgs have been found in streptococcal M proteins (145) and *Yersinia enterocolitica* cell walls (141), and other SAgs not yet shown to be soluble or bound have been found in *Yersinia pseudotuberculosis* (3) and *Mycobacterium tuberculosis* (117). A group of SAgs previously referred to as minor lymphocyte-stimulating antigens (Mls) are now known to be products of murine retroviral tumor viruses and are referred to as endogenous SAgs (5). In addition, SAg activity has been identified in rabies virus (96) and in the virus responsible for murine AIDS (76). Indirect evidence that a SAg may play a role in human AIDS has also been presented (98).

The mechanisms by which these SAgs may benefit their source organisms are not clear but likely include diversion or suppression of host immune responses, direct toxicity, and induction of immune tolerance. This chapter concentrates on the bacterial SAgs and their role in disease. Many recent reviews have highlighted the staphylococcal and streptococcal SAgs (79, 92, 95), so particular use of the mycoplasmal SAg, *Mycoplasma arthritidis* mitogen (MAM), will be made in illustrating general principles, since this model SAg has been studied extensively in our laboratories.

CHARACTERIZATION OF SAgs

SAgs are potent T-cell mitogens capable of T-cell activation by unique mechanisms (Fig. 1A). Unlike conventional antigens, SAgs bind directly to major histocompatibility complex (MHC) molecules outside of the antigen groove (42) without needing to be processed

Allen D. Sawitzke, Kevin L. Knudtson, and Barry C. Cole Department of Internal Medicine, Division of Rheumatology, University of Utah Medical Center, 50 North Medical Drive, Salt Lake City, Utah 84132.

Virulence Mechanisms of Bacterial Pathogens, 2nd ed., Edited by J. A. Roth et al.
© 1995 American Society for Microbiology, Washington, D.C.

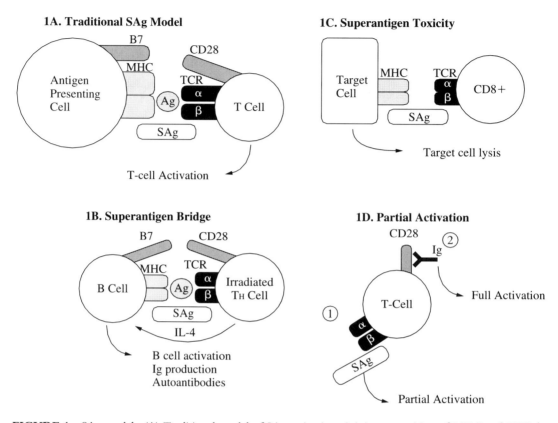

FIGURE 1 SAg models. (A) Traditional model of SAg activation. A joint recognition of MHC and TCR by SAg results in activation, irrespective of the antigen present in the groove of the class II molecule. Accessory molecules are present, but their interaction is not explicitly required. Several modifications of the model have been proposed for specific cases, as reviewed in the text. (B) SAg bridge. Similar recognition of MHC and TCR by the SAg occurs, resulting in B-cell activation with antibody production. Again, accessory molecules and antigen are not required. The T cell functions as the presenting cell. (C) SAg-directed cell lysis. Bridging by SAg of a cytolytic T cell to a class II-positive cell results in cell lysis. The target cell can even be an activated T cell. This type of T-cell–T-cell interaction could contribute to diseases with lymphopenia, including systemic lupus erythematous and AIDS. (D) Partial activation. 1, SAg can bind to the TCR of T cells and result in partial activation such as cytolysis; 2, a second signal, such as Ig against CD28, can complete the activation, thereby allowing proliferation.

by antigen-presenting cells (APC) (31). SAgs recognize the Vβ region of the TCR largely irrespective of other elements such as Vα, Dβ, and Jβ (156). Each SAg interacts with a characteristic set of Vβ chains, although unrelated SAgs may activate T cells bearing similar Vβ chains. As a result, all T cells bearing the Vβ chains that interact with a particular SAg are activated. A special interaction of B and T cells mediated by SAgs has been termed the "SAg bridge," as described for toxic shock syndrome

toxin (TSST) (112) and for MAM (149). In this circumstance, the SAg cross-links T_H cells via their Vβ TCRs with resting B cells via their MHC molecules, a process that signals B-cell differentiation and leads to polyclonal proliferation and immunoglobulin (Ig) secretion (Fig. 1B). T-cell help in the form of interleukin-4 (IL-4) (30) and macrophage production of IL-6 (73) further contribute to B-cell activation.

Recently, SAgs have been shown to activate

other novel immune pathways. It is not known whether these pathways can be generalized to all SAgs or are true only for specific examples. However, each may have special ramifications for interaction with the immune response of the host. T cells expressing the CD8 molecule and primed as antigen-specific cytotoxicity cells are capable of lysing class II–expressing cells when they are bound by SAgs, irrespective of their antigen specificities (69) (Fig. 1C). Additionally, partial T-cell activation has been observed. For example (Fig. 1D), T cells can be activated to cytotoxicity as a result of interaction with SAg even in the absence of MHC binding. In fact, antibodies against SAg have in some instances been able to substitute for class II molecules (61). However, in most cells, full activation of cells leading to proliferation requires a second signal. In vitro, cell adhesion pairs, including CD28/B7 and LFA3/CD2, are capable of providing the second signal (97).

A further property of SAgs is their ability to modify the T-cell repertoire in vivo. Endogenous retroviral SAgs in mouse strains that possess the appropriate presenting MHC class II molecules delete all T cells expressing the same SAg-reactive Vβ chains (126). A similar phenomenon occurs when bacterial SAgs are injected into neonatal or adult mice, especially when repeated doses are given (107, 156). It has been suggested that either deletion of T cells or a "hole" in the T-cell repertoire could benefit the host organism by eliminating T-cell subsets that might ultimately participate in specific diseases (104).

Interaction of SAgs with MHC Molecules

Although conventional antigens are restricted in their presentation by specific MHC molecules, SAgs are able to use a broad range of MHC sites. Thus far, all SAgs are thought to associate with class II MHC molecules. Class II MHC molecules are heterodimeric integral proteins composed of α and β chains. According to the X-ray crystallography model of HLA class I, the chains are folded in a way that forms an antigen-binding groove capable of holding a 10- to 12-amino-acid peptide antigen (18). The chains associate in the endoplasmic reticulum and are protected from endogenous antigen loading by their invariant chain, which occupies the antigen groove until it is cleaved and removed in the endosome (18). Expression of class II molecules is inducible by several cytokines, including gamma interferon and granulocyte-macrophage colony-stimulating factor (GM-CSF) on diverse cell types, including B cells, activated T cells, and renal tubular cells (101). The combination of the class II molecule, the antigen, and the specific TCR is known as the ternary complex. Formation of the ternary complex in association with a coreceptor (CD4, CD8) results in normal T-cell activation. SAgs are able to bind with both the TCR and the class II molecule with or without a traditional antigen.

Several approaches have been used to examine the sites of SAg and class II MHC interaction. Most make use of direct binding studies to test the affinities of SAgs for particular MHC molecules. Additionally, transgenic animals expressing chimeric or mutant molecules can be designed to identify specific binding sites. Other groups have used peptide inhibition studies to define binding regions. More recently, binding regions have been determined by X-ray crystallography as performed for staphylococcal enterotoxin B (SEB) and TSST-1 (6, 125, 142). Each SAg molecule thus far crystallized has two dominant domains, each composed primarily of beta sheets connected by α-helical bridges. While homology between these proteins at the primary level is only about 24%, marked similarity is shown by crystallographic analysis (6). It remains to be seen whether similar structural homologies will be observed in all SAgs. Some evidence of a common SAg-MHC binding region has been shown by competitive binding analysis using peptides derived from retroviral SAgs to inhibit bacterial SAg binding (147).

Thus far, isotypic, allotypic, and xenotypic differences in class II interactions with SAgs have been shown to affect binding. In fact, these differences have been used to help identify the active regions. The HLA-DR alleles vary in their presentations of SEA, and comparison of the HLA-DR variable sequences allowed prediction of a putative binding site involving histidine 81. Such predictions were subsequently tested by examination of mutants expressing nonreactive and reactive alleles with or without the histidine residue at position 81, for their abilities to bind SEA and induce lymphocyte proliferation (68, 88). Synthetic peptides were constructed and used to inhibit the proliferation of appropriate cells (130). Together, these approaches have defined a major MHC binding region for SEA on MHC class II molecules to be an α helix of the β protein chain; this helix borders the antigen-binding site at those residues surrounding position 81. Similar approaches have shown that the N-terminal portion of SEA (residues 1 through 45) is also a region bound by the MHC (59).

Early examinations of SAg binding used protein digests of staphylococcal toxins (51). When this approach was used, SEB was shown to compete for the same sites as SEA on the MHC class II molecule, as indicated by inhibition of lymphocyte proliferation. Mutations of the SEB molecule were made by Kappler and coworkers, and the effects of the mutations on SAg binding to MHC were noted (87). Two regions were shown to be involved in MHC class II recognition. More recently, crystallographic analysis has shown two principal domains surrounding the probable TCR and MHC binding sites. These domains lie in a shallow groove that is consistent with the findings of prior studies of binding (142).

TSST-1 binds nearly equally to human DR, DQ, and DP molecules, but murine isotypes are more variable (133). Braunstein and coworkers (21) made use of this observation and prepared recombinant murine IA molecules for binding studies. They used the fluorescence-activated cell sorter (FACS) to detect binding. Residues in both the α and the β chains were involved in TSST-1 binding, and again, the α-helical borders of the antigen-binding groove were primarily responsible for binding (21).

Similarly, mutant HLA-DR molecules in which the amino acids of the DRα chain were selectively replaced were made, and the effects of these changes on binding were analyzed. Positions 36 and 39 of DRα protein were found to be critical for TSST-1 presentation (119).

MAM is produced by an organism that is a natural pathogen of mice and induces a much stronger proliferative response in murine than in human cells. The many available MHC congenic mouse strains provide a unique opportunity to examine the effect of isotype and allotype on presentation of MAM to T cells. MAM is selective in its response to class II isotypes and yet seems to engage class II molecules primarily by the α chain. Furthermore, Eα presents MAM more effectively than DRα (its human homolog), since L cells expressing Eα-DRβ chimeric molecules are more effective in stimulating human peripheral blood cells than are fibroblasts expressing DRα-DRβ molecules (26a). Molecules expressing H-2Eα or HLA-DRα class II molecules also present MAM much better than those expressing H-2Aα when each is paired with identical H-2Eβ proteins (11, 33). Cells from mice transgenic for the human DQβ molecule (DQB1*0601) when expressed with H-2Aα are also quite effective in presenting MAM to murine T cells (unpublished observations). Interestingly, the DQB1*0601 molecule appears to behave like H-2E, since Eα^- B10.M mice transgenic for DQB1*0601 delete T cells bearing Vβ6, Vβ7, and Vβ8.1 during thymic ontogeny owing to the interaction of their MHC with the Mls1^a retroviral SAg (161). Cells displaying H-2Eα^f-DQB1*0601 hybrid molecules differ from those displaying H-2Eα^f-DQB1*0302 hybrid in their abilities to respond to MAM (unpublished observation). This implies that both α and β class II chains contribute to optimal recognition, as has previously been demonstrated for TSST-1 and SEA (21, 130). It remains to be determined whether allelic differences at HLA-DRβ influence presentation by MAM.

Interaction of SAgs with TCRs

The binding specificity of a particular TCR (α/β) to conventional peptides is generated by the interaction of the various Vα and Jα allelic combinations of the α chain with the Vβ, Dβ, and Jβ allelic combinations of the β chain (41). Because there are potentially millions of different allelic combinations, the frequency of T cells that respond to a conventional antigen is very low (1 in 10^4 to 10^6 T cells) (67). Conversely, the interaction of both endogenous (viral) and exogenous (viral, bacterial, and mycoplasmal) SAgs with T cells is dictated primarily by the particular Vβ chain of the TCR, with minor contributions from Dβ, Jβ, Vα, or Jα chains (79, 80). Thus, because whole subsets of T cells possessing a particular Vβ are activated by a SAg and because SAgs can activate more than one Vβ subset, the percentage of T cells responding to a particular SAg can be very high (5 to 25%) (67, 79).

The notion that the Vβ chain is recognized by SAgs was demonstrated by showing that *Staphylococcus aureus* toxin SEB could specifically stimulate all T cells bearing Vβ3, Vβ8.1, Vβ8.2, and Vβ8.3 but not other Vβ TCR subsets (156). Callahan and colleagues expanded this study to examine the ability of a panel of different staphylococcal toxins to stimulate specific murine Vβ TCRs in vitro (24). They showed that each toxin stimulated a unique subset of murine Vβ TCRs (Table 1) (24). Gascoigne and Ames demonstrated that the β chain alone was sufficient to bind the SAg MHC complex: they showed that soluble murine Vβ3 produced in B cells in the absence of the α chain was specifically bound by SEA (55). Thus, the Vβ-SAg interaction is central to T-cell stimulation by SAg.

The genomic composition of the mouse strain used as well as the allelic polymorphisms of the Vβ chain affect the pattern of Vβ usage by MAM (32). MAM activates Vβ5.1, Vβ6, Vβ8.1, Vβ8.2, and Vβ8.3 in Vβ^b haplotype mice. However, Vβ1, Vβ3.1, Vβ7, and Vβ16 are also activated using V$_β^a$ (C57BR) and/or V$_β^c$ (RIIIS) haplotype mice, which have genomic deletions of the Vβ TCRs activated by MAM in Vβ^b mice. Significantly enhanced expansion of Vβ6 is observed in Vβ^a mice, which lack the Vβ8.1-Vβ8.2-Vβ8.3 subfamily. Dominant expansion of Vβ1 and Vβ7 was observed in Vβ^c mice, which lack both Vβ8 and Vβ6 genes. Alignment of amino acid sequences of MAM-reactive and non-MAM-reactive Vβ TCRs shows that the arginine at position 69 and the phenylalanine at position 75 are conserved in the hypervariable 4 (HV4) domains of Vβ6, Vβ8.1, Vβ8.2, and Vβ8.3 as well as in the human Vβ17, which is also preferentially expanded by MAM (11, 53, 123). Thus, it was predicted that these amino acids would be involved in a MAM-TCR interaction. However, murine Vβ16 and Vβ7, which also possess these amino acids at the same positions, are not expanded by MAM unless Vβ6, Vβ8.1, Vβ8.2, and Vβ8.3 are absent. Conversely, Vβ1, Vβ3.1, and Vβ5.1, which lack the conserved residues in the HV4 region, can be expanded by MAM. Therefore, the presence of the conserved residues in the HV4 region cannot fully account for the differing degrees of MAM reactivity. The structural polymorphisms in the coding regions of Vβ1, Vβ3.1, and Vβ6 between Vβ^b and Vβ^a (138) could account for some of the differences in the ability to be activated by MAM (32). A more comprehensive explanation suggests that MAM interacts with the Vβ at other sites or that other TCR elements are recognized by MAM (123, 124). There is evidence that not all Vβ TCRs exhibit the same MHC requirements for recognition of SAgs. Thus, whereas MAM expanded Vβ5.1, Vβ6, Vβ8.1, Vβ8.2, and Vβ8.3 TCRs in cultures from C3H mice (H-2E$^+$), expansion of Vβ6 and Vβ8.3 was absent in cultures from the congenic H-2E$^-$ mouse strain. Selective expansion of Vβ5.1, Vβ8.1, and Vβ3 was unaffected by the absence of H-2E (32).

Other elements besides the Vβ chain of the TCR can be involved with T-cell interaction with SAg. Hodtsev and coworkers (71) reported that MAM reactivity was absent in human T-cell clones expressing Jβ2.1 but present in other clones expressing other Jβ chains with identical Vβ sequences. Similar involvement of the Jβ chain and/or Dβ chains has been reported for viral SAg interactions with T cells (25, 158), and influences of the α chain on T-cell interaction with SAg have been observed (79). However, the apparent involvement of the TCR α chain and D/Jβ chains does not seem to be important unless the affinity of the Vβ interaction with the SAg-MHC is low (79, 145).

The strength of the MHC-Vβ-SAg interaction is emphasized by the lack of a CD4 and CD8 coreceptor requirement for SAg stimulation of T cells (65). CD4 or CD8 coreceptors bind specific regions of the MHC class II or class I molecule, respectively, during activation of T cells recognizing conventional peptide antigens bound to MHC (46, 115). The adhesion function of CD4-CD8 appears to be necessary for activation of an antigen-specific T-cell response (16). The notion that SAg activation of T cells is CD4 or CD8 coreceptor independent arose when it was shown that MAM could activate T cells expressing CD8 or CD4 or neither receptor (105). Similarly, it was found that SEB could activate CD8$^+$ as well as CD4$^+$ T cells (50). In addition, the activation of T cells by SEB could not be blocked by anti-CD8 monoclonal antibody (MAb) (69). Sekaly and coworkers showed that CD4$^-$ T-cell hybridomas responded to low concentrations of SEA, SEB, or TSST-1 as well as CD4$^+$ cells (135). Although these studies suggested that SAg can provide the adhesion normally supplied by the CD4-CD8 coreceptors in the APC–T-cell interaction, a costimulatory signal such as that provided by CD28-B7 may still be required for full SAg induction of T cells.

Kotb and colleagues showed that proliferation of human T cells is dependent on immobilization or cross-linking of the streptococcal M protein SAg and a costimulatory signal provided by phorbal myristate acetate, IL-1, and IL-6 (93). The observation that the interaction of CD28 and B7 can generate a costimulatory signal in amplifying IL-2 production (56, 100) allowed Fraser and coworkers to use a mutant IL-2 promoter construct to show that the CD28 signal transduction pathway specifically induces IL-2 production following SED presentation by B7-positive APC (52). In addition, anti-B7 MAb affected IL-2 levels of the transfectants containing the wild-type promoter, while IL-2 levels of transfectants containing the mutant promoter were unaffected. Similarly, Ohnishi and coworkers showed that anti-B7–BB1 MAb inhibits the responses of T cells when the streptococcal M protein SAg is presented by APC (118). Thus, it appears as though SAg activation of T cells is dependent on APC–derived soluble factors and membrane-associated costimulatory molecules (102). However, T-cell hybridomas can be activated with MAM presented by formalinized APC or by liposomes containing class II molecules in the absence of soluble factors (14, 31). Therefore, further work needs to be done to define the role of APCs and their soluble mediators in the induction of T cells for each SAg.

The configuration of the complex of MHC, TCR, and SAg has become a subject of controversy. Early reports predicted HV4, which is also referred to as CDR4 or the D-E loop and lies on the side of the TCR away from the antigen-MHC-binding site of the Vβ chain, to be important for SAg recognition. Choi et al. used mutagenesis to replace amino acids 67 to 77 of the SEC2 responder Vβ13.2 into the SEC2 nonresponder Vβ13.1 (29) to generate a TCR responsive to SEC2 induction. Because the amino acids 67 to 77 were predicted to lie in the HV4 region, it was proposed that SAg bound the TCR outside the antigen-MHC-binding site (81). Similarly, Pontzer et al. showed that from a panel of overlapping peptides of the Vβ3 chain, only a peptide corresponding to HV4 blocked SEA binding and T-cell activation (122). As expected, a peptide corresponding to the HV4 region of Vβ8.2, a Vβ element not recognized by SEA, did not block SEA-induced activity.

There is now evidence that the complementarity-determining regions CDR1 and CDR2 of the TCR β chain, which are predicted to interact with the α helices of the MHC (83), are also involved in SAg activation.

Patten et al. showed that SEA and SEB reactivity could be transferred by replacing the CDR1 or CDR2 of a nonreactive Vβ TCR with the CDR1 or CDR2 of a responding Vβ (121). Moreover, those authors suggested that because CDR1 and CDR2 are also involved in antigen-MHC recognition, these SAgs and MHC molecules bind the same sites on Vβ. They propose a model in which the SAg lies in the region between the MHC and the TCR.

The recent solution of the crystal structure of SEB bound to a human class II MHC molecule (HLA-DR1) offers yet another model for the ternary complex. Jardetzky et al. (82) showed that SEB bound to DR1 blocks the residues of the α1 domain that would usually interact with the TCR. This finding supports the observations of Patten et al. in that the SAg would block the interaction between MHC and TCR (121). However, on the basis of the model that SAg binds the HV4 region of Vβ, Jardetzky et al. propose a juxtaposition of the HV4 region during TCR, MHC, and SAg interaction that would place CDR1, CDR2, and CDR3 over the MHC-peptide-binding site, with Vβ bound to SEB and the TCR Vα domain lying above the β1 domain of the MHC (82). This model permits both TCR α-chain and MHC polymorphisms to influence SAg activity. The recent proliferation of models to explain SAg-based activities highlights the many exciting discoveries made over the last couple of years. It also suggests that several different mechanisms are likely to be used by specific SAgs.

Polyclonal B-Cell Activation: the SAg Bridge

Although T-cell activation (as in Fig. 1A) is the best known outcome of SAg interactions with lymphocytes, polyclonal-B-cell activation has been described for some SAgs (Fig. 1B). When irradiated human T cells are added to tonsillar resting B cells to which TSST-1 had been bound, the B cells undergo proliferation and secrete Ig (112). The process, as for SAg-induced T-cell activation, is not MHC restricted. MAM induces human resting-B-cell activation in the presence of irradiated T cells. The re-

action is dependent on CD4$^+$ T$_H$ cells (149). Studies using the murine system show that MAM-reactive T-cell lines can provide help for B cells in the presence of MAM but not in the presence of TSST-1. The same is true for TSST-1-reactive T cells. Thus, the T-cell–B-cell interaction seen is SAg specific and functionally not cross-reactive. Also, MAM and TSST-1 can promote an enhanced in vitro immune response to T-cell-dependent antigens such as sheep erythrocytes in the presence of T cells, B cells, and antigen (148). This observation suggests that SAgs might have an adjuvant effect on an immune response to T-cell-dependent antigens. Recent studies imply that the SAg properties of mycobacteria might be responsible for the adjuvant properties of *M. tuberculosis* (117). As we discuss below, there is now evidence that SAgs can activate the SAg bridge in vivo.

IN VIVO MODULATION BY SAgs

Humoral Effects

The bacterial toxins now classified as SAgs have been known for many years to exert immunomodulatory effects in vivo. In general, SEA, SEB, and TSST-1 suppress primary antibody responses to foreign antigens when they are given prior to immunization but enhance the immune response when they are given after antigen challenge (54, 78, 98, 120). Studies with MAM have shown similar changes. Thus, mice injected intravenously first with sheep erythrocytes and 2 days later with MAM showed fourfold increases in the numbers of antibody-secreting cells over the number in controls. However, when MAM was injected prior to sheep erythrocytes, there was either no effect or a slight decrease in the number of antibody-producing cells (30). Enzyme-linked immunosorbent assay antibodies to albumen (OVA) were enhanced whether MAM was given prior to or after injection of mice with OVA. Similarly, the lymphocyte proliferative response to OVA was increased irrespective of whether the antigen was given before or after administration of MAM. Further evidence of MAM-induced suppression of T-cell functions was the finding that MAM could prolong skin transplants in mice. Recipient C3H mice that had received two or three intravenous injections of MAM 2 days apart prior to transplantation showed increased survival of BALB/c skin transplants compared with control mice that were given phosphate-buffered saline (PBS) (30). In addition, inhibition of T-cell function was indicated by MAM-induced suppression of skin contact sensitivity. Mice were given MAM or PBS 1 and 2 days prior to sensitization with dinitrofluorobenzene on the hind footpads. When the mice were challenged with dinitrofluorobenzene 4 days later, animals receiving MAM showed a 50% decrease in footpad swelling compared with those receiving PBS (30).

Modulation of Cytokine Responses

As expected, SAgs induce a wide variety of cytokines in vitro commensurate with their activation of a variety of lymphoid cells, including CD4$^+$ and CD8$^+$ T lymphocytes, B lymphocytes, macrophages (43), and, as learned more recently, NK cells (45). Only limited studies have been conducted on cytokine induction in vivo. Recently, Gonzalo and coworkers (57) showed that 90 min after intravenous injection of SEB into mice, significant levels of tumor necrosis factor alpha (TNF-α), GM-CSF, IL-1, IL-2, IL-4, and IL-10 could be detected. In contrast, SEA preferentially induced IL-2. As part of a larger study to document the long-term immunomodulatory properties of MAM in vitro, studies were undertaken to determine the effect of this SAg on the cytokine profiles of mice after intravenous injection. Splenocytes collected from mice injected with MAM not only showed decreased abilities to undergo proliferation in response to in vitro challenge with MAM but also exhibited marked suppression in their abilities to produce IL-2. Suppression was maximal 1 to 3 days postinjection of MAM but was still significant after 15 days (30). IL-2 production in response to concanavalin A was also inhibited but much less so. In marked contrast, lymphocytes from mice injected with MAM ex-

hibited a fivefold increase in IL-4 in response to in vitro challenge with MAM or concanavalin A. The response was maximal 3 days postinjection with MAM and returned to normal after 5 days. A similar enhancement of IL-6 responses occurred, although the effect was more transient and was seen only at 1 day postinjection of MAM.

The changes observed in proliferation and cytokine profiles following injection of the MAM SAg into mice suggest a change in T$_H$1-like functions to T$_H$2-like functions, which favor B-cell differentiation and activation. These findings are consistent with the observed increase in responses to lipopolysaccharides of lymphocytes from mice injected with MAM (30) and the increased IgG and IgM secretion of lymphocytes from MAM-injected mice. These observations support the notion that MAM can activate B cells in vivo via the SAg bridge. Prior observations (49) of MAM-induced secretion of rheumatoid factor from human peripheral blood lymphocytes also suggest that autoantibodies can indeed be induced by SAgs, thus supporting a role for these proteins in autoimmune disease, as discussed below.

Anergy, Apoptosis, and Deletion

The abilities of SAgs to cause clonal deletions of specific Vβ TCR-bearing T cells in vivo was first observed by Kappler and coworkers (86) for the endogenous retroviral SAgs of mice previously known as the Mls antigens (5). During thymic development, each SAg in the presence of its MHC receptor is engaged by its reactive Vβ chain segment(s), resulting in the negative selection of T-cell clones bearing that specific receptor(s). Since most mice possess endogenous or exogenous retroviruses that contain SAgs, the majority of mouse strains exhibit holes in their Vβ TCR repertoires.

Deletions of Vβ TCR-bearing cells by bacterial SAgs were first described by White et al. (156), who demonstrated that repeated injections of neonatal mice with SEB virtually eliminated specific T-cells bearing Vβ8.2 and Vβ8.3 in adult mice. When mature mice were injected with SAgs, a state of anergy was in-

duced. In addition, the number of T lymphocytes expressing Vβ8 was reduced by 50% at 2
to 4 weeks postinjection, indicating a partial
deletion (89, 128). When MAM is injected systemically into mice, there is an initial clonal
expansion of Vβ6- and Vβ8-bearing T cells
(32). Also, lymphocytes taken from MAM-injected mice at 1 to 2 days postinjection show
a marked decrease in their abilities to proliferate in response to MAM in vitro and are still
50% suppressed at 10 days following MAM injection. The suppressive effects of MAM and
other SAgs in vivo appear to be Vβ specific (30,
89), since only those Vβ TCR–bearing cells
that respond to a particular SAg show inhibition or apoptosis. In the case of MAM, the
suppressive state may not be due to anergy or
apoptosis, as is seen for other SAgs, since the
state appears to be mediated by CD4$^+$ T lymphocytes that can actively confer lack of responsiveness to MAM on healthy untreated
lymphocytes (40). There is evidence that the
transfer of suppression to healthy cells is also
Vβ specific (29a). It remains to be determined
whether larger doses of MAM are in fact capable of inducing anergy and apoptosis.

Restrictions on the T-cell repertoire mediated by SAgs may render the host vulnerable
to certain pathogenic agents. As we discuss below, these restrictions may also impart some
benefit to the host by protecting against development of autoimmunity mediated by specific Vβ-chain-bearing T cells that are reactive
to self antigens.

ROLE IN BACTERIAL DISEASE

Toxic Effects of SAgs

Long before the immunomodulatory properties of SAgs were discovered, their toxic properties associated with disease pathogenesis were
recognized (reviewed in references 70 and
132). Toxins play a role in human and animal
diseases ranging from food poisoning to shock
(132). In addition, many of the organisms
known to produce toxins can cause arthritis.
Bacterial toxins with demonstrated SAg activities (Table 1) are produced by both gram-pos-

itive and gram-negative bacteria and mycoplasmas. *S. aureus* enterotoxin (SE) serotypes A,
B, C1, C2, C3, D, and E; TSST-1; and streptococcal pyrogenic exotoxin (SPE) serotypes
A, B, and C are considered to be members of
a family of pyrogenic toxins because they have
the abilities to induce fever, enhance the host's
susceptibility to lethal endotoxin shock, and
induce nonspecific T-cell proliferation (19,
132). It remains to be determined whether all
bacterial SAgs fulfill all three criteria defining
pyrogenic toxins.

The mechanisms by which these toxins induce disease remain to be determined. However, all share the basic abilities to bind MHC
class II molecules and to stimulate T cells. Thus
far, two mechanisms for disease induction
through SAg properties have been proposed
(103). In one mechanism, by virtue of the interaction of class II molecules with the toxin,
macrophages and mast cells would be stimulated to release their soluble mediators such as
IL-1, TNF-α, and leukotrienes. These soluble
mediators are known to cause many of the features of the host acute-phase response, including fever, weight loss, and somnolence, that are
associated with the disease states induced by
these toxins. The other proposed mechanism
predicts release of large amounts of soluble mediators such as IL-2, gamma interferon, and
TNF-α from the large number of T cells stimulated by the SAg. Each of these mechanisms
was tested by examining the effects of SEB on
weight loss and immunosuppression in mice
that do not normally respond to SEB, including TCR knockout mice, nude mice, and cy-
closporin A-treated mice (103). Weight loss
and immunosuppression were greatly reduced
in these mice. Moreover, because the T-cell-
deficient mice still had functional class II-
bearing cells, the effect of SEB on weight loss
observed in healthy mice was not due to the
hdirect induction of macrophages by SAg.
Consequently, the T cells are thought to be
responsible for the disease process, because
mice that lack the Vβ TCRs that react with SE
but that still contain normal numbers of T cells

TABLE 1 Vβ specificities of bacterial and mycoplasmal SAgs

Bacterium SAg	Vβ		Reference(s)
	Mouse	Human	
Clostridium perfringens			
CPE	ND[a]	6.9, 22	20
Mycobacterium tuberculosis	ND	8	117
Mycoplasma arthritidis			
MAM	1,[b] 3.1,[b] 5.1, 6, 7,[b] 8.1–3, 16[b]	3.1, 5.1, 7.1–2, 8.1–2, 10, 11.2, 12.2, 15.1, 17, 20	11, 15, 32, 36, 53, 123
Pseudomonas aeruginosa			
Exotoxin A	8.2	ND	45
Staphylococcus aureus			
SEA	1, 3, 10, 11, 12, 17	1.1, 5.3, 6.3, 6.4, 6.9, 7.3, 7.4, 9.1	24, 75, 144
SEB	7, 8.1, 8.2, 8.3	3, 12, 14, 15, 17, 20	24, 27, 67, 156
SEC1	7, 8.2, 8.3, 11	12	24, 85
SEC2	8.2, 10	12, 13.2, 14, 15, 17, 20	24, 27, 85
SEC3	7, 8.2	5, 12, 13.2	24, 29, 67, 85
SED	3, 7, 8.3, 11, 17	5, 12	24, 67, 85
SEE	11, 15, 17	5.1, 6.1–4, 6.9, 8.1, 18	24, 27, 75
Exfoliating toxins A and B	10, 11, 15	2	24, 27
TSST-1	15	2	24, 27, 28
Streptococcus pyogenes			
M protein	ND	2, 4, 8	145, 146
SPE A	8.2	2, 12, 14, 15	77, 146
SPE B	ND	2, 8	2, 146
SPE C	ND	1, 2, 5.1, 10	146
Yersinia enterocolitica	3, 6, 7, 8.1, 9, 11	ND	141
Yersinia pseudotuberculosis	ND	3, 9, 13.1, 13.2	3

[a]ND, not determined.
[b]Activation of Vβ1, Vβ3.1, Vβ7, and Vβ16 by MAM is demonstratable only in Vβ[a] and Vβ[c] haplotype mice.

are asymptomatic following toxin administration.

The family of pyrogenic toxins produced by *S. aureus* and group A streptococci is the best characterized of the bacterial SAgs. All of the SE, TSST-1, exfoliating toxins A and B, and SPE A and SPE C toxins have been sequenced and shown to be monomeric proteins of 22 to 28 kDa (70). Alignment of the amino acid sequences of these proteins shows that many of the proteins are closely related and that they can be divided into two groups: (i) SEA, SEE, and SED, and (ii) SEB, SEC1, and SEC3 (103). In addition, all the members of this toxin family have conserved regions of primary structure (72). The acute and chronic disease associations of these toxins have been reviewed elsewhere

(132). Briefly, toxins produced by members of this family contribute to a variety of disease states, including toxic shock syndrome, food poisoning, erysipelas, rheumatic fever, guttate psoriasis, arthritis, atopic dermatitis, and Kawasaki syndrome (132).

According to polyacrylamide gel electrophoresis (PAGE) analysis, MAM is a protein of 213 amino acids with a molecular mass of 27 kDa (10, 91), a mass similar to the molecular mass of 25.3 kDa that is predicted from the amino acid sequence. Structurally, MAM is unrelated to the SE or SPE SAgs. MAM lacks cysteine residues (9a), which precludes the formation of the disulfide loop thought to be required by SE to provide the stability necessary for toxin function (70). Note that TSST-1 also lacks a disulfide loop and that a linear peptide of TSST-1 is able to induce a mitogenic response similar to that observed with the native toxin (48). Thus, it is possible that linear rather than conformational domains are important for MAM activity. In addition, because both MAM and TSST-1 are capable of SAg bridge formation (54, 112, 149), the structural requirements for MAM activity may be similar to those for TSST-1 activity. While the arthritogenic properties of *M. arthritidis* are well known, the role of MAM in disease is less well defined, as discussed below.

The *Clostridium perfringens* enterotoxin (CPE) is a 34-kDa protein that specifically stimulates human Vβ6.9- and Vβ22-bearing T cells (20). *C. perfringens* is a known etiologic agent of food poisoning, and the abdominal pain and diarrhea associated with the disease are thought to be induced by CPE (106). In addition, CPE has been implicated in antibiotic-induced diarrhea, infantile diarrhea, and sudden infant death syndrome (20, 106). *Pseudomonas aeruginosa* exotoxin A is a unique SAg in that it requires proteinase cleavage in order to induce lymphoproliferation (108). The lymphoproliferative activity of this 66-kDa ADP-ribosylating toxin is associated with its carboxy-terminal end (99).

The toxins that possess SAg properties and that are produced by *Yersinia* spp. are not very well characterized. The SAg component of *Y. enterocolitica* is found in both culture supernatant and cytoplasmic and membrane fractions (141), whereas the SAg activity of *Y. pseudotuberculosis* is found in the culture supernatant (3). Some of the symptoms associated with infection by *Y. pseudotuberculosis* include fever, rash, erythema nodosum, and arthritis, but these symptoms have not been clearly associated with the toxin. Culture supernatants selectively activate T cells bearing human Vβ3, Vβ9, Vβ13.1, and Vβ13.2. Antigen processing is not required, but the response requires class II antigen without classic MHC restriction. Antibodies to SEs have no ability to neutralize *Yersinia* supernatant-induced T-cell proliferation (3). In addition, *Y. enterocolitica* and *Y.*

pseudotuberculosis have been implicated in the development of Reiter's syndrome and reactive arthritis, but it remains to be determined whether their SAgs are involved (3).

SAg-Associated Diseases

Kawasaki syndrome is a systemic vasculitis that on the basis of epidemiologic evidence is suspected to be caused by an infectious agent. It affects principally the young, occurs in outbreaks, shows a seasonality of occurrence, and recurs only rarely. An increase in human TCR Vβ2 and Vβ8.1 has been associated with active disease, as have fever, hypotension, and rash. Importantly, elevated T-cell-clone levels return to normal during the convalescent period. Sequencing of the abnormal clones reveals polyclonality of the junctional region (CDR3) that is consistent with SAg activation and not suggestive of exposure to a traditional antigen (3). Some of the diseases easily confused with Kawasaki syndrome are scarlet fever, staphylococcal scalded-skin syndrome, leptospirosis, Rocky Mountain spotted fever, and juvenile rheumatoid arthritis (136). Each of these diseases has been considered as having possible SAg involvement.

Kawasaki syndrome is treated with aspirin and intravenous gamma globulin (IVIG). The role of IVIG in this therapy is not yet defined, but regulation of idiotypes and blockade of Fc receptors are among the leading hypotheses. Another possibility is that the pooled Ig has anti-SAg activity (143). This possibility is supported by the improved results with early treatment using IVIG and the requirement for very high doses, typically 2 g/kg of body weight, of IVIG, which may well be enough to supply neutralizing quantities of specific antibody.

Toxic shock syndrome became a well-known entity in the 1970s, when its association with the use of highly absorbent feminine tampons was described. The disease is known to occur in many other circumstances and affects men as well. It is classically associated with fever, erythematous rash, and hypotension, but several so-called minor manifestations, including evidence of organ involvement such as elevated liver function tests, decreased renal function, and mental status changes, can also be present. The disease has been associated with several staphylococcal and streptococcal SAgs, including TSST-1, SEB, and SEC. The TSST-1-associated disease is associated also with elevated levels of circulating Vβ2-bearing cells, and increased levels of IL-1 and TNF-α, consistent with a SAg-mediated process, are observed in the acute phases (95).

Recently, evidence for a SAg in *Mycobacterium tuberculosis* has been obtained (117). Pleural fluid from patients with tuberculous pleuritis was analyzed by FACS and found to contain 2- to 13-fold more T cells bearing Vβ8. In addition, healthy human lymphocytes stimulated in vitro with *Mycobacterium tuberculosis* showed a similar skewing of response without a need for antigen processing or classic MHC restriction. MHC class II-bearing cells were required for presentation, and no dependence on CD4 or CD8 expression was present. Sequence analysis of the TCRs from reactive clones showed them to be polyclonal, which is consistent with SAg activation and not suggestive of the traditional antigen pathway. The adjuvant properties of tubercle bacilli are well known and are commonly utilized as complete Freund's adjuvant. The proposed mycobacterial SAgs may in fact be responsible for some of these adjuvant properties, perhaps via a SAg bridge mechanism as described above.

Role of the MAM SAg in Mycoplasma-Induced Disease

M. arthritidis, which commonly causes a latent infection in rodent colonies, can induce a naturally occurring acute to chronic arthritis in these animals (39). The chronic disease in mice closely resembles histologically the joint lesions seen in human rheumatoid arthritis. The synovial lining becomes heavily infiltrated with lymphocytes, and lymphoid follicles and plasma cells are common. Hyperplasia of the synovial membrane with villus formation, pannus, and eventual progressive destruction of articular cartilage and bone develops. The organ-

isms become difficult to isolate in the chronic phase of the disease. Interestingly, periods of remission and exacerbation are common. Other symptoms associated with this disease are an early toxic shock that can be lethal, especially in mice, and paralysis of the hind limbs complicated by fecal impaction, urethritis, and uveitis.

Earlier studies to determine the effect of MAM in *M. arthritidis*-induced disease compared symptoms in inbred and congenic H-2E$^+$ mouse strains whose lymphocytes responded highly to MAM (H-2^k, H-2^d) with those in H-2E$^-$ mice whose lymphocytes were considered to be nonresponsive to MAM (H-2^b). Following administration of MAM, mice that expressed H-2E showed a high incidence of lethal shock. The animals that survived exhibited massive peritoneal adhesions indicative of resolving peritonitis. Lymphocytes from non-MAM-responsive or weakly responsive mice (H-2E$^-$ mice) exhibited a much lower incidence of toxicity and death, and no peritoneal adhesions were present (38). Mice injected subcutaneously with MAM displayed symptoms that were also MHC dependent. Although all mice developed suppurative abscesses, necrotizing abscesses occurred predominantly in those strains that expressed H-2E and that had lymphocytes highly responsive to MAM (37).

Both H-2E$^+$ and H-2E$^-$ mouse strains developed an acute arthritis in response to *M. arthritidis*. Long-term studies on the role of MAM must now be undertaken. In addition, the H-2E$^-$ mouse strains tested are known to respond weakly to MAM. Additional work with strains whose lymphocytes are totally nonresponsive to MAM even at high doses is needed. There can be little doubt that MAM at least contributes to the arthritic response, since live organisms locate and multiply within the joints. In fact, strain DA rats that receive a single intra-articular injection of MAM develop a transient arthritis (26). The pathology is characterized by an early edema and by polymorphonuclear lymphocyte and local lymphoid-cell infiltration below the synovial surface. These symptoms are followed by hypertrophy and hyperplasia with infiltration of the subsynovium by macrophages and fibroblasts and loss of the surface layer of the synovium. The lesions are virtually resolved by 7 days. Interestingly, BN rats whose lymphocytes respond poorly to MAM (and to other T-cell mitogens) develop a less severe arthritis when injected with MAM (25a).

ROLE IN AUTOIMMUNE DISEASE

For many years, the paradigm for autoimmune disease has been that a combination of genetic susceptibility and environmental factors triggers disease. Although many studies have been performed, in most cases neither the genetic predisposition nor the trigger is identifiable. Principal among the candidate genes for predisposition are the MHC products, in particular those of the class II type (H-2A and H-2E in mice and DR, DQ, and DP in humans). These molecules are expressed on selected cells of the host and are responsible for the presentation of linear peptide antigens principally of exogenous origin. MHC molecules are capable of limiting the responsiveness of the host to specific peptides and may account for select autoantibodies in particular diseases (129). Largely on the basis of family and twin studies of systemic immune diseases that show relatively low concordance even among monozygous twins, the disease trigger is believed to be environmental (153). Microbial agents are favorite candidate triggering agents, but proof has been elusive.

Several recent theories of autoimmune pathogenesis make use of SAg in their models. These possible mechanisms include V-region disease as described by Heber-Katz and Acha-Orbea (4, 64), the SAg bridge (54), and molecular mimicry. Importantly, they are not mutually exclusive. The V-region hypothesis suggests that some autoreactive clones escape negative selection and are subsequently activated such that the T cells express a limited number of TCR Vβ types. Ultimately, this could result in oligoclonal self-reactive clones such as are observed in some autoimmune diseases, including multiple sclerosis and experimental allergic encephalomyelitis (EAE). EAE is perhaps the best-known example at present, as is discussed below. The SAg bridge as discussed above provides a means for SAgs to activate B cells as well as T cells and may result in the preferential formation of autoantibodies. A third mechanism is molecular mimicry. This is perhaps best illustrated for the streptococcal M5 peptide, which expresses sites that mimic human cardiac myosin and is a SAg capable of stimulating those reactive clones.

V-Region Disease

EAE is an autoimmune condition of rodents that is characterized by central nervous system inflammation and demyelination. Clinical symptoms include tail or hind limb weakness and paralysis. The disease, which has been extensively used as a model for multiple sclerosis, is induced by immunization of rodents against myelin basic protein or defined epitopes of myelin basic protein (8, 64). Transfer of the disease to healthy recipients can be made by sensitized T cells or T-cell lines bearing specific Vβ chain segments of the TCR. For example, in the PL/J mouse, Vβ8.2-bearing T cells that recognize the N-terminal peptide Ac 1-11 of myelin basic protein are instrumental in development of the disease (4, 150, 159).

SAgs that activate the appropriate myelin basic protein-reactive, Vβ TCR-bearing T cells can activate EAE. Thus, SEB reinduces EAE in PL/J mice that had previously recovered from the disease, and this cycle could be repeated many times. EAE could also be induced by SEB in mice suboptimally immunized with myelin basic protein. Surprisingly, SEA, which does not activate Vβ8.2 TCR-bearing T cells, is also able to reactivate disease. This finding indicates that EAE in these mice might be driven by T cells in addition to those that bear Vβ8.2 chains. Alternatively, production of cytokines as a result of SEA activation of irrelevant Vβ TCRs could be the pathway to reinduction of EAE (131). Similar findings were made by Brocke and coworkers (22), who also showed that EAE transferred by T-cell lines or clones reactive to myelin basic protein can be reactivated by SEB after remission. In that study, SEA was somewhat less effective in triggering disease. The authors hypothesized that TNF-α played a role in exacerbation, since exacerbations in mice treated with anti-TNF-α antibodies were delayed.

As discussed previously, not only do SAgs activate specific Vβ TCR-bearing cells in vivo, but this activated state can also be followed by anergy, in which the responding T cells are deleted owing to apoptosis. Brocke and co-workers (22) showed that injection of mice with SEB before injection with myelin basic protein decreases the subsequent primary and secondary responses of the mice to a myelin basic protein determinant that is recognized by Vβ8+ T cells. Soos et al. (139) supported these findings and further demonstrated that Vβ8 deletion by SEB protects PL/J mice against EAE. The anergy appears to be Vβ specific, since splenocytes from SEB-treated mice fail to respond to SEB but respond normally to SEA. It is clear from these and other studies (127) that the effect of SAg on autoimmune diseases is very dependent on the time of SAg administration.

The human disease that shows striking parallels to EAE is multiple sclerosis, which is characterized by multiple destructive attacks by the immune system on the central nervous system. The repeated attacks result in damage to the myelin coating of the central nervous system, resulting in plaque formation. About 60% of patients are female, and most are young adults when first diagnosed (9). Genetic associations to HLA-DQ and HLA-DR2, especially Dw2, have been reported (140). In addition, consistent with the EAE model, evidence for TCR Vβ skewing has been reported. TCR Vβ5.2 and Vβ6.1 have been found in central nervous system plaques (94). There is a seasonality of occurrence and a selective geography, each of which suggests an infectious etiology. Among the associated immune abnormalities are oligoclonal banding in cerebrospinal fluid analyzed for Ig by PAGE or isoelectric focusing (9). During each attack, the number of circulating CD8+ T cells is decreased. Their levels return to normal following resolution. Interestingly, levels of TNF-α in the cerebrospinal fluids of multiple sclerosis patients are higher during active disease. In addition, peripheral blood lymphocytes from multiple sclerosis patients respond to mitogens with higher TNF-α levels than do controls (137). Recently, human T cells reactive to human myelin basic protein have been shown to be stimulated by appropriate SAgs (23). The combination of genetic predisposition, increase in local production of gamma globulin, increased levels of TNF-α, and probable infection are all suggestive of SAg involvement in this disease.

Molecular Mimicry

Streptococcal M protein is present on the outer cell wall of *Streptococcus pyogenes,* and polymorphisms of M protein relate to the strain differences observed in infectivity as well as the ultimate type of disease associated with infection. Some serotypes are associated with rheumatic fever, while others correlate best with poststreptococcal glomerulonephritis (17). The amino-terminal end of the molecule is now known to contain a SAg referred to as streptococcal M protein. The M5 variant is perhaps the best characterized, as its sequence, binding regions, and specificities are well worked out (154). This SAg is thus far unique in that it has classic molecular mimicry and also the ability to activate huge portions of the resultant autoreactive T cells. Curiously, it does not always require the presence of APC in order to function as a SAg (93). The mimicry is quite impressive, as amino acids 84 to 197 contain segments that mimic myosin, sarcolemma proteins, vimentin, synovium, and cartilage, and several of these segments, including the QKSKQ epitope, which is recognized by sera from rheumatic fever patients, are likely involved in rheumatic fever (154).

Rheumatic fever is a human systemic disease known to occur in a select population that is believed to have a genetic predisposition associated specifically with HLA-DR4 or DR2

(17). In susceptible people exposed to the bacteria, a disease whose major features are fever, rash (erythema marginatum), arthritis, and carditis can develop (17). Some of these problems are thought to be a consequence of mimicry between streptococcal and host proteins. As highlighted above, clones that are autoreactive to the host may then be activated in large numbers by the SAg properties of the M peptide.

SAg Triggering of Experimental Arthritis

As we have discussed, in the V-region disease hypothesis of autoimmune disease, T cells sensitized to autoantigens are seen as the main driving or initiating factor. Autoreactive T cells may occur naturally or be the result of trauma. In the case of arthritis, immunization of mice with type II collagen induces a chronic arthritis (collagen-induced arthritis) that is mediated predominantly by T cells expressing certain Vβ chains of the α/β TCR. Genetic studies imply that in the B10.RIII mouse, T cells expressing Vβ6 and Vβ8 are important in the disease process (12, 13, 62). In addition, there is an enrichment in T cells bearing these same Vβ chains in the joints and draining nodes of arthritic mice. Coincidentally, the MAM SAg activates these same populations of T cells. Since it has been demonstrated that MAM can induce clonal expansion of Vβ6 and Vβ8 in vivo (32), the V-region hypothesis was tested by determining the effect of MAM on collagen-induced arthritis.

MAM injected intraperitoneally or intravenously into mice convalescing from collagen-induced arthritis causes a rapid flare-up of disease (Fig. 2). Furthermore, MAM can also trigger arthritis in nonarthritic mice that have been suboptimally immunized to type II collagen up to 200 days previously (34). MAM increases the severity and shortens the time to onset of arthritis when the antigen is given after suboptimal doses of collagen. There is evidence that the recurrence seen with MAM is Vβ specific, since SEA, a SAg that activates a different set of murine Vβ TCR-bearing cells, fails to trigger disease activity. SEB, like MAM, activates Vβ8-bearing T cells and could also trigger arthritis. Although an increase in antibodies to type II collagen is observed after injection of MAM, it has not been established whether this increase is due to polyclonal production via the SAg bridge or to a specific immune response to cartilage collagen made antigenic by the enhanced inflammatory process (34). Although activation of arthritis appears to be Vβ specific in the collagen-induced arthritis system, nonspecific factors such as cytokines are released as a result of activation of irrelevant Vβ-bearing T-cell subsets or of monocytes and might also contribute to disease. For example, the activation of streptococcal cell wall arthritis by TSST-1 may be a result of cytokine production rather than activation of specific Vβ-bearing T cells (134).

Aged MRL-*lpr/lpr* mice spontaneously develop autoimmunity and arthritis, so these mice have been used to determine the role of SAgs in disease. In a study by Mountz et al. (110), SEB but not SEA was shown to induce an early arthritis after injection into MRL-*lpr/lpr* mice that bore the Vβ8.2 TCR transgene. Although the arthritis only developed 30 days after direct intra-articular injection, the disease was shown to be dependent on expression of both the Vβ8.2 transgene and the *lpr* gene. The *lpr* gene is a mutation of the *fas* gene that is involved in T-cell apoptosis (155). Defects in thymic apoptosis are thought to lead to release of autoimmune T-cell clones. Thus, a preexisting autoreactive T-cell population appears to be required for bacterial SAg-mediated activation of clinical autoimmune disease. Not only can SAgs trigger arthritis in mice that have been presensitized to joint components, but the organisms producing these SAgs might also be able to induce a chronic arthritis that is driven by T cells bearing specific Vβ chain segments. Thus, a TSST-1-producing strain of *S. aureus* induces a chronic erosive arthritis in which the synovium shows a selective increase in T cells bearing Vβ11 (1). Mice that lack Vβ11 because of genomic deletions fail to develop arthritis, as do mice that are Vβ11 depleted by the administration of specific MAb.

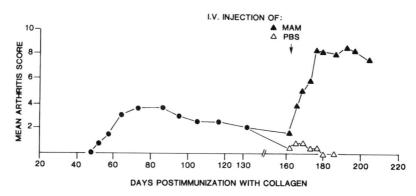

FIGURE 2 SAg-induced activation of collagen-induced arthritis. Mice were immunized with porcine type II collagen on day 0 and day 21. Arthritis peaked at day 80 and had declined by day 160, when the mice were given an intravenous (I.V.) injection of MAM or PBS. MAM-treated mice showed a rapid increase in arthritis that exceeded the level of the previous phase of arthritis mediated by immunization to type II collagen alone.

It is apparent from these observations of collagen-induced arthritis, MRL-*lpr/lpr* mice, and EAE that SAgs can trigger or cause flare-ups in autoimmune diseases by clonally expanding preexisting T cells autoreactive to collagen or to myelin basic protein, respectively.

SAg Involvement in Human Rheumatoid Arthritis

Human rheumatoid arthritis is a disease characterized by symmetrical erosive destruction of the small joints. Its cause remains unknown, but it is believed that a genetic predisposition is triggered into disease expression by environmental agents. The principal candidate triggers are infectious agents. Rheumatoid arthritis is most often accompanied by production of autoantibodies and inflammatory cytokines, resulting in fatigue, chronic anemia, and sometimes other nonarticular symptoms. The synovial lining of the joint contains T and B cells as well as mature plasma cells, macrophages, and dendritic cells. In addition, the joint cytokine profile is dominated by TNF-α, IL-1, and cerebrospinal fluid (101). Mourad and colleagues have shown that rheumatoid arthritis synovial fibroblasts respond with increased levels of gamma interferon, IL-6, and IL-8 following exposure in vitro to SEA (111).

Several reports have suggested that T cells derived from rheumatoid arthritis synovium are oligoclonal (74). Preliminary sequence data from the Vβ and Jβ regions support a SAg-like activation (157). We reiterate that the SAg MAM activates most of the TCR types implicated in rheumatoid arthritis. Further, SAg mechanisms such as the SAg bridge can result in rheumatoid factor production even from healthy lymphocytes, as has been shown for MAM and SED (49, 63).

Allotypic differences in SAg responsiveness, as observed for SEs, suggest a potential mechanism to account for class II associations with autoimmune disease. For example, an individual with an HLA class II allele containing the QKRAAV epitope that is associated with rheumatoid arthritis (58) might selectively respond to a specific SAg and thereby develop rheumatoid arthritis. Amino acids in this region of the β chain affect SAg binding (66). Thus, specific SAgs may trigger disease in genetically predisposed individuals depending on their class II types.

Role of SAgs in AIDS

The profound immunomodulatory and inhibitory effects that SAgs exert on the immune system suggest that these antigens may play a role in immunodeficiency diseases such as AIDS (98). As described above, endogenous or exogenous murine retroviruses code for SAgs that are known to result in the deletion of T cells bearing specific Vβ chains of the TCR. Bacterial SAgs can also induce anergy and apoptosis in T cells that express reactive Vβ chains. Furthermore, the shift toward polyclonal B-cell activation in mice by SAgs such as MAM, which is mediated by an increase in T_H2-like activities and IL-6 production, is consistent with similar findings in AIDS patients. The ability of SAgs to trigger autoimmunity (34) may correlate with the higher incidence of arthritis in AIDS patients (47, 109).

The involvement of SAgs in AIDS may be mediated by the presence of a SAg in the human immunodeficiency virus (HIV) itself and/ or by infection with opportunistic invaders that produce SAgs. It has been suggested that successive infections with SAg-producing organisms would gradually deplete the CD4$^+$ T-cell repertoire (30). Such infections would become more frequent as the T-cell functions of the individual gradually declined, thereby accelerating the disease process. Groux and colleagues have shown that HIV$^+$ cells exposed to SEB enter apoptosis (60). In a related disease, murine AIDS, a SAg has been identified as being part of the virus (76, 84). Furthermore, there is evidence that this SAg is responsible for the progression to anergy, although the response is polyclonal rather than oligoclonal (113).

There is some evidence for involvement of SAgs in human AIDS. Studies by Imberti and coworkers (78) suggest that AIDS patients exhibit a loss of a common set of T cells bearing specific Vβ chain segments, a result that would favor the presence of an HIV-specific SAg rather than the cumulative effects of SAgs resulting from multiple infections. The observations of Laurence et al. (98) are particularly intriguing, since their work indicates that T-cell lines bearing human Vβ12, Vβ17, and Vβ8 TCRs from HIV-infected patients express more HIV proteins than do other Vβ chain-bearing T cells. In addition, irradiated accessory cells from HIV$^+$ donors preferentially activate

T cells bearing Vβ12 chain segments of the TCR. Thus, several mechanisms exist for potential SAg involvement in AIDS, but further work is required to determine if any or all of them are actually used.

FUTURE PROSPECTS

Since the original description of SAgs in 1989, the number of articles in Medline that refer to SAgs has grown from 4 in 1989 to 188 in 1993 and more than 78 in the first 3 months of 1994. We expect that this trend will continue and that many infectious agents will be found to use SAgs in order to gain some advantage over their host. In addition, many of the observations now under investigation are likely to improve modern diagnosis and treatment of autoimmune, infectious, and neoplastic diseases. Genetic and structural information will allow design of DNA-based identification and diagnostic strategies. Therapeutically, the ability to selectively regulate T cells of a specific background may allow selective immunosuppression. Many tumors, such as melanoma, lymphoma, and leukemia, would benefit from the selected immune regulation or apoptosis that could result from the careful application of SAgs reactive to cells of a particular TCR Vβ family. As examples of V-region diseases are described, they, too, might be treated by the deletion of selected subsets. Detailed molecular understanding may allow immunization programs in which certain SAgs are used as adjuvant, antigen, and T-cell restricting elements at the same time. One concern is the toxicity associated with administration of SAg that might accompany any therapeutic regimen. Alber and coworkers have shown that at least for SEB, chemical modification with carboxymethylation dramatically lowers the toxic potential without interfering with mitogenesis (7).

Some efforts to use SAgs in tumor therapy are already under way. They make use of the unique immune activation mechanisms associated with SAg-driven T-cell activation. For example, Newell and coworkers have used SEB to decrease the number of tumors caused by intravenous injection of progressor tumors in mice. They showed that a 50-μg dose of SEB per mouse given at the same time as tumor cells decreased tumor outgrowth from 75% in control animals to 25% in treated animals (114). Wallgren's group used cytotoxic T cells in vitro to kill a large percentage of B cells from a patient with chronic lymphocytic leukemia. They found that administration of tetradecanoyl phorbol acetate followed by T cells and the SAg SEA resulted in increased lysis; specifically, 47% lysis was observed at an effector-to-target ratio of 3:1. They found that tetradecanoyl phorbol acetate activation increased expression of ICAM, LFA1, LFA3, and HLA-DR. Additionally, antibody to ICAM, CD18, or HLA-DR was able to abolish the observed cytotoxicity (152). Thus, selective manipulations of the secondary signal can also alter the SAg therapeutic spectra.

Two trials in the treatment of autoimmune disease have been reported. First, SEB was used to decrease the severity of glomerulonephritis in the MRL-*lpr* mouse model of systemic lupus erythematosus (90). Two injections of SEB were given each week to 6-week-old mice. They developed depletion of Vβ8-bearing T cells, especially those that were negative for both CD4 and CD8 in a dose-dependent fashion. Serologic parameters, including anti-DNA antibody levels, as well as clinical measures improved, although the disease still occurred (90). Second, EAE was effectively treated by the use of SEB in PL/J mice (139). Vandenbark and coworkers (151) showed that peptides derived from T cells reactive to myelin basic protein were able to immunize mice against EAE. Importantly, immunization was mediated by CD8+ T cells and was transferable. The mice also developed an antibody response against the T-cell peptide (151). The peptides were able to treat the extant disease, and their use resulted in decreased severity and duration of illness (116). More recently, multiple sclerosis patients were vaccinated with irradiated T cells reactive to myelin basic protein (160). Of the six patients given T cells, none had side effects, and none showed acute exacerbations. However, no significant improvement was noted either.

Another expected use of SAgs is as adjuvants and/or vaccines. As discussed above, the timing of SAg administration with respect to antigen has a major role in the ultimate immune outcome. In addition, it is likely that different kinds of antigens are affected uniquely. The experiment performed in nature known as rheumatic fever shows a possible negative side of these strategies. Streptococcal M protein results in autoreactive antibodies and causes the pathology associated with rheumatic fever, thereby illustrating the potential pitfalls inherent in this approach if complete understanding is not obtained first.

ACKNOWLEDGMENTS

The work reported here was supported by grants from the National Institute of Allergy and Infectious Diseases (AI-12103), the National Institute of Arthritis and Metabolic Diseases (AR-02255), and the Nora Eccles Treadwell Foundation.

We thank Scott Benson for preparing the illustrations.

REFERENCES

1. **Abdelnour, A., T. Bremell, R. Holmdahl, and A. Tarkowski.** 1994. Clonal expansion of T lymphocytes causes arthritis and mortality in mice infected with toxic shock syndrome toxin-1-producing staphylococci. *Eur. J. Immunol.* **24:** 1161–1166.

2. **Abe, J., J. Forrester, T. Nakahara, J. A. Lafferty, B. L. Kotzin, and D. Y. Leung.** 1991. Selective stimulation of human T cells with streptococcal erythrogenic toxins A and B. *J. Immunol.* **146:**3747–3750.

3. **Abe, J., T. Takeda, Y. Watanabe, H. Nakao, N. Kobayashi, D. Y. M. Leung, and T. Kohsaka.** 1993. Evidence for superantigen production by *Yersinia pseudotuberculosis. J. Immunol.* **151:** 4183–4188.

4. **Acha-Orbea, H., D. J. Mitchell, L. Timmerman, D. C. Wraith, A. S. Tausch, M. K. Waldon, S. S. Zamvil, H. O. McDevitt, and L. Steinman.** 1988. Limited heterogeneity of T cell receptors from lymphocytes mediating autoimmune encephalomyelitis allows specific immune intervention. *Cell* **54:**263–273.

5. **Acha-Orbea, H., and E. Palmer.** 1991. Mls—a retrovirus exploits the immune system. *Immunol. Today* **12:**356–361.

6. **Acharya, K. R., E. F. Passalacqua, E. Y. Jones, K. Harlos, D. I. Stuart, R. D. Brehm, and H. S. Tranter.** 1994. Structural basis of superantigen action inferred from crystal structure of

toxic-shock syndrome toxin-1. *Nature* (London) **367**:94–97.

7. **Alber, G., D. K. Hammer, and B. Fleischer.** 1990. Relationship between enterotoxic and T lymphocyte stimulating activity of staphylococcal enterotoxin B. *J. Immunol.* **144**:4501–4506.

8. **Alvord, E. C., C. M. Shaw, S. Hruby, L. R. Sires, and J. C. Slimp.** 1984. The onset of experimental allergic encephalomyelitis: a useful model for multiple sclerosis, p. 461–466. *In* E. C. Alvord, Jr., M. W. Kies, and A. J. Suckling (ed.), *Progress in Clinical and Biological Research.* Alan R. Liss, Inc., New York.

9. **Antel, J. P., and B. G. W. Arnason.** 1983. Multiple sclerosis and other demyelinating diseases, p. 2098–2104. *In* R. G. Petersdorf, R. D. Adams, E. Braunwald, K. J. Isselbacher, J. B. Martin, and J. D. Wilson (ed.), *Principles of Internal Medicine,* 10th ed. McGraw Hill Book Co., New York.

9a. **Atkin, C. L., and B. C. Cole.** Unpublished observations.

10. **Atkin, C. L., B. C. Cole, G. J. Sullivan, L. R. Washburn, and B. B. Wiley.** 1986. Stimulation of mouse lymphocytes by a mitogen derived from *Mycoplasma arthritidis.* V. A small basic protein from culture supernatants is a potent T-cell mitogen. *J. Immunol.* **137**:1581–1589.

11. **Baccala, R., L. R. Smith, M. Vestberg, P. A. Peterson, B. C. Cole, and A. N. Theofilopoulos.** 1992. *Mycoplasma arthritidis* mitogen Vβ engaged in mice, rats, and humans, and requirement of HLA-DRα for presentation. *Arthritis Rheum.* **35**:434–442.

12. **Banerjee, S., M. A. Behlke, G. Dungeon, D. Y. Loh, J. M. Stuart, H. S. Luthra, and C. S. David.** 1988. Vβ6 genes of T cell receptor may be involved in type II collagen induced arthritis in mice. *FASEB J.* **2**:2120.

13. **Banerjee, S., T. M. Haqqi, H. S. Luthra, J. M. Stuart, and C. S. David.** 1988. Possible role of Vβ T cell receptor genes in susceptibility to collagen-induced arthritis in mice. *J. Exp. Med.* **167**:832–839.

14. **Bekoff, M. C., B. C. Cole, and H. M. Grey.** 1987. Studies on the mechanism of stimulation of T cells by the *Mycoplasma arthritidis*-derived mitogen. Role of class II IE molecules. *J. Immunol.* **139**:3189–3194.

15. **Bhardwaj, N., A. S. Hodtsev, A. Nisanian, S. Kabak, S. M. Friedman, B. C. Cole, and D. N. Posnett.** 1994. Human T-cell responses to *Mycoplasma arthritidis*-derived superantigen. *Infect. Immun.* **62**:135–144.

16. **Bierer, B. E., B. P. Sleckman, S. E. Ratnofsky, and S. J. Burakoff.** 1989. The biologic roles of CD2, CD4, and CD8 in T-cell activation. *Annu. Rev. Immunol.* **7**:579–599.

17. **Bisno, A. L.** 1991. Group A streptococcal infec-

tions and acute rheumatic fever. *N. Engl. J. Med.* **325**:783–793.

18. **Bodmer, H., S. Viville, C. Benoist, and D. Mathis.** 1994. Diversity of endogenous epitopes bound to MHC class II molecules limited by invariant chain. *Science* **263**:1284–1286.

19. **Bohach, G. A., D. J. Fast, R. D. Nelson, and P. M. Schlievert.** 1990. Staphylococcal and streptococcal pyrogenic toxins involved in toxic shock syndrome and related illnesses. *Crit. Rev. Microbiol.* **17**:251–272.

20. **Bowness, P., P. A. Moss, H. Tranter, J. I. Bell, and A. J. McMichael.** 1992. *Clostridium perfringens* enterotoxin is a superantigen reactive with human T cell receptors Vβ6.9 and Vβ22. *J. Exp. Med.* **176**:893–896.

21. **Braunstein, N. S., D. A. Weber, X.-C. Wang, E. O. Long, and D. Karp.** 1992. Sequences in both class II major histocompatibility complex α and β chains contribute to the binding of the superantigen toxic shock syndrome toxin 1. *J. Exp. Med.* **175**:1301–1305.

22. **Brocke, S., A. Gaur, C. Piercy, A. Gautam, K. Gijbels, C. G. Fathman, and L. Steinman.** 1993. Induction of relapsing paralysis in experimental autoimmune encephalomyelitis by bacterial superantigen. *Nature* (London) **365**:642–644.

23. **Burns, J., K. Littlefield, J. Gill, and J. Trotter.** 1992. Bacterial toxin superantigens activate human T lymphocytes reactive with myelin autoantigens. *Ann. Neurol.* **32**:352–357.

24. **Callahan, J. E., A. Herman, J. W. Kappler, and P. Marrack.** 1990. Stimulation of B10.BR T cells with superantigenic staphylococcal toxins. *J. Immunol.* **144**:2473–2479.

25. **Candeias, S., C. Waltzinger, C. Benoist, and D. Mathis.** 1991. The Vβ17+ T cell repertoire: skewed Jβ usage after thymic selection; dissimilar CDR3s in CD4+ versus CD8+ cells. *J. Exp. Med.* **174**:989–1000.

25a. **Cannon, G. W., et al.** Unpublished data.

26. **Cannon, G. W., B. C. Cole, J. R. Ward, J. L. Smith, and E. J. Eichwald.** 1988. Arthritogenic effects of *Mycoplasma arthritidis* T cell mitogen in rats. *J. Rheumatol.* **15**:735–741.

26a. **Chang, M.-d.** Personal communication.

27. **Choi, Y., B. Kotzin, L. Herron, J. Callahan, P. Marrack, and J. Kappler.** 1989. Interaction of *Staphylococcus aureus* toxin "superantigens" with human T cells. *Proc. Natl. Acad. Sci. USA* **86**:8941–8945.

28. **Choi, Y., J. A. Lafferty, J. R. Clements, J. K. Todd, E. W. Gelfand, J. Kappler, P. Marrack, and B. L. Kotzin.** 1990. Selective expansion of T cells expressing Vβ2 in toxic shock syndrome. *J. Exp. Med.* **172**:981–984.

29. **Choi, Y. W., A. Herman, D. DiGiusto, T. Wade, P. Marrack, and J. Kappler.** 1990. Res-

idues of the variable region of the T-cell-receptor beta-chain that interact with *S. aureus* toxin superantigens. *Nature* (London) **346**:471–473.

29a. **Cole, B. C., and E. Ahmed.** Unpublished data.

30. **Cole, B. C., E. Ahmed, B. A. Araneo, J. Shelby, C. Kamerath, S. Wei, S. McCall, and C. L. Atkin.** 1993. Immunomodulation *in vivo* by the *Mycoplasma arthritidis* superantigen, MAM. *Clin. Infect. Dis.* **17**:S163-S169.

31. **Cole, B. C., B. A. Araneo, and G. J. Sullivan.** 1986. Stimulation of mouse lymphocytes by a mitogen derived from *Mycoplasma arthritidis*. IV. Murine T hybridoma cells exhibit differential accessory cell requirements for activation by *M. arthritidis* T cell mitogen, concanavalin A, or eggwhite lysozyme. *J. Immunol.* **136**:3572–3578.

32. **Cole, B. C., R. A. Balderas, E. A. Ahmed, D. Kono, and A. N. Theofilopoulos.** 1993. Genomic composition and allelic polymorphisms influence Vβ usage by the *Mycoplasma arthritidis* superantigen. *J. Immunol.* **150**:3291–3299.

33. **Cole, B. C., C. S. David, D. H. Lynch, and D. R. Kartchner.** 1990. The use of transfected fibroblasts and transgenic mice establishes that stimulation of T cells by the *Mycoplasma arthritidis* mitogen is mediated by Eα. *J. Immunol.* **144**:420–424.

34. **Cole, B. C., and M. M. Griffiths.** 1993. Triggering and exacerbation of autoimmune arthritis by the *Mycoplasma arthritidis* superantigen MAM. *Arthritis Rheum.* **36**:994–1002.

35. **Cole, B. C., D. R. Kartchner, and D. J. Wells.** 1989. Stimulation of mouse lymphocytes by a mitogen derived from *Mycoplasma arthritidis*. VII. Responsiveness is associated with expression of a product(s) of the Vβ8 gene family present on the T cell receptor α/β for antigen. *J. Immunol.* **142**:4131–4137.

36. **Cole, B. C., D. R. Kartchner, and D. J. Wells.** 1990. Stimulation of mouse lymphocytes by a mitogen derived from *Mycoplasma arthritidis* (MAM). VIII. Selective activation of T cells expressing distinct Vβ T cell receptors from various strains of mice by the "superantigen" MAM. *J. Immunol.* **144**:425–431.

37. **Cole, B. C., M. W. Piepkorn, and E. C. Wright.** 1985. Influence of genes of the major histocompatibility complex on ulcerative dermal necrosis induced in mice by *Mycoplasma arthritidis*. *J. Invest. Dermatol.* **85**:357–361.

38. **Cole, B. C., R. N. Thorpe, L. A. Hassell, and J. R. Ward.** 1983. Toxicity but not arthritogenicity of *Mycoplasma arthritidis* for mice associates with the haplotype expressed at the major histocompatibility complex. *Infect. Immun.* **41**:1010–1015.

39. **Cole, B. C., L. R. Washburn, and D. Taylor-Robinson.** 1985. Mycoplasma induced arthritis, p. 107–160. *In* S. Razin and M. F. Bairle (ed.), *The Mycoplasmas,* vol. 4. Academic Press, Inc., New York.

40. **Cole, B. C., and D. J. Wells.** 1990. Immunosuppressive properties of the *Mycoplasma arthritidis* T-cell mitogen in vivo: inhibition of proliferative responses to T-cell mitogens. *Infect. Immun.* **58**:228–236.

41. **Davis, M. M., and P. J. Bjorkman.** 1988. T-cell antigen receptor genes and T-cell recognition. *Nature* (London) **334**:395–402.

42. **Dellabona, P., J. Peccoud, J. Kappler, P. Parrack, C. Benoist, and D. Mathis.** 1990. Superantigens interact with MHC class II molecules outside of the antigen groove. *Cell* **62**:1115–1121.

43. **Dietz, J. N., and B. C. Cole.** 1982. Direct activation of the J774.1 murine macrophage cell line by *Mycoplasma arthritidis*. *Infect. Immun.* **37**:811–819.

44. **Dixon, D. M., R. D. LeGrand, and M. L. Misfeldt.** 1992. Selective activation of murine Vβ8.2 bearing T cells by *Pseudomonas* exotoxin A. *Cell Immunol.* **145**:91–99.

45. **D'Orazio, J. A., B. C. Cole, E. A. Riley, and J. Stein-Streilein.** 1993. Exogenous superantigens augment NK activity of sorted human NK (CD56+) lymphocytes. *J. Immunol.* **150**:301A.

46. **Doyle, C., and J. L. Strominger.** 1987. Interaction between CD4 and class II MHC molecules mediates cell adhesion. *Nature* (London) **330**:256–259.

47. **Edelman, A. S., and S. Zolla-Pazner.** 1989. AIDS: a syndrome of immune dysregulation, dysfunction and deficiency. *FASEB J.* **3**:22–30.

48. **Edwin, C., J. A. Swack, K. Williams, P. F. Bonventre, and E. H. Kass.** 1991. Activation of *in vitro* proliferation of human T cells by a synthetic peptide of toxic shock syndrome toxin-1. *J. Infect. Dis.* **163**:524–529.

49. **Emery, P., G. S. Panyi, K. I. Welsh, and B. C. Cole.** 1985. Rheumatoid factor and HLA-DR4 in RA. *J. Rheumatol.* **12**:217–222.

50. **Fleischer, B., and H. Schrezenmeir.** 1988. T cell stimulation by staphylococcal enterotoxins: clonally variable response and requirements for major histocompatibility complex class II molecules on accessory or target cells. *J. Exp. Med.* **167**:1697–1701.

51. **Fraser, J. D.** 1989. High-affinity of binding of staphylococcal entertoxins A and B to HLA-DR. *Nature* (London) **339**:221–223.

52. **Fraser, J. D., M. E. Newton, and A. Weiss.** 1992. CD28 and T cell antigen receptor signal transduction coordinately regulate interleukin 2 gene expression in response to superantigen stimulation. *J. Exp. Med.* **175**:1131–1134.

53. **Friedman, S. M., M. K. Crow, J. R. Tu-**

mang, M. Tumang, Y. Q. Xu, A. S. Hodtsev, B. C. Cole, and D. N. Posnett. 1991. Characterization of human T cells reactive with the *Mycoplasma arthritidis*-derived superantigen (MAM): generation of a monoclonal antibody against Vβ17, the T cell receptor gene product expressed by a large fraction of MAM-reactive human T cells. *J. Exp. Med.* **174:**891–900.

54. **Friedman, S. M., D. N. Posnett, J. R. Tumang, B. C. Cole, and M. K. Crow.** 1991. A potential role for microbial superantigens in the pathogenesis of systemic autoimmune disease. *Arthritis Rheum.* **34:**468–480.

55. **Gascoigne, N. R., and K. A. Ames.** 1991. Direct binding of secreted T-cell receptor β chain to superantigen associated with class II major histocompatibility complex protein. *Proc. Natl. Acad. Sci. USA* **88:**613–616.

56. **Gimmi, C. D., G. J. Freeman, J. G. Gribben, K. Sugita, A. S. Freedman, C. Morimoto, and L. M. Nadler.** 1991. B-cell antigen B7 provides a costimulatory signal that induces T cells to proliferate and secrete interleukin 2. *Proc. Natl. Acad. Sci. USA* **88:**6575–6579.

57. **Gonzalo, J. A., E. Baixeras, A. Gonzalez-Garcia, and A. George-Chandy.** 1994. Differential *in vivo* effects of a superantigen and an antibody targeted to the same T cell receptor. Activation-induced cell death vs passive macrophage-dependent deletion. *J. Immunol.* **152:**1597–1608.

58. **Gregersen, P. K., J. Silver, and R. J. Winchester.** 1987. The shared epitope hypothesis: an approach to understanding the molecular genetics of susceptibility to rheumatoid arthritis. *Arthritis Rheum.* **30:**1205–1213.

59. **Griggs, N. D., C. H. Pontzer, M. A. Jarpe, and H. M. Johnson.** 1992. Mapping of multiple binding domains of the superantigen staphylococcal enterotoxin A for HLA. *J. Immunol.* **148:**2516–2521.

60. **Groux, H., G. Torpier, D. Monte, Y. Mouton, A. Capron, and J. C. Ameisen.** 1992. Activation-induced death by apoptosis in CD4+ T cells from human immunodeficiency virus-infected asymptomatic individuals. *J. Exp. Med.* **175:**331–340.

61. **Hamad, A. R. A., A. Herman, P. Marrack, and J. W. Kappler.** 1994. Monoclonal antibodies defining functional sites on the toxin superantigen staphylococcal enterotoxin B. *J. Exp. Med.* **180:**615–621.

62. **Haqqi, T. M., G. D. Anderson, S. Banerjee, and C. S. David.** 1992. Restricted heterogeneity in T-cell antigen receptor Vβ gene usage in the lymph nodes and arthritic joints of mice. *Proc. Natl. Acad. Sci. USA* **89:**1253–1255.

63. **He, X., J. J. Goronzy, and C. M. Weyland.**

1993. The repertoire of rheumatoid factor-producing B cells in normal subjects and patients with rheumatoid arthritis. *Arthritis Rheum.* **36**(8):1061–1069.

64. **Heber-Katz, E., and H. Acha-Orbea.** 1989. The V region disease hypothesis: evidence from autoimmune encephalomyelitis. *Immunol. Today* **10:**164–169.

65. **Heeg, K., T. Miethke, P. Bader, S. Bendigs, C. Wahl, and H. Wagner.** 1991. CD4/CD8 coreceptor-independent costimulator-dependent triggering of SEB-reactive murine T cells. *Curr. Top. Microbiol. Immunol.* **174:**93–106.

66. **Herman, A., G. Croteau, R. P. Sekaly, J. Kappler, and P. Marrack.** 1990. HLA-DR alleles differ in their ability to present staphylococcal enterotoxins to T cells. *J. Exp. Med.* **172:**709–717.

67. **Herman, A., J. W. Kappler, P. Marrack, and A. M. Pullen.** 1991. Superantigens: mechanism of T-cell stimulation and role in immune responses. *Annu. Rev. Immunol.* **9:**745–772.

68. **Herman, A., N. Labrecque, J. Thibodeau, P. Marrack, J. W. Kappler, and R. P. Sekaly.** 1991. Identification of the staphylococcal enterotoxin A superantigen binding site in the β1 domain of the human histocompatibility antigen HLA-DR. *Proc. Natl. Acad. Sci. USA* **88:**9954–9958.

69. **Herrmann, T., J. L. Maryanski, P. Romero, and B. Fleischer.** 1990. Activation of MHC class I-restricted CD8+ CTL by microbial T cell mitogens. Dependence upon MHC class II expression of the target cells and Vβ usage of the responder T cells. *J. Immunol.* **144:**1181–1186.

70. **Hewitt, C. R. A., J. D. Hayball, J. R. Lamb, and R. E. O'Hehir.** 1992. The superantigenic activities of bacterial toxins, p. 149–172. *In* C. E. Hormaeche, C. W. Penn, and C. J. Smyth (ed.), *Molecular Biology of Bacterial Infection. Current Status and Future Perspectives,* vol. 49. Cambridge University Press, Cambridge.

71. **Hodtsev, A., N. Bhardwaj, and D. Posnett.** 1994. TCR Jβ product influences superantigen reactivity. *J. Cell. Biochem. Suppl.* **18D:**188A.

72. **Hoffman, M. L., L. M. Jablonski, K. K. Crum, S. P. Hackett, Y. Chi, C. V. Stauffacher, D. L. Stevens, and G. A. Bohach.** 1994. Predictions of T-cell receptor- and major histocompatibility complex-binding sites on staphylococcal enterotoxin C1. *Infect. Immun.* **62:**3396–3407.

73. **Homfeld, J., A. Homfeld, W. Nicklas, and L. Rink.** 1990. Induction of interleukin 6 in murine bone-marrow derived macrophages stimulated by *Mycoplasma arthritidis* mitogen MAM. *Autoimmunity* **7:**317–327.

74. **Howell, M. D., J. P. Diveley, K. A. Lundeen, A. Esty, S. T. Winters, D. J. Carlo, and S.**

W. Brostoff. 1991. Limited T-cell receptor β-chain heterogeneity among interleukin 2 receptor-positive synovial T cells suggests a role for superantigen in rheumatoid arthritis. *Proc. Natl. Acad. Sci. USA* **88**:10921–10925.

75. **Hudson, K. R., H. Robinson, and J. D. Fraser.** 1993. Two adjacent residues in staphylococcal enterotoxins A and E determine T cell receptor Vβ specificity. *J. Exp. Med.* **177:** 175–184.

76. **Hugin, A. W., M. S. Vacchio, and H. C. Morse III.** 1991. A virus-encoded "superantigen" in a retrovirus-induced immunodeficiency syndrome of mice. *Science* **252:**424–427.

77. **Imanishi, K., H. Igarashi, and T. Uchiyama.** 1990. Activation of murine T cells by streptococcal pyrogenic exotoxin type A. Requirement for MHC class II molecules on accessory cells and identification of Vβ elements in T cell receptor of toxin-reactive T cells. *J. Immunol.* **145:**3170–3176.

78. **Imberti, L., A. Sottini, A. Bettinardi, M. Puoti, and D. Primi.** 1991. Selective depletion in HIV infection of T cells that bear specific T cell receptor Vβ sequences. *Science* **254:**860–862.

79. **Irwin, M. J., and N. R. J. Gascoigne.** 1993. Interplay between superantigens and the immune system. *J. Leukocyte Biol.* **54:**495–503.

80. **Irwin, M. J., K. R. Hudson, K. T. Ames, J. D. Fraser, and N. R. J. Gascoigne.** 1993. T-cell receptor β-chain binding to enterotoxin superantigens. *Immunol. Rev.* **131:**61–78.

81. **Janeway, C. A., J. Yagi, P. J. Conrad, M. E. Katz, B. Jones, S. Vroegop, and S. Buxser.** 1989. T-cell responses to Mls and to bacterial proteins that mimic its behavior. *Immunol. Rev.* **107:** 61–88.

82. **Jardetzky, T. S., J. H. Brown, J. C. Gorga, L. J. Stern, R. G. Urban, Y. I. Chi, C. Stauffacher, J. L. Strominger, and D. C. Wiley.** 1994. Three-dimensional structure of a human class II histocompatibility molecule complexed with superantigens. *Nature* (London) **368:**711–718.

83. **Jorgensen, J. L., P. A. Reay, E. W. Ehrich, and M. M. Davis.** 1992. Molecular components of T-cell recognition. *Annu. Rev. Immunol.* **10:** 835–873.

84. **Kanagawa, O., B. A. Nussrallah, M. E. Wiebenga, K. M. Murphy, H. C. Morse III, and F. R. Carbone.** 1992. Murine AIDS superantigen reactivity of the T cells bearing Vβ5 T cell antigen receptor. *J. Immunol.* **149:**9–16.

85. **Kappler, J., B. Kotzin, L. Herron, E. W. Gelfand, R. D. Bigler, A. Boylston, S. Carrel, D. N. Posnett, Y. Choi, and P. Marrack.** 1989. Vβ-specific stimulation of human T cells by staphylococcal toxins. *Science* **244:**811–813.

86. **Kappler, J., N. Roehm, and P. Marrack.**

1987. T cell tolerance by clonal elimination in the thymus. *Cell* **49:**273–280.

87. **Kappler, J. W., A. Herman, J. Clements, and P. Marrack.** 1992. Mutations defining functional regions of the superantigen staphylococcal enterotoxin B. *J. Exp. Med.* **175:**387–396.

88. **Karp, D. R., and E. O. Long.** 1992. Identification of HLA-DR1 β chain residues critical for binding staphylococcal enterotoxins A and E. *J. Exp. Med.* **175:**415–424.

89. **Kawabe, Y., and A. Ochi.** 1991. Programmed cell death and extrathymic reduction of Vβ8+ CD4+ T cells in mice tolerant to *Staphylococcus aureus* enterotoxin B. *Nature* (London) **349:**245–248.

90. **Kim, C., K. A. Siminovitch, and A. Ochi.** 1991. Reduction of lupus nephritis in MRL/*lpr* mice by a bacterial superantigen treatment. *J. Exp. Med.* **174:**1431–1437.

91. **Kirchner, H., G. Brehm, R. Nicklas, R. Beck, and R. Herbert.** 1986. Biochemical characterization of the T cell mitogen derived from *Mycoplasma arthritidis*. *Scand. J. Immunol.* **24:**245–249.

92. **Kotb, M.** 1992. Role of superantigens in the pathogenesis of infectious diseases and their sequelae. *Curr. Opin. Infect. Dis.* **5:**364–374.

93. **Kotb, M., G. Majumdar, M. Tomai, and E. H. Beachey.** 1990. Accessory cell-independent stimulation of human T cells by streptococcal M protein superantigen. *J. Immunol.* **145:**1332–1336.

94. **Kotzin, B., S. Karuturi, L. Cohen, J. M. Forrester, M. Better, G. E. Nedwin, H. Offner, and A. A. Vandenbark.** 1991. Preferential T-cell receptor β-chain variable gene use in myelin basic protein-reactive T-cell clones from patients with multiple sclerosis. *Proc. Natl. Acad. Sci. USA* **88:**9161–9165.

95. **Kotzin, B. L., D. Y. Leung, J. Kappler, and P. Marrack.** 1993. Superantigens and their potential role in human disease. *Adv. Immunol.* **54:** 99–166.

96. **Lafon, M., M. Lafage, A. Martinez-Arends, R. Ramirez, F. Vuillier, D. Charron, V. Lotteau, and D. Scott-Algara.** 1992. Evidence for a viral superantigen in humans. *Nature* (London) **358:**5007–5010.

97. **Lando, P. A., M. Dohlsten, G. Hedlund, T. Brodin, D. Sansom, and T. Kalland.** 1993. Co-stimulation with B7 and targeted superantigen is required for MHC class II–independent T-cell proliferation but not cytotoxicity. *Immunology* **80:** 236–241.

98. **Laurence, J., A. S. Hodtsev, and D. N. Posnett.** 1992. Superantigen implicated in dependence of HIV-1 replication in T cells on TCR Vβ expression. *Nature* (London) **358:**255–259.

99. **Legaard, P. K., R. D. LeGrand, and M. L. Misfeldt.** 1992. Lymphoproliferative activity of *Pseudomonas* exotoxin A is dependent on intracellular processing and is associated with the carboxyl-terminal portion. *Infect. Immun.* **60:**1273–1278.

100. **Linsley, P. S., W. Brady, L. Grosmaire, A. Aruffo, N. K. Damle, and J. A. Ledbetter.** 1991. Binding of the B cell activation antigen B7 to CD28 costimulates T cell proliferation and interleukin 2 mRNA accumulation. *J. Exp. Med.* **173:**721–730.

101. **Lipsky, P. E., L. S. Davis, J. J. Cush, and N. Oppenheimer-Marks.** 1989. The role of cytokines in the pathogenesis of rheumatoid arthritis. *Springer Semin. Immunopathol.* **11:**123–162.

102. **Majumdar, G., H. Ohnishi, M. A. Tomai, A. M. Geller, B. Wang, M. E. Dockter, and M. Kotb.** 1993. Role of antigen-presenting cells in activation of human T cells by the streptococcal M protein superantigen: requirement for secreted and membrane-associated costimulatory factors. *Infect. Immun.* **61:**785–790.

103. **Marrack, P., M. Blackman, E. Kushnir, and J. Kappler.** 1990. The toxicity of staphylococcal enterotoxin B in mice is mediated by T cells. *J. Exp. Med.* **171:**455–464.

104. **Marrack, P., G. M. Winslow, Y. Choi, M. Scherer, A. Pullen, J. White, and J. W. Kappler.** 1993. The bacterial and mouse mammary tumor virus superantigens; two different families of proteins with the same functions. *Immunol. Rev.* **131:**79–92.

105. **Matthes, M., H. Schrezenmeier, J. Homfeld, S. Fleischer, B. Malissen, H. Kirchner, and B. Fleischer.** 1988. Clonal analysis of human T cell activation by the *Mycoplasma arthritidis* mitogen (MAS). *Eur. J. Immunol.* **18:**1733–1737.

106. **McClane, B. A., P. C. Hanna, and A. P. Wnek.** 1988. *Clostridium perfringens* enterotoxin. *Microb. Pathog.* **4:**317–323.

107. **McCormack, J. E., J. E. Callahan, J. Kappler, and P. C. Marrack.** 1993. Profound deletion of mature T cells *in vivo* by chronic exposure to exogenous superantigen. *J. Immunol.* **150:**3785–3792.

108. **Misfeldt, M. L.** 1990. Microbial "superantigens." *Infect. Immun.* **58:**2409–2413.

109. **Morrow, W. J. W., D. A. Isenberg, R. E. Sobol, R. B. Stricter, and T. Kieber-Emmons.** 1991. AIDS virus infections and autoimmunity: a perspective of the clinical, immunological and molecular origins of the autoallergic pathologies associated with HIV disease. *Clin. Immunol. Immunopathol.* **58:**163–180.

110. **Mountz, J. D., T. Zhou, R. E. Long, H. Bluethmann, W. J. Koopman, and C. K. Edwards III.** 1994. T cell influence on superantigen-induced arthritis in MRL-*lpr*/*lpr* mice. *Arthritis Rheum.* **37:**113–124.

111. **Mourad, W., K. Mehindate, T. J. Schall, and S. R. McColl.** 1992. Engagement of major histocompatibility complex class II molecules by superantigen induces inflammatory cytokine gene expression in human rheumatoid fibroblast-like synoviocytes. *J. Exp. Med.* **175:**613–616.

112. **Mourad, W., P. Scholl, A. Diaz, R. Geha, and T. Chatila.** 1989. The staphylococcal toxic shock syndrome toxin 1 triggers B cell proliferation and differentiation via major histocompatibility complex-unrestricted cognate T/B cell interaction. *J. Exp. Med.* **170:**2011–2022.

113. **Muralidhar, G., S. Koch, M. Haas, and S. L. Swain.** 1992. CD4 T cells in murine acquired immunodeficiency syndrome: polyclonal progression to anergy. *J. Exp. Med.* **175:**1589–1599.

114. **Newell, K. A., J. D. I. Ellenhorn, D. S. Bruce, and J. A. Bluestone.** 1991. *In vivo* T-cell activation by staphylococcal enterotoxin B prevents outgrowth of a malignant tumor. *Proc. Natl. Acad. Sci. USA* **88:**1074–1078.

115. **Norment, A. M., R. D. Salter, P. Parham, V. M. Englehard, and D. R. Littman.** 1988. Cell-cell adhesion mediated by CD8 and MHC class I molecules. *Nature* (London) **336:**79–81.

116. **Offner, H., G. A. Hashim, and A. A. Vandenbark.** 1991. T cell receptor peptide therapy triggers autoregulation of experimental encephalomyelitis. *Science* **251:**430–432.

117. **Ohmen, J. D., P. F. Barnes, C. L. Grisso, B. R. Bloom, and R. L. Modlin.** 1994. Evidence for a superantigen in human tuberculosis. *Immunity* **1:**35–43.

118. **Ohnishi, H., T. Tanaka, J. Takahara, and M. Kotb.** 1993. CD28 delivers costimulatory signals for superantigen-induced activation of antigen-presenting cell-depleted human T lymphocytes. *J. Immunol.* **150:**3207–3214.

119. **Panina-Bordignon, P., X. T. Fu, A. Lanzavecchia, and R. W. Karr.** 1992. Identification of HLA-DRα chain residues critical for binding of the toxic shock syndrome toxin superantigen. *J. Exp. Med.* **176:**1779–1784.

120. **Pantaleo, G., J. F. Demarest, H. Soudeyns, C. Graziosi, F. Denis, J. W. Adelsberger, P. Borrow, M. S. Saag, G. M. Shaw, R. F. Sekaly, and A. S. Fauci.** 1994. Major expansion of CD8 + T cells with a predominant Vβ usage during the primary immune response to HIV. *Nature* (London) **370:**463–467.

121. **Patten, P. A., E. P. Rock, T. Sonoda, B. F. de St. Groth, J. L. Jorgensen, and M. M. Davis.** 1993. Transfer of putative complementarity-determining region loops of T cell receptor V domains confers toxin reactivity but not

peptide/MHC specificity. *J. Immunol.* **150:** 2281–2294.

122. **Pontzer, C. H., M. J. Irwin, N. R. Gascoigne, and H. M. Johnson.** 1992. T-cell antigen receptor binding sites for the microbial superantigen staphylococcal enterotoxin A. *Proc. Natl. Acad. Sci. USA* **89:**7727–7731.

123. **Posnett, D. N.** 1993. Do superantigens play a role in autoimmunity? *Semin. Immunol.* **5:**65–72.

124. **Posnett, D. N., S. Kabak, A. S. Hodtsev, E. A. Goldberg, and A. Asch.** 1993. T-cell antigen receptor Vβ subsets are not preferentially deleted in AIDS. *AIDS* **7:**625–631.

125. **Prasad, G. S., C. A. Earhart, D. L. Murray, R. P. Novick, P. M. Schlievert, and D. H. Ohlendorf.** 1993. Structure of toxic shock syndrome toxin 1. *Biochemistry* **32:**13761–13766.

126. **Pullen, A. M., P. Marrack, and J. Kappler.** 1988. The T-cell repertoire is heavily influenced by tolerance to polymorphic self antigens. *Nature* (London) **335:**796–801.

127. **Racke, M. K., L. Quigley, B. Cannella, C. S. Raine, D. E. McFarlin, and D. E. Scott.** 1994. Superantigen modulation of experimental allergic encephalomyelitis: activation of anergy determines outcome. *J. Immunol.* **152:**2051–2059.

128. **Rellahan, B. L., L. A. Jones, A. M. Kruisbeek, A. M. Fry, and L. A. Matis.** 1990. *In vivo* induction of anergy in peripheral Vβ8 + T cells by staphylococcal enterotoxin. *Br. J. Exp. Med.* **172:**1091–1100.

129. **Reveille, J. D., M. J. Macleod, K. Whittington, and F. C. Arnett.** 1991. Specific amino acid residues in the second hypervariable region of HLA-DQA1 and DQB1 chain genes promote the Ro (SS-A)/LA (SS-B) autoantibody responses. *J. Immunol.* **146:**3871–3876.

130. **Russell, J. K., C. H. Pontzer, and H. M. Johnson.** 1991. Both a-helices along the major histocompatibility complex binding cleft are required for staphylococcal enterotoxin A function. *Proc. Natl. Acad. Sci. USA* **88:**7228–7232.

131. **Schiffenbauer, J., H. M. Johnson, E. J. Butfiloski, L. Wegrzyn, and J. M. Soos.** 1993. Staphylococcal enterotoxins can reactivate experimental allergic encephalomyelitis. *Proc. Natl. Acad. Sci. USA* **90:**8543–8546.

132. **Schlievert, P. M.** 1993. Role of superantigens in human disease. *J. Infect. Dis.* **167:**997–1002.

133. **Scholl, P. R., A. Diez, R. Karr, R. P. Sekaly, J. Trowsdale, and R. S. Geha.** 1990. Effect of isotypes and allelic polymorphism on the binding of staphylococcal exotoxins to MHC class II molecules. *J. Immunol.* **144:**226–230.

134. **Schwab, J. H., R. R. Brown, S. K. Anderle, and P. M. Schlievert.** 1993. Superantigen can reactivate bacterial cell wall-induced arthritis. *J. Immunol.* **150:**4151–4159.

135. **Sekaly, R. P., G. Croteau, M. Bowman, P. Scholl, S. Burakoff, and R. S. Geha.** 1991. The CD4 molecule is not always required for the T cell response to bacterial enterotoxins. *J. Exp. Med.* **173:**367–371.

136. **Shackelford, P. G., and A. W. Strauss.** 1991. Kawasaki syndrome. *N. Engl. J. Med.* **324:**1664–1666.

137. **Sharief, M. K., M. Phil, and R. Hentges.** 1991. Association between tumor necrosis factor-a and disease progression in patients with multiple sclerosis. *N. Engl. J. Med.* **325:**467–472.

138. **Smith, L. R., D. H. Kono, M. E. Kammuller, R. S. Balderas, and A. N. Theofilopoulos.** 1992. Vβ repertoire in rats and implications for endogenous superantigens. *Eur. J. Immunol.* **22:**641–645.

139. **Soos, J. M., J. Schiffenbauer, and H. M. Johnson.** 1993. Treatment of PL/J mice with the superantigen, staphylococcal enterotoxin B, prevents development of experimental allergic encephalomyelitis. *J. Neuroimmunol.* **43:**39–43.

140. **Spurkland, A., K. Ronningen, B. Vandvik, E. Thorsby, and F. Vartdal.** 1991. HLA-DQA1 and HLA-DQB1 genes may jointly determine susceptibility to develop multiple sclerosis. *Hum. Immunol.* **30:**65–70.

141. **Stuart, P., and J. G. Woodward.** 1992. *Yersinia enterocolitica* produces superantigenic activity. *J. Immunol.* **148:**225–233.

142. **Swaminathan, S., W. Furey, J. Pletcher, and M. Sax.** 1992. Crystal structure of staphylococcal enterotoxin B, a superantigen. *Nature* (London) **359:**801–806.

143. **Takei, S., Y. K. Arora, and S. M. Walker.** 1993. Intravenous immunoglobulin contains specific antibodies inhibitory to activation of T cells by staphylococcal toxin superantigens. *J. Clin. Invest.* **91:**602–607.

144. **Takimoto, H., Y. Yoshikai, K. Kishihara, G. Matsuzaki, H. Kuga, T. Otani, and K. Nomoto.** 1990. Stimulation of all T cells bearing Vβ1, Vβ3, Vβ11 and Vβ12 by staphylococcal enterotoxin A. *Eur. J. Immunol.* **20:**617–621.

145. **Tomai, M. A., J. A. Aelion, M. E. Dockter, G. Majumdar, D. G. Spinella, and M. Kotb.** 1991. T cell receptor V gene usage by human T cells stimulated with superantigen streptococcal M protein. *J. Exp. Med.* **174:**285–288.

146. **Tomai, M. A., P. M. Schlievert, and M. Kotb.** 1992. Distinct T-cell receptor Vβ gene usage by human T lymphocytes stimulated with the streptococcal pyrogenic exotoxins and pep M5 protein. *Infect. Immun.* **60:**701–705.

147. **Torres, B. A., N. D. Griggs, and H. M. Johnson.** 1993. Bacterial and retroviral super-

antigens share a common binding region on class II MHC antigens. *Nature* (London) **364:**152–154.

148. **Tumang, J. R., E. P. Cherniack, D. M. Gietl, B. C. Cole, C. Russo, M. K. Crow, and S. M. Friedman.** 1991. T helper cell-dependent, microbial superantigen-induced murine B cell activation: polyclonal and antigen-specific antibody responses. *J. Immunol.* **147:**432–438.

149. **Tumang, J. R., D. N. Posnett, B. C. Cole, M. K. Crow, and S. M. Friedman.** 1990. Helper T cell-dependent human B cell differentiation mediated by a mycoplasmal superantigen bridge. *J. Exp. Med.* **171:**2153–2158.

150. **Urban, J. L., V. Kumar, D. H. Kono, C. Gomez, S. J. Horvath, J. Clayton, D. G. Ando, E. E. Sercaz, and L. Hood.** 1988. Restricted use of T cell receptor V genes in murine autoimmune encephalomyelitis raises possibilities for antibody therapy. *Cell* **54:**577–592.

151. **Vandenbark, A. A., G. Hashim, and H. Offner.** 1989. Immunization with a synthetic T-cell receptor V-region peptide protects against experimental autoimmune encephalomyelitis. *Nature* (London) **341:**541–544.

152. **Wallgren, A. C., R. Festin, C. Gidlof, M. Dohlsten, T. Kalland, and T. H. Totterman.** 1993. Efficient killing of chronic B-lymphocytic leukemia cells by superantigen-directed T cells. *Blood* **82:**1230–1238.

153. **Walport, M. J., W. E. R. Ollier, and A. J. Silman.** 1992. Immunogenetics of rheumatoid arthritis and the arthritis and rheumatism council's national repository. *Br. J. Rheumatol.* **31:**701–705.

154. **Wang, B., P. M. Schlievert, A. O. Gaber, and M. Kotb.** 1993. Localization of an immu-nologically functional region of the streptococcal superantigen pepsin-extracted fragment of type 5 M protein. *J. Immunol.* **151:**1419–1429.

155. **Watanabe-Fukunaga, R., C. I. Brannan, A. N. C. Copeland, N. A. Jenkens, and S. Nagata.** 1992. Lymphoproliferation disorder in mice explained by defects in Fas antigen that mediates apoptosis. *Nature* (London) **356:**314–317.

156. **White, J., A. Herman, A. M. Pullen, R. Kubo, J. W. Kappler, and P. Marrack.** 1989. The Vβ-specific superantigen staphylococcal enterotoxin B: stimulation of mature T cells and clonal deletion in neonatal mice. *Cell* **56:**27–35.

157. **Williams, W. V., T. Kieber-Emmons, Q. Fang, J. Von Feldt, B. Wang, T. Ramanujam, and D. B. Weiner.** 1993. Conserved motifs in rheumatoid arthritis synovial tissue T-cell receptor β chains. *DNA Cell Biol.* **12:**425–434.

158. **Woodland, D. L., H. P. Smith, S. Surman, P. Le, R. Wen, and M. A. Blackman.** 1993. Major histocompatibility complex-specific recognition of Mls-1 mediated by multiple elements of the T cell receptor. *J. Exp. Med.* **177:**433–442.

159. **Zamvil, S. S., and L. A. Steinman.** 1990. The T lymphocytes in experimental allergic encephalomyelitis. *Annu. Rev. Immunol.* **8:**579–621.

160. **Zhang, J., R. Medaer, P. Stinissen, D. Hafler, and J. Raus.** 1993. MHC-restricted depletion of human myelin basic protein-reactive T cells by T cell vaccination. *Science* **261:**1451–1454.

161. **Zhou, P., G. D. Anderson, S. Savarirayan, H. Inoko, and C. S. David.** 1991. Human HLA.DQ beta chain presents minor lymphocyte stimulating locus gene products and clonally deletes TCR Vβ6+, Vβ8.1+ T cells in single transgenic mice. *Hum. Immunol.* **31:**47–56.

BACTERIAL TOXINS IN DISEASE PRODUCTION

IV

LIPOPOLYSACCHARIDE AND BACTERIAL VIRULENCE

Richard A. Proctor, Loren C. Denlinger, and Paul J. Bertics

11

Gram-negative bacteria differ from non-gram-negative bacteria and mammalian cells by having both an inner membrane and an outer membrane that consist of lipid bilayers and transmembrane proteins (Fig. 1). The outer leaflet of the outer membrane contains a unique lipid, specifically, a large amphophilic molecule, lipopolysaccharide (LPS). Functionally, LPS provides a barrier to heavy metals, lipid–disrupting agents (e.g., bile salts), and larger molecules (e.g., lytic enzymes, DNA). LPS also contains a highly variable carbohydrate (O antigen) that reduces complement binding and presents the host with multiple antigenic structures in various strains, thus rendering the organisms more resistant to serum-mediated host defenses. Several of these functions and/or characteristics are also provided to gram-positive organisms by lipoteichoic acids.

The precise evolutionary advantage to gram-negative bacteria of using LPS as opposed to other types of lipids is unclear; however, LPS seems to be vital to gram-negative bacteria. The genes for LPS biosynthesis evolved at about the same time as those for lipoteichoic acid biosynthesis and long before the existence of mammalian host defenses. LPS biosynthetic genes have been highly conserved evolutionarily for over 1 billion years (120). They are essential for bacterial viability in that mutations affecting LPS biosynthesis are lethal (42). Finally, LPS is found in a wide variety of bacterial genera and species (113, 120), suggesting that it is important to prokaryotic cells.

Bacteria with bulky O-antigen polysaccharides are in general more resistant to complement-mediated lysis and/or phagocytosis, have a vast array of antigenic variants, and may exhibit molecular mimicry (43, 65, 82). Some nonenteric gram-negative organisms, which have smaller or nonexistent O antigens, use other carbohydrate structures that serve a similar function. For example, phase variation of surface antigens is a common way for some bacteria to evade the immune system and has been documented for the LPSs of *Bordetella pertussis* (117), *Haemophilus influenzae* (97), *Neisseria meningitidis* (143), and *Neisseria gonorrhoeae* (148). Recently, van Putten used isogenic strains of *N. gonorrhoeae* to show that the degree of LPS sialation correlated positively with resistance to complement-mediated lysis but inversely with epithelial-cell invasion (148), suggesting that an unsialated bacterium responsible for the initial mucosal invasion might switch on expression of the bacterial sialotransferase to

Richard A. Proctor, Loren C. Denlinger, and Paul J. Bertics Departments of Medical Microbiology/Immunology, Medicine, and Biomolecular Chemistry, 407 SMI, Madison, Wisconsin 53706.

Virulence Mechanisms of Bacterial Pathogens, 2nd ed., Edited by J. A. Roth et al.

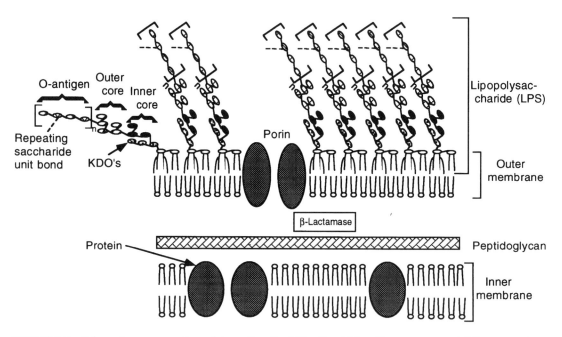

FIGURE 1 Schematic representation of the cell wall of *E. coli,* which will serve as a model for comparisons with lipid A proteins from other genera. LPS forms the outer leaflet of the outer membrane. Phospholipids resembling the phospholipids of mammalian cells make up the other three leaflets of the inner and outer membranes. The inner core LPS is composed of two to three KDOs (◐) and three heptoses (●), which are decorated with phosphate and phosphoethanolamine groups. These phosphate groups bind calcium and magnesium, which help maintain the structural integrity of the outer membrane. The outer core region is usually composed of hexoses (◒) and is shown with five residues. The O antigen also contains hexoses (◒), which are arranged as repeating tri- to hexasaccharides. This diagram shows a repeating tetrasaccharide; the dashed bond is the attachment site for the repeating units. The number of repeats (*n*) may vary within the same culture and often reaches 20 to 30 repeating units.

help itself become established. In this sense, LPS (or maybe more appropriately the sialo-transferase) might be considered a virulence factor, because it helps the bacterium establish a significant infection. Nevertheless, even these non-O-antigen-containing organisms retain the core and lipid A structures (see below) of LPS, which can induce endotoxic shock.

While some of the functions of LPS suggest that it is a virulence factor, an unequivocal demonstration that LPS is able to fulfill the molecular Koch's postulates (35) would involve the generation of knockout mutants, which is difficult to achieve, because knockout mutations are lethal. Although enteric pathogens require bulky O antigens for resistance to bile acids, there are no convincing data that suggest that differences in the LPSs of various enteric bacteria contribute to virulence. Gut patho-

gens usually possess a variety of other virulence factors (e.g., pili, motility, exotoxins, and penetrance factors), making LPS seem less significant to the development of gastroenteritis and other diseases caused by this group of gram-negative organisms. In contrast, LPSs from normal flora as well as from virulent bacteria can be highly pathogenic if released into the bloodstream. The resulting high mortality rate probably does not benefit the bacterium. Therefore, even though the title of this chapter refers to LPS and bacterial virulence, a better way to think of the role of LPS in disease is to consider LPS as a pathogenic factor (92). Such a consideration is the subject of the rest of this chapter.

More than a century ago, the toxic activity of LPS was first suggested (157). Because the LPS was thought to be an integral part of the organism, it was named "endotoxin" in contrast to toxins that were freely released (i.e., proteinaceous "exotoxins"). When filtrates of dead gram-negative organisms were infused into animals, they produced shock and death, thus fulfilling Koch's fifth postulate (38) whereby a toxin harvested from a bacterial filtrate must reproduce the disease. The active principle in these filtrates was both boiling and protease resistant. Koch's fifth postulate has never been widely used because bacteria often have multiple pathogenic factors; for example, infusion of purified *Escherichia coli* hemolysin as well as of *E. coli* LPS can produce shock and death. Nevertheless, LPS reproduces most of the symptoms and signs of gram-negative sepsis.

Within the past decade, major progress in several areas of LPS research has broadened our understanding of this potent biological molecule. Specifically, several groups have (i) identified several genes and enzymes involved in LPS biosynthesis (113–116); (ii) elucidated the covalent structure of bacterial LPS (60, 76, 113, 114, 120); (iii) discovered LPS-binding proteins on macrophages, allowing initial studies regarding signal transduction (84, 95, 116); (iv) characterized the cytokines and other mediators released from macrophages during endotoxic shock (10, 16, 27, 61, 96, 149); and (v) developed novel therapeutic approaches based on our new understanding of LPS structure, the pathways of LPS-mediated macrophage activation, and the mediators involved in endotoxin shock (11, 48, 50, 59, 61, 70–72, 80, 84, 98, 99, 101, 108, 166). Because differences in LPS structure correlate with the abilities to activate macrophages, release mediators, and cause septic shock, this chapter focuses on the information in points ii to v, whereas the identification of LPS biosynthetic genes is reviewed elsewhere in detail (for excellent reviews, see references 113, 114, and 115) and will not be discussed here.

STRUCTURE OF ENDOTOXIN
The lipid portion of LPS allows the LPS to pack neatly to form a planar membrane (69, 112). The lipid component of the outer leaflet

of the outer membrane is entirely LPS (113, 120), and one bacterial cell contains approximately 3.5×10^6 LPS molecules (120). In *E. coli*, approximately 75% of the outer leaflet is composed of LPS (113, 120), while the other 25% is supplied by transmembrane proteins. As can be seen in Fig. 1, the orientation of the LPS in intact bacteria hides the lipid domains beneath the polysaccharide components of the molecule. However, upon bacterial death and autolysis of the cell wall, LPS is released, and the lipid portions of the molecule are then able to interact with other hydrophobic molecules, including mammalian membrane lipids and hydrophobic domains of proteins. Released LPS is thought to play a major role in the development of the septic shock that accompanies gram-negative-bacterium-caused infectious diseases (23, 46, 96).

Research over the past 10 years has elucidated the structure and biosynthetic pathway of endotoxin (60, 76, 112–116, 120). The lipid A (Fig. 2) portion and the inner-core regions of LPS show surprising structural similarities across widely diverse genera (60, 113, 120). At the same time, there is some heterogeneity within a strain that results from a variable number of repeating sugars being added to the O

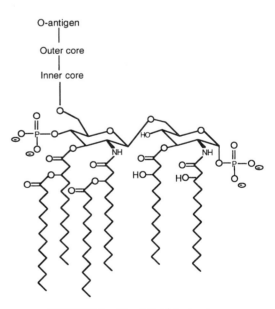

FIGURE 2 *E. coli* lipid A structure.

antigen (65, 119). Much greater heterogeneity between strains occurs in the outer core and the O-antigen hexose residues (65, 119) (Table 1). Nevertheless, an overall structural pattern is seen in most LPS molecules: (i) a negatively charged hydrophilic heteropolysaccharide; (ii) a core oligosaccharide that shows much less intergeneral variation; and (iii) a lipid A region containing two glucosamines that are decorated at the 2, 3, 2′, and 3′ positions with fatty acids. The O antigen is formed by repeating groups of three to five sugars. The sugars in the O antigen vary between strains, and as one of the outermost components of the bacterial cell, they confer antigenic differences within a species. In contrast, the core polysaccharides are less variable (60, 113). The core region is divided into the inner and the outer core. The outer core is made up of hexoses: primarily glucose, galactose, and *N*-acetylglucosamine. The inner core contains (i) heptoses that are decorated with phosphates and phosphoethanolamines and (ii) unique sugars that form the bridge between lipid A and the polysaccharide regions of LPS: 3-deoxy-D-*manno*-octulosonic acid (KDO). The inner core is highly conserved across diverse genera.

Figures 2 and 3 show the LPS structures of several common pathogens as well as some un-

usual LPS structures (60, 113, 120). The LPSs from members of the family *Enterobacteriaceae* and from *Pseudomonas aeruginosa* each contain a large O antigen, a complete outer and inner core, two or three KDOs, and lipid A with four

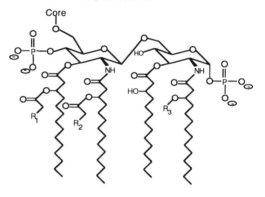

A. Four plus two acylation pattern

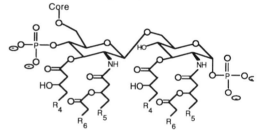

B. Three plus three acylation pattern

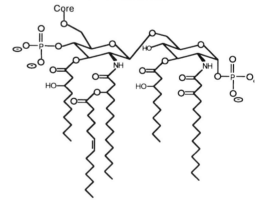

C. *Rhodobacter sphaeroides* lipid A

FIGURE 3 Lipid A structures. R$_1$, 11 to 13 carbons; R$_2$, 9 to 11 carbons; R$_3$, 16 carbons (*Salmonella* spp.) or hydrogen (*E. coli*, *H.influenza*, *P. aeruginosa*, *Salmonella minnesota*); R$_4$, 5 to 7 carbons; R$_5$, 5 to 9 carbons; R$_6$, 7 carbons.

TABLE 1 Comments on the structure of the polysaccharide portion of LPS

1. O antigens are highly variable.
2. The outer core varies between genera and within species but usually contains four to six hexoses.
3. The inner core is composed of three heptoses and two or three *manno*- or *keto*-octulosonic acids, and it is generally conserved across a wide range of genera.
4. Heptoses in the inner core are decorated with charged groups (phosphates and aminoethylphosphates) that can bind Ca^{2+} and Mg^{2+}.
5. Oral-pharyngeal pathogens lack O chains and the usual outer core, but they maintain the usual inner-core composition. *N*-Acetyl neuraminic acid replaces O antigen in *Neisseria* spp. and confers serum resistance.
6. An intracellular pathogen, chlamydia, contains three KDO residues but no core or O-antigen polysaccharides.

hydroxymyristic acids and two fatty acids linked as acyloxyacyl groups to the fatty acids at the 2' and 3' positions of the disaccharide. In contrast, the LPS of oral-pharyngeal pathogens, such as *Neisseria spp., H. influenzae,* and *B. pertussis,* lack the O antigen typical of lower gastrointestinal tract organisms. Instead of an O antigen, other groups, e.g., the *N*-acetyl neuraminic acid found in *Neisseria* species, confer serum resistance. The outer-core hexoses typical of enteric bacteria are also missing on these oral-pharyngeal organisms, but an inner core reminiscent of that *E. coli* and *Salmonella* spp. is present. Heptoses in the inner cores of many gram-negative bacteria are often decorated with negatively charged groups (phosphates and aminoethyl phosphates) that tightly bind Ca^{2+} and Mg^{2+}. Chlamydial LPS completely lacks an O antigen and core polysaccharides except for a KDO trisaccharide.

Lipid A from the common enteric and some oral-pharyngeal organisms usually contains six fatty acids (113, 120), four of them on one glucosamine and two on the other glucosamine (Fig. 3A, Table 2). In contrast, lipid A from *N. meningitidis, N. gonorrheae,* and *P. aeruginosa* has three fatty acids on each glucosamine (Fig. 3B). An exception to the six fatty-acid pattern is the *Salmonella* LPS, which contains a seventh fatty acid attached at R_3. However, this seventh fatty acid is controversial: some workers have found only six fatty acids, while others have claimed to find seven fatty acids in some *Salmonella* species (Fig. 3A) (113). Most fatty acids in enteric organisms contain 12- to 16-carbon fatty acids

(Fig. 3A), whereas the LPSs from oral-pharyngeal pathogens generally have fatty acids that are 10 to14 carbons long. The lipid A from *Rhodobacter* spp. contains both short (10-carbon) and unsaturated fatty acids (Fig. 3C).

STRUCTURE-ACTIVITY RELATIONSHIPS

The precise structural requirements for LPS activation of macrophages are now under intense investigation because such information may allow the development of LPS antagonists as well as provide nontoxic LPS derivatives for use as potent immunostimulants. Several generalizations can be made, but these must be viewed as first attempts toward structural rules rather than finalized statements (Table 3). Testing of lipid A species from many genera and lipid A substructures will further refine these structure-activity rules.

By altering the structure of lipid A or its substructures, the toxic potential of lipid A can be at least partially dissociated from the immunostimulatory activity of LPS. A good deal of evidence has accumulated that the phosphate at the 1 position is important for toxicity (98). Monophosphoryl lipid A (MPL), a disaccharide substructure of lipid A that has the phosphate at the 1 position removed, retains immunostimulatory activity, including the ability to induce small quantities of tumor ne-

TABLE 2 General concepts concerning the lipid A portion of LPS

1. Every toxic lipid A contains disaccharides made up of D–glucosamines or 2,3–diamino-2,3-deoxy-D-glucose.
2. Lipid A contains phosphoryl groups at the 1 and 4' positions.
3. Hydroxylated fatty acids are present at the 2, 3, 2', and 3' positions.
4. Fatty acids are added to the 3-hydroxyl group of the disaccharide-linked fatty acids either asymmetrically or symmetrically.

TABLE 3 Structure-activity relationships for bacterial LPS

Changes in nonlipid portions of lipid A
1. A dephosphorylated lipid A (MPL) is approximately 1,000-fold less toxic but retains immunostimulatory activity.
2. Monosaccharides are 10^7-fold less toxic and nonimmunostimulatory unless a third fatty acid is added synthetically.

Changes in fatty acids
1. Removal of the acyloxyacyl fatty acids reduces toxicity by 10^7-fold.
2. Hepta- and pentaacyl lipid A molecules are less toxic than hexaacyl lipid A.
3. Reduction in fatty acid length (≤12 carbons) reduces toxic activity.
4. Presence of unsaturated fatty acids reduces toxicity.

crosis factor alpha (TNF-α), but it is at least 1,000-fold less toxic than LPS (98, 120, 146, 147). Indeed, MPL has been successfully used as immunotherapy in the treatment of melanoma and as an adjuvant (146, 147). Monosaccharide substructures of lipid A (e.g., lipid X) are more than 10^7-fold less toxic than LPS and show very little immunostimulatory activity (77, 120). However, monosaccharides with a third hydroxymyristic acid added synthetically at the 4 position of the glucosamine are able to induce cytokine release, e.g., granulocyte colony-stimulating factor, interleukin-6 (IL-6), and TNF-α (but not IL-1) and enhance nonspecific immunity against *P. aeruginosa* and *Staphylococcus aureus* (81) infections. Similarly, 4-phospho-D-glucosamines carrying two hydroxymyristic acids and one acyloxyacyl-linked hydroxymyristic acid induce TNF-α release, macrophage activation, tumor regression, and nonspecific resistance to bacterial infections in mice (75, 83, 100, 125, 126, 128).

The number and configuration of the fatty acids attached to the disaccharide also alter the bioactivity of the lipid A (84, 120, 135, 137). Synthetic lipid A substructures containing only the four hydroxymyristic acids attached to the glucosamine disaccharides are as much as 10^7-fold less toxic than the parent hexaacyl synthetic lipid A (120, 137). Another tetraacyl endotoxin derivative is produced by an enzyme found in the lysosomes of phagocytes, 3-acyloxyacyl-hydroxylase, which cleaves the ester-linked fatty acid at the 3 positions of the hydroxymyristic acids (34, 71). Not only is the resulting tetraacyl LPS nontoxic, but this degradation product is able to antagonize the actions of LPS in human cells (34, 71). Even LPS with one fatty acid removed or added, i.e., pentaacyl or heptaacyl lipid A, is less toxic than the hexaacyl forms (120, 137). There is also some indication that when a fatty acid contains ≤12 carbons, then the toxic activity of the LPS is reduced (84, 120, 137). Of interest, not only are the LPSs from *Rhodobacter sphaeroides* and *Rhodobacter capsulatus* nontoxic, but these LPSs are able to antagonize the toxic effects of *E. coli* and *Salmonella* LPSs (48, 84, 120). The *Rho-*

dobacter LPSs vary from the LPSs of enteric and pseudomonal organisms in that both short (10-carbon fatty acids) and unsaturated fatty acids are present (Fig. 3C). These unusual LPSs in *Rhodobacter* spp. may prove to be therapeutically valuable for treating endotoxin shock. Thus, these observations suggest that both polar and hydrophobic groups on lipid A are important for full activity.

While these structure-activity relationships are interesting, some caution should be exercised before they are accepted as rules. First, the responses of mammals with respect to LPS and LPS substructure vary between species. For example, the tetraacyl-disaccharide form of lipid A is an antagonist of LPS in human cells, but it retains many toxic activities in murine cells (36, 48, 73, 86). Second, minor contaminants of LPS substructures produced via chemical synthesis or harvested from mutant bacteria have proven to be biologically active and initially led to some erroneous structure-activity conclusions (77, 135). Third, comparisons of LPSs from different species and genera are difficult, because bacterial LPS is a naturally heterogeneous product that can be further altered during purification procedures because of relatively labile linkages at the KDO-lipid A bond, ester-linked fatty acids, and the 4' phosphoryl group (60, 113, 120). Fourth, the actual physical state of the lipid A or LPS can change dramatically depending on method of preparation, concentration, and storage conditions (26, 41). Recently, methods for producing monomeric, nonaggregated Re LPS from *E. coli* have been reported, and early indications suggest that the monomeric form may be much more active than the aggregated form (129, 139). Finally, the experimental designs of animal and in vitro models of endotoxemia have varied widely between reports, making comparisons between LPSs and LPS substructures difficult. Thus, while questions can be raised concerning the structure activity relationships offered in Table 3, there is substantial experimental support for these ideas, and these conclusions should serve as a stimulus for producing further chemically synthesized compounds to test these structure–activity relationships.

PATHOPHYSIOLOGY OF ENDOTOXIN

Paralleling the considerable progress in defining the covalent structure and biosynthetic pathway of LPS has been a concomitant increase in information about the pathophysiology of endotoxemia (23, 46, 96, 103). Microgram quantities can account for most of the pathophysiology seen in gram-negative septic shock. The release of endotoxin into the bloodstream can come from almost any site of infection; however, most cases of gram-negative septic shock seen in developed countries are due to a major breach in the host defenses (e.g., perforated bowel, major burn, obstructed urinary tract, stone in the biliary tract, immunosuppressive therapy). These infections are frequently caused by normal host flora (e.g., *E. coli, Klebsiella pneumoniae, P. aeruginosa*). Many of these infections are "diseases of medical progress": that is, massive immunosuppression and placement of medical devices (e.g., urinary tract catheters, respiratory tract intubation) allow opportunistic organisms to invade and release their LPSs into the host (163).

The dramatic and often lethal response to LPS makes one question the evolutionary benefit of this host response. Clearly, the host derives some benefit from interactions with endotoxin in that LPS enhances antibody production (5, 6, 146, 151), leads to the release of acute-phase reactants (28), primes phagocytes for more effective bactericidal activity (1), and enhances the tumoricidal activities of macrophages (1). Hence, the host may benefit from a limited release of LPS that activates the host's immune system, especially when the LPS is absorbed in small quantities from intestinal flora that have a symbiotic relationship with the host or when LPS release alerts the host to the danger of invading bacteria. Pain, swelling, and fever are the modest price that is paid for this protective response (28). In contrast, the beneficial effects of LPS provide small solace to patients who develop serious gram-negative

sepsis, that leads to shock and death (17, 21, 74, 102, 103, 163). More than 250,000 U.S. citizens develop gram-negative bacteremia, and mortality is 10 to 90% (21, 74, 102, 103, 163), making endotoxemia the most common fatal toxemia in the United States. While this might be an evolutionary situation wherein the good of the many allows the sacrifice of a few, an equally compelling thought is that LPS may imitate a host-signaling molecule or act upon a host-signaling pathway that is fundamental and indispensable to the cell. Thus, confusion about the evolutionary benefit of the host response to LPS may relate more to our ignorance of the details of LPS action at the cellular level than to an evolutionary mistake.

Classically, diseases caused by bacterial toxins could be sorted into four physiologic mechanisms of action: cytolytic (e.g., the α toxins of *S. aureus* and *Clostridium perfringens,* which form pores or digest membranes, respectively), cytotoxic (e.g., protein synthesis inhibition by diphtheria toxin), enterotoxic (e.g., stimulation of secretion by cholera toxin), and neurotoxic (e.g., inhibition of acetylcholine release by botulinum toxin) (93). The toxicity induced by LPS or by superantigens does not fit neatly into any of these traditional groupings. The disease syndromes caused by endotoxin and the staphylococcal toxic shock syndrome toxin (TSST) (as an example of a superantigen) do have many clinical signs in common, and they act by similar pathophysiologic mechanisms. This finding allows us to add a fifth general mechanism for toxin action, immunotoxicity (Fig. 4). When sufficient toxin enters the circulation, TSST-1 or LPS can act on macrophages to cause the unbridled release of cytokines, especially TNF-α, IL-1α, and IL-6, which act as critical initiators of the "mediator storm" (10, 11, 16, 27, 39, 46, 61, 80, 96, 101, 103, 114, 116, 149, 163, 164). These cytokines trigger the release of bioactive lipids, cascades of other cytokines, and reactive oxygen species from a wide variety of cells (46, 96, 103, 164). These mediators cause increased vascular permeability, decreased cardiac contractility, vasodilation, pulmonary hypertension, and disseminated intra-

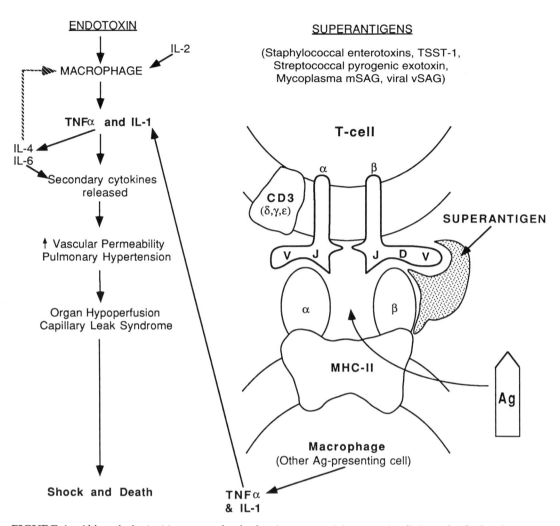

FIGURE 4 Although the inciting agents that lead to immunotoxicity are quite distinct, the final pathways are similar. The cascade of interactions that leads to the release of TNF-α, IL-1, and other mediators is shown by solid arrows. Inhibitory or regulatory responses are indicated by a broken arrow. The T-cell antigen receptor and macrophage major histocompatibility complex (MHC) antigen-presenting receptor are bridged and activated by the superantigen. A "normal" antigen (Ag) is shown to the side of the diagram, and its placement for a normal interaction between a T cell and a macrophage is indicated by the curved arrow. In contrast, endotoxin can act directly on macrophages to release mediators without interacting with T cells.

vascular coagulopathy. Ultimately, these lead to shock and death. Current data show that the macrophage is sufficient to account for the initiation of all endotoxin-stimulatable events by the release of the early cytokines (TNF-α, IL-1α, IL-6) (39, 91, 149, 150). Most convincing is the evidence contributed by Freudenberg et al. (39), who found that the transfer of C3H/HeN macrophages into LPS-resistant C3H/

HeJ mice confers susceptibility. This is not to exclude the possibility that other cells, such as platelets, neutrophils, and endothelial cells, also have an effect (2, 105, 106), but macrophages do appear to play a central role. Similarly, T cells plus macrophages should be able to account for the events seen following the release of a superantigen such as TSST-1. Hence, the interaction of LPS or TSST-1 with macrophages can be viewed as an unbalanced immunologic event; thus the term immunotoxicity.

One troubling aspect of the pathophysiology of endotoxemia is that the disease occurs in spite of and may be made worse by the use of appropriate antibiotic therapy. Some β-lactam antibiotics cause a greater release of endotoxin into the cerebrospinal fluid and blood than aminoglycosides, fluoroquinolones, or chloramphenicol do (3, 14, 32, 52, 62, 63, 90, 122, 127, 141). With some β-lactam antibiotics, this endotoxin release is accompanied by higher levels of TNF-α in animal and tissue culture models of endotoxemia (13, 132), and in the cerebrospinal fluids of children with meningitis (3). A direct role for antibiotic-induced endotoxin release in increased mortality was clearly demonstrated when Morrison et al. (94) used a galactosamine-sensitized CF-1 and C3H/HeJ mouse peritonitis model to show (i) that the release of endotoxin by β-lactam antibiotics directly contributes to mortality (LPS-hyporesponsive mice had lower mortality) and (ii) that ceftazidime releases more LPS than imipenem. Larger amounts of LPS are released by antibiotics that act on penicillin-binding protein (PBP) 3, compared to a PBP2-active drug, and this release follows the development of filamentous organisms (63). Taken together, these studies suggest that antibiotics that release less LPS or compounds that neutralize LPS at the initiation of therapy may reduce the mortality of bacterial sepsis.

ENDOTOXIN AND HOST CELL ACTIVATION

The minute quantities of LPS that are needed to activate a large number of processes in target cells suggest an amplification system such as that used by the signal transduction pathways of hormones and other extracellular modulators. This observation has led to the search for LPS and lipid A receptors in various cells (84, 95, 116). At least four distinct mammalian membrane proteins that can bind LPS have been identified. These include CD14 (55 kDa), CD18 (95 kDa), a 70- to 73-kDa protein, and

the acetyl-low-density lipoprotein receptor (Table 4) (33, 49, 55, 56, 79, 159–162). The first three proteins are candidates for a receptor that may directly mediate LPS signal transduction. Conversely, the interaction of LPS with the acetyl-low-density lipoprotein receptor appears to be involved in an LPS detoxification pathway (49, 56, 84).

Recent studies have centered on the possibility that CD14 serves as an important receptor for LPS. CD14 was first discovered by virtue of its reactivity with monoclonal antibodies directed against a specific differentiation antigen, but its function remained unknown until it was shown to interact with LPS and a plasma LPS-binding protein (LBP) (161, 162). Because CD14 is anchored to the membrane via a glycosylphosphatidylinositol moiety (84), it is less likely that CD14 alone initiates LPS signaling events; rather, the evidence suggests that CD14 interacts with other membrane components to promote signal transduction (84, 134). In addition, the cobinding of LPS to CD14 and LBP may be involved in LPS import into the host cell, and this internalization could be necessary for LPS to induce various intracellular processes. Regardless of the precise function of CD14 and LBP, they appear to play an important role in endotoxemia, because anti-CD14 antibodies can block the LPS-induced release of TNF-α in vitro (162), whereas the depletion of LBP reduces TNF-α production in vivo (130). In addition, B-cell and fibroblast cell lines transfected with CD14 show a markedly increased susceptibility to LPS-mediated activation (51, 78). Nevertheless, additional components are likely to be involved in LPS signaling, given that CD14-negative cells can respond to high concentrations of LPS (85) and that LPS-mediated signal transduction can be blocked even when LBP-CD14-mediated uptake occurs (71).

As indicated above, several other proteins, including CD18 and a 70- to 73-kDa protein, have been implicated in the cellular binding of LPS. CD18 allows particulate-bound LPS (i.e., gram-negative bacteria or LPS-coated erythrocytes) to interact with macrophages, al-

TABLE 4 Cellular signal transduction in response to bacterial LPS

Signature event	Cell type	Assay system	Comment
G-protein activation			
G_i	J774A.1	Pertussin toxin inhibition of IL-1 release	B fragment was active
G_i	U937	Phorbol ester-induced upregulation of G_i correlates with increased LPS-induced excretion	No direct evidence for linkage between events
Non-G_i	RAW264.7	GTPase activity; LPS-stimulatable activity dependent on purines that stimulate P2 receptors; insensitive to pertussis toxin	Direct measure of G-protein activation, but G-protein involvement not ensured until components are characterized
ADP-ribosylation (ARFs)		Ribosylation of G protein	Heat treatment of LPS not performed; therefore, *E. coli* enterotoxin effects are not ruled out
Tyrosine kinase		Tyrosine phosphorylation	Protein tyrosine phosphorylation is *not* specific for membrane receptor tyrosine kinase
MEKs-MAP kinases, Ca^{2+} influx	Macrophages	IL-1 release	Data based on inhibitors that are not specific
PKC	Platelets	Lipid X activation	Lipid X inactive on PKC of other cell types
	RAW264.7	PKC inhibitors block LPS response	LPS substructures that block LPS activity do not act directly on PKC; PKC inhibitors are not specific.
PKA	Lymphocytes	Translocation of one PKA isozyme to nucleus	No direct evidence that this affects signaling
MT bundling	Murine macrophages	Taxol-induced release of TNF-α; *R. sphaeroides* LPS blocks effect of taxol	In addition to MT bundling (slow response), taxol causes major Ca^{2+} influx; massive alterations in cytoskeleton causes many (nonspecific) changes in membrane and intracellular organization
Transforming factors: NF-kB–IkB and OTF-2	70Z/3 cells (lymphocytes)	Kappa-chain production	LPS acts independently of PKA and PKC

though CD18-negative human neutrophils are able to respond to LPS (159). Furthermore, anti-CD18 monoclonal antibodies that block gram-negative bacterial adhesion to macrophages fail to block LPS-induced release of cytokines (159). Thus, CD18 appears to facilitate nonopsonic recognition of gram-negative organisms via LPS binding, but CD18 probably does not play a primary role in phagocyte activation by LPS. In terms of the 70- to 73-kDa LBP, it was first identified by using a radiolabeled, photoactivatable LPS probe (33, 79, 95).

Antibodies that cross-link this LBP cause the release of the same mediators as are induced by LPS (18), suggesting that aggregation of this membrane-associated LBP may cause cellular activation. Altogether, these studies on the 70- to 73-kDa protein together with those on CD14 and CD18 indicate that multiple proteins in macrophage membranes can bind LPS, and these proteins appear to play a role in LPS actions such as the promotion of endotoxic shock. However, in each case, no direct role in LPS signal transduction has been clearly established, and it is not clear if any or all of these binding proteins truly act as classic receptors.

As the search for an LPS receptor proceeded, studies of the events following the initial interaction of LPS with host lymphoreticular cells yielded important information. In many signal transduction systems, GTP-binding proteins (G proteins) are activated as one of the earliest events following ligand–receptor interaction (12, 142). That LPS might also act through G proteins was suggested by several indirect lines of evidence: first, LPS or lipid A can cause the activation of phospholipase A_2, phospholipase C, and protein kinase C (PKC), all of which are known to be activated downstream from G proteins in other signal transduction systems (9). Second, pertussis toxin treatment of J774A.1 macrophages decreases LPS-induced release of IL-1 (64). The fact that pertussis toxin inactivates G_i (GTP binding protein that inhibits adenylate cyclase) via ADP-ribosylation provides presumptive data that G_i is involved. However, further studies have shown that the binding fragment of pertussis toxin (not the active ribosylating enzyme fragment) is able to induce this response in macrophages (136). Since the pertussis toxin binding fragment can enhance adenylate cyclase activity (68), the question became whether its effects on LPS-stimulated IL-1 secretion were the result of a direct action on G_i or whether the effects were the result of an alternative mechanism for modulating intracellular cyclic AMP levels (136). A third line of evidence implicating G proteins in LPS-mediated signal transduction is the observation that phorbol ester-induced increases in G_i expression in U937 cells correlates with the increased responsivity of these cells to LPS (22).

Besides these indirect lines of study implicating G proteins in LPS action, more direct evidence has been provided by studies of GTPase activity in RAW264.7 macrophage cell membranes (140). As G proteins cycle from active to inactive states, GTP is hydrolyzed, thus making GTPase activity a general indicator of G-protein activation (140). In fact, LPS stimulates a macrophage membrane GTPase, and this activity is inhibitable by lipid X (140), which is a lipid A substructure that exhibits antiendotoxin activity in vitro and in vivo (50, 77). The inhibitory activity of lipid X has been enigmatic, because this lipid fails to block LPS or tetraacyl lipid A binding to macrophages (86). One possible explanation is that lipid X blocks LPS activities by acting intracellularly rather than at the cell surface. Synthetic lipid IVa and deacylated LPS also act intracellularly by antagonizing LPS-induced NF-κB binding activity and the production of IL-1β in THP-1 cells (71). This inhibition occurs without inhibiting LPS uptake by the CD14-mediated pathway. Because LPS can be transported into cells and released into the cytoplasm (67), perhaps via specific membrane receptors, including CD14-LBP, it is conceivable that the cellular target for LPS is a G protein. Previous observations consistent with this concept include the following: (i) lipid interactions that alter G-protein activity are well described (144, 154); (ii) endotoxin tolerance is associated with decreased macrophage membrane GTPase activity (19); and (iii) G proteins are common targets for bacterial toxins (93). In sum, both direct and indirect evidence suggests that G proteins are involved in LPS-induced signal transduction.

Studies of the LPS-stimulatable membrane GTPase in macrophage membranes led to the unanticipated observation that GTPase activity is dependent on the presence of adenine nucleotides (9, 108, 140). The profile of active adenine nucleotides suggests a P2 type of purinoreceptor (15), and an adenine nucleotide analog that failed to act synergistically with LPS in stimulating membrane GTPase activity proved efficacious at preventing endotoxic death in a murine model (108). Although these data have shown specificity for both LPS and selected adenine nucleotides (108), the role and precise mechanism(s) for G proteins and a purinoreceptor in endotoxin shock are not completely established and await the further isolation and characterization of the components involved in this system.

Initial attempts to establish the identity of the adenine nucleotide-dependent LPS-stimulatable GTPase have shown that the activity of this GTPase is resistant to pretreatment with three ADP-ribosylating toxins, namely, pertussis toxin, botulinum C3 exoenzyme, and exoenzyme S from *P. aeruginosa* (25). These toxins are active under the conditions used, and their individual substrates are present in the macrophage membrane preparations. Additionally, we noted that *E. coli* serotype O111B4 LPS alone is able to induce membrane ADP-ribosylation of an 83-kDa band in a 15-min reaction at 22°C with 20 μM [^{32}P]NAD. However, this activity is susceptible to heat inactivation (100°C for as little as 1 min), a property that is consistent with the presence of a heat-labile enterotoxin in the commercially obtained LPS preparation. Heat-labile enterotoxin is capable of inactivating heterotrimeric G proteins by ADP-ribosylation (54), suggesting that the labeling of the 83-kDa band may not be a property of LPS but is a feature of the contaminating enterotoxin, since many ADP-ribosylating toxins lack strict substrate specificity. In contrast, the stimulation of GTPase activity by LPS is resistant to heat inactivation (100°C for 10 min) and is reproducible with endotoxins from multiple genera and species (25), confirming that GTPase activity is truly LPS responsive.

Recently, it was reported that pretreatment of bone marrow-derived BALB/c macrophages with 100 ng of *E. coli* LPS per ml for 24 h changes the ADP-ribosyltransferase activities of the cytosolic fractions of these cells, resulting in decreased ADP-ribosylation of a cy-

tosolic 33-kDa protein (58). This effect was not seen if the LPS was coincubated with polymyxin B or if the cells were treated instead with the nonendotoxic LPS from *Rhodopseudomonas palustris*. In addition, LPS from *E. coli* did not decrease the ADP-ribosylation of the 33-kDa cytosolic protein if macrophages derived from the LPS-hyporesponsive mouse strain C3H/HeJ were used. Interestingly, the dose and timing of LPS treatment correlated with an increase in nitrite production by these macrophages. Therefore, Hauschildt et al. concluded that the diminished ADP-ribosylation of this cytosolic protein may be associated with a late event in LPS-induced macrophage activation (58).

In recent years, many investigations of diverse types of cellular stimuli (hormones, neurotransmitters, extracellular matrix components, etc.) have shown that the tyrosine phosphorylation of cellular proteins often follows receptor occupation and serves as a key signaling event (47, 145). LPS rapidly induces the tyrosine phosphorylation of several murine macrophage proteins, some of which have been identified as members of the mitogen-activated protein (MAP) kinase family (24, 155). Although these data are consistent with a membrane receptor possessing intrinsic tyrosine kinase activity, it is possible that an LPS receptor(s) complexes with intracellular tyrosine kinases, thereby regulating their activity. In this regard, LPS has been reported to activate the CD14-associated protein tyrosine kinase p56[lyn] in human monocytes (133). The activation of p56[lyn] also appears to occur concomitantly with the activation of two other protein-tyrosine kinases, namely, p58[hck] and p59[c-fgr] (133). Therefore, there is evidence for direct protein-tyrosine kinase activation by LPS, although it is conceivable that the tyrosine phosphorylation of certain substrates (such as the MAP kinases) is also a downstream signaling event of other systems, including the activation of G proteins, phospholipase C, PKC, Raf-1, or various MEKs, all of which can lead to MAP kinase phosphorylation and activation (44, 104). In fact, these events, together with additional G-protein-sensitive endpoints such as the cyclic-AMP-dependent PKA, phospholipase A$_2$, and calcium flux-calmodulin activation, occur in response to LPS (1, 7, 24, 110). These studies relied heavily on the use of inhibitors, which are able to provide suggestive but not definitive data because of the possible lack of specificity of these various agents. In one case, LPS was reported to directly activate platelet PKC (123), but this appears to be an isolated example, as most of the data favor PKC activation following phospholipase A$_2$ activation in lymphocytes, macrophages, or endothelial cells (1, 111). Also, direct activation of purified PKC by LPS occurs only at high calcium concentrations (158).

Microtubules (MT) have also been implicated in LPS signal transduction. The data are based upon the use of taxol, an MT-bundling agent. Taxol causes the release of TNF-α from murine macrophages and can also promote the internalization of TNF-α receptors in these same cells (30, 116, 152). Moreover, taxol is much less effective at inducing the release of TNF-α from macrophages harvested from the mice less responsive to LPS (C3H/HeJ [152]), and R. sphaeroides lipid A was able to block taxol-induced TNF-α release from murine macrophages (152). Although these data are consistent with MT reorganization playing a role in LPS-mediated TNF-α production, this response may be a nonspecific effect of major cytoskeletal alteration, which could generate widespread effects on many interacting cell systems. Taxol also induces a large calcium influx (66). Thus, taxol may influence many independent systems simultaneously, and perhaps only a subset of these is involved in LPS signal transduction.

Nuclear transcriptional factors are further downstream in the signal transduction pathway than any of the elements considered thus far. LPS results in the activation of NF-κB in THP-1 macrophages (71). LPS also increases kappa-chain biosynthesis in the 70Z/3 murine B-lymphoma cell line via NF-κB and OTF-2 (4, 116, 131). This effect is independent of CD14, as B cells do not express this marker; however, 70Z/3 cells do become more sensitive to LPS when they are transfected with a plasmid expressing CD14 (78). Endothelial cells also do not express CD14, but LPS can activate endothelial NF-κB via interaction with soluble CD14 in serum (118). In some cells, activation of NF-κB comes via PKC phosphorylation I-κB, and indeed, in THP-1 cells, LPS stimulation results in the phosphorylation of MAD3 (an I-κB-like protein) (20). The LPS-induced activation of the 70Z/3 cell line is independent of PKC and PKA (131), but this does not preclude the effects of other kinases, as the tyrosine kinase inhibitor herbimycin A can prevent the LPS-mediated activation of NF-κB in 70Z/3-hCD14 cells (57). Thus, the LPS-stimulatable kinase(s) or other pathways involved in NF-κB and OTF-2 activation are yet to be defined in 70Z/3 cells. Finally, AP-1 is also activated by LPS stimulation, and it was recently shown in J774 macrophages that the activation of junB is via PKA, in contrast to that of c-jun, which is through PKC (40).

In summary, considerable information concerning LPS-stimulated signal transduction is now available for this potent biological molecule: membrane receptors that bind LPS have been identified, G-protein activities have been directly and indirectly measured, protein kinase activities (tyrosine kinase activity, PKA, PKC) appear to be regulated, purinoreceptors have been shown to influence endotoxin lethality, MT interactions have been inferred, and activation of nuclear transcription factors has been found to follow LPS challenge. Nonetheless, the details of how these various systems interact are not yet available; therefore, data providing a cohesive model for endotoxin-mediated signal transduction and cellular activation will necessarily require future work.

NOVEL THERAPEUTIC AGENTS
Paralleling and arising from a greater understanding of the structure and pathophysiology of LPS has been the development of several novel therapeutic approaches. Knowing the covalent structure of lipid A has spurred the

search for mono- and disaccharide derivatives of LPS that can act as nontoxic adjuvants. MPL, a hexaacyl lipid A without the phosphate at the 1 position, has a >1,000-fold decrease in toxicity, yet it retains adjuvant activity (5, 98, 120, 121, 146, 147). MPL reduces the number of suppressor T cells when mice are challenged with pneumococcal polysaccharide antigens, thus leading to higher antibody titers (5). MPL is able to increase the titer of antibodies to protein antigens as well (98, 146). The immunostimulatory activity of MPL is not limited to antibody responses. MPL also has shown promise in the treatment of melanoma (146, 147) and as a nonspecific enhancer of host resistance to gram-positive and gram-negative bacterial challenge (98, 146). More recently, a triacyl monosaccharide derivative of lipid X, SDZ MRL953, has been described. This derivative enhances nonspecific resistance to bacterial challenge and more rapid recovery of the bone marrow in drug-induced leukopenia (81). The protection afforded by SDZ MRL953 may be explained by its abilities to prime neutrophils for enhanced oxidative responses and to induce the production of cytokines (81). Thus, LPS derivatives and substructures have the promise of retaining the immunostimulatory effects of endotoxin with greatly reduced toxicity.

LPS substructures and derivatives are also being studied as inhibitors of endotoxin action. Lipid X was the first LPS-related compound to show utility against endotoxin challenge (77, 109). Studies using tetraacyl *Salmonella typhimurium* Rc LPS, produced via acyloxyacyl hydroxylase, showed inhibition of LPS–induced neutrophil and monocyte adhesion to human vascular endothelial cells (34). Both the naturally occurring (lipid IV$_A$) and synthetic tetraacyl lipid A precursors are similarly able to inhibit several LPS actions (34, 71, 84, 116, 120). A pentaacyl synthetic lipid A-like compound (114), E5531, decreases TNF-α, IL-1, and IL-6 release from LPS-challenged human macrophages; reduces the mortality rate in *Mycobacterium bovis* BCG-primed mice challenged with *E. coli* (70); and decreases LPS binding to LBP

(124). The LPSs of *R. sphaeroides* and *R. capsulatus* are nontoxic (48, 84, 86, 88, 120). The lipid A from each of these bacteria inhibits cytokine release from murine and human phagocytes (48, 84, 86, 88, 120), decreases TNF-α release from LPS-challenged mice (84, 120), and protects animals from endotoxin challenge (84, 120). Therefore, LPS derivatives and nontoxic LPSs show promise in the fight against endotoxemia.

Another approach to the problem of LPS toxicity has been the attempt to bind the toxin and prevent its interaction with host effector cells. Monoclonal antibodies against LPS were developed and tested in humans (107, 166). Although studies of animal models showed very promising results, the human trials gave only modest results that were met with some skepticism (153, 156). Of interest, another monoclonal antibody that binds to the inner core of the LPS, SDZ 219-800, has been reported to be very effective at blocking the following LPS responses: (i) release of IL-6 and TNF-α, (ii) pyrogenic response in rabbits, and (iii) lethality in galactosamine-sensitized mice (31). While this monoclonal antibody appears to show the greatest promise of anti-LPS antibodies developed to date, the recent disappointments in other clinical trials may limit its clinical testing. Endotoxin can also be neutralized by binding to bactericidal permeability-increasing protein (45, 72), lysozyme (138), CAP neutrophil proteins (59), or *Limulus* amebocyte proteins (99). Animal models show promising results (72, 138). A recombinant amino-terminal product of bactericidal permeability-increasing protein that shows as much activity as the holoprotein and appears to be quite nontoxic has been developed (72). Thus, a number of endotoxin-neutralizing agents are being developed, but the major question is whether these agents will work once the endotoxin has begun to induce the signs and symptoms of sepsis, i.e., once the endotoxin has activated host cells.

A targeted reduction in the levels of LPS-induced cytokines is another putative therapeutic modality. Animal and tissue culture models of endotoxemia that use anti-TNF-α

antibodies have shown promise, but antibodies in human trials have been less effective, perhaps because the host was partially immunoparalyzed by the total removal of TNF-α, a stimulator of the immune system (80). IL-1 receptor antagonist, which also gave positive results in animal studies (27, 29, 101) and phase II clinical trials (37), did not reduce the IL-1 response below baseline, yet clinical trials in humans did not show clear-cut efficacy (53). Immunoparalysis may again be part of the explanation. IL-1 does help protect animals from bacterial challenge (87). Perhaps using natural modulators of cell activation, e.g., IL-6 and transforming growth factor β, will prove useful in treating endotoxemia (8, 89). Finding the proper balance between cytokine reduction and immunoparalysis may be a critical and difficult aspect of anticytokine therapy.

A final novel approach that uses adenine nucleotide analogs has arisen from the studies of LPS signal transduction in macrophages (9, 140). An ATP derivative, 2-methylthio-ATP, protects mice challenged with LPS (108). This protection is accompanied by a decrease in release of TNF-α and IL-1 but not IL-6 (108). TNF-α and IL-1 are reduced to basal levels, suggesting that non-LPS-mediated pathways remain intact and maintain basal cytokine production. The selective reduction of particular cytokines shows that 2-methylthio-ATP acts as an immunomodulator, perhaps via a pathway that is relatively specific for the immunotoxic responses. Pentoxyfylline also reduces cytokine release and protects mice challenged with LPS (165). Of interest, while pentoxyfylline is several orders of magnitude less active on a molarity basis, its basic structure is that of a purine ring.

CONCLUSIONS

Over the past decade, major strides have been made in the understanding of endotoxin shock. The covalent structure of lipid A has been defined, and this has allowed the development of lipid A substructures and derivatives for testing as nontoxic immunostimulants and as antiendotoxins. The biosynthetic pathway has also been defined, and this knowledge may lead to the discovery of new antibiotics, because blockage of LPS biosynthesis results in bacterial cell death. With regard to currently available antibiotics, selection of antibiotics that release less LPS as the bacteria are killed may result in lower mortality rates. This may be particularly important for initial doses, when the bacterial burden is high. Neutralization of endotoxin by antibodies or cationic proteins may also reduce mortality, especially if the antibodies or proteins are given prophylactically or shortly after the onset of symptoms. Elucidation of the major cytokines and other mediators released in response to LPS has allowed the development of a number of substances with potential for the prevention of endotoxic shock. Prednisone has already been shown to reduce morbidity and decrease the levels of cytokines in the spinal fluids of children with gram-negative meningitis. Similarly, glucocorticoids reduce mortality in typhoid fever patients, but the pathogenesis of typhoid fever is probably more complex than that of gram-negative meningitis. Finally, some early results with an adenine nucleotide derivative suggest that drugs that interact with purinoreceptors may provide another approach to the treatment of endotoxemia. Therefore, our increasing knowledge about endotoxin and its pathogenesis is now beginning to provide therapeutic dividends.

REFERENCES

1. **Adams, D. O.** 1992. LPS-initiated signal transduction pathways in macrophages, p. 285–309. *In* D. C. Morrison and J. L. Ryan (ed.), *Bacterial Endotoxic Lipopolysaccharides,* vol. I. CRC Press, Inc., Boca Raton, Fla.
2. **Aida, Y., and M. J. Pabst.** 1990. Priming of neutrophils by lipopolysaccharide for enhanced release of superoxide. Requirement for plasma but not for tumor necrosis factor-α. *J. Immunol.* **145:** 3017–3025.
3. **Arditi, M., L. Ables, and R. Yogen.** 1989. Cerebrospinal fluid endotoxin levels in children with *Haemophilus influenzae* meningitis before and after administration of intravenous ceftriaxone. *J. Infect. Dis.* **160:** 1005–1011.
4. **Baeuerle, P. A., and D. Baltimore.** 1988. Activation of DNA-binding activity in an apparently

cytoplasmic precursor of the NF-κB transcription factor. *Cell* **53:**211–217.

5. **Baker, P. J.** 1993. Effect of endotoxin on suppressor T cell function. *Immunobiology* **187:**372–381.

6. **Balish, E.** 1985. Endotoxin effects on germfree animals, p. 338–358. *In* L. B. Hinshaw (ed.), *Handbook of Endotoxin,* vol. II. *Pathophysiology of Endotoxin.* Elsevier, Amsterdam.

7. **Bandekar, J. R., R. Castagna, and B. M. Sultzer.** 1992. Roles of protein kinase C and G proteins in activation of murine resting B lymphocytes by endotoxin-associated proteins. *Infect. Immun.* **60:**231–236.

8. **Barton, B. E., and J. V. Jackson.** 1993. Protective role of interleukin 6 in the lipopolysaccharide-galactosamine septic shock model. *Infect. Immun.* **61:**1496–1499.

9. **Bertics, P. J., J.-W. van de Loo, L. Denlinger, P. S. Leventhal, J. Tomachek, T. Tanke, S. Daugherty, D. T. Golenbock, and R. A. Proctor.** 1993. The purine analog, 2-methylthio-ATP (2-MeS-ATP), protects mice from endotoxin death, p. 233–241. *In* J. Levin, C. R. Alving, R. S. Munford, and P. L. Stütz (ed.), *Bacterial Endotoxins: Recognition and Effector Mechanisms.* Elsevier Science Publishing BV, Amsterdam.

10. **Beutler, B., and A. C. Cerami.** 1988. Tumor necrosis, cachexia, shock, and inflammation: a common mediator. *Annu. Rev. Biochem.* **57:**505–518.

11. **Beutler, B., J. Milsark, and A. C. Cerami.** 1985. Passive immunization against cachectin/tumor necrosis factor protects mice from lethal effect of endotoxin. *Science* **229:**869–871.

12. **Bourne, H. R., D. A. Sanders, and F. McCormick.** 1990. The GTPase superfamily: a conserved switch for diverse cell functions. *Nature* (London) **348:**125–132.

13. **Brandtzaeg, P., and P. Kierulf.** 1992. Endotoxin and meningococcemia: intravascular inflammation induced by native endotoxin in man, p. 327–346. *In* J. L. Ryan and D. C. Morrison (ed.), *Bacterial Endotoxic Lipopolysaccharides,* vol II. CRC Press, Inc., Boca Raton, Fla.

14. **Brandtzaeg, P., P. Kierulf, P. Gaustad, A. Skulberg, J. N. Bruun, S. Halvorsen, and E. Sorensen.** 1989. Plasma endotoxin as a predictor of multiple organ failure and death in systemic meningococcal disease. *J. Infect. Dis.* **159:**195–204.

15. **Burnstock, G., and C. Kennedy.** 1985. Is there a basis for distinguishing two types of P2-purinoceptor? *Gen. Pharmacol.* **16:**433–440.

16. **Cannon, J. G.** 1992. Endotoxin and cytokine responses in human volunteers, p. 311–326. *In* D. C. Morrison and J. L. Ryan (ed.), *Bacterial Endotoxic Lipopolysaccharides,* vol. I. CRC Press, Inc., Boca Raton, Fla.

17. **Centers for Disease Control.** 1990. Increase in national hospital discharge survey rates for septicemia—United States, 1979–1987. *Morbid. Mortal. Weekly Rep.* **39:**31–34.

18. **Chen, T., S. W. Bright, J. L. Pace, S. W. Russell, and D. C. Morrison.** 1990. Induction of macrophage mediated tumor cytotoxicity by a hamster monoclonal antibody with specificity for LPS receptor. *J. Immunol.* **145:**8–12.

19. **Coffee, K. A., P. V. Halushka, S. H. Ashton, G. E. Tempel, W. C. Wise, and J. A. Cook.** 1992. Endotoxin tolerance is associated with altered GTP-binding protein function. *J. Appl. Physiol.* **73:**1008–1013.

20. **Cordle, S. R., R. Donald, M. Read, and J. Hawiger.** 1993. Lipopolysaccharide induces phosphorylation of MAD3 and activation of c-Rel and related NF-kappa B proteins in human monocytic THP-1 cells. *J. Biol. Chem.* **268:** 11803–11810.

21. **Cunnion, R. E., and J. E. Parrillo.** 1989. Myocardial dysfunction in sepsis: recent insights. *Chest* **95:**941–945.

22. **Daniel-Issakani, S., A. M. Spiegel, and B. Strulovici.** 1989. Lipopolysaccharide response is linked to the GTP binding protein, Gi2, in the promonocytic cell line U937. *J. Biol. Chem.* **264:** 20240–20247.

23. **Danner, R. L., R. J. Elin, J. M. Parker, J. M. Hosseini, R. A. Wesley, J. M. Reilly, and J. E. Parillo.** 1991. Endotoxemia in human septic shock. *Chest* **99:**169–175.

24. **DeFranco, A. L., S. L. Weinstein, J. S. Sanghera, S. L. Pelech, and C. H. June.** 1993. LPS stimulation of protein tyrosine phosphorylation in macrophages, p. 255–265. *In* J. Levin, C. R. Alving, R. S. Munford, and P. L. Stütz (ed.), *Bacterial Endotoxin: Recognition and Effector Mechanisms.* Elsevier Science Publishing BV, Amsterdam.

25. **Denlinger, L. C., R. A. Proctor, and P. J. Bertics.** Unpublished data.

26. **Din, Z. Z., P. Mukerjee, M. Kastowsky, and K. Takayama.** 1993. Effect of pH on solubility and ionic state of lipopolysaccharide obtained from the deep rough mutant of *Escherichia coli.* *Biochemistry* **32:**4579–4586.

27. **Dinarello, C. A.** 1989. Interleukin 1 and its biologically related cytokines. *Adv. Immunol.* **44:** 153–205.

28. **Dinarello, C. A.** 1991. Interleukin-1 and interleukin-1 antagonism. *Blood* **77:**1627–1652.

29. **Dinarello, C. A.** 1993. Blocking interleukin-1 in disease, p. 473–479. *In* J. Levin, C. R. Alving, R. S. Munford, and P. L. Stütz (ed.), *Bacterial Endotoxin: Recognition and Effector Mechanisms.* Elsevier Science Publishing BV, Amsterdam.

30. **Ding, A. H., F. Porteu, E. Sanchez, and C.**

F. Nathan. 1990. Shared actions of endotoxin and taxol: induction of TNF release and down/regulation of TNF receptors. *Science* **248:**370–372.

31. **DiPadova, F. E., H. Gram, R. Barclay, B. Kleaser, E. Liehl, and E. T. Rietschel.** 1993. New anticore LPS monoclonal antibodies with clinical potential, p. 325–335. *In* J. Levin, C. R. Alving, R. S. Munford, and P. L. Stütz (ed.), *Bacterial Endotoxin: Recognition and Effector Mechanisms.* Elsevier Science Publishing BV, Amsterdam.

32. **Dofferhoff, A. S. M., J. H. Nijland, H. G. de Vries-Hospers, P. O. M. Mulder, J. Weits, and V. J. J. Bom.** 1991. Effects of different types and combinations of antimicrobial agents on endotoxin release from gram-negative bacteria: an *in vitro* and *in vivo* study. *Scand. J. Infect. Dis.* **23:**745–754.

33. **Dziarski, R.** 1991. Peptidoglycan and lipopolysaccharide bind to the same binding site on lymphocytes. *J. Biol. Chem.* **266:**4719–4725.

34. **Erwin, A. L., and R. S. Munford.** 1992. Processing of LPS by phagocytes, p. 405–434. *In* D. C. Morrison and R. L. Ryan (ed.), *Bacterial Endotoxic Lipopolysaccharides,* vol. I. CRC Press, Inc., Boca Raton, Fla.

35. **Falkow, S.** 1988. Molecular Koch's postulates applied to microbial pathogenicity. *Rev. Infect. Dis.* **10:**S274-S276.

36. **Feist, W., A. J. Ulmer, M.-H. Wang, J. Musehold, C. Schluter, J. Gerdes, H. Herzbeck, H. Brade, S. Kusumoto, T. Diamanstein, E. T. Rietschel, and H.-D. Flad.** 1992. Modulation of lipopolysaccharide-induced production of tumor necrosis factor, interleukin 1, and interleukin-6 by synthetic precursor Ia of lipid A. *FEMS Microbiol. Immunol.* **89:**73–89.

37. **Fisher, C. J. J., G. J. Slotman, S. Opal, J. Pribble, D. Stibs, and M. Catalano.** 1991. Interleukin-1 receptor antagonist reduces mortality in patients with sepsis syndrome. *Am. Coll. Chest Physicians Meet.*

38. **Freeman, B. A.** 1979. *Burrow's Textbook of Microbiology,* p. 221. The W. B. Saunders Co., Philadelphia.

39. **Freudenberg, M. A., D. Keppler, and C. Galanos.** 1986. Requirement for lipopolysaccharide-responsive macrophages in galactosamine-induced sensitization to endotoxin. *Infect. Immun.* **51:**891–895.

40. **Fujihara, M., J. Masashi, Y. Muroi, N. Ito, and T. Suzuki.** 1993. Mechanism of lipopolysaccharide-triggered junB activation in a mouse macrophage-like cell line (J774). *J. Biol. Chem.* **268:**14898–14905.

41. **Galanos, C., and O. Lüderitz.** 1975. Electrodialysis of lipopolysaccharides and their conversion to uniform salt forms. *Eur. J. Biochem.* **54:**603–610.

42. **Galloway, S. M., and C. R. H. Raetz.** 1990. A mutant of *Escherichia coli* defective in the first step of endotoxin biosynthesis. *J. Biol. Chem.* **265:**6394–6402.

43. **Gamian, A., A. Romanowska, and E. Romanowska.** 1992. Immunochemical studies on sialic acid-containing lipopolysaccharides from enterobacterial species. *FEMS Microbiol. Immun.* **4:**323–328.

44. **Gardner, A. M., R. R. Vaillancourt, and G. L. Johnson.** 1993. Activation of mitogen-activated protein kinase/extracellular signal-regulated kinase by G protein and tyrosine kinase oncoproteins. *J. Biol. Chem.* **268:**17896–17901.

45. **Gazzano-Santoro, H., K. Meszaros, C. Birr, S. F. Carroll, G. Thesfan, A. H. Horwitz, E. Lim, S. Aberle, H. Kasler, and J. B. Parent.** 1994. Competition between rBPI$_{23}$, a recombinant fragment of bactericidal/permeability-increasing protein, and lipopolysaccharide (LPS)-binding protein for binding to LPS and gram-negative bacteria. *Infect. Immun.* **62:**1185–1191.

46. **Glauser, M. P., G. Zanetti, J.-D. Baumgartner, and J. Cohen.** 1991. Septic shock: pathogenesis. *Lancet* **338:**732–736.

47. **Glenney, J. R., Jr.** 1992. Tyrosine-phosphorylated proteins: mediators of signal transduction from the tyrosine kinases. *Biochim. Biophys. Acta* **1134:**113–127.

48. **Golenbock, D. T., R. Y. Hampton, N. Qureshi, K. Takayama, and C. R. H. Raetz.** 1991. Lipid A-like molecules that antagonize the effects of endotoxins on human monocytes. *J. Biol. Chem.* **226:**19490–19498.

49. **Golenbock, D. T., R. Y. Hampton, C. R. H. Raetz, and S. D. Wright.** 1990. Human phagocytes have multiple lipid A-binding sites. *Infect. Immun.* **58:**4069–4075.

50. **Golenbock, D. T., J. E. Leggett, W. A. Craig, C. R. H. Raetz, and R. A. Proctor.** 1988. Lipid X protects mice against fatal *Escherichia coli* infection. *Infect. Immun.* **56:**779–784.

51. **Golenbock, D. T., Y. Liu, F. H. Millham, M. W. Freeman, and R. A. Zoeller.** 1993. Surface expression of human CD14 in Chinese hamster ovary fibroblasts imparts macrophage-like responsiveness to bacterial endotoxin. *J. Biol. Chem.* **268:**22055–22059.

52. **Goto, H., and S. Nakamura.** 1980. Liberation of endotoxin from *Escherichia coli* by addition of antibiotics. *Jpn. J. Exp. Med.* **50:**35–43.

53. **Granowitz, E. V., R. Porat, J. W. Mier, S. F. Orencole, M. V. Callahan, J. G. Cannon, E. A. Lynch, K. Ye, D. D. Poutsiaka, E. Vannier, L. Shapiro, J. B. Pribble, D. M. Stiles, M. A. Catalano, S. M. Wolff, and C. A. Dinarello.** 1993. Hematologic and immunomodulatory effects of an interleukin-1 receptor antag-

onist coinfusion during low-dose endotoxemia in healthy humans. *Blood* **82:**2985–2990.

54. **Gyles, C. L.** 1992. *Escherichia coli* cytotoxins and enterotoxins. *Can. J. Microbiol.* **38:**734–746.

55. **Hampton, R. Y., D. T. Golenbock, M. Penman, M. Krieger, and C. R. H. Raetz.** 1991. Recognition and plasma clearance of endotoxin by scavenger receptors. *Nature* (London) **352:** 342–344.

56. **Hampton, R. Y., and C. R. H. Raetz.** 1991. Macrophage catabolism of lipid A is regulated by endotoxin stimulation. *J. Biol. Chem.* **266:**19499–19509.

57. **Han, J., J. D. Lee, P. S. Tobias, and R. J. Ulevitch.** 1993. Endotoxin induces rapid protein tyrosine phosphorylation in 70Z/3 cells expressing CD14. *J. Biol. Chem.* **268:**25009–25014.

58. **Hauschildt, S., P. Scheipers, and W. G. Bessler.** 1994. Lipopolysaccharide-induced change of ADP-ribosylation of a cytosolic protein in bone-marrow-derived macrophages. *Biochem. J.* **297:** 17–20.

59. **Hirata, M., Y. Shimomura, M. Yoshida, J. G. Morgan, I. Palings, D. Wilson, M. H. Yen, S. C. Wright, and J. W. Larrick.** 1994. Characterization of a rabbit cationic protein (CAP18) with lipopolysaccharide-inhibitory activity. *Infect. Immun.* **62:**1421–1426.

60. **Holst, O., and H. Brade.** 1992. Chemical structure of the core region of lipopolysaccharides, p. 135–170. *In* J. L. Ryan and D. C. Morrison (ed.), *Bacterial Endotoxic Lipopolysaccharides,* vol. II. CRC Press, Inc., Boca Raton, Fla.

61. **Hosford, D., and P. Braquet.** 1992. Interactions between platelet-activating factor and lipopolysaccharide: consequences in endotoxemia and sepsis, p. 57–74. *In* J. L Ryan and D. C. Morrison (ed.), *Bacterial Endotoxic Lipids,* vol. II. CRC Press, Inc., Boca Raton, Fla.

62. **Hurley, J. C.** 1992. Antibiotic-induced release of endotoxin: a reappraisal. *Clin. Infect. Dis.* **15:**840–854.

63. **Jackson, J. J., and H. Kropp.** 1992. Beta-lactam antibiotic-induced release of free endotoxin: *in vitro* comparison of penicillin-binding protein (PBP) 2-specific imipenem and PBP 3-specific ceftazidime. *J. Infect. Dis.* **165:**1033–1041.

64. **Jakway, J. P., and A. L. DeFranco.** 1986. Pertussis toxin inhibition of B cell and macrophage responses to bacterial lipopolysaccharide. *Science* **234:**743–746.

65. **Jann, K., and B. Jann.** 1984. Structure and biosynthesis of O-antigens, p. 138–186. *In* E. T. Rietschel (ed.), *Handbook of Endotoxin.* Elsevier, Amsterdam.

66. **Johnson, B. D., and L. Byerly.** 1993. A cytoskeletal mechanism for Ca2+ channel metabolic dependence and inactivation by intracellular Ca2+. *Neuron* **10:**797–804.

67. **Kang, Y.-H., R. S. Dwivedi, and C.-H. Lee.** 1990. Ultrastructural and immunocytochemical study of the uptake and distribution of bacterial lipopolysaccharide in human monocytes. *J. Leukocyte Biol.* **48:**316–332.

68. **Kaslow, H. R., and D. L. Burns.** 1992. Pertussis toxin and target eukaryotic cells: binding, entry, and activation. *FASEB J.* **6:**2684–2690.

69. **Kastowsky, M., T. Gutherlet, and H. Brodaczek.** 1992. Molecular modelling of the three-dimensional structure and conformational flexibility of bacterial lipopolysaccharide. *J. Bacteriol.* **174:**4798–4806.

70. **Kawata, T., J. Bristol, L. McGuigan, D. Rossignol, W. Christ, A. Robidoux, M. Perez, O. Asano, G. Dubuc, L. Hawkins, Y. Wang, and Y. Kishi.** 1992. Anti-endotoxin activities of E5531, a novel synthetic derivate of lipid A, abstr. 1360. *Program Abstr. 32nd Intersci. Conf. Antimicrob. Agents Chemother.*

71. **Kitchens, R. L., R. J. Ulevitch, and R. S. Munford.** 1992. Lipopolysaccharide (LPS) partial structures inhibit responses to LPS in human macrophage cell line without inhibiting LPS uptake by a CD14-mediated pathway. *J. Exp. Med.* **176:** 485–494.

72. **Kohn, F. R., W. S. Ammons, A. Horwitz, L. Grinna, G. Theofan, J. Weickmann, and A. Kung.** 1993. Protective effect of a recombinant amino-terminal fragment of bactericidal/permeability-increasing protein in experimental endotoxemia. *J. Infect. Dis.* **168:**1307–1310.

73. **Kovach, N. L., E. Yee, R. S. Munford, C. R. H. Raetz, and J. M. Harlan.** 1990. Lipid IV$_A$ inhibits synthesis and release of tumor necrosis factor induced by lipopolysaccharide in human whole blood ex vivo. *J. Exp. Med.* **172:**77–84.

74. **Kreger, B. E., D. E. Craven, and W. R. McCabe.** 1980. Gram-negative bacteremia. IV. Re-evaluation of clinical features and treatment in 612 patients. *Am. J. Med.* **68:**344–355.

75. **Kumazawa, Y., M. Matsuura, T. Maruyama, J. Y. Homma, M. Kiso, and A. Hasegawa.** 1986. Structural requirements for inducing *in vitro* B lymphocyte activation by chemically synthesized derivatives related to the nonreducing D-glucosamine subunit of lipid A. *Eur. J. Immunol.* **16:** 1099–1103.

76. **Kusumoto, S.** 1992. Chemical synthesis of lipid A, p. 81–106. *In* D. C. Morrison and J. L. Ryan (ed.), *Bacterial Endotoxic Lipopolysaccharides,* vol. I. CRC Press, Inc., Boca Raton, Fla.

77. **Lam, C., J. Hildebrandt, E. Schütze, B. Rosewirth, R. A. Proctor, E. Liehl, and P. Stütz.** 1991. Immunostimulatory, but not antiendotoxic, activity of lipid X is due to small amounts

of contaminating N,O-acylated disaccharide-1-phosphate: in vitro and in vivo reevaluation of the biological activity of synthetic lipid X. *Infect. Immun.* **59**:2351–2358.

78. **Lee, J. D., K. Kato, P. S. Tobias, T. N. Kirkland, and R. J. Ulevitch.** 1992. Transfection of CD14 into 70Z/3 cells dramatically enhances the sensitivity to complexes of lipopolysaccharide (LPS) and LPS binding protein. *J. Exp. Med.* **175**:1697–1705.

79. **Lei, M., S. A. Stimpson, and D. C. Morrison.** 1991. Specific endotoxic lipopolysaccharide-binding receptors on murine splenocytes. III. Binding specificity and characterization. *J. Immunol.* **147**:1925–1932.

80. **Lesslauer, W., H. Tabuchi, R. Gentz, M. Brockhaus, E. J. Schlaeger, G. Grau, P. F. Piguet, P. Pointaire, P. Vassalli, and H. Loetscher.** 1991. Recombinant soluble tumor necrosis factor receptor proteins protect mice from lipopolysaccharide-induced lethality. *Eur. J. Immunol.* **21**:2883–2886.

81. **Liehl, E., C. Lam, P. Mayer, E. Schütze, G. Bahr, F. Grossmuller, P. Stütz, and J. Hildebrandt.** 1993. SDZ MRL 953, a new cytokine inducing agent and stimulant for nonspecific immunity, p. 389–412. *In* J. Levin, C. R. Alving, R. S. Munford, and P. L. Stütz (ed.), *Bacterial Endotoxin: Recognition and Effector Mechanisms.* Elsevier Science Publishing BV, Amsterdam.

82. **Loos, M., B. Euteneuer, and F. Clas.** 1988. Interaction of bacterial endotoxin (LPS) with fluid phase and macrophage membrane associated C1q, the Fc-recognizing component of the complement system. *Adv. Exp. Med. Biol.* **256**:301–317.

83. **Loppnow, H., H. Brade, I. Durrbaum, C. A. Dinarello, S. Kusumoto, E. T. Rietschel, and H.-D. Flad.** 1989. IL-1 induction-capacity of defined lipopolysaccharide partial structures. *J. Immunol.* **142**:3229–3238.

84. **Lynn, W. A., and D. T. Golenbock.** 1992. Lipopolysaccharide antagonists. *Immunol. Today* **13**:271–276.

85. **Lynn, W. A., Y. Liu, and D. T. Golenbock.** 1993. Neither CD14 nor serum is absolutely necessary for activation of mononuclear phagocytes by bacterial lipopolysaccharide. *Infect. Immun.* **61**:4452–4461.

86. **Lynn, W. A., C. R. H. Raetz, N. Qureshi, and D. T. Golenbock.** 1991. Lipopolysaccharide-induced stimulation of CD11b/CD18 expression on neutrophils. Evidence of specific receptor-based response and inhibition by lipid A antagonists. *J. Immunol.* **147**:3072–3079.

87. **Mancilla, J., P. Garcia, and C. A. Dinarello.** 1993. The interleukin-1 receptor antagonist can either reduce or enhance the lethality of *Klebsiella*

pneumoniae sepsis in newborn rats. *Infect. Immun.* **61**:926–932.

88. **Manthey, C. L., P. Y. Perra, N. Qureshi, P. L. Stütz, T. A. Hamilton, and S. N. Vogel.** 1993. Modulation of lipopolysaccharide-induced macrophage gene expression by *Rhodobacter sphaeroides* lipid A and SDZ880.431. *Infect. Immun.* **61**:3518–3526.

89. **Massague, J.** 1992. Receptors for the TGF-β family. *Cell* **69**:1067–1070.

90. **McConnell, J. S., and J. Cohen.** 1986. Release of endotoxin from *Escherichia coli* by quinolones. *J. Antimicrob. Chemother.* **18**:765–766.

91. **Michalek, S. M., R. N. Moore, J. R. McGhee, D. L. Rosenstreich, and S. E. Mergenhagen.** 1980. The primary role of lymphoreticular cells in the mediation of host responses to bacterial endotoxin. *J. Infect. Dis.* **141**:55–63.

92. **Miles, A. A.** 1955. The meaning of pathogenicity. *Symp. Soc. Gen. Microbiol.* **5**:1–16.

93. **Mims, C. A., J. H. L. Playfair, I. Roitt, D. Wakelin, R. Williams, and R. A. Anderson.** 1993. *Medical Microbiology,* p. 16.1–16.5. Mosby, London.

94. **Morrison, D. C., S. Bucklin, M. Leeson, and Y. Fujihara.** 1994. Bacteremia vs endotoxemia in antibiotic treatment of experimental gram-negative sepsis. *Intensive Care Med.* **20**:20.

95. **Morrison, D. C., M.-G. Lei, T. Kirikae, and T.-Y. Chen.** 1993. Endotoxin receptors on mammalian cells. *Immunobiology* **187**:212–226.

96. **Morrison, D. C., and J. L. Ryan.** 1987. Endotoxins and disease mechanisms. *Annu. Rev. Med.* **38**:417–432.

97. **Moxon, E. R., and D. Maskell.** 1992. *Haemophilus influenzae* lipopolysaccharide: the biochemistry and biology of a virulence factor, p. 75–96. *In* C. E. Hormaeche, C. W. Penn, and J. C. Smith (ed.), *Molecular Biology of Bacterial Infection.* Cambridge University Press, Cambridge.

98. **Myers, K. R., A. T. Truchot, J. Ward, Y. Hudson, and J. T. Ulrich.** 1990. A critical determinant of lipid A endotoxic activity, p. 145–156. *In* A. Nowotny, J. J. Spitzer, and E. J. Ziegler (ed.), *Cellular and Molecular Aspects of Endotoxin Reactions.* Elsevier Science Publishing BV, Amsterdam.

99. **Nakamura, T., H. Furunaka, T. Miyata, F. Tokunaga, T. Muta, S. Iwanaga, M. Niwa, T. Takao, and Y. Shimonishi.** 1988. Tachyplesin, a class of antimicrobial peptides from the hemocytes of the horseshoe crab (*Tachypleus tridentatus*). *J. Biol. Chem.* **263**:16709–16713.

100. **Nakatsuka, M., S. Ikeda, Y. Kumazawa, M. Matsuura, C. Nishimura, J. Y. Homma, M. Kiso, and A. Hasegawa.** 1990. Enhancement of nonspecific resistance to microbial infections of a synthetic lipid A—subunits analogues of

GLA-27 modified at the C1 position of the glucosamine backbone. *Int. J. Immunopharmacol.* **12:** 599–603.

101. **Ohlsson, K., P. Bjork, M. Bergenfeldt, R. Hagemann, and R. C. Thompson.** 1990. Interleukin 1 receptor antagonist reduces mortality from endotoxin shock. *Nature* (London) **348:** 550–552.

102. **Parker, M. M., and J. E. Parrillo.** 1983. Septic shock. *JAMA* **250:**3324–3327.

103. **Parrillo, J. E.** 1990. Septic shock in humans: advances in the understanding of pathogenesis, cardiovascular dysfunction and therapy. *Ann. Intern. Med.* **113:**227–242.

104. **Pelech, S. L., and J. S. Sanghera.** 1992. MAP kinases: charting the regulatory pathways. *Science* **257:**1355–1356.

105. **Piguet, P. F., C. Vesin, J. E. Ryser, G. Senaldi, G. E. Grau, and F. Tacchini-Cottier.** 1993. An effector role for platelets in systemic and local lipopolysaccharide-induced toxicity in mice, mediated by a CD11a- and CD54-independent interaction with endothelium. *Infect. Immun.* **61:**4182–4187.

106. **Pohlman, T. H., and J. M. Harlan.** 1992. Endotoxin-endothelial cell interactions, p. 347–371. *In* D. C. Morrison and J. L. Ryan (ed.), *Bacterial Endotoxic Lipopolysaccharides,* vol. I. CRC Press, Inc., Boca Raton, Fla.

107. **Pollack, M.** 1992. Specificity and function of lipopolysaccharide antibodies, p. 347–374. *In* J. L. Ryan and D. C. Morrison (ed.), *Bacterial Endotoxic Lipopolysaccharides,* vol. II. CRC Press, Inc., Boca Raton, Fla.

108. **Proctor, R. A., L. C. Denlinger, P. S. Leventhal, S. K. Daugherty, J.-W. van de Loo, T. Tanke, G. S. Firestein, and P. J. Bertics.** 1994. Protection of mice from endotoxic death by 2-methylthio-ATP. *Proc. Natl. Acad. Sci. USA* **91:**6017–6020.

109. **Proctor, R. A., J. A. Will, K. E. Burhop, and C. R. H. Raetz.** 1986. Protection of mice against lethal endotoxemia by a lipid A precursor. *Infect. Immun.* **52:**905–907.

110. **Prpic, V., J. E. Weiel, S. D. Somers, S. L. Gonias, S. V. Pizzo, J. DiGusieppi, T. A. Hamilton, B. Herman, and D. O. Adams.** 1987. The effects of bacterial entotoxin on the hydrolysis of phosphatidyl inositol-4,5-biophosphate in murine pentoneal macrophages. *J. Immunol.* **139:**526–533.

111. **Pyzdrowski, K., K. E. Goad, A. E. Schorer, R. A. Proctor, and C. F. Moldow.** 1986. Endotoxin induced activation of protein kinase C is not sufficient for stimulation of endothelial cell tissue factor production. *Clin. Res.* **34:**678A.

112. **Raetz, C. R. H.** 1984. The enzymatic synthesis of lipid A: molecular structure and biologic function of monosaccharide precursors. *Rev. Infect. Dis.* **6:**463–471.

113. **Raetz, C. R. H.** 1990. Biochemistry of endotoxins. *Annu. Rev. Biochem.* **59:**129–170.

114. **Raetz, C. R. H.** 1993. Bacterial endotoxins: extraordinary lipids that activate eucaryotic signal transduction. *J. Bacteriol.* **175:**5745–5753.

115. **Raetz, C. R. H.** 1993. The enzymatic synthesis of lipid A, p. 39–48. *In* J. Levin, C. R. Alving, R. S. Munford, and P. L. Stütz (ed.), *Bacterial Endotoxins: Recognition and Effector Mechanisms.* Elsevier Science Publishing BV, Amsterdam.

116. **Raetz, C. R. H., R. J. Ulevitch, S. D. Wright, C. H. Sibley, A. Ding, and C. F. Nathan.** 1991. Gram-negative endotoxin: an extraordinary lipid with profound effects on eukaryotic signal transduction. *FASEB J.* **5:**2652–2660.

117. **Ray, A., K. Redhead, S. Selkirk, and S. Poole.** 1991. Variability in LPS composition, antigenicity and reactogenicity of phase variants of *Bordetella pertussis. FEMS Microbiol. Lett.* **79:** 211–218.

118. **Read, M., S. R. Cordle, R. A. Veach, C. D. Carlisle, and J. Hawiger.** 1993. Cell-free pool of CD14 mediates activation of transcription factor NF-kappa B by lipopolysaccharide in human endothelial cells. *Proc. Natl. Acad. Sci. USA* **90:** 9887–9891.

119. **Rick, P. D.** 1987. Lipopolysaccharide biosynthesis, p. 648–662. *In* F. C. Neidhardt, J. Ingraham, K. B. Low, B. Magasanik, M. Schaechter, and H. E. Umbarger (ed.), *Escherichia coli* and *Salmonella typhimurium: Cellular and Molecular Biology,* vol. I. American Society for Microbiology, Washington, D.C.

120. **Rietschel, E. T., T. Kirikae, F. U. Schade, U. Mamat, G. Schmidt, H. Loppnow, A. J. Ulmer, U. Zähringer, U. Seydel, F. DiPadova, M. Schreir, and H. Brade.** 1994. Bacterial endotoxin: molecular relationships of structure to activity and function. *FASEB J.* **8:** 217–225.

121. **Rietschel, E. T., T. Kirikae, F. U. Schade, A. J. Ulmer, O. Holst, H. Brade, G. Schmidt, U. Mamat, H.-D. Grimmecke, S. Kusumoto, and U. Zähringer.** 1993. The chemical structure of bacterial endotoxin in relation to bioactivity. *Immunobiology* **187:**169–190.

122. **Rokke, O., A. Revhaug, B. Osterud, and K. E. Giercksky.** 1988. Increased plasma levels of endotoxin and corresponding changes in circulatory performance in a porcine sepsis model: the effect of antibiotic administration. *Prog. Clin. Biol. Res.* **272:**247–262.

123. **Romano, M., and J. Hawiger.** 1990. Interaction of endotoxic lipid A and lipid X with pu-

rified human platelet protein kinase C. *J. Biol. Chem.* **265:**1765–1770.

124. **Rossignal, D., K. Achermann, T. Kawata, W. Christ, A. Robidoux, M. Perez, O. Asano, G. Dubuc, L. Hawkins, Y. Wang, and Y. Kishi.** 1992. Role of lipopolysaccharide binding protein (LBP) in lipopolysaccharide binding and activation of murine macrophage cells: inhibition by E5531, abstr. 1361. *Program Abstr. 32nd Intersci. Conf. Antimicrob. Agents Chemother.*

125. **Saiki, I., H. Maeda, J. Murata, T. Takahashi, S. Sekiguchi, M. Kiso, A. Hasegawa, and I. Azama.** 1990. Production of interleukin 1 from human monocytes stimulated by synthetic lipid A subunit analogues. *Int. J. Immunopharmacol.* **12:**297–305.

126. **Saiki, I., H. Maeda, J. Murata, N. Yamamoto, M. Kiso, A. Hasegawa, and I. Azuma.** 1989. Antimetastatic effect of endogenous tumor necrosis factor induced by the treatment of recombinant interferon λ followed by an analogue (GLA-60) to synthetic lipid A subunit. *Cancer Immunol. Immunother.* **30:**151–157.

127. **Shenep, J. L., R. P. Barton, and K. A. Morgan.** 1985. Role of antibiotic class in the rate of liberation of endotoxin during therapy for experimental gram-negative bacterial sepsis. *J. Infect. Dis.* **151:**1012–1018.

128. **Shimizu, T., T. Masuzawa, Y. Yanagihara, S. Nakamoto, H. Itoh, and K. Achiwa.** 1988. Antitumor activity, mitogenicity, and lethal toxicity of chemically synthesized monosaccharide analogs of lipid A. *J. Pharmacobio-Dyn.* **11:**512–518.

129. **Shnyra, A., K. Hultenby, and A. L. Lindberg.** 1993. Role of the physical state of *Salmonella* lipopolysaccharide in expression of biological and endotoxic properties. *Infect. Immun.* **61:**5351–5360.

130. **Shumann, R. R., S. R. Leong, G. W. Flaggs, P. W. Gray, S. D. Wright, J. C. Mathison, P. S. Tobias, and R. J. Ulevitch.** 1990. Structure and function of lipopolysaccharide binding protein. *Science* **249:**1429–1431.

131. **Sibley, C. H.** 1993. Introduction to the 70Z/3 model system, p. 211–220. *In* J. Levin, C. R. Alving, R. S. Munford, and P. L. Stütz (ed.), *Bacterial Endotoxin: Recognition and Effector Mechanisms.* Elsevier Science Publishing BV, Amsterdam.

132. **Simon, D. M., G. Koenig, and G. M. Trenholme.** 1991. Differences in release of tumor necrosis factor from THP-1 cells stimulated by filtrates of antibiotic-killed *Escherichia coli. J. Infect. Dis.* **164:**800–802.

133. **Stefanova, I., M. L. Corcoran, E. M. Horak, L. M. Wahl, J. B. Bolen, and I. D. Horak.** 1993. Lipopolysaccharide induces activation of CD14-associated protein tyrosine kinase p53/56^lyn. *J. Biol. Chem.* **268:**20725–20728.

134. **Stefanova, I., V. Horejsi, I. J. Ansotegui, W. Knapp, and H. Stockinger.** 1991. GPI-anchored cell-surface molecules complexed to protein tyrosine kinases. *Science* **254:**1016–1019.

135. **Stuetz, P. L., H. Aschauer, J. Hildebrandt, C. Lam, H. Loibner, I. Macher, D. Scholz, E. Schuetze, and H. Vyplel.** 1990. Chemical synthesis of endotoxin analogues and some structure activity relationships, p. 129–144. *In* A. Nowotny, J. J. Spitzer, and E. J. Ziegler (ed.), *Cellular and Molecular Aspects of Endotoxin Reactions.* Elsevier Science Publishing BV, Amsterdam.

136. **Sultzer, B. M., J. Bandekar, and R. Castagna.** 1993. Signal transduction in LPS responder and nonresponder cells, p. 285–291. *In* J. Levin, C. R. Alving, R. S. Munford, and P. L. Stütz (ed.), *Bacterial Endotoxin: Recognition and Effector Mechanisms.* Elsevier Science Publishing BV, Amsterdam.

137. **Takada, H., and S. Kotani.** 1992. Structure-function relationships of lipid A, p. 107–134. *In* D. C. Morrison and J. L. Ryan (ed.), *Bacterial Endotoxic Lipopolysaccharides,* vol. I. CRC Press, Inc., Boca Raton, Fla.

138. **Takada, K., N. Ohno, and T. Yadomae.** 1994. Binding of lysozyme to lipopolysaccharide suppresses tumor necrosis factor production in vivo. *Infect. Immun.* **62:**1171–1175.

139. **Takayama, K., D. H. Mitchell, Z. Z. Din, P. Mukerjee, C. Li, and D. L. Coleman.** 1994. Monomeric Re lipopolysaccharide from *Escherichia coli* is more active than the aggregated form in the *Limulus* amebocyte lysate assay and in inducing Egr-1 mRNA in murine peritoneal macrophages. *J. Biol. Chem.* **269:**2241–2244.

140. **Tanke, T., J.-W. van de Loo, H. Rhim, P. S. Leventhal, R. A. Proctor, and P. J. Bertics.** 1991. Bacterial lipopolysaccharide-stimulated GTPase activity in RAW264.7 macrophage membranes. *Biochem. J.* **277:**379–385.

141. **Täuber, M. G., A. M. Shibl, C. J. Hackbarth, J. W. Larrick, and M. A. Sande.** 1987. Antibiotic therapy, endotoxin concentration in cerebrospinal fluid, and brain edema in experimental *Escherichia coli* meningitis in rabbits. *J. Infect. Dis.* **156:**456–462.

142. **Taylor, C. W.** 1990. The role of G-proteins in transmembrane signalling. *Biochem. J.* **272:**1–13.

143. **Tsai, C. M., R. Boykins, and C. E. Frasch.** 1983. Heterogeneity and variation among *Neisseria meningitidis* lipopolysaccharides. *J. Bacteriol.* **155:**498–504.

144. **Tsai, M.-H., A. Hall, and D. W. Stacey.** 1989. Inhibition by phospholipids of the interaction between R-ras, Rho, and their GTPase-activating proteins. *Mol. Cell. Biol.* **9:**5260–5264.

145. **Ullrich, A., and J. Schlessinger.** 1990. Signal

transduction by receptors with tyrosine kinase activity. *Cell* **61**:203–212.

146. **Ulrich, J. T., J. L. Cantrell, G. L. Gustafson, K. R. Myers, J. A. Rudbach, and J. R. Hiernaux.** 1991. The adjuvant activity of monophosphoryl lipid A., p. 133–143. *In* D. P. Spriggs and W. C. Koff (ed.), *Topics in Vaccine Adjuvant Research.* CRC Press, Inc., Boca Raton, Fla.

147. **Van Eschen, K. B.** 1992. Monophosphoryl lipid A and immunotherapy, p. 411–428. *In* J. L. Ryan and D. C. Morrison (ed.), *Bacterial Endotoxic Lipopolysaccharides,* vol. II. CRC Press, Inc., Boca Raton, Fla.

148. **van Putten, J. P.** 1993. Phase variation of lipopolysaccharide directs interconversion of invasive and immuno-resistant phenotypes of *Neisseria gonorrhoeae. EMBO J.* **12**:4043–4051.

149. **Van Snick, J.** 1990. Interleukin-6: an overview. *Annu. Rev. Immunol.* **8**:253–278.

150. **Vogel, S. N., and B. E. Henricson.** 1990. Role of cytokines as mediators of endotoxin-induced manifestations: a comparison of cytokines by doses of LPS and MPL that elicit comparable early endotoxin tolerance, p. 465–474. *In* A. Nowotny, J. J. Spitzer, and E. J. Ziegler (ed.), *Cellular and Molecular Aspects of Endotoxin Reactions.* Elsevier Science Publishing BV, Amsterdam.

151. **Vogel, S. N., and M. M. Hogan.** 1990. Role of cytokines in endotoxin-mediated host responses, p. 238–258. *In* J. J. Openhiem and E. M. Shevach (ed.), *Immunophysiology—the Role of Cells and Cytokines in Immunity and Inflamation.* Oxford University Press, New York.

152. **Vogel, S. N., C. L. Manthey, M. E. Brandes, P. Y. Perera, and C. A. Salkowski.** 1993. LPS mimetic effect of taxol on LPS-inducible gene expression, glucocorticoid receptor expression and tyrosine phosphorylation in murine macrophages, p. 243–253. *In* J. Levin, C. R. Alving, R. S. Munford, and P. L. Stütz (ed.), *Bacterial Endotoxin: Recognition and Effector Mechanisms.* Elsevier Science Publishing BV, Amsterdam.

153. **Warren, H. S., R. L. Danner, and R. S. Munford.** 1992. Anti-endotoxin monoclonal antibodies. *N. Engl. J. Med.* **326**:1153–1156.

154. **Wedegaertner, P. B., D. H. Chu, P. T. Wilson, M. J. Levis, and H. R. Bourne.** 1993. Palmitoylation is required for signaling functions and membrane attachment of Gqa and Gsa. *J. Biol. Chem.* **268**:25001–25008.

155. **Weinstein, S. L., M. R. Gold, and A. L. DeFranco.** 1991. Bacterial lipopolysaccharide stimulates protein tyrosine phosphorylation in macrophages. *Proc. Natl. Acad. Sci. USA* **88**:4148–4152.

156. **Wenzel, R. P.** 1992. Anti-endotoxin monoclonal antibodies—a second look. *N. Engl. J. Med.* **326**:1151–1152.

157. **Westphal, O., U. Westphal, and T. Sommer.** 1978. The history of pyrogen research, p. 221–238. *In* D. Schlessinger (ed.), *Microbiology—1977.* American Society for Microbiology, Washington, D.C.

158. **Wightman, P. D., and C. R. H. Raetz.** 1984. The activation of protein kinase C by biologically active lipid moieties of lipopolysaccharide. *J. Biol. Chem.* **259**:10048–10052.

159. **Wright, S. D., P. A. Detmers, Y. Aida, R. Adamowski, D. C. Anderson, Z. Chad, L. G. Kabbash, and M. J. Pabst.** 1990. CD18-deficient cells respond to lipopolysaccharide *in vitro. J. Immunol.* **144**:2566–2571.

160. **Wright, S. D., S. M. Levin, M. T. Jong, Z. Chad, and L. G. Kabbash.** 1989. CR3 (CD11b/CD18) expresses one binding site for Arg-Gly-Asp-containing peptides and a second site for bacterial lipopolysaccharide. *J. Exp. Med.* **169**:175–183.

161. **Wright, S. D., R. A. Ramos, A. H. Hermanowski-Vosatka, P. Rockwell, and P. A. Detmers.** 1991. Activation of the adhesive capacity of CR3 on neutrophils by endotoxin: dependence on lipopolysaccharide binding and CD14. *J. Exp. Med.* **173**:1281–1286.

162. **Wright, S. D., R. A. Ramos, P. S. Tobias, R. J. Ulevitch, and J. C. Mathison.** 1990. CD14, a receptor for complexes of lipopolysaccharide (LPS) and LPS binding protein. *Science* **249**:1431–1433.

163. **Young, L. S.** 1990. Gram-negative sepsis, p. 611–636. *In* G. L. Mandell, R. G. Douglas, Jr., and J. E. Bennett (ed.), *Principles and Practice of Infectious Disease.* Churchill Livingstone, New York.

164. **Young, L. S., R. A. Proctor, B. Beutler, W. R. McCabe, and J. N. Sheagren.** 1991. University of California/Davis Interdepartmental Conference on gram-negative septicemia. *Rev. Infect. Dis.* **13**:666–687.

165. **Zabel, P., F. U. Schade, and M. Schlaak.** 1993. Modulating effects of pentoxyfylline on cytokine release syndromes, p. 413–421. *In* J. Leven, C. R. Alving, R. S. Munford, and P. L. Stütz (ed.), *Bacterial Endotoxin: Recognition and Effector Mechanisms.* Elsevier Science Publishing BV, Amsterdam.

166. **Ziegler, E. J., C. J. Fisher, C. L. Sprung, R. C. Straube, J. C. Sadoff, G. E. Foulke, C. H. Wortel, M. P. Fink, R. P. Dellinger, N. N. H. Teng, I. E. Allen, H. J. Berger, G. L. Knatterud, A. F. L. Buglio, C. R. Smith, and the HA-1A Sepsis Study Group.** 1991. Treatment of gram-negative bacteremia and septic shock with HA-1A human monoclonal antibody against endotoxin. *N. Engl. J. Med.* **324**:429–436.

PHYLOGENETIC ANALYSES OF THE
RTX TOXIN FAMILY

Rodney A. Welch

12

In this brief review, I describe the members and genetic organization of the repeats in toxin (RTX) exoprotein family, which is found among numerous genera of gram-negative bacteria. A short summary of the biology and structural elements of the RTX toxins is presented. I also derive phylogenies for the pore-forming-toxin branch of the RTX exoprotein family that are based on amino acid sequence alignments of the respective proteins encoded by the RTX operons. From these phylogenetic analyses I derive several new hypotheses to explain the vertical as well as the apparent horizontal evolutionary events that have led to the present-day RTX exotoxin family.

RTX EXOPROTEIN FAMILY

Several different families of exoproteins among gram-negative eubacteria are organized on the basis of their clear protein sequence homologies and their similarity in operon arrangement and secretion pathways (45). One of the best-characterized exoprotein families is the RTX exoprotein family, which includes cytolytic toxins with either broad or narrow target cell reactivities (hemolysins versus leukotoxins)

(69), metallo–dependent proteases (15), lipases (60), nodulation-associated proteins (16), and a meningococcal exoprotein of unknown function (64). Table 1 provides a list of the known RTX exoproteins. The RTX family is distinguished from other exoprotein groups by the following common traits: (i) there is no cleavage of an amino-terminal leader peptide from the exported protein during secretion of the protein across the cytoplasmic membrane (17); (ii) a structure within the carboxy terminus of the exported RTX protein "targets" that protein for export (49); (iii) a specific RTX secretory apparatus consists of a peripheral inner membrane protein (termed the RTX B protein), which provides energy for RTX export through ATP hydrolysis (33, 66) and a transmembrane channel consisting of an RTX D protein (55) and an outer membrane TolC-like protein (65); (iv) there is no detectable pool of RTX A in the periplasm during export (19, 61); and (v) tandem repeats consisting of the signature RTX nonamer are amino proximal to the export targeting structure (18, 71).

The members of the RTX exoprotein family that are distinguished by the diverse activities described in Table 1 are differentiated from one another by the absence of amino acid sequence similarity in the amino-terminal regions proximal to the signature repeats. In

Rodney A. Welch   Department of Medical Microbiology and Immunology, University of Wisconsin—Madison Medical School, 481 Medical Sciences Building, 1300 University Avenue, Madison, Wisconsin 53706.

Virulence Mechanisms of Bacterial Pathogens, 2nd ed., Edited by J. A. Roth et al.
© 1995 American Society for Microbiology, Washington, D.C.

TABLE 1 Members of the RTX exoprotein family

Function, genus, and species	Activity	RTX *A* gene product	Reference
Toxins			
Escherichia coli	Hemolysin	HlyA	18
Pasteurella haemolytica	Leukotoxin	LktA	11, 29, 43
Actinobacillus actinomycetemcomitans	Leukotoxin	AaltA	31, 37
Bordetella pertussis	Adenylate cyclase-hemolysin	CyaA	23
Actinobacillus pleuropneumoniae	Hemolysin	ApxIA	22
	Leukotoxin	ApxIIA	13
	Leukotoxin	ApxIIIA	46
Actinobacillus suis	Leukotoxin	AshA	8
Proteus vulgaris	Hemolysin	PvxA	32
Morganella morganii	Hemolysin	MmxA	68
Pasteurella haemolytica-like	Leukotoxin	PlktA	10
Moraxella bovis	Hemolysin	?	44
Escherichia coli	Enteroaggregative exotoxin	?	2
	Enterohemorrhagic hemolysin	?	54
Proteases			
Serratia marcescens	Metalloprotease	PrtSM	40, 58
Erwinia chrysanthemi	Metalloprotease	PrtB, PrtC	39
Pseudomonas aeruginosa	Alkaline protease	AprA	50
Nodulation			
Rhizobium leguminosarum	?	NodO	16
Lipases			
Pseudomonas fluorescens	Lipase	LipA	60
Serratia marcescens	Lipase	LipA	1
Unknown			
Neisseria meningitidis	?	FrpA, FrpC	62, 63

most instances, however, the conserved *B* and *D* RTX export genes are located 3′ to the RTX *A* genes, with a rho-independent terminator separating the *A* gene from the *B* and *D* genes (18). Therefore, the fundamental RTX genetic framework in 5′-to-3′ order as shown in Fig. 1 is as follows: *A*-gene repeats–*A*-gene targeting sequence–rho-independent terminator–*B* gene–*D* gene. On either the 5′ or the 3′ side of this scaffold, other genes are added or gene fusions of distinct functional cassettes are brought to the 5′ sides of the *A*-gene repeats.

RTX B EXPORTER PROTEINS ARE A DISTINCT BRANCH OF THE ABC SUPERFAMILY OF MEMBRANE TRANSPORTERS

During the past decade, the discovery was made that prokaryotic and eukaryotic cells contain a widespread, ubiquitous plasma membrane protein family of probably a single homologous origin that shares the motifs of amino-terminal transmembrane domains and a carboxy-terminal cytoplasmic region containing an ATP-binding consensus sequence (28). All of the members are involved in transport of hydrophilic molecules across the plasma membrane. The members can be grouped on the basis of the type of molecule they transport and whether the molecules are imported or exported from the cell. For the RTX exoprotein branch of the ATP-binding cassette (ABC) superfamily, the cytoplasmic-membrane RTX B protein facilitates extracellular secretion of the RTX A protein, presumably by providing energy for the secretory process (34). What specific event along the export pathway is accomplished by this expenditure of energy is unknown. There is genetic evidence that RTX B proteins directly interact with the A-protein

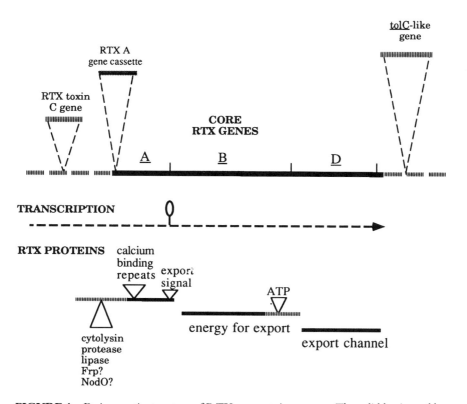

FIGURE 1 Basic genetic structure of RTX exoprotein operons. The solid horizontal bar at the top represents the cores of homologous RTX *A, B,* and *D* gene sequences and their transcriptional order. This arrangement is common to nearly all of the RTX exoproteins listed in Table 1 (the exception is ApxII). Shaded, dashed lines to the left and right of the solid horizontal bar depict the regions where different genes or fragments of genes are fused to the core RTX operon. Insertions of genes into the RTX operon are depicted as shaded lines lying above the genetic map. The dashed line below the gene map shows the direction of transcription and the presence of a predicted RNA stem-loop structure between the *A* and *B* genes common to the RTX operons. Labeled below as RTX proteins are the products of the core RTX genes. Their functions are listed above or below the individual bars representing the different RTX proteins. The question marks beside Frp and NodO indicate that the functions of the amino acid sequences upstream of the repeat region in the *Neisseria menigitidis* Frp and the *Rhizobium leguminosarum* NodO proteins are unknown.

targeting structure; i.e., extragenic suppressor mutants in the *Escherichia coli* hemolysin B protein that relieve an export defect in the carboxy terminus of HlyA have been isolated (72).

RTX D EXPORTER PROTEINS ARE MEMBERS OF THE MFP FAMILY

Dinh and colleagues recently reported the discovery of the membrane fusion protein (MFP) family of proteins, within which the RTX D transport proteins represent a distinct subcluster (14). These proteins putatively span the inner and outer membranes of gram-negative bacteria. Besides the RTX D-protein transporters, this family includes drug and mutagen transporters in *E. coli* (EmrA and AcrA, respectively), the putative lipooligosaccharide transporter NolF of *Rhizobium meliloti,* and the toxic metal transporters CzcB and CnrB of *Alcaligenes* species. Although the overall amino acid sequence identity within the family is low (the most distantly related proteins, HlyD and

EmrA, have only 18% identity), certain structural and functional features suggest that these proteins are homologous. For example, all of the proteins possess a highly hydrophobic stretch of amino acids from approximately positions 80 through 150 and a second, less hydrophobic region from residues 320 through 420. In the amino acid sequence alignments of MFP family members, these two regions also have the greatest scores for sequence similarity. In addition, each member of the MFP family is functionally associated with an inner membrane protein believed to have the ability to hydrolyze ATP. These inner membrane proteins share a common ATP-binding motif within a carboxy-terminal cytoplasmic domain but are not necessarily all members of the aforementioned ABC family. Thus, Dinh and colleagues propose that the MFPs span the inner and outer membranes via a membrane fusion-like process. The MFPs then transport hydrophilic molecules within a direct MFP pore, with an inner membrane protein providing energy for the vectorial transport process. Their hypothesis is consistent with the data presently available concerning secretion of the E. coli hemolysin by HlyB and HlyD (33). If a third gene product, TolC or its equivalent, is directly a part of the RTX export complex, it would be interesting to know whether the non-RTX members of the MFP family also require TolC-like proteins for their respective transport functions.

BIOLOGY OF THE RTX EXOTOXINS
The RTX toxins in general are >300 kDa in molecular size (70). They are composed of an unknown number of a single species of polypeptide (RTX A gene products). The RTX toxin A polypeptides range in length from 953 to 1,706 amino acids (23, 42). The purification and chemical analysis of homogeneously pure and active material have not been conclusively accomplished, but for at least one member of the RTX toxin family, the E. coli hemolysin, such attempts suggest that besides a polypeptide component to the toxin, lipopolysaccharide (LPS) may be present (6, 7, 51). Evidence that

lends support to the hypothesis that hemolysin is a protein-LPS complex comes from the observation that E. coli and Salmonella typhimurium mutants in the synthesis of inner and outer portions of LPS are affected in the synthesis and activity of hemolysin (9, 57). The RTX C gene, which is always linked 5′ to the A gene, encodes a product that participates in the activation of the RTX A toxin polypeptide. In the case of the E. coli hemolysin, the activation amounts to an acylation, possibly at one or more lysine residues (25). The last known components of the E. coli hemolysin cytolytic complex are calcium ions, which bind in an unknown stoichiometry to the tandem nonameric repeats that are common to all of the RTX toxins (5). Calcium appears to be a common requirement for the cytotoxic activities of RTX exotoxins (69).

One of the intriguing evolutionary questions about the RTX toxin family is how they came to have different patterns of host and cell type cytotoxic specificity. The RTX toxins with erythrolytic capabilities, such as the E. coli hemolysin and the Actinobacillus pleuropneumoniae ApxIA toxin, are also cytotoxic to phagocytic cells (22, 69). The E. coli hemolysin is active against such cells regardless of whether they are murine, ruminant, or primate in origin (69). That broad target cell reactivity stands in stark contrast to that of the homologous RTX toxins, the Pasteurella haemolytica and Actinobacillus actinomycetemcomitans leukotoxins. These have little erythrolytic activity but a potent cytotoxic activity toward phagocytic cells of a particular animal origin, with ruminant and some primate cells being the respective susceptible hosts (69). The structural basis for toxin cellular specificity was examined through the construction of hybrid toxin genes by Forestier and Welch (20) and McWhinney et al. (48). The results from these experiments are not directly comparable, because different pairs of RTX toxin genes were used by the two laboratories. The hybrid-gene approach successfully demonstrated that discrete exchanged portions of hemolysins expressed their erythrolytic activities within hybrids that were mostly leukotoxin gene sequences. An unresolved problem with the studies was that each laboratory concluded that different portions of the respective RTX hemolysins conferred erythrolytic activity (20, 48). Forestier and Welch showed that a class of hybrid toxins had erythrolytic activity but no cytotoxic activity against either the nucleated BL3 (bovine) or the Raji (human) cell targets (20). The activities of these hybrids indicated that RTX A-protein structures responsible for lysis of erythrocytes are potentially separable from those needed for cytotoxicity against nucleated cells. More recent studies also followed that line of reasoning. E. coli hemolysin mutants altered in the repeat region were constructed, because hybrids with the erythrolytic-positive, leukotoxic-negative phenotype (Ery⁺ Lko⁻) have the common trait of a hybrid gene fusion joint within the repeats (53). Two different Ery⁺ Lko⁻ mutants with in-frame linker insertions at two different points in the repeat area were isolated. Analysis of the calcium requirement for lytic activity revealed that compared to organisms with the wild-type hemolysin, the insertion mutants needed 10 to 50 times more calcium in the assay medium (>1 mM) for maximal erythrolytic or leukotoxic activity. These results meant that the Ery⁺ Lko⁻ phenotype previously described was misleading (20). The 10 mM calcium in the standard erythrolytic assays was adequate for maximal erythrolytic activity for the Ery⁺ Lko⁻ mutants, but in the chromium release assays for leukotoxicity, the calcium concentration ranged from 0.2 to 0.25 mM. This was sufficient calcium for wild-type-toxin leukotoxic activity, but it was insufficient for these particular mutants. In a subsequent reexamination of the calcium requirements of the hybrid toxins that possessed the Ery⁺ Lko⁻ phenotype (52), several of the hybrids had increased leukotoxicity with 1 mM calcium present, but several others did not. This observation leaves some uncertainty as to the complete explanation for the Ery⁺ Lko⁻ mutant phenomenon.

There remain the issues of what the RTX export machinery has as an absolute requirement for recognition in RTX A proteins and what is also needed for an efficient export process. It is apparent that as few as 27 to 53 residues of the carboxy terminus are required for detectable E. coli HlyA export into the medium (26, 47, 56). The carboxy terminus was confirmed to be required for extracellular secretion of HlyA, but an internal deletion of HlyA that removed 11 of the 13 tandem nonameric repeats caused a significant decrease in the secretion rate of HlyA into the medium (19), which suggests that the intact tandem array of RTX repeats aids the secretion process in some unknown way. Data supporting this contention came from work involving the E. coli HlyA and the Bordetella pertussis CyaA proteins in gene fusion systems for export of heterologous proteins (41). The removal of the encoding portion of the repeats in the gene fusion system remarkably impeded the extracellular secretion of different fusion proteins.

These data may be interpreted from an evolutionary viewpoint as well. As previously mentioned and as depicted in Fig. 1, there is an apparent cassettelike or mosaic pattern of the different RTX A proteins. All of them have retained both the repeat region and a less well conserved carboxy-terminal sequence absolutely required for export. There is little homology among the different classes of RTX exoproteins immediately amino terminal to the RTX repeats. This fact suggests that RTX A protein interaction with the products of the linked B and D transport genes requires retention of both conserved RTX A-protein features. The early genetic dissection of the HlyA export signal possibly caused an erroneous presumption that the RTX repeats represent a protein domain separable from the very carboxy-terminal export sequence. This error has been perpetuated by the recognition that a common requirement for all RTX exoproteins is calcium and that the repeats represent the region in which calcium binds to the protein (4, 5). It is also suggested that besides the calcium-binding function, a selection pressure for maintenance of the RTX repeats exists, because without its repeats, an RTX exoprotein will not be adequately expressed at the site where it is needed.

EVOLUTION OF THE RTX TOXIN BRANCH OF THE RTX EXOPROTEIN FAMILY

The predicted amino acid sequences for all of the available RTX toxin C, A, B, and D proteins were aligned by the Clustell algorithm for multiple sequence alignments present in the DNASTAR package (DNASTAR, Madison, Wisc.). Maximum parsimony analyses of the aligned amino acid sequences that used a heuristic tree search and bootstrap analysis were performed with the PAUP package (59). The monophylogeny of derived trees was assessed by 100 replications of the bootstrap method. The nodes present in all of the phylogenetic trees possessed bootstrap values greater than 50%. The trees for each RTX protein are shown in Fig. 2. The trees for the RTX C and A proteins are superimposable on one another, but despite the common linkage of *CABD* genes among most RTX toxin determinants and the invariant linkage of *B* and *D* genes, the toxin B and D phylogenetic trees are not similar to the toxin C and A trees. Among the RTX C, A, B, and D trees, the different apparent phylogenies arise from shifts in branch nodes involving the different degrees of relatedness of the three *A. pleuropneumoniae* Apx determinants to one another and to the other RTX toxin proteins. Judging from the RTX toxin C and A trees, there appear to be two major trunks and a solitary branch to the RTX

toxin family tree, with *E. coli* hemolysin, *A. pleuropneumoniae* ApxI and ApxIII, and *A. actinomycetemcomitans* leukotoxin composing one of the trunks and *Pasteurella haemolytica* leukotoxin, *P. haemolytica*-like leukotoxin, *Actinobacillus suis* Ash, and *A. pleuropneumoniae* ApxII determinants on the other trunk. The *B. pertussis* adenylate cyclase-hemolysin represents a single branch that is nearly equally divergent from the two major branches. If the phylogenetic analyses are based on the toxin B or D tree, a different conclusion about the clustering of family members can be made. The branches of the B tree are relatively shorter in genetic distance than those of the C, A, and D trees, which means that there is less confidence in the validity of the branch nodes for that tree. The bootstrapping of that tree nevertheless indicates that the tree represents a robust phylogeny. Therefore, considering the patterns for the toxin C and A trees, it is remarkable that the ApxIIIB sequence appears to be more closely related to the *P. haemolytica* LktB protein sequence than the ApxIIIC and ApxIIIA sequences are related to the LktC and LktA sequences, respectively. The most striking shift occurs with the ApxID sequences. According to the RTX toxin C and A trees, ApxI is closely related to the Hly and Aalt cluster, but in the RTX toxin D tree, ApxI almost appears to be the RTX determinant most distantly related to that particular grouping.

Frey et al. summarized the occurrence of the different full and partial *apx* operons among the type isolates for each of the *A. pleuropneumoniae*

FIGURE 2 Phylogeny of RTX toxin C, A, B, and D proteins. Phylograms were derived from the PAUP software package (59), using Clustell alignments of the predicted amino acid sequences of the RTX toxin C, A, B, and D proteins. Numbers above the branches in each tree represent the mean genetic distance as calculated by the PAUP package. Lengths of the branches are proportional to the genetic distances in each individual phylogeny. Gene product designations are identical to those listed in Table 1. (A) Toxin C. (B) Toxin A. In addition to the RTX toxin proteins listed in Table 1, the LktA1, LktA11, and LktA3 proteins, which represent the LktA sequences present in GenBank that were derived from *P. haemolytica* serotypes 1, 11, and 3, respectively, are shown. CyaA-N term. del. indicates that the amino-terminal 300 amino acids of the predicted CyaA sequence (23) were removed in the RTX A toxin alignment that resulted in the RTX A phylogeny. (C) Toxin B. PvxB represents the predicted amino acid sequence of the *Proteus vulgaris* B gene product (35). (D) Toxin D. In panels C and D, Hly-plasmid and Hly-chromosome indicate that the *E. coli* sequences used in these phylogenies were derived from *hly* operons present on a plasmid in a murine fecal isolate (27) and on the chromosome of a human uropathogenic strain (18), respectively.

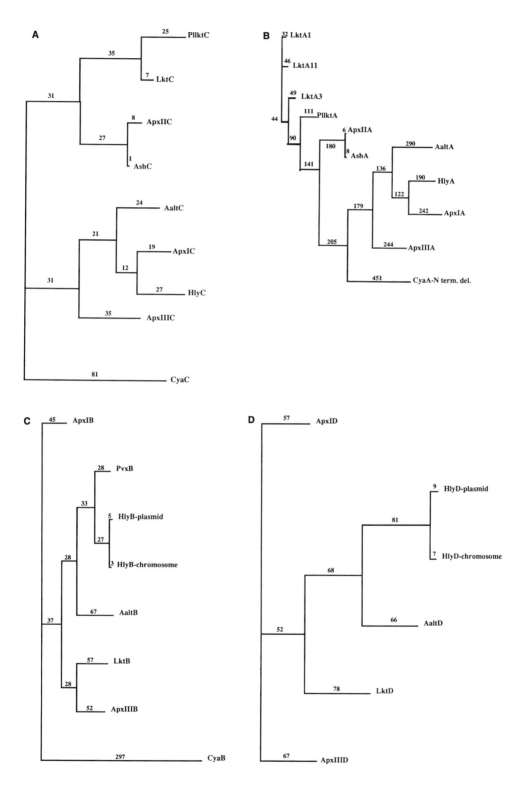

serotypes (21). In summary, the *apxICABD* and *apxIIICABD* operons are exclusive of one another, and the *apxIICA* determinant, which is suspected to have arisen by deletion of the *apxIIBD* genes, occurs in all but one serotype. A new point to be made is that in the form of either *apxICABD* or *apxIBD,* the *apxIBD* genes are in all but one of the *A. pleuropneumoniae* serotypes. Therefore, the most common *rtx* genes in *A. pleuropneumoniae* are *apxIICA* and *apxIBD*. The phylogenetic data indicate that the ApxID sequence is more related to LktD and ApxIIID than to HlyD and AaltD. This suggests that the full-length *apxICABD* operon arose from a recombination event that substituted an *hlyCA*-like sequence into an ancient progenitor of the *apxCABDII-apxCABDIII* cluster. The phylogenies imply that the full-length *apxICABD* operon may be a hybrid operon.

Chang et al. suggested that the *apxIBD* configuration arose by deletion of the upstream *apxIC* and *A* sequences (13). The evidence for the deletion leaving the *apxIBD* determinant was that the region 5′ to *apxIB* possessed fragments of an *apxIA* 3′ sequence that could encode amino acid sequences more similar to HlyA than to LktA.

Perhaps a single, primordial full-length RTX determinant in *A. pleuropneumoniae* was the progenitor of the *apxIICABD-apxIIICABD* operons. A duplication of the determinant led to a second *apx* locus that subsequently diverged to yield distinct *apxIICABD* and *apxIIICABD* operons. For the *hlyCABD* determinant in *E. coli,* the precedent in RTX biology that second copies of *rtx* operons do exist in the chromosome has been established (30). These multiple copies probably occurred by either straightforward gene duplications or horizontal acquisition of second *hly* operons. The *apxIICABD* operon that arose in this scenario of duplication and divergence must have undergone deletional loss of the *B* and *D* genes that resulted in the present-day *apxIICA* determinant. Because the *apxIICA* locus is so common, these events would have had to oc-

cur very early in the evolution of *A. pleuropneumoniae,* before the divergence of that organism into many different serotypes.

The general pattern of the *B. pertussis cya* genes, representing the outlying, most distantly related RTX family member, is consistent in each RTX toxin gene product phylogeny. That is expected, given that the *cya* locus is different from the other RTX toxin determinants in at least five distinct ways. First, the guanosine-plus-cytosine content of the *cyaA* gene is 66.6% instead of the general pattern of 38 to 40% for the other RTX toxin *A* genes (69). Second, as mentioned above, the *B. pertussis* equivalent of *tolC, cyaE,* is physically linked 3′ to the cotranscribed *cyaD* gene (24). For the other RTX toxin operons, the *tolC* equivalent is at a distant unlinked locus. Third, unlike the other RTX toxin operons, in which the *C* and *A* genes are linked in the 5′-to-3′ direction, the *cyaC* gene is 5′ to *cyaA* but is transcribed from the DNA strand opposite to *cyaA* (3). Fourth, the RTX-like *cyaA* gene is disrupted at its 5′ end by the fusion of a gene sequence that encodes a calmodulin-dependent adenylate cyclase activity (23). Last, the *cyaABDE* genes are under the regulatory control of a global effector *bvg* (38). There does not appear to be a similar form of regulation for the other RTX toxin loci.

The *E. coli* Hly and *A. actinomycetemcomitans* AaltC, -A, -B, and -D sequences cluster together in each of the four phylogenetic trees. This fact indicates that they share with each other a more recent common ancestor than they share with the members of the *P. haemolytica* leukotoxin branch of the RTX toxin family. This fact is probably most significant in efforts to understand the evolution of RTX toxin target specificity. Aside from the *B. pertussis* adenylate cyclase-hemolysin, only two RTX toxins have unambiguous broad cell reactivity: the *E. coli* hemolysin and the *A. pleuropneumoniae* ApxIA hemolysin (69). Because they are found clustered with the *A. actinomycetemcomitans* AaltA, with its narrow specificity toward human and Old World ape leukocytes, the questions arise of how and when narrow versus broad target cell and different host cell specificities occurred in the RTX toxin family. The differences in types of hosts differentiate the narrow specificity of the *P. haemolytica* leukotoxin group of toxins from the specificity of the *A. actinomycetemcomitans* leukotoxin. The application of Occam's razor to this situation is difficult, because the simplest explanation is elusive. The branching of the RTX A protein phylogeny does not provide adequate clues about the ordering of events that led to the specificity for leukocytes or to different host type specificities. The broad cell reactivity of the *E. coli* hemolysin and the *A. pleuropneumoniae* ApxIA hemolysin was probably a relatively recent event because this activity does not occur in the *P. haemolytica* leukotoxin branch of the RTX toxin family. Hybrid gene construction and mutant analysis will perhaps tell us just how many substitutions are needed to derive toxins that exchange target cell specificities. This knowledge will lead to an understanding of how much sequence divergence or recombination would have had to occur to make a leukotoxin become a hemolysin or vice versa.

Early during the exploration of the RTX family, investigators perceived that B-protein sequences are considerably less variable than RTX C-, A-, and D-protein sequences (10, 12, 36). This fact suggested that there are greater functional constraints on RTX B than on other RTX proteins (69). Alternatively, there may have been immune selection against the surface-expressed and -secreted RTX toxin A proteins. Thus, negative selection caused greater divergence among RTX A-protein sequences than among RTX B protein sequences. This finding presumes that there is little selection against RTX toxin B proteins because they are not expressed on the cell surface. That possibility, when combined with some unknown B functional constraints, may explain the relatively greater B-sequence conservation compared to that of the A proteins. The RTX toxin C and D proteins have sequence differences intermediate in degree compared with the extremes of the A and B proteins. Perhaps some of the differences in the RTX toxin C proteins are driven by differences in the RTX toxin A proteins. This possibility would be logical if C proteins have to directly interact with A proteins in a complex during acylation and activation. It will be interesting to test this hypothesis by observing whether second-site intergenic suppressor mutants of HlyC that relieve mutations in HlyA that are negatively affected in the acylation process can be isolated.

SUMMARY

The RTX toxin family represents the largest and most ubiquitous bacterial exotoxin family that is known. There is little question that research on the different RTX toxins will yield significant progress toward understanding the pathogenesis of important human and animal diseases. We can look forward to the phylogenetic analyses of additional RTX toxin genes, which for the first time have been found in human diarrheal pathogens. Sequence information for the new RTX determinants discovered in enterohemorrhagic *E. coli* O157 strains (54) and enteroaggregative *E. coli* strains should soon be available (2). Neither of these two types of *E. coli* is overtly beta-hemolytic when cultured on sheep erythrocyte agar plates (67). This lack may create the interesting situation of discovering *E. coli* RTX toxins that have narrower target cell specificities than the *E. coli* hemolysin. If such a discovery is made, the resulting data when combined with results from additional RTX *A* hybrid gene constructions and mutant isolations will give us a better understanding of the structure, function, and evolution of the RTX toxins.

ACKNOWLEDGMENTS

The work described in this review was funded by USDA grant 91-37204-6735, Public Health Service grant AI20323, and a Romnes Fellowship from the University of Wisconsin.

I thank Margaret Bauer for her helpful suggestions with the manuscript. I also thank Peg Riley for her infectious energy and instruction with the phylogenetic analyses using the PAUP software.

REFERENCES

1. **Akatsuka, H., E. Kawai, K. Omori, S. Komatsubara, T. Shibatini, and T. Tosa.** 1994. The *lipA* gene of *Serratia marcescens*. *J. Bacteriol.* **176:**1949–1956.

2. **Baldwin, T. J., S. Knutton, L. Sellers, H. A. M. Hernandez, A. Aitken, and P. H. Williams.** 1992. Enteroaggregative *Escherichia coli* strains secrete a heat-labile toxin antigenically related to *E. coli* hemolysin. *Infect. Immun.* **60:**2092–2095.

3. **Barry, E. M., A. A. Weiss, I. E. Ehrmann, M. C. Gray, E. Hewlett, and M. S. Goodwin.** 1991. *Bordetella pertussis* adenylate cyclase toxin and hemolytic activities require a second gene, *cyaC,* for activation. *J. Bacteriol.* **173:**720–726.

4. **Baumann, U., S. Wu, K. M. Flaherty, and D. McKay.** 1993. Three-dimensional structure of the alkaline protease of *Pseudomonas aeruginosa*: a two-domain protein with a calcium binding parallel beta roll motif. *EMBO J.* **12:**3357–3364.

5. **Boehm, D. F., R. A. Welch, and I. S. Snyder.** 1990. Calcium is required for binding of *Escherichia coli* hemolysin (HlyA) to erythrocyte membranes. *Infect. Immun.* **58:**1951–1958.

6. **Bohach, G., and I. S. Snyder.** 1986. Composition of affinity-purified alpha-hemolysin of *Escherichia coli*. *Infect. Immun.* **53:**435–437.

7. **Bohach, G. A., and I. S. Snyder.** 1985. Chemical and immunological analysis of the complex structure of *Escherichia coli* alpha-hemolysin. *J. Bacteriol.* **164:**1071–1080.

8. **Burrows, L. L., and R. Y. C. Lo.** 1992. Molecular characterization of an RTX toxin determinant from *Actinobacillus suis*. *Infect. Immun.* **60:**2166–2173.

9. **Camprubi, S., J. Tomas, F. Munoa, C. Madrid, and A. Juarez.** 1990. Influence of lipopolysaccharide on external hemolytic activity of *Salmonella typhimurium* and *Klebsiella pneumoniae*. *Curr. Microbiol.* **20:**1–3.

10. **Chang, Y., D. Ma, J. Shi, and M. M. Chengappa.** 1993. Molecular characterization of a leukotoxin gene from a *Pasteurella haemolytica*-like organism, encoding a new member of the RTX toxin family. *Infect. Immun.* **61:**2089–2095.

11. **Chang, Y.-F., R. Young, D. Post, and D. K. Struck.** 1987. Identification and characterization of the *Pasteurella haemolytica* leukotoxin. *Infect. Immun.* **55:**2348–2354.

12. **Chang, Y.-F., R. Young, and D. K. Struck.** 1989. Cloning and characterization of a hemolysin gene from *Actinobacillus (Haemophilus) pleuropneumoniae*. *DNA* **8:**635–647.

13. **Chang, Y.-F., R. Young, and D. K. Struck.** 1991. The *Actinobacillus pleuropneumoniae* hemolysin determinant: unlinked *appCA* and *appBD* loci flanked by pseudogenes. *J. Bacteriol.* **173:**5151–5158.

14. **Dinh, T., I. T. Paulsen, and M. H. Saier.** 1994. A family of extracytoplasmic proteins that allow transport of large molecules across the outer membranes of gram-negative bacteria. *J. Bacteriol.* **176:**3825–3831.

15. **Duong, F., A. Lazdunski, B. Cami, and M. Murgier.** 1992. Sequence of a cluster of genes controlling synthesis and secretion of alkaline protease in *Pseudomonas aeruginosa*: relationships to other secretory pathways. *Gene* **121:**47–54.

16. **Economou, A., W. D. O. Hamilton, A. W. B. Johnston, and J. A. Downie.** 1990. The *Rhizobium* nodulation gene *nodO* encodes a Ca^{2+}-binding protein that is exported without N-terminal cleavage and is homologous to haemolysin and related proteins. *EMBO J.* **9:**349–354.

17. **Felmlee, T., S. Pellett, E. Y. Lee, and R. A. Welch.** 1985. The *Escherichia coli* hemolysin is released extracellularly without cleavage of a signal peptide. *J. Bacteriol.* **163:**88–93.

18. **Felmlee, T., S. Pellett, and R. A. Welch.** 1985. The nucleotide sequence of an *Escherichia coli* chromosomal hemolysin. *J. Bacteriol.* **163:**94–105.

19. **Felmlee, T., and R. A. Welch.** 1988. Alterations of amino acid repeats in the *Escherichia coli* hemolysin affect cytolytic activity and secretion. *Proc. Natl. Acad. Sci. USA* **85:**5269–5273.

20. **Forestier, C., and R. A. Welch.** 1991. Identification of RTX toxin target cell specificity domains by use of hybrid genes. *Infect. Immun.* **59:**4212–4220.

21. **Frey, J., J. T. Bosse, Y. F. Chang, J. M. Cullen, B. Fenwick, G. F. Gerlach, D. Gygi, F. Haesebrouck, T. J. Inzana, R. Jansen, E. M. Kamp, J. MacDonald, J. I. MacInnes, K. R. Mittal, J. Nicolet, A. Rycroft, R. P. Segars, M. A. Smits, E. Stegenbaek, D. K. Struck, J. F. V. D. Bosch, P. J. Willison, and R. Young.** 1993. *Actinobacillus pleuropneumoniae* RTX-toxins: uniform designation of haemolysins, cytolysins, pleurotoxin and their genes. *J. Gen. Microbiol.* **139:**1723–1728.

22. **Frey, J., R. Meier, D. Gygi, and J. Nicolet.** 1991. Nucleotide sequence of the hemolysin I gene from *Actinobacillus pleuropneumoniae*. *Infect. Immun.* **59:**3026–3032.

23. **Glaser, P., D. Ladant, O. Sezer, F. Pichot, A. Ullman, and A. Danchin.** 1988. The calmodulin-sensitive adenylate cyclase of *Bordetella pertussis*: cloning and expression in *Escherichia coli*. *Mol. Microbiol.* **2:**19–30.

24. **Glaser, P., H. Sakamoto, J. Bellalou, A. Ull-**

mann, and A. Danchin. 1988. Secretion of cyclolysin, the calmodulin-sensitive adenylate cyclase-haemolysin bifunctional protein of *Bordetella pertussis. EMBO J.* 7:3997–4004.

25. Hardie, K. R., J. P. Issartel, E. Koronakis, C. Hughes, and V. Koronakis. 1991. *In vitro* activation of *Escherichia coli* prohaemolysin to the mature membrane-targeted toxin requires HlyC and a low molecular weight cytosolic polypeptide. *Mol. Microbiol.* 5:1669–1679.

26. Hess, J., I. Gentschev, W. Goebel, and T. Jarchau. 1990. Analysis of the haemolysin secretion system by PhoA-HlyA fusion proteins. *Mol. Gen. Genet.* 224:201–208.

27. Hess, J., W. Wels, M. Vogel, and W. Goebel. 1986. Nucleotide sequence of a plasmid-encoded hemolysin determinant and its comparison with a corresponding chromosomal hemolysin sequence. *FEMS Microbiol. Lett.* 34:1–11.

28. Higgins, C. F. 1992. ABC transporters: from microorganism to man. *Annu. Rev. Cell Biol.* 8:67–113.

29. Highlander, S. K., M. Chidambaram, M. J. Engler, and G. M. Weinstock. 1989. DNA sequence of the *Pasteurella haemolytica* leukotoxin gene cluster. *DNA* 8:15–28.

30. Knapp, S., J. Hacker, I. Then, D. Muller, and W. Goebel. 1984. Multiple copies of hemolysin genes and associated sequences in the chromosomes of uropathogenic *Escherichia coli* strains. *J. Bacteriol.* 159:1027–1033.

31. Kolodrubetz, D., T. Dailey, J. Ebersole, and E. Kraig. 1989. Cloning and expression of the leukotoxin gene from *Actinobacillus actinomycetemcomitans. Infect. Immun.* 57:1465–1469.

32. Koronakis, V., M. Cross, B. Senior, E. Koranakis, and C. Hughes. 1987. The secreted hemolysins of *Proteus mirabilis, Proteus vulgaris,* and *Morganella morganii* are genetically related to each other and to the alpha-hemolysin of *Escherichia coli. J. Bacteriol.* 169:1509–1515.

33. Koronakis, V., C. Hughes, and E. Koronakis. 1991. Energetically distinct early and late stages of HlyB/HlyD-dependent secretion across both *Escherichia coli* membranes. *EMBO J.* 10:3263–3272.

34. Koranakis, V., C. Hughes, and E. Koranakis. 1993. ATPase activity and ATP/ADP-induced conformational change in soluble domain of the bacterial protein translocator. *Mol. Microbiol.* 8:1163–1175.

35. Koronakis, V., E. Koronakis, and C. Hughes. 1988. Comparison of the haemolysin secretion protein HlyB from *Proteus vulgaris* and *Escherichia coli;* site-directed mutagenesis causing impairment of export function. *Mol. Gen. Genet.* 213:551–555.

36. Kraig, E., T. Dailey, and D. Kolodrubetz.

1990. Nucleotide sequence of the leukotoxin gene from *Actinobacillus actinomycetemcomitans:* homology to the alpha-hemolysin/leukotoxin gene family. *Infect. Immun.* 58:920–929.

37. Lally, E., I. Kieba, D. Demuth, J. Rosenbloom, N. Taichman, and C. Gibson. 1989. Identification and expression of the *Actinobacillus actinomycetemcomitans* leukotoxin gene. *Biochem. Biophys. Res. Commun.* 159:256–262.

38. Laoide, B. M., and A. Ullmann. 1990. Virulence dependent and independent regulation of the *Bordetella pertussis cya* operon. *EMBO J.* 9:999–1005.

39. Letoffe, S., P. Delepelaire, and C. Wandersman. 1990. Protease secretion by *Erwinia chrysanthemi:* the specific secretion functions are analogous to those of *Escherichia coli* alpha-haemolysin. *EMBO J.* 9:1375–1382.

40. Letoffe, S., P. Delepelaire, and C. Wandersman. 1991. Cloning and expression in *Escherichia coli* of the *Serratia marcescens* metalloprotease gene: secretion of the protease from *E. coli* in the presence of the *Erwinia chrysanthemi* protease secretion functions. *J. Bacteriol.* 173:2160–2166.

41. Letoffe, S., and C. Wandersman. 1992. Secretion of CyaA-PrtB and HlyA-PrtB fusion proteins in *Escherichia coli:* involvement of the glycine-rich repeat domain of *Erwinia chrysanthemi* protease B. *J. Bacteriol.* 174:4920–4927.

42. Lo, R. Y. C., P. Shewen, C. Strathdee, and C. N. Greer. 1985. Cloning and expression of the leukotoxin gene of *Pasteurella haemolytica* A1 in *Escherichia coli* K-12. *Infect. Immun.* 50:667–671.

43. Lo, R. Y. C., C. Strathdee, and P. Shewen. 1987. Nucleotide sequence of the leukotoxin genes of *Pasteurella haemolytica* A1. *Infect. Immun.* 55:1987–1996.

44. Lobo, A., and R. A. Welch. Unpublished data.

45. Lory, S. 1992. Determinants of extracellular protein secretion in gram-negative bacteria. *J. Bacteriol.* 174:3423–3428.

46. Macdonald, J., and A. N. Rycroft. 1992. Molecular cloning and expression of *ptxA*, the gene encoding the 120-kilodalton cytotoxin of *Actinobacillus pleuropneumoniae* serotype 2. *Infect. Immun.* 60:2726–2732.

47. Mackman, N., K. Baker, L. Gray, R. Haigh, J. M. Nicaud, and I. B. Holland. 1987. Release of a chimeric protein into the medium from *Escherichia coli* using the C-terminal secretion signal of hemolysin. *EMBO J.* 6:2835–2841.

48. McWhinney, D. R., Y.-F. Chang, R. Young, and D. K. Struck. 1992. Separable domains define target cell specificities of an RTX hemolysin from *Actinobacillus pleuropneumoniae. J. Bacteriol.* 174:291–297.

49. Nicaud, J. M., N. Mackman, L. Gray, and I. B. Holland. 1986. The C-terminal 23 kDa pep-

tide of *E. coli* hemolysin 2001 contains all the information necessary for its secretion by the haemolysin export machinery. *FEBS Lett.* **204:**331–335.

50. **Okuda, K., K. Morihara, Y. Atsumi, H. Takeuchi, S. Kawamoto, H. Kawasaki, K. Suzuki, and J. Fukushima.** 1990. Complete nucleotide sequence of the structural gene for alkaline proteinase from *Pseudomonas aeruginosa* IFO 3455. *Infect. Immun.* **58:**4083–4088.

51. **Ostolaza, H., B. Bartoleme, J. Serra, F. d. l. Cruz, and F. Goni.** 1991. Alpha-hemolysin from *E. coli:* purification and self-aggregation properties. *FEBS Lett.* **280:**195–198.

52. **Regassa, L., and R. A. Welch.** Unpublished data.

53. **Rowe, G. E., S. Pellett, and R. A. Welch.** 1994. Analysis of toxinogenic functions associated with the RTX repeat region and monoclonal antibody D12 epitope of *Escherichia coli* hemolysin (HlyA). *Infect. Immun.* **62:**579–588.

54. **Schmidt, H., H. Karche, and L. Beutin.** 1994. The large-sized plasmids of enterohemorrhagic *Escherichia coli* O157 strains encode hemolysins which are presumably members of the *E. coli* alpha-hemolysin family. *FEMS Microbiol. Lett.* **117:**189–196.

55. **Schulein, R., I. Gentschev, H.-J. Mollenkopf, and W. Goebel.** 1992. A topological model for the haemolysin translocator protein HlyD. *Mol. Gen. Genet.* **234:**155–163.

56. **Stanley, P., V. Koronakis, and C. Hughes.** 1991. Mutational analysis supports a role for multiple structural features in the C-terminal secretion signal of *Escherichia coli* haemolysin. *Mol. Microbiol.* **5:**2391–2403.

57. **Stanley, P. L. D., P. Diaz, M. J. A. Bailey, D. Gygi, A. Juarez, and C. Hughes.** 1993. Loss of activity in the secreted form of *Escherichia coli* haemolysin caused by an *rfaP* lesion in core lipopolysaccharide assembly. *Mol. Microbiol.* **10:**781–787.

58. **Suh, Y., and M. J. Benedik.** 1992. Production of active *Serratia marcescens* metalloprotease from *Escherichia coli* by alpha-hemolysin HlyB and HlyD. *J. Bacteriol.* **174:**2361–2366.

59. **Swofford, D. L.** 1993. *Phylogenetic Analysis Using Parsimony,* version 3.0s, Illinois Natural History Survey, Champaign, Ill.

60. **Tan, Y., and K. J. Miller.** 1992. Cloning, expression and nucleotide sequence of a lipase gene from *Pseudomonas fluorescens* B52. *Appl. Environ. Microbiol.* **58:**1402–1407.

61. **Thomas, W. D., Jr., S. Wagner, and R. A. Welch.** 1992. A heterologous membrane protein domain fused to the C-terminal ATP-binding domain of HlyB can export *Escherichia coli* hemolysin. *J. Bacteriol.* **174:**6771–6779.

62. **Thompson, S., and P. F. Sparling.** 1993. The RTX cytotoxin-related FrpA protein of *Neisseria meningitidis* is secreted extracellularly by meningococci and by HlyBD and *Escherichia coli. Infect. Immun.* **61:**2906–2911.

63. **Thompson, S., L. Wang, and P. F. Sparling.** 1993. Cloning and nucleotide sequence of frpC, a second gene from *Neisseria meningitidis* encoding a protein similar to RTX cytotoxins. *Mol. Microbiol.* **9:**85–96.

64. **Thompson, S., L. Wang, and P. F. Sparling.** 1993. *Neisseria meningitidis* iron-regulated proteins related to the RTX family of exoproteins. *J. Bacteriol.* **175:**811–818.

65. **Wandersman, C., and P. Delepelaire.** 1990. TolC, an *Escherichia coli* outer membrane protein required for hemolysin secretion. *Proc. Natl. Acad. Sci. USA* **87:**4776–4780.

66. **Wang, R., S. J. Seror, M. Blight, J. M. Pratt, J. K. Broome-Smith, and I. B. Holland.** 1991. Analysis of the membrane organization of an *Escherichia coli* protein translocator, HlyB, a member of a large family of prokaryote and eukaryote surface transport proteins. *J. Mol. Biol.* **217:**441–454.

67. **Welch, R. A.** Unpublished data.

68. **Welch, R. A.** 1987. Identification of two different hemolysin determinants in uropathogenic *Proteus* isolates. *Infect. Immun.* **55:**2183–2190.

69. **Welch, R. A.** 1991. Pore-forming cytolysins of Gram-negative bacteria. *Mol. Microbiol.* **5:**521–528.

70. **Welch, R. A.** 1994. Holistic perspective on the *Escherichia coli* hemolysin, p. 351–364. *In* V. L. Miller, J. B. Kaper, D. Portnoy, and R. R. Isberg (ed.), *Molecular Genetics of Bacterial Pathogenesis.* American Society for Microbiology, Washington, D.C.

71. **Welch, R. A., T. Felmlee, S. Pellett, and D. Chenoweth.** 1986. The *E. coli* hemolysin: its gene organization, secretion and interaction with neutrophil receptors, p. 431–438. *In* D. Lark, S. Normark, B. E. Uhlin, and H. Wolf-Watz (ed.), *Protein-Carbohydrate Interactions in Biological Systems.* Academic Press, Inc., New York.

72. **Zhang, F., J. A. Sheps, and V. Ling.** 1993. Complementation of transport-deficient mutants of *Escherichia coli* alpha-hemolysin by second site mutations in the transporter hemolysin B. *J. Biol. Chem.* **268:**19889–19895.

PORE-FORMING TOXINS OF
GRAM-POSITIVE BACTERIA

Rodney K. Tweten

13

In recent years, the study of the regulation, cytolytic mechanism, and role in disease of pore-forming cytolytic proteins from the gram-positive bacteria has provided interesting insights into this group of proteins. These proteins are somewhat enigmatic in that they are generally produced as soluble proteins that ultimately end up as membrane-associated complexes, even though their structures often do not resemble that of a membrane-associated protein. Typically, these proteins aggregate into supramolecular complexes of various sizes on the membrane. This transition from a soluble protein to a membrane-associated complex is probably accomplished in a variety of ways, but many fundamental questions about the transition remain. The role these proteins play in development of disease is only now being elucidated. As more of these toxins are discovered and studied, classes based on similarities in mechanism and/or primary sequence are becoming evident. Some of these cytolytic proteins have clearly evolved from a single progenitor gene, whereas others may share mechanistic features but lack sequence similarities. In some cases, divergent evolution has

forced cytolytic proteins with a common ancestral gene to acquire new attributes that facilitate the pathogenesis of a particular organism. Still others appear to have undergone convergent evolution, with similar mechanisms evolving for one reason or another.

Since these toxins are generally released by the cell into the host tissues, their role in disease is difficult to address. They can have a myriad of effects on many different cell types, although it is not a trivial task to sort out the cellular effects that have a true role in the pathogenic mechanism from those that are merely artifacts of the in vitro system being used. Most diseases caused by bacteria that produce pore-forming toxins are multifactorial, which further complicates the study of the contribution to disease of these toxins.

As the mechanisms and gene sequences for these toxins have been elucidated, it has become apparent that some of these toxins may be grouped by similarities in structure and cytolytic mechanism. I have attempted to group those toxins that exhibit similarities in mechanism or sequence, but at least some of these groupings are tentative and may change as more information becomes available. In this chapter, I examine the cytolytic mechanisms of the gram-positive pore-forming toxins and review their role in pathogenesis.

Rodney K. Tweten   Department of Microbiology and Immunology, University of Oklahoma Health Sciences Center, Oklahoma City, Oklahoma 73190.

Virulence Mechanisms of Bacterial Pathogens, 2nd ed., Edited by J. A. Roth et al.
© 1995 American Society for Microbiology, Washington, D.C.

THIOL-ACTIVATED CYTOLYSINS

Grouped on the basis of sequence and mechanism, the thiol-activated cytolysins make up the largest group of related cytolytic proteins produced by the gram-positive bacteria. Consequently, there is a vast amount of information available on these toxins; it has been reviewed by a number of workers (1, 8, 14, 24, 115). These cytolytic toxins are found in species of *Clostridium, Streptococcus, Listeria,* and *Bacillus.* Their widespread occurrence suggests that they have successfully complemented the pathogenic mechanisms of the various bacterial species that have acquired one of these toxins. A common set of features has been described for these toxins: use of cholesterol as the putative membrane receptor, aggregation of membrane monomers of the toxins into supramolecular complexes, lysis of most cells that contain cholesterol, and sequence similarity among all toxins of this family sequenced to date. These toxins probably originally evolved from a single progenitor gene, since identity in the amino acid sequences of the toxins sequenced to date ranges from approximately 40 to 70%. The thiol-activated toxins cloned and sequenced are perfringolysin O (PFO; from *Clostridium perfringens*) (125), streptolysin O (SLO; from *Streptococcus pyogenes*) (70), listeriolysin O (LLO; from *Listeria monocytogenes*) (81), ivanolysin (IVO; from *Listeria ivanovii*) (49), seeligeriolysin (LSO; from *Listeria seeligeri*) (49), cereolysin (CEO; from *Bacillus cereus*) (87a), pneumolysin (PLY; from *Streptococcus pneumoniae*) (131), and alveolysin (ALV; from *Bacillus alvei*) (46). The dendrogram in Fig. 1 represents the sequence similarity of each toxin. SLO, PFO, CEO, and ALV exhibit the highest degree of sequence identity (≈60 to 70%), whereas *Listeria*-derived toxins (LLO, IVO, and LSO) and PLY exhibit the least identity with the rest (≈40%).

Because of their high degree of similarity, it is tempting to speculate that ALV, SLO, CEO, and PFO are closest to the putative progenitor gene for these toxins. The fact that PLY, LSO, IVO, and LLO exhibit significantly less similarity to SLO, PFO, and ALV may be the result

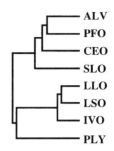

FIGURE 1 Dendrogram of clustering relationships of the thiol-activated toxins with known sequences. ALV, alveolysin (*B. alvei*); PFO, perfringolysin O (*C. perfringens*); CEO, cereolysin (*B. cereus*); SLO, streptolysin O (*S. pyogenes*); LLO, listeriolysin O (*Listeria monocytogenes*); LSO (*Listeria seeligeri*); IVO, ivanolysin (*Listeria ivanovii*); and PLY, pneumolysin. The clustering relationships were generated by the program Pileup from the Wisconsin Genetics Package.

of evolutionary adjustments to the specific pathogenic mechanism of each bacterial species. The roles of these toxins in the pathogenic mechanisms of the bacteria that produce them are varied. Some of the toxins may exert a direct toxic effect on mammalian cells, but in other members of this family, the toxin may facilitate infection by the bacterial species producing it without having a direct toxic effect on host cell tissues.

Cytolytic Mechanism

The hallmark of this family of toxins is the capacity of thiol reagents to restore lost cytolytic activity. It has been known for some time that crude toxin preparations lose activity with time but that the activity can be restored by thiol reagents such as cysteine. Thus, the term "thiol activated" was applied to these toxins. DNA sequence analysis of many of the genes for these toxins has shown that all but IVO have a single cysteine in the mature protein. This cysteine is invariably present in a conserved 11-residue peptide with the sequence ECTGLAW-EWWR. Only LSO varies from this theme and has a phenylalanine substituted for the alanine (49). The contribution of this conserved peptide to the mechanism of the toxins remains controversial. Iwamoto et al. (62) found that modifying the cysteine of PFO with *N*-ethyl-maleimide or 5,5-dithio-*bis*-(2-nitrobenzoic acid) significantly reduced the ability of PFO to bind to erythrocytes. Removal of the 5,5-dithio-*bis*-(2-nitrobenzoic acid) (a reversible modifier of -SH groups) by a reducing reagent restored activity. Their data were complicated, however, by a conformational change that accompanied the modification. The conformational change was reversed when the 5,5-dithio-*bis*-(2-nitrobenzoic acid) was removed with a reducing agent. In contrast, in studies of PLY and SLO (99, 109) in which the cysteine was converted to a serine, no apparent change was observed in the ability to bind the membrane and form aggregates, even though cytolytic activity was reduced 70 to 85%. For PLY (109), the decreased activity of the serine and glycine substitutions for the cysteine could not be explained by a change in binding or aggregation. However, rigorous binding analyses were not carried out in these studies as was done by Iwamoto et al. (62); therefore, it is possible that the binding kinetics of these mutants was altered.

Studies by Pinkney et al. (99) and Saunders et al. (109) clearly showed that the sulfhydryl of the single cysteine was not required for activity. Replacement of the cysteine with alanine resulted in mutants of PLY and SLO with near wild-type activity. It is likely that the side chain hydrophobicity is important, since replacement of the cysteine with the more hydrophilic serine resulted in a 75 to 80% reduction in hemolytic activity. The strong conservation of the cysteine at this site is somewhat puzzling, since alanine could function nearly as well without the associated risk of inactivation of the toxin by oxidation. Whether the cysteine plays a role in the natural infections caused by the bacteria that produce these toxins remains to be elucidated. Although the sulfhydryl function of the essential cysteine is not necessary for the activity of these toxins, the cysteine is clearly in a sensitive site that cannot withstand chemical modification. The term "thiol activated" is no longer a true description of these toxins; however, the fact that this term is so entrenched in the literature makes it unlikely that it will be changed.

Several of the other residues within and outside of the ECTGLAWEWWR peptide in PLY have been altered by Boulnois's group (21) with some interesting results. Conversion of Trp-433 to Phe resulted in loss of >99% of the hemolytic activity, but oligomer formation and cholesterol binding appeared to remain intact. Conversion of Trp-435 to Phe resulted in a loss of about 87% of the hemolytic activity, but conversion of Trp-436 to Phe did not result in any loss of hemolytic activity. Binding and oligomer formation were not reported for the last two mutations. The only mutation that appeared to affect cholesterol binding was Glu-434 to Asp. In this case, cholesterol binding was reduced approximately 50%.

The mutagenesis data indicate that while some mutations (Trp-433 to Phe) can eliminate >99% of the hemolytic activity, cholesterol binding and oligomerization appear to remain unaffected. These observations make it difficult to envision an obvious role for this highly conserved region. Boulnois et al. (21) suggested that this region may interact with the sterol after membrane binding, and they have data to suggest that a site outside the ECTGLAWEWWR peptide may be responsible for initial membrane binding. Chemical modification of PLY with diethylpyrocarbonate resulted in toxin that was cytolytically inactive and unable to bind erythrocytes. Since diethylpyrocarbonate primarily targets histidine, it was hypothesized (21) that a histidine was involved in the binding event. The only histidine found to be conserved among PLY, LLO, PFO, and SLO was His-367 of PLY. Since their report, several additional toxins have been sequenced from this family, and this histidine is conserved in all eight of the toxins (Fig. 2). This site also completely conserves a tyrosine-valine-alanine triplet that endows this region with an obvious hydrophobicity. Replacement of His-367 in PLY with arginine results in an inactive toxin that does not bind erythrocytes, a situation that supports the chemical modification data (21). Boulnois et al. (21) suggested that this site may

```
PLY  LLDHSGAYVAQ
PFO  NLDHSGAYVAQ
SLO  NLSHQGAYVAQ
LLO  NIDHSGGYVAQ
ALV  KLDHSGAYVAQ
CEO  TLDHYGAYVAQ
IVO  NLDHSGAYVAR
ISO  NIDHSGGYVAQ
     . * *.***.
```

FIGURE 2 Sequences of the various thiol-activated toxins that correspond to residues 364 to 374 of PLY. The conserved histine is underlined. Abbreviations for the toxins are the same as in Fig. 1.

be the primary site for membrane binding and that the ECTGLAWEWWR peptide may interact with the sterol after binding is achieved.

Boulnois et al. suggested that the ECT-GLAWEWWR sequence may function in a postbinding event (21) that involves an interaction with the sterol. Their data in conjunction with the chemical modification data of Iwamoto et al. (62) (which showed that a conformational change was associated with cysteine modification) indicate that the ECT-GLAWEWWR sequence may function as a trigger for a postbinding conformational change, perhaps one that readies the protein for oligomerization. However, the interaction of these regions with cholesterol and the membrane will have to be directly measured in order to elucidate the role for the ECTGLAW-EWWR sequence as well as other regions of these toxins.

The receptor-binding function of PFO resides in the carboxy-terminal fragment generated by trypsin cleavage (63, 126). This 22,268-Da fragment starts at Asn-304 of PFO and includes all residues to the carboxy terminus. The fragment includes the ECTGLAW-EWWR sequence and the histidine-containing sequence identified by Boulnois et al. (21) (Fig. 2). This fragment was termed T2 by Ohno-Iwashita et al. (87), who originally identified it. The T2 fragment was subsequently found to contain both receptor-binding function and at least one of the putative

complementary domains responsible for the oligomerization of PFO. Using different approaches, Tweten et al. (126) and Iwamoto et al. (63) showed that the T2 fragment could bind erythrocytes and inhibit hemolysis by native PFO. Although T2 could compete for cell receptors with native PFO, competition for receptor was not the basis for the observed inhibition. At least 500-fold more T2 was required to inhibit the binding of PFO to the membrane than was required to inhibit PFO-induced hemolysis (63). The T2 fragment inhibited PFO cytolytic activity on erythrocytes by the formation of T2-PFO pairs that could not oligomerize into the supramolecular complexes required for activity of the toxin. The T2 fragment was apparently not capable of oligomerizing with itself, since preincubation of T2 on the erythrocytes did not diminish the ability of T2 to inhibit cytolytic activity or oligomerization (126). These data provide evidence that the T2 fragment contains the receptor-binding site of PFO and at least one of the complementary interfacial domains required for oligomerization of PFO.

Oligomerization of these toxins into supramolecular complexes on the membrane is a necessary step in the cytolytic mechanism of these toxins. Complexes composed of polymerized monomers of a variety of thiol-activated toxins have been visualized by electron microscopy (15, 34, 35, 85, 95, 116). These complexes typically form rings and arcs that are composed of a variable number of toxin monomers. For SLO, these complexes are estimated to contain 25 to 100 monomers (15). More recently, electron microscopy–derived projection structures for the membrane-bound complexes of PFO and SLO were reported (89, 112). In both cases, the ring exhibited periodic inner and outer regions of density. On the basis of projection data for PFO, Olofsson et al. (89) estimated that a closed ring comprises approximately 50 monomers. These monomers exhibit a periodic repeat of 2.4 nm, and one monomer was estimated to span the width of the ring. The ring width was estimated to be 6 nm, and the internal diameter of the ring complex was approximately 30 nm. Sekiya et al. (112) estimated the width of the SLO ring to be 4.9 nm and the internal diameter of the ring complex to be 24 nm. In contrast to the conclusions reached by Olofsson et al. (89), Sekiya et al. suggested that SLO formed a double ring on the membrane; both rings were estimated to contain approximately 22 to 24 SLO monomers.

Measurements of the ring structure and the pore diameter for these toxins were reasonably close in the two studies, but quite different interpretations of the data were presented. Sekiya et al. (112) hypothesized, on the basis of extensive measurements, that two SLO molecules span the width of the ring in a staggered conformation (Fig. 3A), whereas Olofsson et al. (89) indicated that a single PFO spans the width of the ring (Fig. 3B). At this time, there is no additional evidence that might discriminate between these two models, although it is possible (though unlikely) that PFO and SLO form different oligomeric complexes on the membrane.

Although many reports have presented electron micrographic structures of the oligomerized toxin on membranes, no direct measurements of the oligomerization process were made until recently. Harris et al. (50) reported a method based on fluorescence energy transfer that could directly measure the oligomerization of PFO on the membranes of erythrocyte ghosts. This approach utilized energy transfer between the fluorescent dyes of fluorophore-labeled PFO molecules. Kinetic measurements of the oligomerization process facilitated the measurement of both the oligomerization kinetics and the activation energy for oligomerization. It was determined that oligomerization of the PFO was mostly complete prior to the onset of hemolysis.

It has been known for some time that these toxins are inactive at low temperature, but the basis for the inactivity was unknown. At 0 to 4°C, these toxins do not lyse erythrocytes; however, if the unbound toxin is removed by washing the erythrocytes and then the temperature is raised to 37°C, lysis proceeds rap-

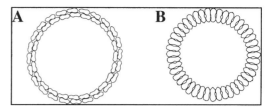

FIGURE 3 Schematic representation of the membrane structure of oligomeric SLO and PFO. Representative structure of the staggered double ring of Sekiya et al. (112) for SLO and the single-ring structure of PFO based on the data of Olofsson et al. (89).

idly. Presumably, the toxin that remained bound to the cells could not proceed to form a lytic complex at the low temperature. Membranes that contain cholesterol in appreciable amounts (>30 mol%), which is typical of most mammalian cells, do not undergo sharp thermal-phase transitions (71); however, the bulk viscosity of the lipid still decreases as the temperature is lowered. Cholesterol apparently abolishes the lipid thermal-phase transitions and also forms cholesterol-rich domains that largely exclude lipid. PFO preferentially binds cholesterol in these cholesterol-rich domains surrounded by C_{18} phospholipids (86). Using fluorescence energy transfer between fluorescent-probe-labeled PFO, Harris et al. (50) showed that oligomerization of PFO was undetectable at 4°C, which confirmed that the inability of PFO to oligomerize at low temperature was probably responsible for the lack of cytolytic activity at this temperature. However, the reason that aggregation does not proceed at low temperature remains unclear, though it is possible that PFO itself has a temperature-dependent conformation step that does not occur at low temperature. Harris et al. (50) also determined that activation energies were 18.7 kcal/mol for oligomerization and 23.2 kcal/mol for hemolysis, which suggests that oligomerization drives the cytolytic process.

Pathogenesis
The roles of these toxins in pathogenesis are varied and not yet fully understood; they appear to differ in each of the studied pathogens

that produces a thiol-activated cytolysin. Since these toxins are efficient in the lysis of eukaryotic cells, it is assumed that at sufficient concentrations, they will simply lyse the target cell. Sublytic amounts of these toxins on the membrane of a eukaryotic cell can in some cases cause significant perturbations in the cell's physiology. These perturbations may be especially important when cells of the immune system are affected.

The thiol-activated toxin with the best studied role in pathogenesis is LLO. *L. monocytogenes* most often causes meningitis or sepsis in neonates and immunocompromised individuals. Food-borne transmission of *L. monocytogenes* is an important route of exposure to the organism. *L. monocytogenes* is a facultative intracellular parasite that invades and grows within macrophages as well as other cell types such as epithelial cells and fibroblasts. It gains entry to the cell cytoplasm by a mechanism involving phagocytosis and a mechanism in which LLO plays an intregral part. *L. monocytogenes* is phagocytosed by the cell and then escapes the phagosome and replicates within the cell cytoplasm. Although many nuances of this process remain to be understood, it is clear that LLO plays a central role in the escape of *L. monocytogenes* from the phagosome. The observation that LLO functions in such a limited, though important, role probably reflects the fact that its contribution to pathogenesis may be the best understood of all contributions by thiol-activated cytolysins.

The loss of LLO synthesis or its cytolytic function results in an avirulence phenotype in the mouse model (33, 83). The actual contribution of LLO to virulence was worked out by Portnoy's laboratory group in a series of elegant studies in which they showed that LLO facilitates the escape of *L. monocytogenes* from the phagocytic vesicle (16, 100, 101). LLO is so well suited to this task that its placement and expression in *Bacillus subtilis* facilitate the intracellular growth of *B. subtilis* in J774 cells (16), a macrophagelike cell line used for the *Listeria* studies. The key feature that allows LLO to function in this manner is the pH optimum for

its cytolytic activity. LLO is most active at pH around 5.5, the approximate pH of a phagosome after its formation. Little cytolytic activity for LLO is detected at pH at or above 7.0, whereas related cytolysins such as SLO and PFO are active at pH 5.5 as well as at pH 7.0 (101). It appears that once the *Listeria* cell escapes the phagosome, the activity of the LLO must be reduced or the eukaryotic cell membrane may be lysed. It would not benefit the bacterial cell to lyse the host cell if further replication of the bacterium is to be achieved within the protected environment of the eukaryotic cell cytoplasm. Additional support for this scenario came from studies in which the PFO gene was placed and expressed in *B. subtilis* (101) and was used to replace the LLO gene in *L. monocytogenes* (66a). PFO facilitated the escape of the bacilli and listeriae from J774 cells, but unlike the situation with LLO, no significant intracellular growth of the bacterium was achieved. It appeared that the bacterial cells escaped from the endosome but that the PFO killed the macrophage, probably because PFO remains active at neutral pH (101). The basis for the pH optimum of LLO remains unknown; it is not clear whether LLO is unable to bind to its receptor (cholesterol) or whether it is inhibited at one of the later stages of the cytolytic mechanism (e.g., oligomerization). Except for studies in which various residues in the conserved cysteine-containing undecapeptide have been altered, no work has identified any residues that may contribute to the pH optimum of LLO.

PLY also contributes to the pathogenesis of *S. pneumoniae,* but its role in the disease is less clear than that of LLO. Noted that, as stated earlier, PLY is not secreted from the *S. pneumoniae* cell (it lacks a signal peptide) but is released only upon autolysis of the bacterial cell. Immunization of mice to PLY (96) or inactivation of the PLY gene in *S. pneumoniae* by transposon mutagenesis (9) results in a significantly increased survival time in mice challenged by the intranasal route; however, complete protection is not afforded. PLY is also the soluble factor that injures pulmonary endothelial cells in vitro and may be responsible for alveolar hemorrhage during infection (106). PLY can also directly bind the Fc portion of the immunoglobulin G (IgG) molecule, and this ability has been suggested as the basis for PLY activation of the classic complement pathway in the absence of anti-PLY antibody (84). Residues 121 to 162 (domain 1) and 368 to 397 (domain 2) of PLY exhibit homology (approximately 27 and 42% identity, respectively) with human C-reactive protein, an acute-phase protein known to bind the C polysaccharides of the pneumococci and to activate complement (68). Boulnois et al. (22) showed that conversion of Tyr-384 to Phe results in an 86% reduction in the ability of PLY to activate complement and a 75% reduction in Fc-binding activity. No alteration in hemolytic activity was observed, suggesting that the putative Fc-binding site may not be important to the hemolytic activity of the toxin. Curiously, this region overlaps the region identified by Boulnois et al. (22) as the putative receptor-binding site (Fig. 4), which suggests that cell-bound PLY would not be available for binding IgG and that IgG-bound PLY should not bind cells. It was not determined whether IgG-bound PLY was inhibited for cytolytic activity, although it seems likely that if PLY was bound with an IgG molecule, it would most likely occlude this putative binding site.

```
PLY  SGAYVAQYYITWDELSYDHQGKEVLTPKAW

PFO  SGAYVAQFEVAWDEVSYDKEGNEVLTHKTW

SLO  QGAYVAQYEILWDEINYDDKGKEVITKRRW

LLO  SGGYVAQFNISWDEVNYDPEGNEIVQHKNW

ALV  SGAYVAQFEVYWDEFSYDADGQEIVTRKSW

CEO  YGAYVAQFDVSWDGFTFDQNGKEILTHKTW

IVO  SGAYVARFNVTWDEVSYDANGNEVVEHKKW

IVO  SGGYVAQFNISWDEVSYDENGNEIKVHKKW

     *.***.. . **    .*  *.*.    . *
```

FIGURE 4 Domain 2 of the putative Fc-binding domain of PLY (22). Residues of the Fc-binding domain that overlap the putative receptor-binding domain of PLY are underlined (see Fig. 2).

C. perfringens type A is the main causative agent of gas gangrene and produces both alpha toxin (phospholipase A) and PFO (theta toxin) as well as several other extracellular enzymes and proteins. Because of the large number of extracellular factors produced by each clostridial species, it has been difficult to sort out the various contributions by these factors to the development of the disease. One striking characteristic of a gangrenous lesion is the absence of infiltrating polymorphonuclear lymphocytes (PMNLs). An early histologic study (105) showed that PMNLs do not infiltrate the necrotic tissue of a gangrenous lesion; instead, they appear to pile up at the outer zones of the necrotic tissue, suggesting that one or more diffusible substances affect the PMNLs. The fact that PFO can lyse most eukaryotic cells indicates that it probably has a direct toxic effect on the PMNLs; however, there may be other effects on the PMNLs when PFO concentration is decreased. Bremm et al. (25, 27, 28) showed that at sublytic levels, PFO decreases leukotriene production by PMNLs and increases the rate of conversion of leukotriene B_4 to less chemotactic derivatives. Stevens et al. (118) showed that sublytic levels of PFO cause marked morphologic changes in PMNLs and reduce the zymosan-stimulated chemiluminescent burst and the directed and random migrations of the PMNLs. In contrast, alpha toxin (phospholipase C) does not significantly affect PMNLs. The effects of PFO on PMNLs were consistent with the absence of PMNLs in the gangrenous lesion, and it is possible that one role for PFO in natural infection is to prevent infiltration of the PMNLs, thus facilitating the fulminant growth of bacterial cells.

In contrast to the data of Stevens et al. (118), gene knock-out experiments which eliminated PFO production in a type A *C. perfringens* (2a) did not appear to support the role of PFO in restricting the influx of PMNL into the gangreneous lesion. Awad et al. (2a) have shown in the mouse model that when the PFO gene was eliminated, tissue necrosis still occurred in mice infected with the PFO-negative strain of *C. perfringens*. In the same animals PMNLs were not present in the necrotic tissue to any significant degree, although in the alpha toxin knockout mutants significant PMNL infiltration was observed. It was concluded by the authors that PFO contributes somewhat to tissue necrosis, but that alpha toxin was largely responsible for restricting the infiltration of PMNLs into the lesion and the majority of the tissue necrosis. These observations show that in vitro observations cannot always be applied to the actual in vivo effects of a toxin.

The role of the remaining thiol-activated cytolysins in disease has not been explored. Curiously, there is little information on the contribution to disease of SLO, the prototype toxin for the family of thiol-activated toxins. Only recently has the SLO gene been knocked out in *S. pyogenes,* and work to examine the effect of losing its production in a mouse model system is currently under way (35a).

STAPHYLOCOCCUS AUREUS ALPHA TOXIN, LEUKOCIDIN, GAMMA HEMOLYSIN, AND DELTA TOXIN AND C. PERFRINGENS BETA TOXIN

The *S. aureus* alpha and delta toxins are the best-characterized toxins in this group; the remaining toxins, *S. aureus* leukocidin and gamma hemolysin and the *C. perfringens* beta toxin, are mentioned here only briefly, since it is not clear whether they are pore-forming toxins. However, they do exhibit sequence similarity to the *S. aureus* alpha toxin (a pore-forming toxin); hence, they may also be pore-forming toxins. *S. aureus* alpha toxin was the first cytolytic protein to be identified as a pore-forming toxin (43). Since then, a considerable amount of work has been devoted to the study of the mechanism of alpha toxin. Recently, alpha toxin was the subject of an excellent review by Bhakdi and Tranum (13), and therefore, only the salient points and the information published since that review are addressed here. The Paton-Valentine leukocidin and gamma hemolysin appear to be related to each other and to a lesser extent to alpha toxin (31, 36, 67, 120). Both gamma hemolysin and leukocidin are two-component systems that are

contained on the same operon, and the two share a common protein that is required for activity (67). Finck-Barbancon et al. (37) suggested that the leukocidin is a pore-forming protein, since it makes the membranes of polymorphonuclear leukocytes permeable to various divalent cations. No evidence is available to indicate whether the gamma hemolysin is also a pore-forming toxin. *S. aureus* delta toxin is quite different from most of the other cytolytic bacterial toxins in that it is a short polypeptide of 26 amino acids (40) that forms channels in lipid bilayers (39). Recently, the gene for the beta toxin from *C. perfringens* was cloned and sequenced (58). Beta toxin also exhibits sequence similarity to *S. aureus* alpha toxin (28% identity); however, little is known about its mechanism except that this toxin appears to oligomerize easily during the purification of native toxin from *C. perfringens* and the recombinant form from *Escherichia coli* (58). The fact that beta toxin easily oligomerizes suggests that it may also be a pore-forming toxin, though no direct evidence is available to support this possibility. Since there is little in the published literature to suggest that *S. aureus* leukocidin and gamma hemolysin and *C. perfringens* beta toxin are pore-forming toxins, this section mainly discusses *S. aureus* alpha and delta toxins, both well-studied pore-forming toxins with quite different physical characteristics.

S. aureus Alpha Toxin

The gene for staphylococcal alpha toxin was cloned more than a decade ago by Gray and Kehoe (47). At that time, no proteins that exhibited similarity to alpha toxin had been identified. Since then, several proteins that share a limited degree of sequence similarity with alpha toxin have been identified (Fig. 5). Alpha toxin may be distantly related to the leukocidin S (18% identity in the carboxy-terminal half of each protein) and F (27% identity) proteins, which are apparently nearly identical to proteins encoded by the gamma hemolysin *B* and *C* genes from *S. aureus* (Fig. 5) (31, 120). The beta hemolysin of *C. perfringens* also exhibits

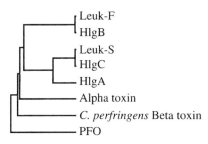

FIGURE 5 Dendrogram of clustering relationships showing the sequence relatedness of the leukotoxin proteins with *S. aureus* alpha toxin. Leukotoxins S and F are the same proteins as the gamma C (HlgC) and B (HlgB) proteins. Clustering relationships were generated by the program Pileup from the Wisconsin Genetics Package. PFO was included as an unrelated protein.

similarity to *S. aureus* alpha toxin, gamma hemolysin HlgB, and leukocidin F (28 to 29% identity) and to a lesser extent to *S. aureus* leukocidin S and gamma hemolysin HlgA and HlgC proteins (18 to 22% identity). As mentioned above, whether this sequence similarity to these other toxins indicates that they, too, are pore-forming toxins remains to be proven by direct physical methods such as planar membrane analysis.

CYTOLYTIC MECHANISM

The first step in alpha-toxin action on a cell is membrane binding. The fact that alpha toxin may bind to protein-free liposomes indicates that no specific receptor is required. Although the receptor has been variously identified as the band 3 protein of erythrocytes (erythrocyte anion antiporter) (77) or gangliosides (69), the latter may actually inhibit toxin binding. No confirmation that band 3 is the alpha-toxin receptor has been published. Cassidy and Harshman (30) originally identified what appeared to be both high- and low-affinity binding sites for alpha toxin. The high-affinity receptors were apparently present on susceptible erythrocytes (rabbit erythrocytes), whereas only low-affinity sites were present on erythrocytes (i.e., human) that were resistant to the action of alpha toxin. A caveat of this study was that the alpha toxin, which was labeled to high specific activity with

[125]I, retained only 20% of its activity. The major features of Cassidy and Harshman's work were subsequently confirmed by Hildebrand et al. (56), who showed that 1,500 to 2,000 high-affinity receptors were present on rabbit erythrocytes and that once these sites were saturated, alpha toxin bound via a nonspecific absorptive mechanism. Sufficient toxin added to either resistant or susceptible cells results in lysis and the formation of hexameric complexes on the membrane.

Several workers, using a variety of approaches, have shown that alpha toxin forms a hexameric oligomer of the alpha-toxin monomers on membranes (10, 20, 44, 52, 60, 61, 91, 123). These oligomers form a pore in the membrane that is approximately 1 to 1.5 nm in diameter (6, 80). Following a theme of the thiol-activated cytolysins, alpha toxin apparently binds to the membrane, diffuses laterally, and oligomerizes with other toxin monomers to form the pore. Several investigators have examined the ultrastructure of the oligomerized membrane complex of alpha toxin (55, 88, 90–92, 132), and most have reaffirmed the hexameric structure of the toxin on the membrane. However, studies that have examined alpha-toxin structure in crystalline layers show the formation of oligomers with either 4- or 8-subunit structures (88, 90) that do not appear to be fully inserted into the bilayer. The significance of these findings remains unclear.

One of the earliest studies to identify regions potentially important in toxin function was that by Cassidy and Harshman (30), who determined that a single tyrosine of alpha toxin was uniquely iodinated during ^{125}I labeling and resulted in a 90% reduction in hemolytic activity. Their peptide sequence work and the DNA-derived primary structure of alpha toxin indicated that this tyrosine residue is Tyr-27 (47). Harshman et al. (53) later determined that a monoclonal antibody that binds to alpha toxin does not inhibit toxin binding to membranes or interfere with hexamer formation. CNBr digestion of alpha toxin showed that the monoclonal antibody binds to the CNBr-derived carboxy-terminal peptide VII, which corresponds to residues 235 to 291 of alpha toxin. In contrast, polyclonal rabbit antibody to alpha toxin, which inhibits alpha-toxin-induced hemolysis, binds to CNBr fragments IV, V, and VII. These peptides correspond to residues 1 to 33, 204 to 234, and 235 to 291, respectively, of the alpha-toxin sequence. Harshman et al. (51) generated synthetic peptides derived from the sequences of CNBr fragments IV (peptide PI; KTGDLVTYDKENG) and V (peptide PII; PNKASSLLSSGFS). Both peptides marginally slow the rate of hexamer formation by alpha toxin. The amino-terminal-derived PI is the most effective in slowing aggregation, although the results were not definitive, since aggregation proceeded nearly as well as in the untreated alpha toxin.

Blomqvist and Thelestam (19) suggested that a naturally occurring amino-terminal proteolytic fragment of alpha toxin is hemolytic but not lethal to Y1 adrenal cells. These findings as well as a subsequent report by Blomqvist et al. (18) have led to some confusion in the assignment of function to various domains of alpha toxin. In a recent review, Bhakdi and Tranum (13) suggested that the naturally occurring fragment identified by Blomqvist and Thelestam (19) contains important regions that are not present on the trypsin-derived 17,000-Da amino-terminal fragment, since the latter was devoid of any hemolytic or binding activity. However, Blomqvist et al. (18) subsequently showed that the naturally occurring proteolytic fragment (NF-I) originally identified by them (19) is actually the carboxy-terminal fragment (Fig. 6). In fact, the trypsin-derived carboxy-terminal fragment (TF-1) is slightly larger than the naturally generated fragment. Therefore, the suggestion made by Bhakdi and Tranum (13) is not valid, since the two peptides are actually derived from opposite ends of the molecule.

Blomqvist and Thelestam (19) did not rigorously demonstrate that the NF-I fragment had been separated from the rest of the toxin. It was likely that the hemolytic activity was due to nicked but intact alpha toxin, since native polyacrylamide gel electrophoresis and gel filtration showed the nicked fragment to nearly comigrate with native alpha toxin. They suggested that the NF-I fragment aggregated to form this complex, but subsequent findings by Palmer et al. (94) did not support the results of Blomqvist amd Thelestam (19). Palmer et al. found that proteinase K-nicked alpha toxin resulted in two peptides of similar sizes that could be dissociated only by 6 M urea. The nicked toxin retained hemolytic activity, but the separated fragments exhibited no activity. Also, the nicked toxin could not permeabilize nucleated cells, and the pore formed by this variant was smaller (0.6 to 0.9 nm) than that formed by the native toxin. Therefore, it was likely (as mentioned above) that Blomqvist and Thelestam (19) did not separate the NF-I and NF-II fragments and were actually working with nicked but intact alpha toxin. Tomita et al. (124) recently showed that another naturally nicked form of alpha toxin is cleaved between Glu-71 and Gly-72. The nicked toxin retains the ability to oligomerize on the membrane, although channel-forming activity decreases 20-fold. The inevitable conclusion one reaches from these studies is that the proteolytic nicking of alpha toxin at various sites appears to alter the nature of the pore formed. Presumably, proteolytic cleavage of the protein relaxes the structural constraints placed on the alpha toxin such that the pore structure is altered. This concept has some support from reports (88, 90) of structural differences in the oligomers formed by nicked and native toxin that were found when these oligomers were examined by electron microscopy and image processing on lipid bilayers.

Walker et al. (130) showed that elimination of up to 22 residues from the amino terminus of alpha toxin does not affect oligomerization of alpha toxin on erythrocytes. These amino-terminal truncation mutants were, however, not hemolytic. The data of Walker et al. provide an intriguing insight into the mechanism of alpha toxin. Their observation indicates that a step that facilitates pore formation occurs after oligomerization. The oligomer may have a transient, preinsertion state that is frozen in its

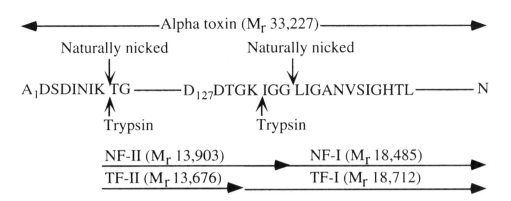

FIGURE 6 Naturally occurring and trypsin-generated peptide fragments of *S. aureus* alpha toxin.

amino-terminal truncation mutants. In the same study, it was found that alpha-toxin-induced hemolysis is sensitive to the loss of three or five residues from the carboxy-terminal end. Hemolysis proceeded to nearly the same endpoint as with wild-type toxin but at a significantly reduced rate. The reduced rate of hemolysis was correlated with a reduced rate of oligomerization.

Modification of the histidine residues of alpha toxin with diethylpyrocarbonate (97) results in decreased membrane binding and oligomerization of the toxin. Recently, Menzies and Kernodle (82) altered histidines 35, 48, 144, and 259 of the alpha toxin gene by site-directed mutagenesis. Substitution of leucine for His-35 resulted in a toxin that lacked hemolytic and lethal activities. Replacement of His-48, His-144, and His-259 with leucine resulted in derivatives that retained 7, 16, and 46%, respectively, of wild-type hemolytic activity. The lethal activity of the His-48 mutant alpha toxin was decreased approximately 10-fold over that of the wild-type, whereas replacement of His-259 did not detectably change the toxin's lethal activity. Hexamer formation was considerably decreased by the replacement of His-35, although it was not totally abolished. Menzies and Kernodle (82) suggested that the His-35 mutant may not be able to penetrate the membrane efficiently or its pore may remain closed. Either prospect is reasonable and correlates with the previous ob-

servations of Walker et al. (130), who found that removal of residues from the amino terminus resulted in hemolytically inactive alpha toxin that could still oligomerize. However, the dramatic change in side chain attributes resulting from a substitution of leucine for histidine cannot be ignored; a more conservative change may not have had such a dramatic effect on activity.

PATHOGENESIS

Alpha toxin, as well as other cytolytic toxins, stimulates the generation and release of inflammatory mediators (i.e., lipoxygenase factors and leukotrienes) from PMNLs (26). It also stimulates phosphatidylinositol turnover and arachidonic acid generation in undifferentiated cultures of pheochromocytoma PC12 cells (38). In monocytes, sublytic doses stimulate the release of tumor necrosis factor alpha, and cytocidal doses stimulate the release of interleukin-1β (11). Alpha-toxin treatment of platelets results in assembly of the prothrombinase complex (2). Clearly, alpha toxin exhibits a myriad of effects on various cell types in vitro; however, it remains unclear whether these effects are important in the development of the various *S. aureus* disease syndromes. In addition, many of these effects are also observed with other cytolytic toxins (25, 27, 48, 104, 110) and may be a general effect resulting from the actions of cytolytic toxins on the membranes of susceptible cells. It is easy to envision how these various physiologic effects on cells could contribute to disease, but their significance in the disease process is difficult to prove directly.

Bramley et al. (23) used the mouse mastitis model to show that intramammary inoculation with alpha-toxin-negative mutants (generated by allelic replacement of the toxin gene) causes acute mastitis but not death in the lactating mouse. In contrast, the alpha-toxin-positive wild-type strain causes acute mastitis and death of 60% of the mice within 48 h. Both mutant and wild-type strains are phagocytosed similarly up to 8 h postinfection. At 24 h, the neutrophils in the glands infected with the wild-type alpha-toxin-positive strain become degenerate, and there is a paucity of fresh neutrophils. The severity of the infection is clearly altered by loss of alpha-toxin production, which may be related to the effects of alpha toxin on neutrophils (107).

S. aureus Delta Toxin

S. aureus delta toxin is a 26-residue secreted peptide (MAQDIISTIGDLVKWIIDTVDK-FTKK) (40) that shares structural and some mechanistic traits with the bee venom melittin. Delta toxin is a relatively weak toxin: its 50% lethal dose is near 120 mg/kg of body weight for mice (73) and 30 mg/kg for guinea pigs. These values indicate that delta toxin is about 10,000 to 100,000 times less toxic than *S. aureus* alpha toxin and the thiol-activated toxins. If the sequence of delta toxin is analyzed by a helical wheel program (Fig. 7), an amphipathic structure with a hydrophobic face and a hydrophilic face is apparent (41). This is consistent with previous findings (39) that showed an appreciable amount of alpha-helical structure in delta toxin. The helical nature of delta toxin was confirmed by nuclear magnetic resonance studies (76) of delta toxin associated with dodecylphosphocholine micelles. Residues 5 to 23 form an extended helix with an amphipathic structure.

CYTOLYTIC MECHANISM

Delta toxin may form a pore in the membrane by the aggregation of monomers into a barrel stave structure (42). The size of delta toxin in

FIGURE 7 Helical wheel analysis of *S. aureus* delta toxin. Hydrophobic residues (outlined letters) are on the left, and charged or hydrophilic residues are on the right (solid letters).

solution has been estimated to be 210 kDa (17); however, in the presence of Tween 80, this complex is reduced to approximately 20 kDa, which corresponds to the size of a hexameric complex (41). Some evidence for the oligomeric structure has been obtained from both patch clamp (79) and molecular modeling (108) studies. It has been suggested that delta toxin may aggregate on the membrane surface and then insert into the bilayer, forming antiparallel, hexameric barrels in the membrane (103, 122). Although these studies suggest that delta toxin forms these barrel stave-type pores in the membrane, the oligomeric structure (if formed) of the toxin in the membrane still remains ambiguous.

Although the majority of the work on delta toxin concerns its interaction with membranes, other intriguing aspects of this toxin are also not understood. Delta toxin is secreted by the staphylococci, although secretion does not appear to follow a signal peptide-dependent process, simply because the delta toxin gene (*hld*) does not appear to encode a signal peptide (66). Two ribosome-binding sites occur within this region (shown in Fig. 8 as SD-1 and SD-2).

```
SD-1                                                                      SD-2
GATGGAAAATAGTTG ATG AGT TGT TTA ATT TTA AGA ATT TTT ATC TTA ATT AAG GAA GGA GTG
                Ser Met ser cys leu ile leu arg ile phe ile leu ile lys glu gly val

ATT TCA ATG GCA CAA GAT ATC ATT TCA ACA ATC GGT GAC TTA GTA AAA TGG ATT ATC GAC
ile ser met ala gln asp ile ile ser thr ile gly asp leu val lys trp ile ile asp

ACA GTG AAC AAA TTC ACT AAA AAA TAA
thr val asn lys phe thr lys lys OCH
```

FIGURE 8 Gene sequence of the delta toxin gene (*hld*). The sequence of the extracellular form is shown in italics. Translation starts at the second ribosome-binding site (SD-2) and accounts for at least 90% of the extracellular delta toxin (40).

The sequence of the extracellular form of delta toxin corresponds to the sequence that starts with the methionine following the SD-2 ribosome-binding site. It is thought that the majority of delta toxin is synthesized from SD-2, since 90% of the extracellular form contains a formylmethionine (40). Therefore, delta toxin is secreted without a signal peptide, unless the toxin itself acts as a signal peptide. It is possible that the remaining 10% of the delta toxin, which lacks a formylmethionine, is synthesized from the SD-1 promoter. However, the additional coding sequence still does not resemble the structure of a normal signal peptide, and this explanation cannot account for the majority of the extracellular toxin. Whether delta toxin is secreted by a *sec*-dependent secretion system in *S. aureus* or another, as yet undefined mechanism, is not known.

The gene for delta toxin occurs upstream from and is transcribed in the opposite direction of the *agr* gene complex, which comprises three genes (*agrA, agrB, and agrC*) (66, 98) that are part of a regulatory system involved in controlling the expression of at least 14 different extracellular proteins of *S. aureus* (65). Delta toxin is transcribed from its promoter, which is located immediately upstream of the *agr* locus, and its expression is controlled by the *agr* locus (65). However, the *hld* transcript appears to be important in the *agr* regulation of the other extracellular proteins controlled by *agr*, since mutations that delete the *hld* gene result in the same phenotype as that of *agrA* deletion mutants (i.e., depressed levels of expression) (65). The expression of the delta-toxin protein does not appear to be important to this regulation, except for regulation of the expression of extracellular proteases. Therefore, the *hld* transcript, and to a lesser extent the delta toxin itself, appear to participate in the *agr* regulatory system.

PATHOGENESIS
Because of the low toxicity of delta toxin, it is difficult to envision a dominant role for it in *S. aureus* pathogenesis. Comparatively little information is available on the contribution of delta toxin to pathogenesis. Of the *S. aureus* strains isolated from animals with bovine mastitis, only 12% were delta toxin positive (78). Delta-toxin production may be correlated with *S. aureus* infant necrotizing enterocolitis (93); however, only three patients were examined in that study. Also, it was not proven that *S. aureus* is actually the etiologic agent of infant necrotizing enterocolitis. Delta toxin may be important in the disease involving coagulase-negative staphylococci (54, 75, 111, 128), but no direct evidence supports a role of delta toxin in disease associated with either *S. aureus* or the coagulase-negative staphylococci. The reports mainly consist of correlative data that show delta-toxin production in staphylococcal isolates from patients with various disease syndromes, and in most cases it was not clear that the staphylococcal species were the etiologic agents of the disease.

Clostridium septicum Alpha Toxin
C. septicum alpha toxin was originally described in 1944 by Bernheimer (7), who noted that when proteins from the culture supernatants of

C. septicum were partially fractionated by then-available means, the lethal and hemolytic activities appeared to be the same protein. Since that time, little has been written on the alpha toxin until quite recently, when Ballard et al. (3) purified the lethal activity from *C. septicum* BX96, one of the original Bernheimer strains. The results of that work proved the original hypothesis of Bernheimer that alpha toxin was both hemolytic and lethal. Alpha toxin was determined to be a protein of approximately 48-kDa mass that exhibited approximately 106 hemolytic units per mg. The 50% lethal dose for mice was estimated to be approximately 10 μg/kg. When antibody to the pure toxin was made and used to probe an immunoblot of crude supernatant proteins from other *C. septicum* isolates, all exhibited a single band with the same molecular weight as the purified alpha toxin. To date, we have not found a *C. septicum* isolate that does not produce alpha toxin (unpublished observations).

CYTOLYTIC MECHANISM

During the storage of alpha toxin, we noted that with time, a smaller fragment of approximately 44 kDa was generated, apparently by proteolytic cleavage. In subsequent work, Ballard et al. (4) demonstrated that alpha toxin requires proteolytic cleavage for activity and that this cleavage occurs in the carboxy terminus of the toxin and removes approximately 4 kDa of the toxin. The activation of protoxin could be achieved with a variety of serine proteases (trypsin, subtilisin, chymotrypsin, proteinase K), with trypsin giving a single product that resembles the native nicked toxin. All of the proteases gave slightly different cleavage products, but none were smaller than the trypsin-generated product or the native nicked product, which indicates that the 44-kDa form is the smallest active fragment that can be generated.

In a subsequent study of the mechanism of alpha toxin, Ballard et al. (4) determined that erythrocytes treated with alpha toxin prelytically release potassium, which is typical of cells treated with a pore-forming toxin. Planar membrane studies confirmed that proteolytically activated alpha toxin forms channels in bilayers composed of cholesterol and phospholipid. These pores are approximately 1.5 nm in size and exhibit weak anion selectivity. The protoxin form of alpha toxin is not capable of lysing cells and does not exhibit any of the pore-forming characteristics of the proteolytically activated toxin. Why the protoxin is cytolytically inactive remains to be completely understood; however, the proteolytically activated toxin is able to oligomerize on erythrocyte membranes, but the protoxin remains monomeric. This finding suggests that the carboxy-terminal tail obscures one of the domains involved in oligomerization of the toxin and therefore prevents the oligomerization process. Whether the carboxy-terminal tail that is cleaved by the protease remains attached to the rest of the toxin after its cleavage remains unclear. Ballard et al. (4) suggested that the carboxy-terminal tail acts as an intramolecular chaperone sequence and may be lost after cleavage.

The characteristics of the alpha-toxin mechanism do not resemble those of any known clostridial toxin and are not similar to those of any characterized toxin from the gram-positive bacteria. Surprisingly, the cytolytic mechanism of alpha toxin does bear a remarkable resemblance to the mechanism of the cytolytic protein aerolysin, which is produced by *Aeromonas hydrophila* (29, 45, 133, 134). Both toxins require a proteolytic activation in the carboxy-terminal end of the protein that can be accomplished by several proteases, the most effective being trypsin. Both toxins oligomerize into structures resembling hexamers or heptamers on membranes, and the oligomers of both toxins are resistant to dissociation by sodium dodecyl sulfate, thiol, and heat. The amino acid sequences of *C. septicum* alpha toxin (2b) and aerolysin (57, 59) exhibit approximately 27% identity, suggesting that they may be distantly related. The primary structure of alpha toxin does not contain any extended stretches of hydrophobic amino acids that would suggest the presence of a transmembrane domain. Although no sequence relatedness to the thiol-activated cystolysins and *S. aureus* alpha toxin is observed, *C. septicum* alpha toxin, like these toxins, does not exhibit any extended regions of hydrophobic residues in its structure. Thus, all of the toxins addressed to this point, except for *S. aureus* delta toxin, appear to have in common two distinct characteristics: a relatively hydrophilic primary structure lacking extended regions of hydrophobic residues and an oligomerization-driven cytolytic process.

PATHOGENESIS

Comparatively little is known about the role of alpha toxin in disease. *C. septicum* causes a fulminant nontraumatic gas gangrene, mostly in individuals with various predisposing diseases, including colon cancer, leukemia, diabetes, and neutropenia (5, 32, 64, 114, 119, 135). However, *C. septicum* disease can also occur in healthy individuals for unclear reasons. The likely source of the *C. septicum* is thought to be the bowel, but this is unproven. The incidence of *C. septicum* is highest in persons suffering from a colonic malignancy (72). A patient who survives *C. septicum* gangrene is invariably checked for a colinic malignancy, since the association rate of *C. septicum* infection with colonic cancer is high (72). The mortality rate from *C. septicum* infection is 60 to 100%, a rate largely due to the fulminant nature of the disease and the difficulty in diagnosing it in its early stages. Typically, patients who have developed myonecrosis go into a deep shock from which most do not recover, even if the infection is rapidly cleared by the use of antibiotics (102). The basis for the development of this shock is unclear, but alpha toxin is a suspect in this aspect of the disease. Mice injected with pure alpha toxin typically undergo cardiopulmonary distress that gives the appearance of shock (3). Also, mice immunized with alpha toxin appear to be offered partial protection against challenge with *C. septicum* (3), but that study did not determine if all mice had mounted a successful response to the alpha toxin. It is possible that protective levels of neutralizing antibody for alpha toxin are not easily achieved, and therefore, many of the mice may not have had the necessary level of protection against the alpha toxin that was produced during the infection.

The full role of alpha toxin in *C. septicum* disease is only now beginning to be explored, and many questions remain. One curious aspect of the system is why alpha toxin is produced as a protoxin. Many cytolytic proteins are produced in fully functional forms by the gram-positive bacteria. Therefore, why alpha toxin is produced as an inactive precursor is unclear. These and other questions concerning the contribution of alpha toxin to disease, the structure-function relationships of alpha toxin, and the early stages of *C. septicum* nontraumatic gangrene will have to be investigated to fully understand this disease.

DISCUSSION

Although the literature abounds with work on cytolytic pore-forming toxins, some of the more intriguing aspects of these toxins, particularly those events that take place after the toxins bind to the membrane, remain enigmatic. One of the central problems in understanding pore formation by these toxins is learning how these water-soluble toxins undergo their transition into the membrane bilayer. This transition is remarkable in view of the fact that these toxins do not exhibit extended sequences of hydrophobic residues within their structures (Fig. 9). These toxins must reorganize their structures to facilitate their transition into the membrane, presumably by organizing a hydrophobic surface that can interact with the lipid bilayer. The reorganization of the structures of these toxins may be driven by the processes of membrane binding and oligomerization. Upon membrane binding, these toxins may undergo a conformational change that facilitates insertion of the monomer into the membrane, or perhaps interfacial domains that facilitate oligomerization are exposed. It is not known whether the monomeric forms of these toxins insert into the bilayer and then oligomerize or whether oligomerization precedes insertion of the complex.

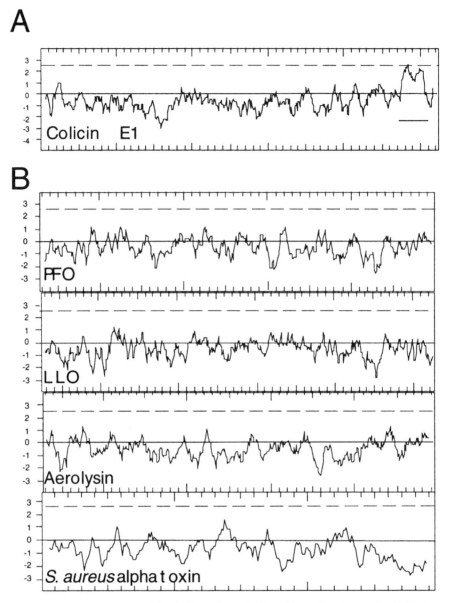

FIGURE 9 Comparison of the hydropathies of selected pore-forming toxins. (A) Hydropathy plot for colicin E1, which contains a well-defined transmembrane domain at its carboxy terminus (solid bar). (B) Hydropathy analyses of three thiol-activated cytolysins and of *S. aureus* alpha toxin and *A. hydrophila* aerolysin, the latter being an example of a pore-forming toxin from a gram–negative bacterium. In the last examples, the relative hydrophobicity of the colicin transmembrane domain is denoted by the dashed line in each plot. Note the the absence of hydrophobic domains in PFO, LLO, aerolysin, and *S. aureus* alpha toxin that are comparable to the one in colicin E1.

A recent report on the mechanism of aerolysin indicated that it oligomerizes either in solution and/or on the cell surface prior to its insertion into the membrane (129). The work on *S. aureus* alpha toxin by Walker et al. (130) and Menzies and Kernodle (82) also suggests that oligomerization may precede insertion. Oligomerization of the monomers on the membrane makes sense, since this process could be the driving force behind the rearrangement of the monomers to form the amphipathic complex. Perhaps the activation energy of the oligomerization process (50) drives the reorganization of the monomers into an amphipathic complex that then inserts into the membrane. Since most of these proteins do not contain extended hydrophobic sequences within their primary structures, workers have not been able to identify domains of these proteins that may interact with the lipid bilayer by inspecting their amino acid sequences. The questions surrounding the process of membrane insertion should be amenable to approaches that are currently being applied to the study of the colicins (74, 113, 117), a group of cytolytic proteins that apparently do not aggregate and have large transmembrane domains within their primary structures. In these studies, environmentally sensitive fluorescent dyes or spin label probes have been used to modify strategically placed cysteine residues that have been inserted into the structures of the toxins by in vitro mutagenesis. These probes are then used to follow the polar to nonpolar transitions of these residues during the process of membrane insertion. Similar approaches may reveal the nature of the toxin reorganization described here that facilitates insertion of these toxins into the membrane.

The role of these toxins in pathogenesis has been explored to a limited extent. The best-understood role in pathogenesis is that of LLO, probably because *L. monocytogenes* is a facultative intracellular parasite and the activity of LLO is limited to the interior of the phagosome. In contrast, the role of a cytolytic toxin in disease becomes increasingly difficult to assess if the toxin is secreted by an extracellular pathogen; in these situations, the toxin can potentially affect many cell types both close to and distant from the foci of infection. Most cytolytic toxins described here are capable of stimulating the arachidonic pathway, which results in the generation of a variety of by-products (12, 25–27, 110, 121, 127). These types of effects are often mediated by the influx of Ca^{2+} through an unregulated pore, and hence, it is not surprising that these phenomena are observed in eukaryotic cells treated with pore-forming toxins. It is reasonable to assume that these cellular effects are important in disease, but a direct role for these phenomena in disease has yet to be demonstrated. Most, if not all, of these cytolytic toxins probably contribute to disease, but ultimately proving the specific nature and extent of that contribution will pose a significant challenge to investigators.

The study of pore-forming cytolysins has been going on for many years, and yet they still sequester within their structures the secrets to how these hydrophilic proteins disrupt membranes. Although we are beginning to understand how these toxins affect the membranes of eukaryotic cells and how they may contribute to disease, some of the most challenging problems remain to be solved by present and future workers in the field of cytolytic proteins.

REFERENCES

1. **Alouf, J. E., and C. Geoffrey.** 1984. Structure activity relationships in the sulfhydryl-activated toxins, p. 165–171. *In* J. E. Alouf, F. J. Fehrenbach, J. H. Freer, and J. Jeljaszewicz (ed.), *Bacterial Protein Toxins.* Academic Press, London.

2. **Arvand, M., S. Bhakdi, B. Dahlback, and K. T. Preissner.** 1990. *Staphylococcus aureus* alphatoxin attack on human platelets promotes assembly of the prothrombinase complex. *J. Biol. Chem.* **265:**14377–14381.

2a. **Awad, M. M., A. E. Bryant, D. L. Stevens, and J. I. Rood.** Virulence studies on chromosomal α-toxin and Θ-toxin mutants constructed by allelic exchange provide genetic evidence for the essential role of α-toxin in *Clostridium perfringens*-mediated gas gangrene. *Mol. Microbiol.,* in press.

2b. **Ballard, J., J. Crabtree, B. A. Roe, and R. K. Tweten.** 1995. The primary structure of *Clostridium septicum* alpha toxin exhibits similarity with

Aeromonas hydrophila aerolysin. *Infect. Immun.* **63:** 340–344.

3. **Ballard, J., A. Bryant, D. Stevens, and R. K. Tweten.** 1992. Purification and characterization of the lethal toxin (alpha toxin) of *Clostridium septicum*. *Infect. Immun.* **60:**784–790.

4. **Ballard, J., Y. Sokolov, W.-L. Yuan, B. L. Kagan, and R. K. Tweten.** 1993. Activation and mechanism of *Clostridium septicum* alpha toxin. *Mol. Microbiol.* **10:**627–634.

5. **Barza, M., M. Pins, R. Sacknoff, R. E. Scully, E. T. Ryan, R. Mangrulkar, and D. Hamer.** 1993. An 81-year-old man with pain and crepitus in the shoulder—*Clostridium septicum* myonecrosis (gas gangrene)—adenocarcinoma, cecum, with metastasis to right kidney and pericolic and peripancreatic lymph nodes. *N. Engl. J. Med.* **328:**340–346.

6. **Belmonte, G., L. Cescatti, B. Ferrari, T. Nicolussi, M. Ropele, and G. Menestrina.** 1987. Pore formation by *Staphylococcus aureus* alpha-toxin in lipid bilayers. Dependence upon temperature and toxin concentration. *Eur. Biophys. J.* **14:**349–358.

7. **Bernheimer, A. W.** 1944. Parallelism in the lethal and hemolytic activity of the toxin of *Clostridium septicum*. *J. Exp. Med.* **80:**309–320.

8. **Bernheimer, A. W., and B. Rudy.** 1986. Interactions between membranes and cytolytic peptides. *Biochim. Biophys. Acta* **864:**123–141.

9. **Berry, A. M., J. C. Paton, and D. Hansman.** 1992. Effect of insertional inactivation of the genes encoding pneumolysin and autolysin on the virulence of *Streptococcus pneumoniae* type-3. *Microb. Pathog.* **12:**87–93.

10. **Bhakdi, S., R. Fussle, and J. J. Tranum.** 1981. Staphylococcal alpha-toxin: oligomerization of hydrophilic monomers to form amphiphilic hexamers induced through contact with deoxycholate detergent micelles. *Proc. Natl. Acad. Sci. USA* **78:** 5475–5479.

11. **Bhakdi, S., M. Muhly, S. Korom, and F. Hugo.** 1989. Release of interleukin-1 beta associated with potent cytocidal action of staphylococcal alpha-toxin on human monocytes. *Infect. Immun.* **57:**3512–3519.

12. **Bhakdi, S., N. Suttorp, W. Seeger, R. Fussle, and J. J. Tranum.** 1984. Molecular basis for the pathogenicity of *Staphylococcus aureus* alpha-toxins. *Immun. Infect.* **12:**279–285.

13. **Bhakdi, S., and J. J. Tranum.** 1991. Alpha-toxin of *Staphylococcus aureus*. *Microbiol. Rev.* **55:** 733–751.

14. **Bhakdi, S., and J. Tranum-Jensen.** 1988. Damage to cell membranes by pore-forming bacterial cytolysins. *Prog. Allergy* **40:**1–43.

15. **Bhakdi, S., J. Tranum-Jensen, and A. Sziegoleit.** 1984. Structure of streptolysin-O in target membranes, p. 173–180. *In* J. E. Alouf, F. J. Fehrenbach, J. H. Freer, and J. Jeljaszewicz (ed.), *Bacterial Protein Toxins.* Academic Press, London.

16. **Bielecki, J., P. Youngman, P. Connelly, and D. A. Portnoy.** 1990. *Bacillus subtilis* expressing a haemolysin gene from *Listeria monocytogenes* can grow in mammalian cells. *Nature* (London) **345:** 175–176.

17. **Birkbeck, T. H., and D. D. Whitelaw.** 1980. Immunogenicity and molecular characterisation of staphylococcal delta haemolysin. *J. Med. Microbiol.* **13:**213–221.

18. **Blomqvist, L., T. Bergman, M. Thelestam, and H. Jornvall.** 1987. Characterization of domain borders and of a naturally occurring major fragment of staphylococcal alpha-toxin. *FEBS Lett.* **211:**127–132.

19. **Blomqvist, L., and M. Thelestam.** 1986. A staphylococcal alpha-toxin fragment. Its characterization and use for mapping biologically-active regions of alpha-toxin. *Acta Pathol. Microbiol. Immunol. Scand. Sect. B Microbiol.* **94:**277–283.

20. **Blomqvist, L., and M. Thelestam.** 1988. Oligomerization of ^{3}H-labelled staphylococcal alpha-toxin and fragments on adrenocortical Y1 tumour cells. *Microb. Pathog.* **4:**223–229.

21. **Boulnois, G. J., T. J. Mitchell, F. K. Saunders, F. J. Mendez, and P. W. Andrew.** 1990. Structure and function of pneumolysin, the thiol-activated toxin of *Streptococcus pneumoniae,* p. 43–51. *In* R. Rappuoli, J. E. Alouf, P. Falmagne, F. J. Fehrenbach, J. Freer, R. Gross, J. Jeljaszewicz, C. Montecucco, M. Tomasi, T. Wadström, and B. Witholt (ed.), *Bacterial Protein Toxins.* Gustav Fischer Verlag, Stuttgart, Germany.

22. **Boulnois, G. J., J. C. Paton, T. J. Mitchell, and P. W. Andrew.** 1991. Structure and function of pneumolysin, the multifunctional, thiol-activated toxin of *Streptococcus pneumoniae. Mol. Microbiol.* **5:**2611–2616.

23. **Bramley, A. J., A. H. Patel, M. O'Reilly, R. Foster, and T. J. Foster.** 1989. Roles of alpha-toxin and beta-toxin in virulence of *Staphylococcus aureus* for the mouse mammary gland. *Infect. Immun.* **57:**2489–2494.

24. **Braun, V., and T. Focareta.** 1991. Pore-forming bacterial protein hemolysins (cytolysins). *Crit. Rev. Microbiol.* **18:**115–158.

25. **Bremm, K. D., H. J. Brom, J. E. Alouf, W. König, B. Spur, A. Crea, and W. Peters.** 1984. Generation of leukotrienes from human granulocytes by alveolysin from *Bacillus alvei. Infect. Immun.* **44:**188–193.

26. **Bremm, K. D., J. Brom, W. Konig, B. Spur, A. Crea, S. Bhakdi, F. Lutz, and F. J. Fehrenbach.** 1983. Generation of leukotrienes and lipoxygenase factors from human polymorphonuclear granulocytes during bacterial phagocytosis

and interaction with bacterial exotoxins. *Zentralbl. Bakteriol. Mikrobiol. Hyg.* **254**:500–514.

27. **Bremm, K. D., W. König, P. Pfeiffer, I. Rauschen, K. Theobald, M. Thelestam, and J. E. Alouf.** 1985. Effect of thiol-activated toxins (streptolysin O, alveolysin, and theta toxin) on the generation of leukotrienes and leukotriene-inducing and -metabolizing enzymes from human polymorphonuclear granulocytes. *Infect. Immun.* **50**:844–851.

28. **Bremm, K. D., W. König, M. Thelestam, and J. E. Alouf.** 1987. Modulation of granulocyte functions by bacterial exotoxin and endotoxins. *Immunology* **62**:363–371.

29. **Buckley, J. T.** 1991. Secretion and mechanism of action of the hole-forming toxin aerolysin. *Experientia* **47**:418–419.

30. **Cassidy, P., and S. Harshman.** 1976. Biochemical studies on the binding of staphylococcal 125-I labeled alpha-toxin to rabbit erythrocytes. *Biochemistry* **15**:2348–2355.

31. **Cooney, J., Z. Kienle, T. J. Foster, and P. W. O'Toole.** 1993. The gamma-hemolysin locus of *Staphylococcus aureus* comprises three linked genes, two of which are identical to the genes for the F and S components of leukocidin. *Infect. Immun.* **61**:768–771.

32. **Corey, E. C.** 1991. Nontraumatic gas gangrene: case report and review of emergency therapeutics. *J. Emerg. Med.* **9**:431–436.

33. **Cossart, P., M. F. Vincente, J. Mengaud, F. Baquero, J. C. Perez-Diaz, and P. Berche.** 1989. Listeriolysin O is essential for the virulence of *Listeria monocytogenes*: direct evidence obtained by gene complementation. *Infect. Immun.* **57**:3629–3639.

34. **Cowell, J. L., K. Kim, and A. W. Bernheimer.** 1978. Alteration by cereolysin of the structure of cholesterol-containing membranes. *Biochim. Biophys. Acta* **507**:230–241.

35. **Duncan, J. L., and R. Schlegel.** 1975. Effect of streptolysin O on erythrocyte membranes, liposomes, and lipid dispersions. A protein-cholesterol interaction. *J. Cell Biol.* **67**:160–174.

35a. **Ferretti, J.** Personal communication.

36. **Finck-Barbancon, V., G. Duportail, O. Meunier, and D. A. Colin.** 1993. Pore formation by a two-component leukocidin from *Staphylococcus aureus* within the membrane of human polymorphonuclear leukocytes. *Biochim. Biophys. Acta* **1182**:275–282.

37. **Finck-Barbancon, V., G. Prevost, and Y. Piemont.** 1991. Improved purification of leukocidin from *Staphylococcus aureus* and toxin distribution among hospital strains. *Res. Microbiol.* **142**:75–85.

38. **Fink, D., M. L. Contreras, P. I. Lelkes, and P. Lazarovici.** 1989. *Staphylococcus aureus* alpha-toxin activates phospholipases and induces a Ca^{2+} influx in PC12 cells. *Cell. Signal.* **1**:387–393.

39. **Fitton, J. E.** 1981. Physicochemical studies on delta haemolysin, a staphylococcal cytolytic polypeptide. *FEBS Lett.* **130**:257–260.

40. **Fitton, J. E., A. Dell, and W. V. Shaw.** 1980. The amino acid sequence of the delta haemolysin of *Staphylococcus aureus*. *FEBS Lett.* **115**:209–212.

41. **Freer, J. H., and T. H. Birkbeck.** 1982. Possible conformation of delta-lysin, a membrane-damaging peptide of *Staphylococcus aureus*. *J. Theor. Biol.* **94**:535–540.

42. **Freer, J. H., T. H. Birkbeck, and M. Bhakoo.** 1984. Interaction of staphylococcal delta-lysin with phospholipid monolayers and bilayers—a short review, p. 181–189. *In* J. E. Alouf, F. J. Fehrenbach, J. H. Freer, and J. Jeljaszewicz (ed.), *Bacterial Protein Toxins*. Academic Press, London.

43. **Füssle, R., S. Bhakdi, A. Sziegoleit, J. Tranum-Jensen, T. Kranz, and H. L. Wellensiek.** 1981. On the mechanism of membrane damage by *S. aureus* alpha toxin. *J. Cell Biol.* **91**:83–94.

44. **Fussle, R., and A. Sziegoleit.** 1988. Incorporation of staphylococcal alpha-toxin in glutaraldehyde fixed erythrocytes. *Zentralbl. Bakteriol. Mikrobiol. Hyg. Ser. A* **269**:346–354.

45. **Garland, W. J., and J. T. Buckley.** 1988. The cytolytic toxin aerolysin must aggregate to disrupt erythrocytes, and aggregation is stimulated by human glycophorin. *Infect. Immun.* **56**:1249–1253.

46. **Geoffroy, C., J. Mengaud, J. E. Alouf, and P. Cossart.** 1990. Alveolysin, the thiol-activated toxin of *Bacillus alvei,* is homologous to listeriolysin-O, perfringolysin-O, pneumolysin, and streptolysin-O and contains a single cysteine. *J. Bacteriol.* **172**:7301–7305.

47. **Gray, G. S., and M. Kehoe.** 1984. Primary sequence of the alpha-toxin gene from *Staphylococcus aureus* wood 46. *Infect. Immun.* **46**:615–618.

48. **Grimminger, F., C. Scholz, S. Bhakdi, and W. Seeger.** 1991. Subhemolytic doses of *Escherichia coli* hemolysin evoke large quantities of lipoxygenase products in human neutrophils. *J. Biol. Chem.* **266**:14262–14269.

49. **Haas, A., M. Dumbsky, and J. Kreft.** 1992. Listeriolysin genes: complete sequence of *ilo* from *Listeria ivanovii* and of *lso* from *Listeria seeligeri*. *Biochim. Biophys. Acta* **1130**:81–84.

50. **Harris, R. W., P. J. Sims, and R. K. Tweten.** 1991. Kinetic aspects of the aggregation of *Clostridium perfringens* theta toxin on erythrocyte membranes: a fluorescence energy transfer study. *J. Biol. Chem.* **266**:6936–6941.

51. **Harshman, S., J. E. Alouf, O. Siffert, and F. Baleux.** 1989. Reaction of staphylococcal alpha-toxin with peptide-induced antibodies. *Infect. Immun.* **57**:3856–3862.

52. **Harshman, S., P. Boquet, E. Duflot, J. E. Alouf, C. Montecucco, and E. Papini.** 1989. Staphylococcal alpha-toxin: a study of membrane penetration and pore formation. *J. Biol. Chem.* **264:**14978–14984.

53. **Harshman, S., N. Sugg, B. Gametchu, and R. W. Harrison.** 1986. Staphylococcal alpha-toxin: a structure-function study using a monoclonal antibody. *Toxicon* **24:**403–411.

54. **Hebert, G. A.** 1990. Hemolysins and other characteristics that help differentiate and biotype *Staphylococcus lugdunensis* and *Staphylococcus schleiferi. J. Clin. Microbiol.* **28:**2425–2431.

55. **Hebert, H., A. Olofsson, M. Thelestam, and E. Skriver.** 1992. Oligomer formation of staphylococcal alpha-toxin analyzed by electron microscopy and image processing. *FEMS Microbiol. Immunol.* **5:**5–12.

56. **Hildebrand, A., M. Pohl, and S. Bhakdi.** 1991. *Staphylococcus aureus* alpha-toxin. Dual mechanism of binding to target cells. *J. Biol. Chem.* **266:**17195–17200.

57. **Howard, S. P., W. J. Garland, M. J. Green, and J. T. Buckley.** 1987. Nucleotide sequence of the gene for the hole-forming toxin aerolysin of *Aeromonas hydrophila. J. Bacteriol.* **169:**2869–2871.

58. **Hunter, S. E., J. E. Brown, P. C. Oyston, J. Sakurai, and R. W. Titball.** 1993. Molecular genetic analysis of beta-toxin of *Clostridium perfringens* reveals sequence homology with alpha-toxin, gamma-toxin, and leukocidin of *Staphylococcus aureus. Infect. Immun.* **61:**3958–3965.

59. **Husslein, V., B. Huhle, T. Jarchau, R. Lurz, W. Goebel, and T. Chakraborty.** 1988. Nucleotide sequence and transcriptional analysis of the CaerA region of *Aeromonas sobria* encoding aerolysin and its regulatory region. *Mol. Microbiol.* **2:**507–517.

60. **Ikigai, H., and T. Nakae.** 1985. Conformational alteration in alpha-toxin from *Staphylococcus aureus* concomitant with the transformation of the water-soluble monomer to the membrane oligomer. *Biochem. Biophys. Res. Commun.* **130:**175–181.

61. **Ikigai, H., and T. Nakae.** 1987. Assembly of the alpha-toxin-hexamer of *Staphylococcus aureus* in the liposome membrane. *J. Biol. Chem.* **262:**2156–2160.

62. **Iwamoto, M., Y. Ohno-Iwashita, and S. Ando.** 1987. Role of the essential thiol group in the thiol-activated cytolysin from *Clostridium perfringens. Eur. J. Biochem.* **167:**425–430.

63. **Iwamoto, M., Y. Ohno-Iwashita, and S. Ando.** 1990. Effect of isolated C-terminal fragment of theta-toxin (perfringolysin-O) on toxin assembly and membrane lysis. *Eur. J. Biochem.* **194:**25–31.

64. **Jamison, J. P., and F. M. Ivey.** 1986. Nontraumatic clostridial myonecrosis, a case report. *Orthop. Rev.* **15:**658–663.

65. **Janzon, L., and S. Arvidson.** 1990. The role of the delta-lysin gene (*hld*) in the regulation of virulence genes by the accessory gene regulator (*agr*) in *Staphylococcus aureus. EMBO J.* **9:**1391–1399.

66. **Janzon, L., S. Löfdahl, and S. Arvidson.** 1989. Identification of the delta-lysin gene, *hld,* adjacent to the accessory gene regulator (*agr*) of *Staphylococcus aureus. Mol. Gen. Genet.* **219:**480–485.

66a. **Jones, S., and D. A. Portnoy.** Personal communication.

67. **Kamio, Y., A. Rahman, H. Nariya, T. Ozawa, and K. Izaki.** 1993. The two staphylococcal bi-component toxins, leukocidin and gamma-hemolysin, share one component in common. *FEBS Lett.* **321:**15–18.

68. **Kaplan, M. H., and J. E. Volanakis.** 1974. Interaction of C-reactive protein complexes with the complement system. I. Consumption of human complement associated with the reaction of C-reactive protein with pneumococcal C-polysaccharide and choline phosphatides, lecithin and sphingomyelin. *J. Immunol.* **112:**2135–2147.

69. **Kato, I., and M. Naiki.** 1975. Ganglioside and rabbit erythrocyte membrane receptor of staphylococcal alpha-toxin. *Infect. Immun.* **13:**289–291.

70. **Kehoe, M. A., L. Miller, J. A. Walker, and G. J. Boulnois.** 1987. Nucleotide sequence of the streptolysin O (SLO) gene: structural homologies between SLO and other membrane-damaging, thiol-activated toxins. *Infect. Immun.* **55:**3228–3232.

71. **Keough, K. M. W., and P. J. Davis.** 1984. Thermal analysis of membranes, p. 55–97. *In* M. Kates and L. A. Manson (ed.), *Membrane Fluidity.* Plenum Press, New York.

72. **Kornbluth, A. A., J. B. Danzig, and L. H. Bernstein.** 1989. *Clostridium septicum* infection and associated malignancy. Report of 2 cases and review of the literature. *Medicine* (Baltimore) **68:**30–37.

73. **Kreger, A. S., K.-S. Kim, F. Zaboretzky, and A. W. Bernheimer.** 1971. Purification and some properties of staphylococcal delta hemolysin. *Infect. Immun.* **3:**444–465.

74. **Lakey, J. H., D. Duché, J.-M. González-Manas, D. Baty, and F. Pattus.** 1993. Fluorescence energy transfer distance measurements. The hydrophobic helical hairpin of colicin A in the membrane bound state. *J. Mol. Biol.* **230:**1055–1067.

75. **Lambe, D. J., K. P. Ferguson, J. L. Keplinger, C. G. Gemmell, and J. H. Kalbfleisch.** 1990. Pathogenicity of *Staphylococcus lugdunensis, Staphylococcus schleiferi,* and three other coagulase-negative staphylococci in a mouse model and pos-

sible virulence factors. *Can. J. Microbiol.* **36:**455–463.

76. **Lee, K. H., J. E. Fitton, and K. Wuthrich.** 1987. Nuclear magnetic resonance investigation of the conformation of delta-haemolysin bound to dodecylphosphocholine micelles. *Biochim. Biophys. Acta* **911:**144–153.

77. **Maharaj, I., and H. B. Fackrell.** 1980. Rabbit erythrocyte band 3: a receptor for staphylococcal alpha toxin. *Can. J. Microbiol.* **26:**524–531.

78. **Matsunaga, T., S. Kamata, N. Kakiichi, and K. Uchida.** 1993. Characteristics of *Staphylococcus aureus* isolated from peracute, acute and chronic bovine mastitis. *J. Vet. Med. Sci.* **55:**297–300.

79. **Mellor, I. R., D. H. Thomas, and M. S. Sansom.** 1988. Properties of ion channels formed by *Staphylococcus aureus* delta-toxin. *Biochim. Biophys. Acta* **942:**280–294.

80. **Menestrina, G.** 1986. Ionic channels formed by *Staphylococcus aureus* alpha-toxin: voltage-dependent inhibition by divalent and trivalent cations. *J. Membr. Biol.* **90:**177–190.

81. **Mengaud, J., M. F. Vicente, J. Chenevert, J. M. Pereira, C. Geoffroy, S. B. Gicquel, F. Baquero, D. J. Perez, and P. Cossart.** 1988. Expression in *Escherichia coli* and sequence analysis of the listeriolysin O determinant of *Listeria monocytogenes. Infect. Immun.* **56:**766–772.

82. **Menzies, B. E., and D. S. Kernodle.** 1994. Site-directed mutagenesis of the alpha toxin gene of *Staphylococcus aureus:* role of the histidines in toxin activity and in a murine model. *Infect. Immun.* **62:**1843–1847.

83. **Michel, E., K. A. Reich, R. Favier, P. Berche, and P. Cossart.** 1990. Attenuated mutants of the intracellular bacterium *Listeria monocytogenes* obtained by single amino acid substitutions in listeriolysin O. *Mol. Microbiol.* **4:**2167–2178.

84. **Mitchell, T. J., P. W. Andrew, F. K. Saunders, A. N. Smith, and G. J. Boulnois.** 1991. Complement activation and antibody binding by pneumolysin via a region of the toxin homologous to a human acute-phase protein. *Mol. Microbiol.* **5:**1883–1888.

85. **Mitsui, K., T. Sekiya, S. Okamura, Y. Nozawa, and J. Hase.** 1979. Ring formation of perfringolysin O as revealed by negative stain electron microscopy. *Biochim. Biophys. Acta* **558:**307–313.

86. **Ohno-Iwashita, Y., M. Iwamoto, K. Mitsui, S. Ando, and S. Iwashita.** 1991. A cytolysin, theta-toxin, preferentially binds to membrane cholesterol surrounded by phospholipids with 18-carbon hydrocarbon chains in cholesterol-rich region. *J. Biochem.* (Tokyo) **110:**369–375.

87. **Ohno-Iwashita, Y., M. Iwamoto, K. Mitsui, H. Kawasaki, and S. Ando.** 1986. Cold-labile hemolysin produced by limited proteolysis of theta-toxin from *Clostridium perfringens. Biochemistry* **25:**6048–6053.

87a. **Okamura, K., et al.** Unpublished data.

88. **Olofsson, A., H. Hebert, and M. Thelestam.** 1991. The structure of *Staphylococcus aureus* alpha-toxin: effects of trypsin treatment. *J. Struct. Biol.* **106:**199–204.

89. **Olofsson, A., H. Hebert, and M. Thelestam.** 1993. The projection structure of perfringolysin-O (*Clostridium perfringens* omicron-toxin). *FEBS Lett.* **319:**125–127.

90. **Olofsson, A., U. Kaveus, I. Hacksell, M. Thelestam, and H. Hebert.** 1990. Crystalline layers and three-dimensional structure of *Staphylococcus aureus* alpha-toxin. *J. Mol. Biol.* **214:**299–306.

91. **Olofsson, A., U. Kaveus, M. Thelestam, and H. Hebert.** 1988. The projection structure of alpha-toxin from *Staphylococcus aureus* in human platelet membranes as analyzed by electron microscopy and image processing. *J. Ultrastruct. Mol. Struct. Res.* **100:**194–200.

92. **Olofsson, A., U. Kaveus, M. Thelestam, and H. Hebert.** 1992. The three-dimensional structure of trypsin-treated *Staphylococcus aureus* alpha-toxin. *J. Struct. Biol.* **108:**238–244.

93. **Overturf, G. D., M. P. Sherman, D. W. Scheifele, and L. C. Wong.** 1990. Neonatal necrotizing enterocolitis associated with delta toxin-producing methicillin-resistant *Staphylococcus aureus. Pediatr. Infect. Dis. J.* **9:**88–91.

94. **Palmer, M., U. Weller, M. Messner, and S. Bhakdi.** 1993. Altered pore-forming properties of proteolytically nicked staphylococcal alpha-toxin. *J. Biol. Chem.* **268:**11963–11967.

95. **Parrisius, J., S. Bhakdi, M. Roth, J. J. Tranum, W. Goebel, and H. P. Seeliger.** 1986. Production of listeriolysin by beta-hemolytic strains of *Listeria monocytogenes. Infect. Immun.* **51:**314–319.

96. **Paton, J. C., R. A. Lock, and D. J. Hansman.** 1983. Effect of immunization with pneumolysin on survival time of mice challenged with *Streptococcus pneumoniae. Infect. Immun.* **40:**548–552.

97. **Pederzolli, C., L. Cescatti, and G. Menestrina.** 1991. Chemical modification of *Staphylococcus aureus* alpha-toxin by diethylpyrocarbonate: role of histidines in its membrane-damaging properties. *J. Membr. Biol.* **119:**41–52.

98. **Peng, H.-L., R. P. Novick, B. Kreiswirth, J. Kornblum, and P. Schlievert.** 1988. Cloning, characterization, and sequencing of an accessory gene regulator *(agr)* in *Staphylococcus aureus. J. Bacteriol.* **170:**4365–4372.

99. **Pinkney, M., E. Beachey, and M. Kehoe.** 1989. The thiol-activated toxin streptolysin O

does not require a thiol group for activity. *Infect. Immun.* **57**:2553–2558.

100. **Portnoy, D., P. S. Jacks, and D. Hinrichs.** 1988. The role of hemolysin for intracellular growth of *Listeria monocytogenes. J. Exp. Med.* **167**:1459–1471.

101. **Portnoy, D. A., R. K. Tweten, M. Kehoe, and J. Bielecki.** 1992. The capacity of listeriolysin O, streptolysin O, and perfringolysin O to mediate growth of *Bacillus subtilis* within mammalian cells. *Infect. Immun.* **60**:2710–2717.

102. **Price, C. I., A. B. Hollingsworth, and R. K. Tweten.** 1991. Atraumatic gas gangrene caused by *Clostridium septicum. Abstr. 43rd Annu. Meet. Southwestern Surg. Congr.*

103. **Raghunathan, G., P. Seetharamulu, B. R. Brooks, and H. R. Guy.** 1990. Models of delta-hemolysin membrane channels and crystal structures. *Protein Struct. Funct. Genet.* **8**: 213–225.

104. **Raulf, M., J. E. Alouf, and W. Konig.** 1990. Effect of staphylococcal delta-toxin and bee venom peptide melittin on leukotriene induction and metabolism of human polymorphonuclear granulocytes. *Infect. Immun.* **58**:2678–2682.

105. **Robb-Smith, A. H. T.** 1945. Tissue changes induced by *Cl. whelchii* type A filtrates. *Lancet* **ii**: 362–368.

106. **Rubins, J. B., P. G. Duane, D. Charboneau, and E. N. Janoff.** 1992. Toxicity of pneumolysin to pulmonary endothelial cells *in vitro. Infect. Immun.* **60**:1740–1746.

107. **Russel, R. J., P. C. Wilkinson, R. J. McInroy, S. McKay, A. C. McCartney, and J. P. Arbuthnott.** 1976. Effects of staphylococcal products on locomotion and chemotaxis of human blood neutrophils and monocytes. *J. Med. Microbiol.* **8**:433–449.

108. **Sansom, M. S., I. D. Kerr, and I. R. Mellor.** 1991. Ion channels formed by amphipathic helical peptides. A molecular modelling study. *Eur. Biophys. J.* **20**:229–240.

109. **Saunders, K. F., T. J. Mitchell, J. A. Walker, P. W. Andrew, and G. J. Boulnois.** 1989. Pneumolysin, the thiol-activated toxin of *Streptococcus pneumoniae,* does not require a thiol group for in vitro activity. *Infect. Immun.* **57**:2547–2552.

110. **Scheffer, J., W. König, V. Braun, and W. Goebel.** 1988. Comparison of four hemolysin-producing organisms (*Escherichia coli, Serratia marcescens, Aeromonas hydrophila,* and *Listeria monocytogenes*) for release of inflammatory mediators from various cells. *J. Clin. Microbiol.* **26**:544–551.

111. **Scheifele, D. W., G. L. Bjornson, R. A. Dyer, and J. E. Dimmick.** 1987. Delta-like toxin produced by coagulase-negative staphylococci is associated with neonatal necrotizing enterocolitis. *Infect. Immun.* **55**:2268–2273.

112. **Sekiya, K., R. Satoh, H. Danbara, and Y. Futaesaku.** 1993. A ring-shaped structure with a crown formed by streptolysin-O on the erythrocyte membrane. *J. Bacteriol.* **175**:5953–5961.

113. **Shin, Y.-K., C. Levinthal, F. Levinthal, and W. L. Hubbell.** 1993. Colicin E1 binding to membranes: time-resolved studies of spin labeled mutants. *Science* **259**:960–963.

114. **Sjlin, S. U., and A. K. Hansen.** 1991. *Clostridium septicum* gas gangrene and an intestinal malignant lesion. A case report. *J. Bone Joint Surg.* (America) **73**:772–773.

115. **Smyth, C. J., and J. L. Duncan.** 1978. Thiol-activated (oxygen-labile) cytolysins, p. 129–183, *In* J. Jeljaszewicz and T. Wadstrom (ed.), *Bacterial Toxins and Cell Membranes.* Academic Press, London.

116. **Smyth, C. J., J. H. Freer, and J. P. Arbuthnot.** 1975. Interaction of *Clostridium perfringens* theta-haemolysin, a contaminant of commercial phospholipase C, with erythrocyte ghost membranes and lipid dispersions. A morphological study. *Biochim. Biophys. Acta* **382**:479–493.

117. **Steer, B. A., and A. R. Merrill.** 1994. The colicin E1 insertion-competent state: detection of structural changes using fluorescence resonance energy transfer. *Biochemistry* **33**:1108–1115.

118. **Stevens, D. L., J. Mitten, and C. Henry.** 1987. Effects of alpha and theta toxins from *Clostridium perfringens* on human polymorphonuclear leukocytes. *J. Infect. Dis.* **156**:324–333.

119. **Stevens, D. L., D. M. Musher, D. A. Watson, H. Eddy, R. J. Hamill, F. Gyorkey, H. Rosen, and J. Mader.** 1990. Spontaneous, nontraumatic gangrene due to *Clostridium septicum. Rev. Infect. Dis.* **12**:286–296.

120. **Supersac, G., G. Prevost, and Y. Piemont.** 1993. Sequencing of leucocidin R from *Staphylococcus aureus* P83 suggests that staphylococcal leucocidins and gamma-hemolysin are members of a single, two-component family of toxins. *Infect. Immun.* **61**:580–587.

121. **Suttorp, N., W. Seeger, R. J. Zucker, L. Roka, and S. Bhakdi.** 1987. Mechanism of leukotriene generation in polymorphonuclear leukocytes by staphylococcal alpha-toxin. *Infect. Immun.* **55**:104–110.

122. **Thiaudiere, E., O. Siffert, J. C. Talbot, J. Bolard, J. E. Alouf, and J. Dufourcq.** 1991. The amphiphilic alpha-helix concept. Consequences on the structure of staphylococcal delta-toxin in solution and bound to lipids. *Eur. J. Biochem.* **195**:203–213.

123. **Tobkes, N., B. A. Wallace, and H. Bayley.** 1985. Secondary structure and assembly mecha-

nism of an oligomeric channel protein. *Biochemistry* **24:**1915–1920.

124. **Tomita, T., M. Watanabe, and Y. Yarita.** 1993. Assembly and channel-forming activity of a naturally-occurring nicked molecule of *Staphylococcus aureus* alpha-toxin. *Biochim. Biophys. Acta* **1145:**51–57.

125. **Tweten, R. K.** 1988. Nucleotide sequence of the gene for perfringolysin O (theta-toxin) from *Clostridium perfringens:* significant homology with the genes for streptolysin O and pneumolysin. *Infect. Immun.* **56:**3235–3240.

126. **Tweten, R. K., R. W. Harris, and P. J. Sims.** 1991. Isolation of a tryptic fragment from *Clostridium perfringens* Θ-toxin that contains sites for membrane binding and self-aggregation. *J. Biol. Chem.* **266:**12449–12454.

127. **Umezawa, K., I. B. Weinstein and W. V. Shaw.** 1980. Staphylococcal delta-hemolysin inhibits cellular binding of epidermal growth factor and induces arachidonic acid release. *Biochem. Biophys. Res. Commun.* **94:**625–629.

128. **Vandenesch, F., M. J. Storrs, L. F. Poitevin, J. Etienne, P. Courvalin, and J. Fleurette.** 1991. Delta-like haemolysin produced by *Staphylococcus lugdunensis. FEMS Microbiol. Lett.* **62:**65–68.

129. **Van der Goot, F. G., F. Pattus, K. R. Wong, and J. T. Buckley.** 1993. Oligomerization of the channel-forming toxin aerolysin precedes insertion into lipid bilayers. *Biochemistry* **21:**2636–2642.

130. **Walker, B., M. Krishnasastry, L. Zorn, and H. Bayley.** 1992. Assembly of the oligomeric membrane pore formed by staphylococcal alpha-hemolysin examined by truncation mutagenesis. *J. Biol. Chem.* **267:**21782–21786.

131. **Walker, J. A., R. L. Allen, P. Falmagne, M. K. Johnson, and G. J. Boulnois.** 1987. Molecular cloning, characterization, and complete nucleotide sequence of the gene for pneumolysin, the sulfhydryl-activated toxin of *Streptococcus pneumoniae. Infect. Immun.* **55:**1184–1189.

132. **Ward, R. J., and K. Leonard.** 1992. The *Staphylococcus aureus* alpha-toxin channel complex and the effect of Ca^{2+} ions on its interaction with lipid layers. *J. Struct. Biol.* **109:**129–141.

133. **Wilmsen, H. U., K. R. Leonard, W. Tichelaar, J. T. Buckley, and F. Pattus.** 1992. The aerolysin membrane channel is formed by heptamerization of the monomer. *EMBO J.* **11:** 2457–2463.

134. **Wilmsen, H. U., F. Pattus, and J. T. Buckley.** 1990. Aerolysin, a hemolysin from *Aeromonas hydrophila,* forms voltage-gated channels in planar lipid bilayers. *J. Membr. Biol.* **115:**71–81.

135. **Yeong, M. L., and G. I. Nicholson.** 1988. *Clostridium septicum* infection in neutropenic enterocolitis. *Pathology* **20:**194–197.

BACTERIAL ADP-RIBOSYLATING EXOTOXINS

Kathleen M. Krueger and Joseph T. Barbieri

14

Pathogenic bacteria utilize a variety of virulence factors that contribute to their pathogenesis. One group of virulence factors includes bacterial exotoxins that covalently transfer ADP-ribose to eukaryotic target proteins. While early biochemical studies showed that several of these bacterial ADP-ribosylating exotoxins (bAREs) differ with respect to quaternary organization and the eukaryotic proteins that are targeted for ADP-ribosylation, many bAREs possess a common structural organization as A-B proenzymes (20, 27). The A domain possesses ADP-ribosyltransferase activity, while the B domain is required for efficient delivery of the A domain to the intracellular target protein. Subsequent biochemical and crystallographic studies have organized most bAREs into one of several A-B arrangements (Fig. 1): a single polypeptide protein with the A-B domains covalently linked, a multiprotein complex with the A-B domains noncovalently associated, and A-B domains located on separate proteins in solution.

In addition to their A-B organization, many bAREs are produced as proenzymes that require posttranslational modification to express enzymatic activity. Mechanisms for in vitro activation of bAREs include proteolysis to generate a catalytically active peptide, disulfide bond reduction, and allosteric activation by either nucleotides or eukaryotic accessory proteins. One challenging area of research has been to correlate in vitro mechanisms that activate bAREs with their in vivo intoxication pathways.

This chapter provides an overview of the structure-function properties of bAREs followed by a description of our current understanding of the biochemical and molecular properties of diphtheria toxin (DT) of *Corynebacterium diphtheriae,* exoenzyme S (ExoS) of *Pseudomonas aeruginosa,* and pertussis toxin (PT) of *Bordetella pertussis.*

BACTERIAL TOXINS THAT ADP-RIBOSYLATE EUKARYOTIC PROTEINS

A growing number of bacteria have been shown to produce enzymes that covalently transfer the ADP-ribose moiety of NAD^+ to eukaryotic target proteins (Table 1):

$$NAD^+ + \text{target protein} \rightarrow \text{ADP-ribose-target} \\ \text{protein} + \text{nicotinamide} + H^+$$

These bAREs are termed exotoxins, although it is not clear whether all bAREs are capable

Joseph T. Barbieri   Department of Microbiology, Medical College of Wisconsin, 8701 Watertown Plank Road, Milwaukee, Wisconsin 53226.   *Kathleen M. Krueger* Duke University Medical School, Durham, North Carolina 27710.

Virulence Mechanisms of Bacterial Pathogens, 2nd ed., Edited by J. A. Roth et al.
© 1995 American Society for Microbiology, Washington, D.C.

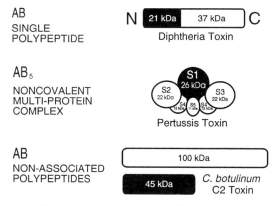

FIGURE 1 A-B organization of bAREs, which are grouped according to the structural organizations of their A and B domains. The A domains represent the catalytic domains and are shown in black, while the B domains are responsible for the delivery of the A domains to their eukaryotic target proteins and are shown in white.

of intoxicating eukaryotic cells under physiologic conditions. Bacteriologists have studied bAREs as virulence factors that contribute to the pathogenesis of the bacterium and as immunogens in vaccine development. Several bAREs ADP-ribosylate eukaryotic proteins that regulate eukaryotic-cell physiology, including both the heterotrimeric G proteins and the low-molecular-weight GTP-binding proteins, a fact that has broadened interest in

bAREs as tools for studying eukaryotic-cell physiology. Though bAREs are arranged into several types based on quaternary organization and though they ADP-ribosylate a diverse group of eukaryotic proteins, advances in defining molecular and structural properties of bAREs predict a structural and functional conservation of their active sites. Key studies that allowed identification of the conservation of the active sites of bAREs were performed on DT, *P. aeruginosa* exotoxin A (ETA), and the heat-labile enterotoxin of *Escherichia coli* (LT).

Although DT and ETA ADP-ribosylate the same eukaryotic target protein, elongation factor 2, early studies indicated that the two toxins did not share structural or molecular properties. The primary amino acid sequences of DT and ETA showed little overall homology (29, 30). The catalytic A domains of DT and ETA were at opposite ends of the primary amino acid sequences of each toxin. Despite this apparent lack of homology, Carroll and Collier (8, 9) showed that Glu-148 of DT and Glu-553 of ETA were homologs within the NAD binding sites of their respective toxins. Both Glu-148 of DT and Glu-553 of ETA were targeted for a novel photolabeling reaction in which the nicotinamide ring of NAD was transferred to the γ carbon of glutamic acid via a UV-dependent mechanism (10). Studies by Domenighini

TABLE 1 Family of bAREs

Exotoxin	Target (residue)	Physiologic response
DT *Pseudomonas* ETA	EF-2 (diphthamide, modified His)	Inhibits protein synthesis
CT, *E. coli* LT	G_s, G_t (Arg)	Inhibits GTPase of Gα
	G_i, G_o, G_t (Cys)	Uncouples G protein and G-protein-coupled receptor interactions
Botulinum C2 and related toxins	Actin (Arg)	Inhibits actin polymerization
Botulinum C3 and related toxins from *Clostridium limosum* and *Bacillus cereus*	Low-mol-wt G proteins, Rho, Rac families (Asn)	Unknown
Pseudomonas ExoS	Vimentin, Ras, others (unknown)	Unknown
EDIN (*S. aureus*)	Rho (unknown)	Disassembles Golgi apparatus
Bacillus sphaericus mosquitocidal toxin	38 kDa, 42 kDa unidentified (unknown)	Unknown

et al. showed that the A domain of DT could be modeled along the α-carbon backbone within the active site of the crystal structure of ETA, a fact that extended our appreciation of the structural conservation between the active sites of DT and ETA (23). The conservation of structure of the active sites in DT and ETA has been verified by recent analysis of the crystal structure of DT (13). Thus, despite the lack of apparent primary amino acid homology, the active sites of DT and ETA show a conservation of structure-function. Two other bAREs, PT (3, 15, 19) and *Clostridium limosum* C3-like ADP-ribosyltransferase (35), also possess glutamic acids that are homologs to Glu-148 of DT, as shown by photolabeling with NAD via the UV-dependent mechanism. Kinetic studies have shown that Glu-148 of DT and Glu-129 of PT are involved in the catalytic portion of the ADP-ribosyltransferase reaction.

Hol and coworkers identified a conservation of structure within the active sites of ETA and LT. These investigators solved the crystal structure of LT and showed that the active site of LT could be superimposed on the active site of ETA, although only 3 of 44 amino acids within the respective active sites were identical (62). Consistent with the catalytic role of glutamic acid within the active site was the alignment of Glu-112 of LT with Glu-553 of ETA. The structural alignment of the active sites of LT and ETA expanded our appreciation of the structural conservation of the active sites of the bAREs, since LT and ETA possess different biochemical properties with respect to their A-B organization and classes of eukaryotic proteins that are ADP-ribosylated. Together, these data provide the foundation for a model that predicts an overall conservation of the structure and function of the active sites of bAREs despite the fact that these exotoxins share little overall primary amino acid homology.

CONTRIBUTION OF ADP-RIBOSYLATING EXOTOXINS TO BACTERIAL PATHOGENESIS

Bacteria often express multiple virulence factors to produce a clinically recognized infection. Although this complicates the identification of the contribution of individual virulence factors to bacterial pathogenesis, several bAREs, including DT and cholera toxin, clearly contribute to the production of clinically relevant infections. Virulent strains of *C. diphtheriae* produce DT as their major virulence factor. Purified DT is responsible for the systemic pathology attributed to diphtheria (20). Cholera toxin is responsible for the secretory diarrhea associated with cholera (4). In contrast, the role of other bAREs in bacterial pathogenesis has not been as clearly defined. PT is one of several virulence factors produced by *B. pertussis* that contributes to both the systemic disease and the ability of *B. pertussis* to colonize the ciliated epithelial cells of the upper respiratory tract (71). *P. aeruginosa* produces numerous virulence factors that contribute to the virulence of this opportunistic pathogen, including two ADP-ribosyltransferases, ETA, and ExoS. Strains of *P. aeruginosa* that did not express ETA or ExoS are less virulent than parental strains (55). Although this fact implicates both ETA and ExoS as virulence factors of *P. aeruginosa,* the absolute contribution of ETA and ExoS to pathogenesis awaits the evaluation of the pathogenic properties of strains of *P. aeruginosa* that lack the structural gene encoding either ETA or ExoS. In addition to neurotoxins, *Clostridium botulinum* produces two ADP-ribosyltransferases, C2 and C3. Although C2 has been shown to be cytotoxic for cultured cells (1), the contribution of either C2 or C3 to the clinical pathogenesis of *Clostridium botulinum* has not been defined. *Staphylococcus aureus* produces the ADP-ribosyltransferase EDIN. Although the contribution of EDIN to the pathogenesis of *S. aureus* has not been defined, Wu and collaborators showed that purified EDIN disrupts the Golgi apparatus in cultured cells through a mechanism that mimics brefeldin A (65). Defining the contributions of these and other bAREs to bacterial pathogenesis will require the development of both animal models that mimic natural infections and genetic systems that allow the construction of bARE knockout strains for the evaluation of pathogenic potential.

MOLECULAR PROPERTIES OF DT

DT is the major virulence determinant of *C. diphtheriae*. Purified DT elicits the systemic pathology recognized as diphtheria. Formalin-treated DT is an effective immunogen for the prevention of diphtheria and is used as the diphtheria component of the diphtheria-pertussis-tetanus (DPT) vaccine (20). The gene encoding DT is contained within the genome of the β phage of *C. diphtheriae* and has been cloned (30). DT is a 535-amino-acid protein that ADP-ribosylates elongation factor 2, inhibiting eukaryotic protein synthesis. The A-B properties of DT were originally predicted by biochemical and genetic analysis and have been confirmed by crystallographic analysis (13). The A domain is composed of residues 1 through 193 and catalyzes the ADP-ribosylation of a posttranslationally modified histidine on elongation factor 2, and the B domain comprises an internal translocation domain (composed of residues 205 through 378) and a caboxyl-terminal receptor-binding domain (composed of residues 386 through 535).

Eidels and coworkers cloned a gene from a Vero cell cDNA library that encoded a protein that functions as a receptor for DT (53). The deduced amino acid sequence of this functional DT receptor possessed 97% amino acid homology to human heparin-binding epidermal growth factor-like growth factor precursor. The binding affinity of DT for the cloned DT receptor was about 10-fold lower than the affinity of DT for Vero cells, which indicated that other eukaryotic proteins contributed to the high-affinity binding of DT to Vero cells. While Mekada and coworkers implicated the CD9 antigen as a component of the DT receptor (48), Eidels and coworkers subsequently showed that the CD9 antigen influenced the number of DT receptors rather than the affinity of DT for the cloned DT receptor (5).

Molecular and structural analyses of the translocation domain of DT (residues 205 through 378) have provided considerable insight into one of the least-defined properties of bAREs: the ability to translocate the A domain across a cell membrane. The following model has been proposed for the translocation of the A domain of DT across the endosomal membrane (13). DT binds to its cell surface receptor, and the DT-receptor complex enters the cell via receptor-mediated endocytosis. The acidic pH of the endosome stimulates the protonation of six acidic residues within a loop region that connects two hydrophobic alpha helixes within the translocation domain. This renders the loop regions membrane soluble and triggers the insertion of the two hydrophobic alpha helixes of the translocation domain into the endosomal membrane. Upon penetration through the endosomal membrane, the acidic amino acids within the loop region deprotonate in the neutral environment of the cytosol and anchor the two hydrophobic alpha helixes within the endosomal membrane. This membrane-anchored structure has been proposed to contribute to the translocation of the catalytic A domain across the endosomal membrane. The catalytic A domain is activated through proteolysis and disulfide bond reduction to release a soluble A domain into the cytosol. One future goal of DT research is to define the specific molecular steps that occur during the translocation of the catalytic fragment across the endosomal membrane.

Carroll and Collier showed that Glu-148 of DT could be photolabeled by NAD via a UV-dependent mechanism (8) in which the nicotinamide ring of NAD was transferred to the γ carbon of glutamic acid (10). DT mutants that possess conservative substitutions at Glu-148 (Glu-148-Asp) possess reduced catalytic activity (68) without a decrease in binding affinities for either NAD or elongation factor 2 (73). This indicates that Glu-148 plays a catalytic role in the ADP-ribosyltransferase reaction. These studies have proven to be a reference for comparative catalytic analyses of other bAREs.

MOLECULAR PROPERTIES OF ExoS OF *P. AERUGINOSA*

P. aeruginosa produces two ADP-ribosyltransferases, ETA and ExoS, that differ in their biochemical properties. ETA ADP-ribosylates elongation factor 2, while ExoS is more promiscuous and ADP-ribosylates several eukaryotic proteins, including vimentin and low-molecular-weight GTP-binding proteins, in vitro (18). The in vivo eukaryotic target protein(s) of ExoS has not been defined. ETA and ExoS also possess other biochemical differences where ExoS requires a eukaryotic accessory protein, termed FAS (factor-activating ExoS), to express ADP-ribosyltransferase activity in vitro (17, 34). The gene encoding FAS has recently been cloned from a bovine brain cDNA library and shown to be a member of the 14-3-3 family of eukaryotic proteins (26). Members of the 14-3-3 family of eukaryotic proteins possess a variety of functions, including phospholipase activity and regulatory functions. By genetic analysis, ExoS has been implicated as a mediator of *P. aeruginosa* pathogenesis in burn wounds and chronic lung infections (56).

Iglewski and coworkers initially described the ADP-ribosyltransferase activity of ExoS (34). ExoS activity was secreted from *P. aeruginosa* 388 as a noncovalent aggregate that co-purified with a 53-kDa protein and a 49-kDa protein. Upon being extracted from sodium dodecyl sulfate-polyacrylamide gels, the 49-kDa protein of ExoS possessed enzymatic activity (17, 56) and was designated the enzymatically active form of ExoS, while the 53-kDa protein of ExoS did not possess ADP-ribosyltransferase activity in vitro (56) and was designated the enzymatically inactive form of ExoS. The 53- and 49-kDa proteins share several properties, including immunologic cross-reactivity (41, 56) and possession of a common amino-terminal amino acid sequence (16). However, the absolute biochemical and genetic relationship between the 53- and 49-kDa forms of ExoS has not been resolved.

Recently, the gene encoding the 49-kDa form of ExoS (*exoS*) was cloned from *P. aeruginosa* 388 (42). The deduced amino acid sequence predicted that *exoS* encoded a 453-amino-acid protein. ExoS did not possess an amino-terminal leader sequence, which indicated that ExoS was not secreted from *P. aeruginosa* via the general secretory pathway. Although the deduced amino acid sequence of ExoS lacked overall amino acid homology with other bAREs, three regions of local primary amino acid homology between ExoS and primary amino acid sequences that constituted the active site of LT were identified. Analysis of the deduced amino acid sequence of ExoS allowed the prediction that either Glu-265 or Glu-408 represents an active-site glutamic acid homolog, since a BESTFIT analysis of ExoS and LT aligned both Glu-265 and Glu-408 of ExoS with Glu-112 of LT. We recently observed that the recombinant 49-kDa form of ExoS expressed in *E. coli* possesses ADP-ribosyltransferase activity with a specific activity that is similar to that of native ExoS isolated from *P. aeruginosa* (41a). This proves that the open reading frame of *exoS* is necessary and sufficient for the expression of the FAS-dependent ExoS ADP-ribosyltransferase activity.

Future studies of ExoS will determine its A-B structure-function properties, define residues that are involved in the ADP-ribosyltransferase reaction, and determine its role in the pathogenesis of *P. aeruginosa*.

MOLECULAR PROPERTIES OF PT OF *B. PERTUSSIS*

PT ADP-ribosylates a subset of heterotrimeric GTP-binding proteins. Much of our understanding of the molecular properties of PT has been derived from research directed at developing an effective acellular pertussis vaccine to prevent infection by *B. pertussis,* the etiologic agent responsible for whooping cough.

Whooping cough is a major cause of morbidity and mortality in children. Children under 10 years of age account for most clinical cases, with children under 6 months of age having the highest mortality rate. Complications of pertussis infections range from mild to severe, with severe complications often involving the respiratory system or the central nervous system (33). Pertussis causes more than half a million deaths worldwide annually, primarily in developing countries where vaccination is not administered (24). Vaccination against pertussis was initiated in the United States in the 1940s and has proved effective in controlling the disease. Prior to vaccination, 115,000 to 270,000 cases of whooping cough, resulting in 5,000 to 10,000 deaths, were reported annually in the United States. Currently, approximately 1,200 to 4,000 cases of whooping cough, resulting in 5 to 10 deaths, occur annually in the United States (24). Recent research on pertussis has focused on the development of less-reactive pertussis vaccines. Acellular vaccine candidates composed of one or more virulence factors of *B. pertussis* have been constructed. Since antibodies generated against PT provide protection against infection (61), the first generation of acellular vaccines included a chemically detoxified form of PT. In the United States, the protocol for vaccination against pertussis has recently been modified such that the whole-cell vaccine is used in the initial three doses, but an acellular pertussis vaccine (DTP$_a$ vaccine) may substitute for the whole-cell pertussis vaccine in the fourth and fifth doses (11). This acellular pertussis vaccine was licensed by the Food and Drug Administration in December 1991 and includes formaldehyde-detoxified PT, filamentous hemagglutinin, pertactin, and type 2 fimbriae (11). Current studies have led to the construction of genetically engineered PT toxoids, which have shown promise as potential components of future acellular pertussis vaccines (45, 46, 58, 60).

Pertussis vaccines are currently administered only to children. However, evidence indicates that the immunity to pertussis that is generated by childhood immunization is insufficient to protect individuals into adulthood (25). The adult population therefore serves as a potential reservoir for *B. pertussis*, and data have implicated infected adults in the transmission of pertussis to children (12, 25). Future studies should determine the efficacy of the acellular vaccines in preventing pertussis infections in the adult population and in reducing the adult reservoir.

PT AS AN ADP-RIBOSYLATING EXOTOXIN

PT is one of the major virulence factors of *B. pertussis*. Isogeneic strains of *B. pertussis* that are defective in the production of PT are impaired in the ability to cause lethal infection in a mouse model system (the 50% lethal dose for a PT-deficient mutant of *B. pertussis* is more than 100-fold higher than that for parental wild-type *B. pertussis*) (72). PT catalyzes the ADP-ribosylation of the α subunit of a subset of heterotrimeric GTP-binding proteins, including G_i, G_o, and G_t. The trimeric form of the G protein, composed of αβγ subunits, is the preferred target for ADP-ribosylation (54, 70). G proteins susceptible to ADP-ribosylation by PT contain a cysteine residue that is located 4 amino acids from the carboxyl terminus of G_α and is the site of ADP-ribosylation (38). The carboxyl terminus of G_α is important for the interaction of the G protein with the seven transmembrane-spanning G-protein-coupled receptors (21). This is consistent with ADP-ribosylation of G_α disrupting the interaction of the G protein with the receptor by effectively uncoupling the receptor from the G protein (52, 59, 67, 69). The ability of ADP-ribosylation to uncouple signal transduction between the G protein and the G-protein-coupled receptor has made PT a useful pharmacologic tool. Recently, peptides composed of the carboxyl-terminal 10 to 20 residues of G_α have been shown to be ADP-ribosylated by PT, indicating that this region of G_α is sufficient for recognition for ADP-ribosylation (28).

PT is a hexameric protein with a molecular mass of 105 kDa that is composed of five different polypeptides: S1 (26,220 Da), S2 (21,920 Da), S3 (21,860 Da), S4 (12,060 Da), and S5 (10,940 Da) (57). These polypeptides are associated via noncovalent interactions within the holotoxin in a 1:1:1:2:1 ratio, respectively (66). The genes of the subunits of PT are located within a single operon that contains five closely linked open reading frames that encode the individual subunits (44, 57). Gene duplication appears to be responsible for the production of genes encoding S2 and S3, since they are 75% homologous at the nucleotide level and 70% homologous at the protein level. The ADP-ribosyltransferase activity, the A domain, of PT resides in S1, while the remaining subunits constitute the B domain, termed the B oligomer (67).

Mutagenesis of S1 (235 amino acids) has been used to characterize the functions of various regions and residues involved in ADP-ribosylation (Fig. 2). Two residues, His–35 and Glu–129, play a catalytic role in the ADP-ribosyltransferase reaction (2, 74), while Trp–26 is responsible for NAD binding (22). Arg–9 plays an important but as yet undefined role in the ADP-ribosyltransferase reaction (6, 14, 58). The recently solved crystal structure of PT showed that, consistent with their proposed functional roles, His–35, Glu–129, Trp–26, and Arg–9 reside within the active site of S1 (63). Deletion mutagenesis has allowed the functional mapping of S1. The amino-terminal 180 amino acids of S1 represent all the residues required for the ADP-ribosyltransferase reaction, including the NAD-binding domain and the one of two binding sites for the heterotrimeric G protein that is defined by the 20 carboxy-terminal amino acids of $G_{i\alpha}$. Residues between 195 and 204 represent a second G-protein-binding site within S1 that interacts with a region of G_{α} that is distinct from the 20 carboxyl-terminal amino acids of G_{α} and is defined by the trimeric conformation of the G protein. This region (residues 195 to 204) confers efficient ADP-ribosyltransferase activity on S1 via high affinity binding of S1 to the heterotrimeric G protein. Residues 219 to 235 are required for the in vitro assembly of S1 with the B oligomer (40). Studies have not identified a role for amino acids 205 through 219, but this region may act as a linker to align the catalytic core of S1 and the carboxyl-terminal B-oligomer-binding domain. This would be analogous to the 20-amino-acid alpha helix that is located near the carboxyl terminus of the A subunit of LT that links the enzymatic portion of the A subunit with the carboxyl-terminal B-oligomer-binding domain (40). The functional assignment of these regions of S1 agrees with

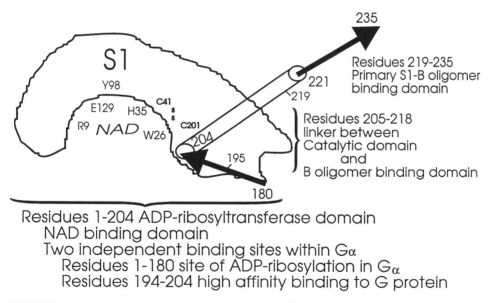

FIGURE 2 Functional organization of the S1 subunit of PT. S1 is composed of several functional domains. Residues 1 to 180 compose the complete NAD glycohydrolase activity and the minimum region that catalyzes the ADP-ribosylation reaction. Residues 195 to 204 compose a region that confers high-affinity binding to transducin and allows for the efficient ADP-ribosylation of transducin. Residues 215 to 219 represent a protease-sensitive site within the holotoxin. Residues 219 to 235 represent the primary B-oligomer-binding region of S1. Biochemical and crystallographic studies have identified additional contacts between amino-terminal regions of S1 and the B oligomer.

their location within the crystal structure of PT (63).

Similar to other ADP-ribosylating exotoxins, *B. pertussis* PT is secreted as a proenzyme that is cytotoxic but not catalytic when assayed in vitro. In vitro activation of the ADP-ribosyltransferase activity of PT requires ATP (37, 39, 43, 47, 70) and the reduction of the Cys-41–Cys-201 disulfide (50). The mechanism of ATP activation of PT has been the subject of considerable investigation. Early studies showed that ATP hydrolysis was not required for activation and that other nucleoside triphosphates, nucleoside diphosphates, and inorganic polyphosphates were able to activate PT. ATP did not stimulate the catalytic activity of purified S1 (7, 39, 51), which implies that ATP stimulation was B oligomer dependent. Subsequent studies showed that ATP binds to PT and purified B oligomer but not to purified S1 (32, 47). Several phospholipids and detergents were also shown to stimulate the ADP-ribosyltransferase and NAD glycohydrolase activities of PT (37, 51). One detergent, the zwitterion CHAPS {3-[(3-cholamidopropyl)-dimethyl-ammonio]-1-propanesulfonate}, stimulated NAD glycohydrolase activity but inhibited ADP-ribosyltransferase activity (37, 51). Burns and Manclark showed that in the presence of CHAPS, ATP stimulates the release of S1 from the B oligomer (7). Together, these data led to a model that proposed that ATP activates PT via the release of S1 from the B oligomer (36). However, since CHAPS inhibits the ADP-ribosyltransferase reaction (51), it was not clear how ATP activated the ADP-ribosyltransferase activity of PT in the absence of CHAPS. Subsequent studies have shown that ATP altered the kinetic and conformational state of S1 within the holotoxin to that observed for isolated S1. ATP was also observed to reduce the affinity of S1 for the holotoxin without concomitant release of S1 from the B oligomer (39).

While the mechanism of entry of PT into susceptible cells has not been defined, recent studies have allowed the development of a working model for the in vivo ADP-ribosyla-tion of G proteins by PT. PT binds to susceptible cells via an interaction of the S2-S4 and S3-S4 dimers of the B oligomer with eukaryotic cells (67). Although a specific PT receptor has not been identified, PT binds carbohydrate moieties, particularly sialic acid, of certain glycoproteins and glycolipids. The interaction of PT with cells has recently been reviewed by Kaslow and Burns (36). Studies by Montecucco et al. (49) that used photoreactive phospholipids indicated that the S4 subunit of PT is labeled by a superficial phospholipid probe, whereas both the S2 and the S3 subunits are labeled by a deeper probe. These results indicate that components of the B oligomer might be involved in membrane translocation. Concentrations of reduced glutathione and ATP similar to those found in intact cells were sufficient to activate the in vitro catalytic activity of PT, indicating that these compounds could activate the toxin in vivo (37). Although ATP and dithiothreitol in the presence of phospholipids or detergents stimulate the release of a significant amount of the S1 subunit from the B oligomer (7, 31, 36), release of S1 from the B oligomer is not required for the in vitro activation of PT. Thus, release of S1 from the B oligomer may not be required for in vivo ADP-ribosylation of G proteins. PT-sensitive G proteins are generally thought to be associated with the plasma membrane. However, a member of the G_i family (G_{i3}) was localized to the Golgi membranes of LLC-PK$_1$ epithelial cells (64). This G protein regulates protein secretion and is a target for PT-mediated ADP-ribosylation in vivo (64). Thus, a mechanism for trafficking PT within the cell should include delivery of PT or S1 to the Golgi.

Future research on PT will include resolution of an efficient mechanism to genetically engineer a toxoid of PT as a component of an acellular pertussis vaccine, determination of how PT is trafficked within the eukaryotic cell to ADP-ribosylate both Golgi- and plasma membrane-associated G proteins, and determination of the contribution of the ADP-ribosylation of subsets of G proteins to the cytopathology of PT.

SUMMARY

bAREs represent one family of virulence factors that exert their toxic effects by transferring the ADP-ribose moiety of NAD onto specific eukaryotic target proteins. bAREs possess little primary amino acid homology and have diverse quaternary structure-function organizations. However, biochemical and crystallographic studies have shown that underlying their apparent diversity, several bAREs have structurally conserved active sites. Future studies into the molecular mechanisms responsible for the ADP-ribosylation of eukaryotic proteins by bAREs will increase our understanding of the role that these proteins play in the bacterial pathogenesis and will also provide insight into the physiologic processes of the eukaryotic proteins that are the targets of ADP-ribosylation.

REFERENCES

1. **Aktories, K., M. Wille, and I. Just.** 1992. Clostridial actin-ADP-ribosylating toxins. *Curr. Top. Microbiol. Immunol.* **175:**97–113.

2. **Antoine, R., and C. Locht.** 1994. The NAD-glycohydrolase activity of the pertussis toxin S1 subunit: involvement of the catalytic His-35 residue. *J. Biol. Chem.* **269:**6450–6457.

3. **Barbieri, J. T., L. M. Mende-Mueller, R. Rappuoli, and R. J. Collier.** 1989. Photolabeling of Glu-129 of the S1 subunit of pertussis toxin with NAD. *Infect. Immun.* **57:**3549–3554.

4. **Birnbaumer, L., J. Codina, R. Mattera, A. Yatani, N. Scherer, M.-J. Toro, and A. M. Brown.** 1990. Recent advances in the understanding of multiple roles of G proteins in coupling of receptors to ionic channels and other effectors, p. 225–266. *In* J. Moss and M. Vaughan (ed.), *ADP-Ribosylating Toxins and G Proteins: Insights into Signal Transduction.* American Society for Microbiology, Washington, D.C.

5. **Brown, J. G., B. D. Almond, J. G. Naglich, and L. Eidels.** 1993. Hypersensitivity to diphtheria toxin by mouse cells expressing both diphtheria toxin and CD9 antigen. *Proc. Natl. Acad. Sci. USA* **90:**8184–8188.

6. **Burnette, W. N., W. Cieplak, V. L. Mar, K. T. Kaljot, H. Sato, and J. M. Keith.** 1988. Pertussis toxin S1 mutant with reduced enzyme activity and a conserved protective epitope. *Science* **242:**72–74.

7. **Burns, D. L., and C. R. Manclark.** 1986. Adenine nucleotides promote dissociation of pertussis toxin subunits. *J. Biol. Chem.* **261:**4324–4327.

8. **Carroll, S. F., and R. J. Collier.** 1984. NAD binding site of diphtheria toxin: identification of a residue within the nicotinamide subsite by photochemical modification by NAD. *Proc. Natl. Acad. Sci. USA* **81:**3307–3311.

9. **Carroll, S. F., and R. J. Collier.** 1988. Active site of *Pseudomonas aeruginosa* exotoxin A: glutamic acid 533 is photolabeled by NAD and shows functional homology with glutamic acid 148 of diphtheria toxin. *J. Biol. Chem.* **262:**8707–8711.

10. **Carroll, S. F., J. A. McCloskey, P. F. Crain, N. J. Oppenheimer, T. M. Marschner, and R. J. Collier.** 1985. Photoaffinity labeling of diphtheria toxin fragment A with NAD: structure of the photoproduct at position 148. *Proc. Natl. Acad. Sci. USA* **82:**7237–7241.

11. **Center for Disease Control.** 1992. Pertussis vaccination: acellular pertussis vaccine for reinforcing and booster use—supplementary ACIP statement. *Morbid. Mortal. Weekly Rep.* **41:**RR-1.

12. **Cherry, J. D.** 1993. Acellular pertussis vaccines—a solution to the pertussis problem. *J. Infect. Dis.* **168:**21–24.

13. **Choe, S., M. J. Bennett, G. Fujii, P. M. G. Curmi, K. A. Kantardjieff, R. J. Collier, and D. Eisenberg.** 1992. The crystal structure of diphtheria toxin. *Nature* (London) **357:**216–222.

14. **Cieplak, W., W. N. Burnette, V. L. Mar, K. T. Kaljot, C. F. Morris, K. K. Chen, H. Sato, and J. M. Keith.** 1988. Identification of a region in the S1 subunit of pertussis toxin that is required for enzymatic activity and that contributes to the formation of a neutralizing antigenic determinant. *Proc. Natl. Acad. Sci. USA* **85:**4667–4671.

15. **Cieplak, W., C. Locht, V. L. Mar, W. N. Burnette, and J. M. Keith.** 1990. Photolabeling of mutant forms of the S1 subunit of pertussis toxin with NAD⁺. *Biochem. J.* **268:**547–551.

16. **Coburn, J.** 1992. *Pseudomonas aeruginosa* exoenzyme S. *Curr. Top. Microbiol. Immunol.* **175:**133–143.

17. **Coburn, J., A. V. Kane, L. Feig, and D. M. Gill.** 1991. *Pseudomonas aeruginosa* exoenzyme S requires a eukaryotic protein for ADP-ribosyltransferase activity. *J. Biol. Chem.* **266:**6438–6446.

18. **Coburn, J., R. T. Wyatt, B. H. Iglewski, and D. M. Gill.** 1989. Several GTP-binding proteins, including p21^cHras, are preferred substrates of *Pseudomonas aeruginosa* exoenzyme S. *J. Biol. Chem.* **264:**9004–9008.

19. **Cockle, S.** 1989. Identification of an active-site residue in subunit S1 of pertussis toxin by photo-crosslinking to NAD. *FEBS Lett.* **249:**329–332.

20. **Collier, R. J.** 1975. Diphtheria toxin: mode of action and structure. *Bacteriol. Rev.* **39**:54–85.

21. **Conklin, B. R., and H. R. Bourne.** 1993. Structural elements of Gα subunits that interact with Gβγ, receptors, and effectors. *Cell* **73**:631–641.

22. **Cortina, G., and J. T. Barbieri.** 1989. Role of tryptophan 26 in the NAD glycohydrolase reaction of the S1 subunit of pertussis toxin. *J. Biol. Chem.* **264**:17322–17328.

23. **Domenighini, M., C. Montecucco, W. C. Ripka, and R. Rappuoli.** 1991. Computer modelling of the NAD binding site of ADP-ribosylating toxins. Active-site structure and mechanism of NAD binding. *Mol. Microbiol.* **5**:23–31.

24. **Edwards, K. M.** 1993. Acellular pertussis vaccines—a solution to the pertussis problem? *J. Infect. Dis.* **168**:15–20.

25. **Edwards, K. M., D. Decker, B. S. Graham, J. Mezzatesta, J. Scott, and J. Hackell.** 1993. Adult immunization with acellular pertussis vaccine. *JAMA* **269**:53–56.

26. **Fu, H., J. Coburn, and R. J. Collier.** 1993. The eukaryotic host factor that activates exoenzyme S of *Pseudomonas aeruginosa* is a member of the 14-3-3 protein family. *Proc. Natl. Acad. Sci. USA* **90**:2320–2324.

27. **Gill, D. M.** 1976. The arrangements of the subunits of cholera toxin. *Biochemistry* **15**:1242–1248.

28. **Graf, R., J. Codina, and L. Birnbaumer.** 1992. Peptide inhibitors of ADP-ribosylation by pertussis toxin are substrates with affinities comparable to those of the trimeric GTP-binding proteins. *Mol. Pharmacol.* **42**:760–764.

29. **Gray, G. L., D. H. Smith, J. S. Baldridge, R. N. Harkins, M. L. Vasil, E. Y. Chen, and H. L. Heyneker.** 1984. Cloning, nucleotide sequence, and expression in *Escherichia coli* of the exotoxin A structural gene of *Pseudomonas aeruginosa*. *Proc. Natl. Acad. Sci. USA* **81**:2645–2649.

30. **Greenfield, L., M. J. Bjorn, G. Horn, D. Fong, G. A. Buck, R. J. Collier, and D. A. Kaplan.** 1983. Nucleotide sequence of the structural gene for diphtheria toxin carried by corynebacteriophage β. *Proc. Natl. Acad. Sci. USA* **80**:6853–6857.

31. **Hausman, S. Z., and D. L. Burns.** 1992. Interaction of pertussis toxin with cells and model membranes. *J. Biol. Chem.* **267**:13735–13739.

32. **Hausman, S. Z., C. R. Manclark, and D. L. Burns.** 1990. Binding of ATP by pertussis toxin and isolated toxin subunits. *Biochemistry* **29**:6128–6131.

33. **Howson, C. P., C. J. Howe, and H. V. Fineberg (ed.).** 1991. *Adverse Effects of Pertussis and Rubella Vaccines,* p. 9–31, p. 320–326. National Academy Press, Washington, D.C.

34. **Iglewski, B. H., J. Sadoff, M. J. Bjorn, and M. S. Maxwell.** 1978. *Pseudomonas aeruginosa* exoenzyme S: an adenosine diphosphate ribosyltransferase distinct from toxin A. *Proc. Natl. Acad. Sci. USA* **75**:3211–3215.

35. **Jung, M., I. Just, J. van Damme, J. Vandekerckhove, and K. Aktories.** 1993. NAD-binding site of the C3-like ADP-ribosyltransferase from *Clostridium limosum. J. Biol. Chem.* **268**:23215–23218.

36. **Kaslow, H. R., and D. L. Burns.** 1992. Pertussis toxin and target eukaryotic cells: binding, entry, and activation. *FASEB J.* **6**:2684–2690.

37. **Kaslow, H. R., L.-K. Lim, J. Moss, and D. D. Lesikar.** 1987. Structure-activity analysis of the activation of pertussis toxin. *Biochemistry* **26**:123–127.

38. **Kaziro, Y., H. Itoh, T. Kozasa, M. Nakafuku, and T. Satoh.** 1991. Structure and function of signal-transducing GTP-binding proteins. *Annu. Rev. Biochem.* **60**:349–400.

39. **Krueger, K. M., and J. T. Barbieri.** 1993. Molecular characterization of the in vitro activation of pertussis toxin by ATP. *J. Biol. Chem.* **268**:12570–12578.

40. **Krueger, K. M., and J. T. Barbieri.** 1994. Functional mapping of the carboxyl terminus of the S1 subunit of pertussis toxin. *Infect. Immun.* **62**:2071–2078.

41. **Kulich, S. M., D. W. Frank, and J. T. Barbieri.** 1993. Purification and characterization of exoenzyme S from *Pseudomonas aeruginosa* 388. *Infect. Immun.* **61**:307–313.

41a. **Kulich, S. M., D. W. Frank, and J. T. Barbieri.** 1995. Expression of recombinant exoenzymes of *Pseudomonas aeruginosa. Infect. Immun.* **63**:1–8.

42. **Kulich, S. M., T. Yahr, L. Mende-Mueller, J. T. Barbieri, and D. W. Frank.** 1994. Cloning the structural gene for the 49 kDa form of exoenzyme S from *Pseudomonas aeruginosa* strain 388. *J. Biol. Chem.* **269**:10431–10437.

43. **Lim, L.-K., R. D. Sekura, and H. R. Kaslow.** 1985. Adenine nucleotides directly stimulate pertussis toxin. *J. Biol. Chem.* **20**:2585–2588.

44. **Locht, C., and J. M. Keith.** 1986. Pertussis toxin gene: nucleotide sequence and genetic organization. *Science* **232**:1258–1264.

45. **Loosmore, S., S. Cockle, G. Zealey, H. Boux, K. Phillips, R. Fahim, and M. Klein.** 1991. Detoxification of pertussis toxin by site-directed mutagenesis: a review of Connaught strategy to develop a recombinant pertussis vaccine. *Mol. Immunol.* **28**:235–238.

46. **Loosmore, S. M., G. R. Zealey, H. A. Boux, S. Cockle, K. Radika, R. E. F. Fahim, G. J. Zobrist, R. K. Yacoob, P. Chong, F.-L. Yao, and M. H. Klein.** 1990. Engineering of genetically detoxified pertussis toxin analogs for devel-

opment of a recombinant whooping cough vaccine. *Infect. Immun.* **58:**3653–3662.

47. **Mattera, R., J. Codina, R. D. Sekura, and L. Birnbaumer.** 1986. The interaction of nucleotides with pertussis toxin. Direct evidence for a nucleotide binding site on the toxin regulating the rate of ADP-ribosylation of N_i, the inhibitory regulatory component of adenylyl cyclase. *J. Biol. Chem.* **261:**11173–11179.

48. **Mitamura, T., R. Iwamoto, T. Umata, T. Yomo, I. Urabe, M. Tsuneoka, and E. Mekada.** 1992. The 27-kd diphtheria toxin receptor-associated protein (DRAP27) from Vero cells is the monkey homologue of human CD9 antigen: expression of DRAP27 elevates the number of diphtheria toxin receptors on toxin-sensitive cells. *J. Cell Biol.* **118:**1389–1399.

49. **Montecucco, C., M. Tomasi, G. Schiavo, and R. Rappuoli.** 1986. Hydrophobic photolabelling of pertussis toxin subunits interacting with lipids. *FEBS Lett.* **194:**301–304.

50. **Moss, J., S. J. Stanley, D. L. Burns, J. A. Hsia, D. A. Yost, G. A. Myers, and E. L. Hewlett.** 1983. Activation by thiol of the latent NAD glycohydrolase and ADP-ribosyltransferase activities of *Bordetella pertussis* toxin (islet-activating protein). *J. Biol. Chem.* **258:**11879–11882.

51. **Moss, J., S. J. Stanley, P. A. Watkins, D. L. Burns, C. R. Manclark, H. R. Kaslow, and E. L. Hewlett.** 1986. Stimulation of the thiol-dependent ADP-ribosyltransferase and NAD glycohydrolase activities of *Bordetella pertussis* toxin by adenine nucleotides, phospholipids, and detergents. *Biochemistry* **25:**2720–2725.

52. **Murayama, T., and M. Ui.** 1984. [³H]GDP release from rat and hamster adipocyte membranes independently linked to receptors involved in activation or inhibition of adenylate cyclase. Differential susceptibility to two bacterial toxins. *J. Biol. Chem.* **259:**761–769.

53. **Naglich, J. G., J. E. Metherall, D. W. Russell, and L. Eidels.** 1992. Expression cloning of a diphtheria toxin receptor: identity with a heparin-binding EGF-like growth factor precursor. *Cell* **69:**1051–1061.

54. **Neer, E. J., J. M. Lok, and L. G. Wolf.** 1984. Purification and properties of the inhibitory guanine nucleotide regulatory unit of brain adenylate cyclase. *J. Biol. Chem.* **259:**14222–14229.

55. **Nicas, T. I., and B. Iglewski.** 1985. The contribution of exoproducts to virulence of *Pseudomonas aeruginosa. Can. J. Microbiol.* 31:387–392.

56. **Nicas, T. I., and B. H. Iglewski.** 1984. Isolation and characterization of transposon-induced mutants of *Pseudomonas aeruginosa* deficient in production of exoenzyme S. *Infect. Immun.* **45:**470–474.

57. **Nicosia, A., M. Perugini, C. Franzini, M. C.**

Casagli, M. G. Borri, G. Antoni, M. Almoni, P. Neri, G. Ratti, and R. Rappuoli. 1986. Cloning and sequencing of the pertussis toxin genes: operon structure and gene duplication. *Proc. Natl. Acad. Sci. USA* **83:**4631–4635.

58. **Pizza, M., A. Covacci, A. Bartoloni, M. Perugini, L. Nencioni, M. T. De Magistris, L. Villa, D. Nucci, R. Manetti, M. Bugnoli, F. Giovannoni, R. Olivieri, J. T. Barbieri, H. Sato, and R. Rappuoli.** 1989. Mutants of pertussis toxin suitable for vaccine development. *Science* **246:**497–500.

59. **Ramdas, L., R. M. Disher, and T. G. Wensel.** 1991. Nucleotide exchange and cGMP phosphodiesterase activation by pertussis toxin inactivated transducin. *Biochemistry* **30:**11637–11645.

60. **Rappuoli, R., M. Pizza, M. T. De Magistris, A. Podda, M. Bugnoli, R. Manetti, and L. Nencioni.** 1992. Development and clinical testing of an acellular pertussis vaccine containing genetically detoxified pertussis toxin. *Immunobiology* **184:**230–239.

61. **Sato, H., A. Ito, J. Chiba, Y. Sato.** 1984. Monoclonal antibodies against pertussis toxin: effect on toxin activity and pertussis infections. *Infect. Immun.* **46:**422–428.

62. **Sixma, T. K., S. E. Pronk, K. H. Kalk, E. S. Wartna, B. A. M. van Zanten, B. Witholt, and W. G. J. Hol.** 1991. Crystal structure of a cholera toxin-related heat-labile enterotoxin from *E. coli. Nature* (London) **351:**371–377.

63. **Stein, E. P., A. Boodhoo, G. D. Armstrong, S. A. Cockle, M. H. Klein, and R. Read.** 1994. The crystal structure of pertussis toxin. *Structure* **2:**45–59.

64. **Stow, J. L., J. B. de Almeida, N. Narula, E. J. Holtzman, L. Ercolani, and D. A. Ausiello.** 1991. A heterotrimeric G protein, $G\alpha_{i-3}$, on Golgi membranes regulates the secretion of a heparin sulfate proteoglycan in LLC-PK$_1$ epithelial cells. *J. Cell Biol.* **114:**1113–1124.

65. **Sugai, M., C.-H. Chen, and H. Wu.** 1992. Bacterial ADP-ribosyltransferase with a substrate specificity of the Rho protein disassembles the Golgi apparatus in Vero cells and mimics the action of brefeldin A. *Proc. Natl. Acad. Sci. USA* **89:**8903–8907.

66. **Tamura, M., K. Nogimori, S. Murai, M. Yajima, K. Ito, T. Katada, and M. Ui.** 1982. Subunit structure of islet-activating protein, pertussis toxin, in conformity with the A-B model. *Biochemistry* **21:**5516–5522.

67. **Tamura, M., K. Nogimori, M. Yajima, K. Ase, and M. Ui.** 1983. A role of the B-oligomer moiety of islet-activating protein, pertussis toxin, in development of the biological effects on intact cells. *J. Biol. Chem.* **258:**6756–6761.

68. **Tweten, R. K., J. T. Barbieri, and R. J. Col-**

lier. 1985. Diphtheria toxin: effect of substituting aspartic acid for glutamic acid 148 on ADP-ribosyltransferase activity. *J. Biol. Chem.* **260:** 10392–10394.

69. **Ui, M.** 1990. Pertussis toxin as a valuable probe for G-protein involvement in signal transduction, p. 45–66. *In* J. Moss and M. Vaughan (ed.), *ADP-Ribosylating Toxins and G Proteins: Insights into Signal Transduction.* American Society for Microbiology, Washington, D.C.

70. **Watkins, P. A., D. L. Burns, Y. Kanaho, T.-Y. Liu, E. L. Hewlett, and J. Moss.** 1985. ADP-ribosylation of transducin by pertussis toxin. *J. Biol. Chem.* **260:**13478–13482.

71. **Weiss, A. A., and E. L. Hewlett.** 1986. Virulence factors of *Bordetella pertussis. Annu. Rev. Microbiol.* **40:**661–686.

72. **Weiss, A. A., E. L. Hewlett, G. A. Myers, and S. Falkow.** 1984. Pertussis toxin and extracytoplasmic adenylate cyclase as virulence factors of *Bordetella pertussis. J. Infect. Dis.* **150:**219–222.

73. **Wilson, B. A., K. A. Reich, B. R. Weinstein, and R. J. Collier.** 1990. Active-site mutations of diphtheria toxin: effects of replacing glutamic acid-148 with aspartic acid, glutamine, or serine. *Biochemistry* **29:**8643–8651.

74. **Xu, Y., V. Barbançon-Finck, and J. T. Barbieri.** 1994. Role of histidine 35 of the S1 subunit of pertussis toxin in the ADP-ribosylation of transducin. *J. Biol. Chem.* **269:**9993–9999.

STRATEGIES
TO OVERCOME
BACTERIAL
VIRULENCE
MECHANISMS

ECOLOGICAL CONCEPTS IN THE CONTROL OF PATHOGENS

Kenneth H. Wilson

15

The body surfaces of humans and animals support complex natural bacterial communities. Despite their adverse effects on the host such as shortened life span and occasional invasion of tissues, these communities overall have a favorable impact. Aside from being required for the proper functioning of the immune system, the biota is itself a major host defense. This chapter concentrates on the gastrointestinal biota in reviewing what is known of natural communities as an ecological barrier to pathogens and evaluates progress in applying this knowledge.

The role of the biota as a host defense was first demonstrated experimentally in the mid-1950s by Bohnhoff et al. (4) and Freter (17). Unfortunately, a large literature in which even the key observation of bacterial antagonism has been "rediscovered" and renamed repeatedly (70, 84) has given the appearance of chaos. Even very recently, the study of microecology of the gut has been described as consisting of "scattered information that needs to be interrelated into a coherent understanding" (41). As this chapter attempts to demonstrate, this depiction is not entirely fair. Granted, much published work has been based on inadequate methodology and has tested nonsensical or vague hypotheses. However, there is a cohesive core of literature that defines the important properties of the colonic ecosystem and establishes a theoretic framework that allows researchers to ask questions that mean something. This chapter reviews the major concepts of that literature. It then contrasts what we know of human biota with what we have attempted to put into practice.

INTESTINAL BIOTA AND ITS HOST DEFENSE FUNCTION

In many animals, the stomach or rumen supports a complex biota. However, more work has been done on the role of these communities in nutrition than on their role as a host defense. Often, as in humans, the stomach is either sterile or supports a very sparse biota. In humans, only bacteria of the genus *Helicobacter* are known to colonize the stomach (45, 74, 88). Other bacteria found there do not form stable populations but are just passing through. The possibility that strains of *Helicobacter* compete has not been investigated but would be an interesting study, given the pathogenic potential of at least some strains of *Helicobacter pylori* (25, 65). It may be possible to protect subjects from pathogenic strains by establishing popu-

Kenneth H. Wilson Division of Infectious Diseases, Veterans Affairs Medical Center and Duke University Medical Center, 508 Fulton Street, Durham, North Carolina 27705.

Virulence Mechanisms of Bacterial Pathogens, 2nd ed., Edited by J. A. Roth et al.
© 1995 American Society for Microbiology, Washington, D.C.

lations of nonpathogenic strains. Although significant bacterial populations have been found in the small bowels of some animals, little work on their role in host defense has been done. Instead, the stomach and small bowel, at least in humans, have other defenses against bacterial pathogens. Gastric acid is lethal to a wide range of bacteria. In both the stomach and the small intestine, digestive enzymes are in the lumen, and the mucosa is covered with a thick layer of mucin. Transit time is fast, so there is no time for bacteria to multiply into large populations.

The colon, on the other hand, contains a large bacterial population. The germfree newborn is rapidly colonized first by facultatively anaerobic organisms such as *Escherichia coli* (42). These are followed by relatively aerotolerant anaerobes such as lactobacilli. During ecologic succession, toxigenic clostridia such as *Clostridium difficile* and *Clostridium botulinum* are relatively common community members (49, 86). However, most of these early populations are later suppressed by the climax-stage biota. While there is overlap in bacterial taxa present, to a large extent the composition of climax-stage biota is host species specific (36, 54). In their collected work on composition of the predominant human colonic biota, Ed and Lillian Moore (personal communication) have identified more than 560 different species of bacteria. The work of Finegold et al. (14, 15) is very similar. Around 100 species make up about 95% of the biomass (36). Anaerobes outnumber facultatively anaerobic species by more than 100 to 1 (14, 36). This ecosystem is composed of hundreds of species in any individual, with no one species predominating. Ethnicity has no influence on the composition of the biota (12, 54) despite the fact that there is considerable person-to-person variation (36). Although a few populations of bacteria change with radical dietary manipulation (12, 53), the overall composition of an individual's biota remains stable over time (36, 98).

The climax-stage biota's host defense role is implicit, as stated above: most bacteria in the biosphere are not community members. The likes of *Yersinia pestis, Bacillus anthracis, Clostridium tetani, C. botulinum,* and *C. difficile* are notably absent. Because most of the pathogenic bacteria tested can colonize the germfree animal (21, 32, 60, 64, 90), it can be concluded that the biota excludes them. Although some specialized pathogens are able to cause disease by adhering to enterocytes or invading them (40, 56, 67), none is able to compete effectively against an intact biota within the lumen to establish a population. However, one of our greatest current medical problems appears to be induced when patients are treated with antibiotics, most of which simplify the biota (13). The role of host defense was originally observed experimentally in rodents that had been treated with the antibiotic streptomycin and then challenged with *Salmonella, Shigella* (4), or *Vibrio cholerae* (17). In the same mode, we are plagued by nosocomial infections caused by *Enterococcus* species, pathogenic *E. coli, Candida albicans, Staphylococcus epidermidis,* and other organisms that reside in the guts of antibiotic-treated humans. Antibiotic treatment of animals has been implicated in the spread of *Salmonella* spp. (37, 82, 83).

C. difficile illustrates the biota's defense function. Not indigenous to the adult hamster, it rapidly makes an appearance after antibiotic treatment and exposure of animals to the organism (Fig. 1). Within a matter of hours, it has reached a population size of 100 million CFU/ml. On the other hand, the untreated hamster is unaffected by even large challenges of either vegetative cells or spores. After spores germinate in the small bowel, they behave as vegetative cells and fail to initiate growth in the cecum (Fig. 2). The host defense role can also be demonstrated in the gnotobiotic mouse. Germfree mice can readily be colonized by *C. difficile,* and they are much less susceptible than hamsters to the toxin (60, 96). In monoassociated mice, the pathogen maintains a population of about 400 million CFU/ml. However, by 2 to 3 weeks after an indigenous mouse biota has been introduced into these monoassociated mice (about the length of time

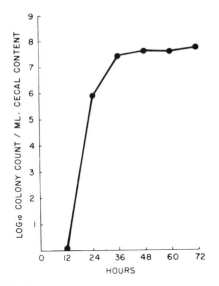

FIGURE 1 Rate of appearance and population size of *C. difficile* in Syrian hamsters treated orogastrically at time zero with clindamycin and caged in a room contaminated with spores (unpublished data).

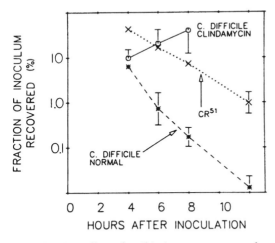

FIGURE 2 Effect of antibiotic treatment on colonization resistance. This graph shows the rate of elimination of *C. difficile* in the ceca of Syrian hamsters orally administered spores along with a ^{51}Cr tracer. In healthy animals, *C. difficile* cells are eliminated faster than the tracer; in antibiotic-treated animals, the pathogen initiates growth. (Reprinted with permission from reference 95a. © 1985 by the University of Chicago. All rights reserved.)

required for ecological succession), *C. difficile* is suppressed to undetectable levels (93).

OVERALL THEORY

To be useful and to fit the pattern of conventional scientific thought in more established fields, concepts should help in understanding cause and effect, be supported by experimental data, and make common sense. The best all-encompassing theoretical framework is Freter et al.'s theory of colonic microecology (22, 23). This theory grew directly from experimental observation in the 1960s that the ability of organisms to compete with *Shigella flexneri* in the guts of gnotobiotic mice was predicted well by their ability to compete in continuous-flow (CF) culture (34). No other in vitro system is able to reproduce in vivo interactions in this way.

This observation was key because of prior work. In 1950, Monod (52) and Novik and Szilard (58) independently described the theory of the chemostat (theoretically perfect CF culture), which holds that two experimentally observed phenomena are useful in mathematical models to predict the behavior of bacteria in chemostats. First, the population size of an organism is proportional to the amount of substrate supplied. Second, one substrate turns out to be growth limiting, and if two organisms are limited by the same substrate, only one will be able to colonize the chemostat long term unless both organisms are precisely equally fit. Fitness can be calculated if the dilution rate of the chemostat, the maximum growth rate of each organism, and the limiting substrate concentration at half the maximal growth rate for each organism are known. Less-fit organisms will be forced to divide at a rate slower than the dilution rate, and their populations will diminish to zero. Extensive experimental data now support the theory of the chemostat; the theory is able to predict accurately the outcome of microbial competition in CF cultures (16, 31).

Thus, the fact that CF cultures succeeded as in vitro models of bacterial competition in the gut suggested that bacteria were competing for nutrients. Further study showed this to be indeed the case for both *Shigella flexneri* (18) and *E. coli* (61). These results have been confirmed repeatedly (30, 94). An intriguing question now arose: How far can the analogy between a CF culture ecosystem and a colonic ecosystem be carried? Studies were done with the biotas of mice (24), rats (85), humans (50), and chickens (57), and all supported the concept that the entire ecosystem could be modeled in this manner. Subsequent studies showed that an environment-simulating medium was better than a standard laboratory broth medium (93). Figure 3 compares a CF culture of mouse biota in veal infusion broth with one using homogenates of fecal pellets from germfree mice as a medium. The tapered gram-negative rods occurring in palisades only in the environment-simulating medium are typical of the predominant mouse cecal biota. This result is

not surprising, because in the cases studied so far, carbon source has been growth limiting. The major carbon sources in the colon are unabsorbed plant carbohydrates and gastrointestinal mucin, neither of which is present in standard laboratory media. In fact, the theory was adequately predictive that adding to CF culture medium the monosaccharide moieties found in mucin that were growth limiting to *C. difficile* fostered the growth of organisms in CF culture that were able to suppress *C. difficile* in vivo (94).

A serious shortcoming of the Monod theory, however, is that it fails to predict some important phenomena that occur in CF cultures and intact animals. First, if a strain of *E. coli* is implanted, it forms a barrier to establishment of a second strain. Furthermore, if implanted first, *E. coli* can maintain a population

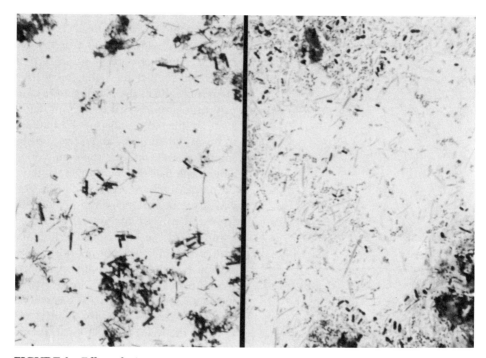

FIGURE 3 Effect of using an environment-simulating medium on composition of mouse cecal biota in CF culture. The Gram stain on the left is of a CF culture containing biota growing on veal infusion broth. On the right, the medium consists of a sterile extract of fecal pellets from germfree mice. The latter medium consistently supported a biota giving an appearance more characteristic of conventional mouse biota, which is usually dominated by tapered gram-negative or gram-variable cells that form palisades (93a).

even if the entire ecosystem is subsequently introduced. Simple chemostat theory does not explain these findings. Nor does it explain why colonization resistance is impaired when the adherent wall growth is removed from a CF culture (24). A major shortcoming of the Monod theory is that the "idealized" CF culture does not have adherent populations. When the original theory was revised to mathematically model competition for adhesion sites as well as for nutrients, the model predicted colonization resistance (23). Further experimental observations have supported the more complex model (20). A relatively unheralded advantage of the newer mathematical model is that it is more in keeping with the extremely complex ecosystem known to exist in the colon. Even though it has relatively poor substrate affinity, an organism could coexist with a more efficient competitor by adhering to the colonic wall to maintain a population. Thus, there could be more than one organism for each growth-limiting nutrient.

It appears that the organisms making up this complex ecosystem interact with each other and the host in ways consistent with the theory described above. In doing so, they create an environment that by its physicochemical nature has inhibitory effects on many other microorganisms. Not only does the biota monopolize growth-limiting substrates, but anaerobiosis itself slows the growth rate of some facultative organisms (10), and a slower growth rate equals a disadvantageous shift of the Monod curve. The high H_2S tension within the gut lumen acts to make *E. coli* compete less effectively for a carbon source (22). High concentrations of volatile fatty acids have an inhibitory effect on many organisms in short-term experiments in vitro (35, 48, 66). Despite the fact that the effect as measured in vitro has often failed to correlate with in vivo antagonism (44, 78), high concentrations of these acids probably lead to a prolonged lag phase (23). Although the importance in vivo of secondary bile acids has not yet been demonstrated, the ability of organisms indigenous to the gut to grow in their presence contrasts sharply with

the suppressive effect of secondary bile acids on oral biota.

Clearly, the mechanisms in favor of selection of certain organisms to be community members and against others to be excluded is multifactorial despite the fact that these mechanisms can be conceptualized in one theory. The control mechanisms are highly redundant (19); this is not surprising in a complex yet stable ecosystem.

APPLICATION OF ECOLOGICAL CONCEPTS

The Concept of "Good" Bacteria

In 1908, Metchnikoff noted that ingestion of putrefied food was often disastrous. On the other hand, foods prepared by fermentation with lactic acid bacilli seldom putrefied and were not unhealthy to eat. He concluded that because putrefaction occurs in the human gut, ingesting lactic acid bacilli could prolong life (47). Since that time, some people have believed that bacteria that happen to produce lactic acid as the major end product of fermentation are especially beneficial to humans or animals. Organisms of particular interest in recent studies have included *Lactobacillus acidophilus* (9), *Lactobacillus casei* (9, 73), *Lactobacillus reuteri* (81), *Enterococcus faecium* (97), and various bifidobacteria (3, 33). Claims have included prevention of traveler's diarrhea (59), treatment of antibiotic-associated diarrhea (72), prevention of colon cancer (28), treatment of lethal irradiation (11), treatment of *C. difficile* colitis (29), and lowering of serum cholesterol (51, 97).

Overall, this literature can still be described as conflicting and uncontrolled. This is not surprising, because the studies usually have no logical rationale. There is no direct evidence that lactic acid-producing bacteria have any effect that is different from those of organisms that produce acetic acid or any other end product. Pickling food with acetic acid also preserves it, so there is an equally good argument in favor of acetic acid producers. Furthermore, lactic acid producers represent an unrelated array of gram-positive bacteria. Thus, it is not clear why the organisms claimed to be beneficial should have any common effect not shared by many bacteria. When administered orally, these organisms can generally be shown to at least partially survive transit through the gastrointestinal tract, but they do not implant. There is little evidence that they divide or carry out any metabolic activity on their way through. Thus, the notion that they would have any effect on the host in the presence of a finely tuned ecosystem consisting of hundreds of well-adapted species seems irrational. The idea that they could function in the place of that ecosystem after antibiotic treatment seems equally unfounded.

Competition between Strains

It is well known that strains of the same bacterial species tend to compete when at the same body site. Given the importance of available adhesion sites for the process of implantation, this phenomenon should not be surprising. Although several examples in the gastrointestinal tract are known (2, 61, 95), this discussion discusses what to date is probably the most successful application of microbial ecology to clinical medicine: the control of staphylococcal colonization.

During the period just prior to the introduction of semisynthetic penicillins, it was difficult to treat staphylococcal infections. *Salmonella aureus* had become commonly resistant to penicillin, and phage type 80/81 in particular caused widespread outbreaks of nosocomial infection. It was especially common and devastating for this organism to spread in newborn nurseries. Staphylococcal epidemiology was studied extensively. *S. aureus* colonizes the anterior nares of around 25% of healthy people. When one strain has colonized the nares, it is difficult to implant a second strain (5). In studying one nursery outbreak, it was found that colonization with a particular strain of the staphylococcus strain, 502A, was negatively correlated with disease (70). It appeared that 502A was relatively nonpathogenic, and it protected against colonization with the 80/81

strain. In a series of studies at various geographic locations, 502A was intentionally implanted in the nares and umbilical stumps of newborn infants during outbreaks of infection due to 80/81 (6, 43, 71). In all eight nurseries where it was used, the introduction of 502A led to a cessation of the outbreak of staphylococcal disease (69).

Further studies addressed the problem of familial furunculosis (7, 76). In this disorder, several members in a household repeatedly develop staphylococcal skin abscesses. Although drainage and treatment with semisynthetic penicillins can cure the abscesses, they recur. Members of affected families were treated with oxacillin to clear the carrier state (an approach that is probably less effective today [91]), and 502A was implanted. Four of 82 patients (total of 4 lesions) compared with 15 of 51 saline-treated controls (total of 27 lesions) developed staphylococcal lesions during approximately 1 year of follow-up. Strain 502A was cultured from lesions in three of the four infections in the treated group. These three lesions consisted of two small skin abscesses and a sty. In comparison, the 27 staphylococcal infections in the control group included 3 ischial abscesses, 8 deep skin abscesses, and 1 case of peritonitis.

Given the effectiveness of interstrain bacterial antagonism, it seems unfortunate that the approach has essentially been abandoned. The advent of more antibiotics, better hygiene in nurseries, and increased numbers of medical tort cases probably all contributed to the abandonment of this promising line of work. However, the appearance of vancomycin- and methicillin-resistant *S. aureus* in the near future seems all but assured, and once again, there will probably be little to fall back on. In such a setting, a resurgence of interest in an ecological approach seems likely. We have a much better understanding now of the molecular basis of virulence in this pathogen and an ability to genetically alter it. Thus, it may be possible to engineer a strain to be even less virulent than 502A. Whether this can be done without decreasing a strain's competitive ability remains to be seen.

The Concept of Biodiversity

As described above, community members of the colonic ecosystem are highly diverse. As it turns out, this diversity itself is critical to the biota's function in host defense. The more strains of anaerobes that are introduced into gnotobiotic mice, the more potentially pathogenic strains of enterobacteriaceae are suppressed (21, 89). By the same token, *C. difficile* is suppressed to a greater degree in gnotobiotic mice as the complexity of the biota increases (93). In humans, antibiotics do not totally eliminate the biota; they simplify it (13). When this process occurs, suppressed populations such as *Enterococcus* species and *Candida albicans* increase, and new organisms such as *C. difficile* are able to colonize. Organisms that are antibiotic resistant and that possess special virulence characteristics are major problems in our hospitals. To a large extent, this problem probably stems from the concentration of antibiotic-treated patients with simplified biotas found in these environments. One would expect that in such patients, not all metabolic substrates have been scavenged, not all adhesion sites have been taken, and the physicochemical environment of the colon is no longer disadvantageous to nonindigenous bacterial strains.

Although it has been hoped that some single organism could replace the entire ecosystem in this situation, such a hope is unrealistic. To again establish normal homeostasis would require establishing biodiversity. Surprisingly, this concept has been applied rarely except in laboratory animals. In humans, fecal enemas have been given to patients with antibiotic-associated colitis (8, 68). Although the results are promising, the studies are uncontrolled.

So far, the problems in establishing biodiversity outside the laboratory setting have not been solved. One problem is logistical. To avoid facing the biosafety problem inherent in using fecal material, one would have to maintain hundreds of fastidious anaerobic species in a viable state long enough to be administered. This is not known to be possible using methods that would be commercially feasible. Further-more, the fact remains that no collection of organisms has yet been shown to be capable of performing all the functions of a complete enteric biota. Ideally, then, the ecosystem should be maintained intact as far as possible. However, collections of life-forms as they occur in nature are not patentable, a fact that discourages commercial development in this area. Hopefully, patents on processes will eventually make development feasible.

Importance of Subpopulations

Although isolates picked at random from the predominant anaerobic biota can reproduce many normal functions, it is clear that subpopulations are also important in suppression of pathogens. For instance, 100 anaerobic strains in gnotobiotic mice suppressed *Shigella flexneri* less than the same strains plus *E. coli* (21). Random isolates from the predominant climax-stage biota of mice suppressed *C. difficile* less than the climax-stage isolates plus isolates that had predominated during ecologic succession (93). Furthermore, the entire collection of isolates picked at random suppressed *C. difficile* less than a mouse's fecal specimen that had been heat shocked and then introduced into germ-free mice (77). Presumably, heat shocking selected for spore-forming bacteria that were particularly suppressive to *C. difficile*. These results suggest that it may be possible to select isolates from indigenous biota that are particularly suppressive to certain pathogens. The mechanisms for such interactions have not been demonstrated, and we are not close to using this approach in a rational way. However, "designer biotas" would have the advantage of potentially being patentable.

Microbes as Pharmacologic Agents

Saccharomyces boulardii is effective in prophylaxis and prevention of relapse in *C. difficile* disease (79, 80). The organism was originally found in investigations of an Indochinese folk remedy for diarrhea (46). Compared with other probiotic agents, this organism stands out as having undergone controlled trials, been shown to be effective, and undergone rational study of the mechanism of action. As with other probiotic agents, it does not implant in the intestinal tract but to some extent survives passage (46). As one would expect of an organism that is probably metabolically inactive at the site of action, it does not affect the population size of *C. difficile*. Rather, it is suspected to either bind the toxin or cleave it with a preformed protease (62). Thus, it could be described more as a living pharmacologic agent than as an agent of ecological manipulation.

Another concept in probiotics has been studied in a controlled manner: that massive numbers of foreign bacterial cells introduced into the small bowel during viral gastroenteritis can serve as an immune adjuvant. At least one controlled study showed that treatment with *L. casei* decreased diarrhea in infants infected with rotavirus (38). The mechanism appeared to be that the bacterium accelerated the development of an immune response (39). It is not known which components of the bacterial cell are important or which other bacteria would cause this effect. It is also not known whether the adjuvant effect is confined to infants who have incompletely developed immune systems. Again, it appears that when an effect of a current probiotic agent can be measured and its mechanism determined, it does not involve bacterial interactions within the ecosystem of the gut.

NEW DIRECTIONS

Since Bohnhoff and Freter discovered bacterial antagonism or colonization resistance in the mid-1950s, a substantial and coherent literature has developed. The theoretical framework in this field is now mathematical and predictive and makes intuitive sense. In many instances, it has been possible to control bacterial pathogens by using ecological manipulations in experimental animals.

However, progress in applying laboratory approaches and scientific theory has been slow. Emerging infectious diseases, especially those caused by antibiotic-resistant nosocomial pathogens, are of increasing concern. Many of these organisms are normally suppressed by our bi-

ota, but in hospital populations inundated with antibiotics, this normal host defense is compromised. Despite real advances in the study of the microbial ecology of the gut, approaches being taken in practice have not changed much during this century. A therapeutic armamentarium consisting of various strains of *Lactobacillus* and related organisms has no rational basis and is ineffective.

Thus, the potential for applying concepts in microbial ecology to the control of pathogens is high. Not only are the concepts relatively well developed compared with their applications, but new tools that promise to revolutionize the field are available. For instance, by using oligonucleotide probes directed at rRNA (27, 41, 63), it should be possible for the first time to quickly determine whether or not a given organism is colonizing a host, even in small numbers. Sequencing of ribosomal DNA makes it possible to detect organisms that cannot be cultured (26, 75, 87, 92). Denaturation gradients make it possible to separate complex amplicons (55), and two-dimensional electrophoresis of DNA may make it possible to see an entire ecosystem of amplified ribosomal DNAs spread before one's eyes. It is now possible to develop recombinant systems for indigenous anaerobes (1). In short, the pieces are in place for rapid advancement in this field.

REFERENCES

1. **Bedzyk, L. A., N. B. Shoemaker, K. E. Young, and A. A. Salyers.** 1992. Insertion and excision of *Bacteroides* conjugative chromosomal elements. *J. Bacteriol.* **174:**166–172.
2. **Berg, R. D.** 1978. Antagonism among the normal anaerobic bacteria of the mouse gastrointestinal tract determined by immunofluorescence. *Appl. Environ. Microbiol.* **35:**1066–1073.
3. **Biavati, B., T. Sozzi, P. Mattarelli, and L. D. Trovatelli.** 1992. Survival of bifidobacteria from human habitat in acidified milk. *Microbiologica* **15:**197–200.
4. **Bohnhoff, M., B. L. Drake, and C. P. Miller.** 1954. Effect of streptomycin on susceptibility of the intestinal tract to experimental salmonella infection. *Proc. Soc. Exp. Biol. Med.* **86:**132–137.
5. **Boris, M., T. F. Sellers, H. F. Eichenwald, J. C. Ribble, and H. R. Shinefield.** 1964. Bacterial interference. *Am. J. Dis. Child.* **108:**252–261.

6. **Boris, M., H. R. Shinefield, J. C. Ribble, H. F. Eichenwald, G. H. Hauser, and C. T. Caraway.** 1963. The Louisiana epidemic. *Am. J. Dis. Child.* **105**:674–682.

7. **Boris, M., H. R. Shinefield, P. Romano, D. P. McCarthy, and A. L. Florman.** 1968. Bacterial interference. Protection against recurrent intrafamilial staphylococcal disease. *Am. J. Dis. Child.* **115**:521–529.

8. **Bowden, T. A., A. R. Mansberger, and L. E. Lykins.** 1981. Pseudomembraneous enterocolitis: mechanism of restoring floral homeostasis. *Am. Surg.* **47**:178–183.

9. **Clements, M. L., M. M. Levine, R. E. Black, R. M. Robins-Browne, L. A. Cisneros, G. L. Drusano, C. F. Lanata, and A. J. Saah.** 1981. *Lactobacillus prophylaxis* for diarrhea due to enterotoxigenic *Escherichia coli*. *Antimicrob. Agents Chemother.* **20**:104–108.

10. **Davis, B. D., R. Dulbecco, H. N. Eisen, and H. S. Ginsberg.** 1980. *Microbiology Including Immunology and Molecular Genetics*. Harper & Row, Philadelphia.

11. **Dong, M.-Y., T.-W. Chang, and S. L. Gorbach.** 1987. Effects of feeding *Lactobacillus* GG on lethal irradiation in mice. *Diagn. Microbiol. Infect. Dis.* **7**:1–7.

12. **Finegold, S. M., H. R. Attebery, and V. L. Sutter.** 1974. Effect of diet on human fecal flora: Comparison of Japanese and American diets. *Am. J. Clin. Nutr.* **27**:1456–1469.

13. **Finegold, S. M., G. E. Mathison, and W. L. George.** 1983. Changes in human intestinal flora related to the administration of antimicrobial agents, p. 356–438. *In* D. J. Hentges (ed.), *Human Intestinal Flora in Health and Disease*. Academic Press, Inc., New York.

14. **Finegold, S. M., V. L. Sutter, and G. E. Mathison.** 1983. Normal indigenous intestinal flora, p. 3–31. *In* D. J. Hentges. (ed.), *Human Intestinal Microflora in Health and Disease*. Academic Press, Inc., New York.

15. **Finegold, S. M., V. L. Sutter, P. T. Sugihara, H. A. Elder, S. M. Lehmann, and R. L. Phillips.** 1977. Fecal microbial flora in Seventh Day Adventist populations and control subjects. *Am. J. Clin. Nutr.* **30**:1781–1792.

16. **Fredrickson, A. G.** 1977. Behavior of mixed cultures of microorganisms. *Annu. Rev. Microbiol.* **31**:63–87.

17. **Freter, R.** 1955. Fatal enteric cholera infection in the guinea pig achieved by inhibition of normal enteric flora. *J. Infect. Dis.* **97**:57–64.

18. **Freter, R.** 1962. *In vivo* and *in vitro* antagonism of intestinal bacteria against *Shigella flexneri*. II. The inhibitory mechanism. *J. Infect. Dis.* **110**:38–46.

19. **Freter, R.** 1974. Interactions between mechanisms controlling the intestinal microflora. *Am. J. Clin. Nutr.* **27**:1409–1416.

20. **Freter, R.** 1988. Mechanisms of bacterial colonization of the mucosal surfaces of the gut, p. 45–60. *In* J. A. Roth (ed.), *Virulence Mechanisms of Bacterial Pathogens*. American Society for Microbiology, Washington, D.C.

21. **Freter, R., and G. D. Abrams.** 1972. Function of various intestinal bacteria in converting germ-free mice to the normal state. *Infect. Immun.* **6**:119–126.

22. **Freter, R., H. Brickner, M. Botney, D. Cleven, and A. Aranki.** 1983. Mechanisms that control bacterial populations in continuous-flow culture models of mouse large intestinal flora. *Infect. Immun.* **39**:676–685.

23. **Freter, R., H. Brickner, J. Fekete, M. M. Vickerman, and K. E. Carey.** 1983. Survival and implantation of *Escherichia coli* in the intestinal tract. *Infect. Immun.* **39**:686–703.

24. **Freter, R., E. Stauffer, D. Cleven, L. V. Holdeman, and W. E. C. Moore.** 1983. Continuous-flow cultures as in vitro models of the ecology of large intestinal flora. *Infect. Immun.* **39**:666–675.

25. **George, L. I., T. J. Borody, and P. Andrews.** 1990. Cure of duodenal ulcer after eradication of *Helicobacter pylori*. *Med. J. Aust.* **153**:145–149.

26. **Giovannoni, S. L., T. B. Britschgi, C. L. Moyer, and K. G. Field.** 1990. Genetic diversity in Sargasso Sea bacterioplankton. *Nature* (London) **345**:60–63.

27. **Giovannoni, S. L., E. F. DeLong, G. J. Olsen, and N. R. Pace.** 1988. Phylogenetic group-specific oligodeoxynucleotide probes for identification of single microbial cells. *J. Bacteriol.* **170**:720–726.

28. **Gorbach, S. L.** 1990. Lactic acid bacteria and human health. *Ann. Med.* **22**:37–41.

29. **Gorbach, S. L., T.-W. Chang, and B. Goldin.** 1987. Successful treatment of relapsing *Clostridium difficile* colitis with *Lactobacillus* GG. *Lancet* **ii**:1519.

30. **Guiot, H. F.** 1982. Role of competition for substrate in bacterial antagonism in the gut. *Infect. Immun.* **38**:887–892.

31. **Hansen, S. R., and S. P. Hubbell.** 1980. Single-nutrient microbial competition: qualitative agreement between experimental and theoretically forcast outcomes. *Science* **207**:1491–1493.

32. **Hazenberg, M. P., M. Bakker, and A. V. Burggraaf.** 1981. Effects of the human intestinal flora on germ-free mice. *J. Appl. Bacteriol.* **50**:95–106.

33. **Hekmat, S., and D. J. McMahon.** 1992. Survival of *Lactobacillus acidophilus* and *Bifidobacterium bifidum* in ice cream for use as a probiotic food. *J. Dairy Sci.* **75**:1415–1422.

34. **Hentges, D. J., and R. Freter.** 1962. *In vivo* and *in vitro* antagonism of intestinal bacteria against *Shigella flexneri*. I. Correlation between various tests. *J. Infect. Dis.* **110:**30–37.

35. **Hentges, D. J., and B. R. Maier.** 1972. Inhibition of *Shigella flexneri* by the normal intestinal flora. III. Interactions with *Bacteroides fragilis in vitro*. *Infect. Immun.* **6:**168–173.

36. **Holdeman, L. V., E. P. Cato, and W. E. C. Moore.** 1976. Human fecal flora: variation in bacterial composition within individuals and a possible effect of emotional stress. *Appl. Environ. Microbiol.* **32:**359–375.

37. **Holmberg, S. D., M. T. Osterholm, K. A. Senger, and M. L. Cohen.** 1984. Drug-resistant salmonella from animals fed antimicrobials. *N. Engl. J. Med.* **311:**617–622.

38. **Isolauri, E., M. Juntunen, T. Rautanen, P. Sillanaukee, and T. Koivula.** 1991. A human lactobacillus strain (*Lactobacillus* GG) promotes recovery from acute diarrhea in children. *Pediatrics* **88:**90–97.

39. **Kaila, M., E. Isolauri, E. Sopi, E. Virtanen, S. Laine, and H. Arvilommi.** 1992. Enhancement of the circulating antibody secreting cell response in human diarrhea by a human *Lactobacillus* strain. *Pediatr. Res.* **32:**141–144.

40. **Karch, H., J. Heesemann, R. Laufs, A. D. O'Brien, C. O. Tacket, and M. M. Levine.** 1987. A plasmid of enterohemorrhagic *Escherichia coli* O157:H7 is required for expression of a new fimbrial antigen and for adhesion to epithelial cells. *Infect. Immun.* **55:**455–461.

41. **Krogfelt, K. A., L. K. Poulsen, and S. Molin.** 1993. Identification of coccoid *Escherichia coli* BJ4 cells in the large intestine of streptomycin-treated mice. *Infect. Immun.* **61:**5029–5034.

42. **Lee, A., and E. Gemmell.** 1972. Changes in the mouse intestinal microflora during weaning: role of volatile fatty acids. *Infect. Immun.* **51:**1–7.

43. **Light, I., R. L. Walton, J. M. Sutherland, H. R. Shinefield, and V. Brackvogel.** 1967. Use of bacterial interference to control a staphylococcal nursery outbreak. *Am. J. Dis. Child.* **113:**291–300.

44. **Maier, B. R., and D. J. Hentges.** 1972. Experimental shigella infection in animals. *Infect. Immun.* **6:**168–173.

45. **Marshall, B. J., and J. R. Warren.** 1984. Unidentified curved bacilli in the stomach of patients with gastritis and peptic ulceration. *Lancet* **i:**1311–1314.

46. **McFarland, L. V., and P. Bernasconi.** 1993. *Saccharomyces boulardii*: a review of an innovative biotherapeutic agent. *Microb. Ecol. Health Dis.* **6:**157–171.

47. **Metchnikoff, E.** 1908. *The Prolongation of Life.*

Optimistic Studies. G. P. Putnam's Sons, New York.

48. **Meynell, G. G.** 1963. Antibacterial mechanisms of the mouse gut. *Br. J. Exp. Pathol.* **44:**209–219.

49. **Midura, T. F., and S. S. Arnon.** 1976. Infant botulism: identification of *Clostridium botulinum* and its toxins in faeces. *Lancet* **ii:**934–936.

50. **Miller, T. L., and M. J. Wolin.** 1981. Fermentation by the human large intestine microbial community in a semicontinuous culture system. *Appl. Environ. Microbiol.* **42:**400–407.

51. **Molin, G., R. Andersson, S. Ahrne, C. Lonner, I. Marklinder, M. L. Johansson, B. Jeppsson, and S. Bengmark.** 1992. Effect of fermented oatmeal soup on the cholesterol level and the *Lactobacillus* colonization of rat intestinal mucosa. *Antonie van Leeuwenhoek* **61:**167–173.

52. **Monod, J.** 1950. La technique de cultur continue: theorie et applications. *Ann. Inst. Pasteur* **79:**390–410.

53. **Moore, W. E., E. P. Cato, I. J. Good, and L. V. Holdeman.** 1981. The effect of diet on the human fecal flora. Banbury report 7. *Gastrointestinal Cancer: Endogenous Factors.* Cold Spring Harbor Laboratory Press, Cold Spring Harbor, N.Y.

54. **Moore, W. E. C., and L. V. Holdeman.** 1974. Human fecal flora: the normal flora of 21 Japanese-Hawaiians. *Appl. Microbiol.* **27:**961–979.

55. **Muyzer, G., E. C. De Waal, and A. G. Uitterlinden.** 1993. Profiling of complex microbial populations by denaturing gradient gel electrophoresis analysis of polymerase chain reaction-amplified genes coding for 16S rRNA. *Appl. Environ. Microbiol.* **59:**695–700.

56. **Nataro, J. P., M. M. Baldini, J. B. Kaper, R. E. Black, N. Bravo, and M. M. Levine.** 1985. Detection of an adherence factor of enteropathogenic *Escherichia coli* with a DNA probe. *J. Infect. Dis.* **152:**560.

57. **Nisbet, D. J., S. C. Ricke, C. M. Scanlan, D. E. Corrier, A. G. Hollister, and J. R. Deloach.** 1994. Inoculation of broiler chicks with a continuous-flow derived bacterial culture facilitates early cecal bacterial colonization and increases resistance to *Salmonella typhimurium*. *J. Food Prot.* **57:**12–15.

58. **Novik, A., and L. Szilard.** 1950. Experiments with the chemostat on spontaneous mutations of bacteria. *Proc. Natl. Acad. Sci. USA* **36:**708–719.

59. **Oksanen, P. J., S. Salminen, M. Saxelin, P. Hamalainen, A. Ihantola-Vormisto, L. Muurasniemi-Isoviita, S. L. Nikkari, T. Oksanen, I. Porsti, and E. Salminen.** 1990. Prevention of travellers' diarrhea by *Lactobacillus* GG. *Ann. Med.* **22:**53–56.

60. **Onderdonk, A. B., R. L. Cisneros, and J. G. Bartlett.** 1980. *Clostridium difficile* in gnotobiotic mice. *Infect. Immun.* **28:**277–282.

61. **Ozawa, A., and R. Freter.** 1964. Ecologic mechanism controlling growth of *Escherichia coli* in continuous-flow culture and in the mouse intestine. *J. Infect. Dis.* **114**:235–242.

62. **Pothoulakis, C., C. P. Kelly, M. A. Joshi, N. Gao, C. J. O'Keane, I. Castagliuolo, and J. T. LaMont.** 1993. *Saccharomyces boulardii* inhibits *Clostridium difficile* toxin A binding and enterotoxicity in rat ileum. *Gastroenterology* **104**:1108–1115.

63. **Poulsen, L. K., G. Ballard, and D. A. Stahl.** 1993. Use of rRNA fluorescence in situ hybridization for measuring the activity of single cells in young and established biofilms. *Appl. Environ. Microbiol.* **59**:1354–1360.

64. **Raibaud, P., R. Ducluzeau, F. Dubos, S. Hudault, H. Bewa, and M. C. Muller.** 1980. Implantation of bacteria from the digestive tract of man and various animals into gnotobiotic mice. *Am. J. Clin. Nutr.* **33**:2440–2447.

65. **Rauws, E. A. J., and G. N. J. Tytgat.** 1990. Cure of duodenal ulcer associated with eradication of *Helicobacter pylori*. *Lancet* **335**:1233–1235.

66. **Rolfe, R. D.** 1984. Role of volatile fatty acids in colonization resistance to *Clostridium difficile*. *Infect. Immun.* **45**:185–191.

67. **Sansonetti, P. J., T. L. Hale, G. J. Dammin, C. Kapfer, H. H. Collins, and S. B. Formal.** 1983. Alterations in the pathogenicity of *Escherichia coli* K-12 after transfer of plasmid and chromosomal genes from *Shigella flexneri*. *Infect. Immun.* **39**:1392–1402.

68. **Schwan, A., S. Sjolin, U. Trottestam, and B. Aronsson.** 1984. Relapsing *Clostridium difficile* enterocolitis cured by rectal infusion of normal faeces. *Scand. J. Infect. Dis.* **16**:211–215.

69. **Shinefield, H. R., J. C. Ribble, and M. Boris.** 1971. Bacterial interference between strains. *Am. J. Dis. Child.* **121**:148–152.

70. **Shinefield, H. R., J. C. Ribble, M. Boris, and H. F. Eichenwald.** 1963. Bacterial interference: its effect on nursery-acquired infection with *Staphylococcus aureus*. I. Preliminary observations on artificial colonization of newborns. *Am. J. Dis. Child.* **105**:146–154.

71. **Shinefield, H. R., J. M. Sutherland, J. C. Ribble, and H. F. Eichenwald.** 1963. The Ohio epidemic. *Am. J. Dis. Child.* **105**:655–662.

72. **Siitonen, S., H. Vapaatalo, S. Salminen, A. Gordin, M. Saxelin, R. Wikberg, and A. L. Kirkkola.** 1990. Effect of *Lactobacillus* GG yogurt in prevention of antibiotic associated diarrhoea. *Ann. Med.* **22**:57–59.

73. **Silva, M., N. V. Jacobus, C. Deneke, and S. L. Gorbach.** 1987. Antimicrobial substance from a human *Lactobacillus* strain. *Antimicrob. Agents Chemother.* **31**:1231–1233.

74. **Solnick, J. V., J. O'Rourke, A. Lee, B. J. Paster, F. E. Dewhirst, and L. S. Tompkins.** 1993. An uncultured gastric spiral organism is a newly identified *Helicobacter* in humans. *J. Infect. Dis.* **168**:379–385.

75. **Stahl, D. A., D. J. Lane, G. J. Olsen, and N. R. Pace.** 1985. Characterization of a Yellowstone hot spring microbial community by 5S rRNA sequences. *Appl. Environ. Microbiol.* **49**:1379–1384.

76. **Strauss, W. G., H. I. Maibach, and H. R. Shinefield.** 1969. Bacterial interference treatment of recurrent furunculosis. *JAMA* **208**:861–863.

77. **Su, W. J., P. Bourlioux, M. Bournaud, M. O. Besnier, and J. Fourniat.** 1986. Mise au point d'un modele experimental animal permettant de la microflore coecale du hamster, antagoniste de *Clostridium difficile*. *Ann. Inst. Pasteur* **137A**:89–96.

78. **Su, W. J., M. J. Waechter, P. Bourlioux, M. Dolegeal, J. Fourniat, and G. Mahuzier.** 1987. Role of volatile fatty acids in colonization resistance to *Clostridium difficile* in gnotobiotic mice. *Infect. Immun.* **55**:1686–1691.

79. **Surawicz, C. M., G. W. Elmer, P. Speelman, L. V. McFarland, J. Chinn, and G. Van Belle.** 1989. Prevention of antibiotic-associated diarrhea by *Saccharomyces boulardii*: a prospective study. *Gastroenterology* **96**:981–988.

80. **Surawicz, C. M., L. V. McFarland, G. Elmer, and J. Chinn.** 1989. Treatment of recurrent *Clostridium difficile* colitis with vancomycin and *Saccharomyces boulardii*. *Am. J. Gastroenterol.* **84**:1285–1287.

81. **Talarico, T. L., and W. J. Dobrogosz.** 1989. Chemical characterization of an antimicrobial substance produced by *Lactobacillus reuteri*. *Antimicrob. Agents Chemother.* **33**:674–679.

82. **Threlfall, E. J., L. R. Ward, A. S. Ashley, and B. Rowe.** 1980. Plasmid-encoded trimethoprim resistance in multiresistant epidemic *Salmonella typhimurium* phage types 204 and 193 in Britain. *Br. Med. J.* **280**:1210–1211.

83. **Threlfall, E. J., L. R. Ward, and B. Rowe.** 1978. Spread of multiresistant strains of *Salmonella typhimurium* phage types 204 and 193 in Britain. *Br. Med. J.* **2**:997.

84. **van der Waaij, D., J. M. Berghuis-de Vries, and J. E. C. Lekkerkerk-van der Wees.** 1971. Colonization of the digestive tract in conventional and antibiotic-treated mice. *J. Hyg.* **69**:405–411.

85. **Veilleux, B. G., and I. Rowland.** 1981. Simulation of the rat intestinal ecosystem using a two-stage continuous culture system. *J. Gen. Microbiol.* **123**:103–115.

86. **Viscidi, R., S. Willey, and J. G. Bartlett.** 1981. Isolation rates and toxigenic potential of *Clostridium difficile* isolates from various populations. *Gastroenterology* **81**:5–9.

87. **Ward, D. M., R. Weller, and M. M. Bateson.** 1990. 16S rRNA sequences reveal numerous un-

cultured microorganisms in a natural community. *Nature* (London) **345**:63–65.

88. **Warren, J. R.** 1983. Unidentified curved bacilli on gastric epithelium in active chronic gastritis. *Lancet* **i**:273–275 (Letter).

89. **Welling, B. W., G. Groen, J. H. M. Tuinte, J. P. Koopman, and H. M. Kennis.** 1980. Biochemical effects on germ-free mice of association with several strains of anaerobic bacteria. *J. Gen. Microbiol.* **117**:57–63.

90. **Wells, C. L., and E. Balish.** 1983. *Clostridium tetani* growth and toxin production in the intestines of germfree rats. *Infect. Immun.* **41**:826–828.

91. **Wheat, L. J., R. B. Kohler, A. L. White, and A. White.** 1981. Effect of rifampin on nasal carriers of coagulase-positive staphylococci. *J. Infect. Dis.* **144**:177.

92. **Wilson, K. H., R. B. Blitchington, R. Frothingham, and J. A. P. Wilson.** 1991. Whipple's disease associated with a bacterium related to *Streptomyces* and *Arthrobacterium,* abstr. 1092. *Program Abstr. 31st Intersci. Conf. Antimicrob. Agents Chemother.*

93. **Wilson, K. H., and R. Freter.** 1986. Interactions of *Clostridium difficile* and *E. coli* with microfloras in continuous-flow cultures and gnotobiotic mice. *Infect. Immun.* **54**:354–358.

93a. **Wilson, K. H., and R. Freter.** Unpublished data.

94. **Wilson, K. H., and F. Perini.** 1988. Role of competition for nutrients in suppression of *Clostridium difficile* by the colonic microflora. *Infect. Immun.* **56**:2610–2614.

95. **Wilson, K. H., and J. N. Sheagren.** 1983. Antagonism of toxigenic *Clostridium difficile* by nontoxigenic *C. difficile.* *J. Infect. Dis.* **147**:733–736.

95a. **Wilson, K. H., J. N. Sheagren, and R. Freter.** 1985. Population dynamics of ingested *Clostridium difficile* in the gut of the Syrian hamster. *J. Infect. Dis.* **151**:355–361.

96. **Wilson, K. H., J. N. Sheagren, R. Freter, L. Weatherbee, and D. Lyerly.** 1986. Gnotobiotic models for study of the microbial ecology of *Clostridium difficile* and *E. coli.* *J. Infect. Dis.* **153**:547–551.

97. **Zacconi, C., V. Bottazzi, A. Rebecchi, E. Bosi, P. G. Sarra, and L. Tagliaferri.** 1992. Serum cholesterol levels in axenic mice colonized with *Enterococcus faecium* and *Lactobacillus acidophilus.* *Microbiologica* **15**:413–417.

98. **Zubrzycki, L., and E. H. Spaulding.** 1962. Studies on the stability of the normal human fecal flora. *J. Bacteriol.* **83**:968–974.

ENHANCEMENT OF NONSPECIFIC RESISTANCE TO BACTERIAL INFECTION BY BIOLOGICAL RESPONSE MODIFIERS

Charles J. Czuprynski

16

The mammalian host has developed a number of defense mechanisms to combat bacterial pathogens. Although these are generally successful, sometimes the pathogen gains the upper hand and goes on to produce disease. Mankind has dealt rather successfully with this situation for the last 50 years by exploiting the antibacterial properties of various natural microbial products (antibiotics). This strategy has been further refined through the synthesis of new generations of drugs that possess advantageous pharmacologic properties not present in the original compounds. The age-old struggle between microbes and humans continues, however, as the microbes devise new means of inactivating or otherwise circumventing the antibacterial activities of many antibiotics. As a result, we are now faced with the specter of multiple-drug-resistant *Mycobacterium tuberculosis,* penicillin-resistant *Streptococcus pneumoniae,* and other antibiotic-resistant pathogenic bacteria that threaten to reverse our hard-earned gains in the battle against infectious disease (10, 63).

A different strategy for combatting these infectious agents is to augment the effectiveness of the body's natural defense mechanisms through administration of a variety of agents (biological response modifiers) that directly or indirectly enhance defense mechanisms. Various types of natural products with this property have been described. These include certain microbes (i.e., *Mycobacterium bovis* BCG, *Propionibacterium acnes*), components of the microbial cell wall (i.e., MDP, β-glucan), and other natural and synthetic compounds. Although these are protective in many types of experimental infections, they have not achieved widespread use in human medicine. As a result, they are not considered further in this review, and the reader is referred elsewhere for more extensive information (4, 78).

As our understanding of the immune response to infectious agents has become more sophisticated, it has become clear that many of the molecules that facilitate communication and interaction among host cells during active infection can also be utilized to improve host defense. In particular, attention has focused on the small proteins (i.e., cytokines) produced by immune and tissue cells in response to bacteria and bacterial products (16, 90, 94). The list of these mediators is large and continues to grow as contemporary immunologic and molecular biological investigations identify new mediators with important biological effects. It is clear

Charles J. Czuprynski   Department of Pathobiological Sciences, School of Veterinary Medicine, University of Wisconsin, Madison, Wisconsin 53706.

Virulence Mechanisms of Bacterial Pathogens, 2nd ed., Edited by J. A. Roth et al.

that these cytokines are numerous and pleiotropic and that they often possess overlapping biological activities (16, 90, 94). The biotechnology revolution has made it possible to obtain large amounts of these proteins in a relatively pure state, and they can then be used for in vitro and in vivo investigation of their effects on host defense. Table 1 provides a partial list of cytokines that are reported to increase resistance to bacterial infection in experimental animals or humans. Although many of these cytokines have a distinct array of biological activities, they share the ability to increase resistance to bacterial infection. Furthermore, at least some cytokines can exert additive or synergistic biological effects when administered in combination (76).

When recombinant cytokines first became available to investigators in the mid-1980s, there was an initial period of euphoria in which it was assumed that one or more of these would prove to be a "magic bullet" that would dramatically reduce our reliance on antibiotics to treat infectious diseases. Unfortunately, that optimism and enthusiasm have not generally been realized. Our increased understanding of these cytokines, their receptor-ligand interactions, and their in vivo biological activities has made it clear that no single cytokine or combination of cytokines will soon replace antibiotics as the primary treatment modality for infectious diseases. Nonetheless, there are several examples of cytokines that have proven to be clinically effective and other examples of cytokines that are useful adjuncts to other types of therapy. Furthermore, as we continue to unravel the complexities of the immune response, new cytokines and regulatory networks that may become targets for future rounds of investigation are being identified.

This chapter does not attempt to provide a comprehensive overview of all types of biological response modifiers. Instead, it provides a brief review of two classes of cytokines, the interferons (IFNs) and the colony-stimulating factors (CSFs), that are now being used routinely as prophylactic or therapeutic agents for certain conditions. It also touches briefly on the use of cytokines as adjuvants for vaccines and as adjuncts to conventional antibiotic therapy. It concludes with a discussion of the opposing biological effects of two recently described cytokines, interleukin-10 (IL-10) and IL-12, whose balance appears to be critical to the generation of protective cellular immunity.

EFFECTS OF IFN ON RESISTANCE TO MICROBIAL INFECTION

The abilities of IFNs to confer resistance to viral infection of cells has long been recognized (16, 36, 78). Over time, it became clear that the IFNs (IFN-α, IFN-β, and IFN-γ) possessed an additional array of biological activities that stimulate host defense mechanisms. When recombinant IFNs first became available in the early 1980s, they were evaluated for their anticipated beneficial effects on resistance to tumors and infectious agents. To a large extent, the IFNs have failed to live up to these overly optimistic expectations. However, IFNs have

TABLE 1 Cytokine protection against experimental infection[a]

Cytokine	Reference	Cytokine	Reference
CSF-1	33, 56	IL-8	2
G-CSF	57	IL-12	41, 52, 67, 87, 93
GM-CSF	17, 25	TNF-α[b]	9, 76
IFN-α	5, 19, 22	IL-1	48, 68, 76
IFN-β	26	IL-2	17, 38, 72
IFN-γ	20	IL-4	85
		IL-6	53

[a]The cytokines listed here are reported to be effective against infection in experimental animals or in humans.
[b]TNF-α, tumor necrosis factor alpha.

proven to be highly efficacious against several human disease problems and are approved for use in the United States and abroad.

Although not directly involving an infectious agent, the use of IFN-α to treat various neoplasms deserves mention. The first cancer shown to be highly responsive to IFN-α was hairy cell leukemia. In many instances, IFN-α treatment resulted in a prolonged remission or even cure that contrasted greatly with the poor prognosis offered by previous therapies (36). Subsequently, it was shown that IFN-α is useful for treatment of chronic myelogenous leukemia (36) and Kaposi's sarcoma, a common affliction of AIDS patients (84). There is little doubt that IFN-α treatment of these conditions has led to a reduction in the opportunistic infections that accompany progressive neoplastic disease or that might result from aggressive treatment with immunosuppressive chemotherapeutic agents. Although IFN-α treatment is not without side effects, these are usually reversible.

In addition to these successes, IFN-α has also proven beneficial for treatment of the chronic viral infections hepatitis C and hepatitis D (19, 22). These severe and progressive liver infections typically lead to significant loss of hepatic function. Treatment with high doses of IFN-α has led to remarkable improvement and even reversal of liver damage in a number of patients (22). There is also some evidence that IFN-α reduces replication of the human immunodeficiency virus (HIV) in human monocytes (84). However, the potential usefulness of IFN-α in HIV-infected individuals is unclear, since other evidence suggests that IFN-α, like other cytokines, can increase HIV replication in later stages of infection. Besides these reported benefits for treatment of human viral infections, there is also evidence that recombinant IFN-α can protect cattle against viral and bacterial infections in the lung (5).

Perhaps even greater expectations were held for the potential clinical usefulness of IFN-γ, the immune IFN produced by activated T lymphocytes and natural killer (NK) cells. When laboratory animal and in vitro studies

identified IFN-γ as the principal macrophage-activating agent (69, 79), it was assumed by some that administration of IFN-γ to enhance macrophage antibacterial activity would soon become a part of the physician's armamentarium. Unfortunately, this has not proven to be the case. However, IFN-γ is a useful therapy in several specific instances, albeit for diseases of low prevalence in the United States.

The first of these diseases is chronic granulomatous disease, an autosomal defect that renders the cytochrome oxidase of phagocytic cells inoperative (21). As a result, leukocytes from these patients are unable to undergo the oxidative burst that produces antimicrobial reactive oxygen intermediates. These patients suffer from recurrent infections to which they typically succumb at an early age. Administration of IFN-γ dramatically reduces the incidence and severity of bacterial infection (21). However, the mechanism by which this occurs is unknown, since it does not correlate with restoration of the oxidative burst (16, 21).

A second disease in which IFN-γ has proven useful in limited clinical trials is lepromatous leprosy, the severe form of leprosy in which the host is seemingly unable to control intracellular multiplication of the bacilli in macrophages. In trials conducted outside the United States, Kaplan and coworkers provided convincing evidence that local administration of IFN-γ increased accumulation of mononuclear cells in skin lesions (50). Histopathologic evaluation suggested that macrophages in these skin lesions were activated to control the intracellular leprosy bacilli (50). These effects were apparent both locally and at distant sites. Similar observations were made previously by these same investigators after intradermal injection of purified protein derivative or IL-2 (49). More recently, Holland and coworkers reported that IFN-γ administration as an adjunct to conventional antimycobacterial therapy resulted in dramatic clinical improvement in *Mycobacterium avium*-infected individuals who were refractory to antimycobacterial therapy alone (44). These observations hold promise that IFN-γ may yet prove to be useful for treatment of certain stubborn intracellular infections that are resistant to conventional therapy.

CLINICAL USE OF CSF

Investigations during the 1960s led to the recognition that hematopoiesis involves the stepwise progression of multipotential bone marrow cells through a series of stages in which they become committed to various lineages (i.e., erythrocytic, myelocytic, lymphocytic) and progressively mature into cells that eventually leave the bone marrow and make their way to the bloodstream (34). A number of distinct growth factors, or CSFs, that drive the progression of bone marrow cells through these maturation pathways were identified. As these CSFs were purified, cloned, and characterized, it became clear that certain factors might influence only one cell type (i.e., erythropoietin and granulocyte CSF [G-CSF]). In contrast, other CSFs influenced cells that were not yet committed to a single lineage or affected committed cells of more than one lineage (i.e., IL-3 and granulocyte-macrophage CSF [GM-CSF]) (34). Investigations with laboratory animals provided solid evidence that these CSFs had potent effects in vivo. In particular, it became clear that after treatment of the animals with radiation or radiomimetic drugs (i.e., the alkylating agents used in cancer chemotherapy), certain CSFs (i.e., G-CSF) could more rapidly restore normal numbers of circulating leukocytes (57). These studies led investigators to determine whether a similar effect occurs in human patients whose circulating leukocytes were depressed, as, for example, after cancer chemotherapy. Treatment of leukopenic patients with G-CSF resulted in more rapid restoration of normal numbers of leukocytes (3, 32, 35, 42, 47, 55, 66, 89). In some but not all trials, this treatment also resulted in fewer episodes of fever and infection. There is evidence that macrophage CSF therapy of granulocytopenic patients significantly reduces the incidence of opportunistic fungal infections (62) that otherwise plague this highly susceptible patient population. Promising results have also been obtained with GM-CSF therapy,

which has the added benefit of increasing numbers of circulating monocytes and eosinophils as well as of neutrophils (3, 32). In addition to restoration of leukocytes in cancer patients undergoing chemotherapy, G-CSF and GM-CSF are also useful for autologous bone marrow transplantation patients (32). Presumably, this reflects the ability of additional CSFs to drive leukocytes through the normal developmental pathway. In addition, it is possible that some of the beneficial effects of CSF therapy result from activation of the effector function of mature leukocytes, since there is ample evidence that G-CSF and GM-CSF can activate neutrophils and that GM-CSF and CSF-1 can activate macrophages in vitro (32, 89).

USE OF CYTOKINES AS ADJUVANTS AND AS ADJUNCTS TO ANTIBIOTIC THERAPY

It is common to use some form of adjuvant for vaccines, particularly when inactivated organisms (i.e., bacterins) or subunit vaccines are being used. A number of compounds have been used, but only a few are approved for use in human beings. Recombinant cytokines have received attention as potential new adjuvants that will augment the host immune response (particularly cellular immunity) without eliciting significant untoward effects. Most of this work has been done with food animal or laboratory animal species. There is evidence that recombinant IL-1 and IL-2 enhance the immunization of cattle against respiratory infection (46, 71, 72). A beneficial effect of IL-1 on immunization of sheep has also been reported (61). Both IL-1 and IL-2 have been reported to increase the protection offered by adoptive immunotherapy of experimental listeriosis in mice (38, 39). A particularly striking recent report indicated that use of IL-12 as an adjuvant for mice immunized with soluble *Leishmania major* antigens resulted in strong protective immunity, whereas the soluble antigens alone were not protective (1). These results are similar to those obtained when β-glucan was used as an adjuvant for immunization against *Leishmania donovani* (65).

There may also be some benefit to combined treatment with cytokines and antibiotics. Recombinant IL-1ß in combination with various antibiotics greatly enhances the resistance of mice to a variety of infectious agents (60). More recent evidence suggests that IL-1β in conjunction with combined antibiotic administration might also decrease the incidence and severity of *Staphylococcus aureus* mastitis (48). Other evidence suggests that IL-2 alone and GM-CSF alone are also of some benefit against experimental mastitis in cattle (17, 64, 73). As described earlier, there is exciting new evidence for the potential use of IFN-γ as an adjunct for drug-resistant mycobacterial infections (44).

OPPOSING ROLES OF IL-10 AND IL-12 IN CELLULAR IMMUNITY

That there is often a dichotomy between humoral and cellular immunity, particularly in chronic infections, has been recognized. Several older studies showed that a vigorous antibody response is correlated with reduced resistance to intracellular pathogens, whereas expression of delayed-type hypersensitivity reflects a greater capacity to resist and eradicate these infections (8). The mechanisms by which this dichotomy is regulated remained obscure for many years. Then, in the 1980s, Mosmann and Coffman made the salient observation that T-cell clones could be differentiated into two broad classes on the basis of the cytokine profiles the clones produced. The first class, which came to be known as TH1 cells, produced predominantly IL-2 and IFN-γ and thus was able to elicit delayed-type hypersensitivity and other manifestations of cellular immunity. The second class, which came to be known as TH2 cells, produced little IFN-γ and IL-2 and larger amounts of IL-4, IL-5, and other cytokines that preferentially lead to B-cell maturation and production of certain immunoglobulin isotypes (most prominently immunoglobulin E) (58). These observations and those by other investigators led to the development of a working model that stated that TH1 cytokines result in increased resistance to facultative intracel-

lular pathogens (40). In contrast, an immune response that elicits a TH2 cytokine profile would be of greater benefit against infections that require an antibody response (58). In particular, a TH2 cytokine response would result in production of immunoglobulin E, which is protective against certain helminth infections (85).

Although this model has since been modified somewhat, it continues to provide a valuable framework for examining immunoregulatory events. A number of studies have confirmed that a TH1 type of cytokine response results in increased resistance to facultative intracellular pathogens. The first and most striking example was provided by mice experimentally infected with *L. major,* a protozoal parasite of macrophages (40). In strains of mice that preferentially elicit a TH1 type of response, self-limiting infection results. In contrast, the infection is severe and progressive in mouse strains in which a TH2 response is elicited (40, 95). If the latter group of susceptible mice is treated with agents that drive the mice toward a TH1 type of immune response (i.e., anti-IL-4 monoclonal antibody [MAb] treatment), the mice become resistant to infection and switch to a TH1 type of immune response (74). Similar results have since been obtained with mice experimentally infected with other intracellular pathogens such as *Listeria monocytogenes* (37). The TH1-versus-TH2 dichotomy is less striking in humans (31, 59, 97) and cattle (12), where T cells that produce both types of cytokines are often observed (i.e., TH0 phenotype). However, the overall evidence suggests that the presence of TH2 cytokines, particularly IL-4, is detrimental to host defense even in the presence of TH1 cytokines like IL-2 and IFN-γ.

Until recently, there were few clues as to how regulation of the immune response would select for either a TH1 or a TH2 cytokine response. Recent evidence provides new insights into some of the ways in which this may occur. Earlier investigation had identified a cytokine that could inhibit production of TH1 cell cytokines. Called IL-10, this cytokine was shown to be produced by macrophages and TH2 cells and to exert its effects in both a paracrine and an autocrine fashion (11, 29, 45, 82, 88). Various lines of evidence showed that IL-10 downregulated the progression of T cells toward the TH1 cytokine profile in vitro (45) and the expression of protective cellular immunity in vivo (70). Mice treated with an anti-IL-10 MAb expressed increased resistance to infection with *L. major* (40, 74), *Listeria monocytogenes* (91), *M. avium* (7, 18), *Candida albicans* (77), and *Trypanosoma cruzi* (83). It has been reported that IL-10 is elevated in the cerebrospinal fluid during gram-positive bacterial meningitis and that this IL-10 impairs macrophage antibacterial activity (24). There is also evidence that humans infected with *Mycobacterium leprae* or *L. donovani* produce increased amounts of IL-10 when they are in the progressive stage of these infections (31, 43, 59, 81).

Our laboratory obtained evidence for the detrimental effects of IL-10 in antibacterial resistance in vivo when we evaluated, by in situ hybridization, the presence of cytokine-expressing cells in the livers of *Listeria monocytogenes*-infected mice. We noted that the number of IL-10-expressing cells increased within 1 day of infection and remained elevated during the early stage of infection, when the listeriae are multiplying in the spleen and liver (91). Other investigators have independently reported similar results (23). These data were consistent with the hypothesis that IL-10 production inhibits or delays the onset of protective immunity. We next tested this hypothesis directly by administering an anti-IL-10 MAb to mice just before challenge with *Listeria monocytogenes*. The results indicated that this treatment reduced plasma IL-10 levels and enhanced resistance to the infection (92). These data, too, were consistent with current thinking that IL-10 is detrimental to host defense. Paradoxically, however, we noted that anti-IL-10-treated mice had a delayed ability to completely clear the *Listeria monocytogenes* infection from the liver. Furthermore, when the anti-IL-10 MAb was given 1 to 3 days after the infec-

tion was initiated, the resistance of the mice was impaired rather than elevated (92). This suggests that IL-10 plays some essential role in the later stages of the infection, when the host is attempting to eliminate the listeriae. Although these data appear to be contrary to current thinking about IL-10, there are other reports that IL-10 has biological effects that might contribute to host defense. For example, IL-10 has been shown to facilitate leukocyte accumulation in vivo (96), downregulate expression of certain inflammatory cytokines (6, 14, 30, 51, 82), increase expression of Fc receptors (86), and reduce the severity of septic shock in vivo (6). Taken as a whole, these data suggest that the actual role of IL-10 in host defense may be more complex than was previously realized. In particular, the net effect of IL-10 may be influenced by the time at which it is present, its concentration, and the presence of other immunoregulatory molecules.

IL-10 is released by macrophages in response to microbes, despite the ability of these same microbes to elicit a strong cellular immune response. It was therefore reasoned that yet another soluble factor must be able to overcome the downregulatory effects of IL-10. The key to understanding these observations became clear in 1993, when several laboratories provided evidence that another cytokine, known as IL-12, could overcome the downregulatory effects of IL-10 (44, 88). IL-12 is a heterodimer produced by macrophages and B cells (13). Originally described for its ability to activate NK-cell cytolytic activity, it has since been shown to stimulate IFN-γ production by NK cells and TH1 cells (27, 88) and to select for progression of T cells along the TH1 cell cytokine pathway (45, 80). In particular, these studies indicate that IL-12 alone is able to overcome the downregulatory effects of IL-10 (45, 88). There is evidence that both mRNA and IL-12 heterodimer are present in the pleural fluid of human tuberculosis patients (98).

Several laboratories have shown that IL-12 has beneficial effects on host defense in vivo. For example, endogenous IL-12 or recombinant IL-12 increases resistance to *L. major* (41,

54, 75), *Toxoplasma gondii* (28, 52), and *Listeria monocytogenes* (87, 93). Evidence suggests that these effects are mediated at least in part by IFN-γ (28, 41, 87, 93), perhaps released by NK cells (27, 88). A recent report that use of recombinant IL-12 as an adjuvant for immunization of mice with soluble antigens of *L. major* resulted in strong protective immunity was particularly impressive (1). These data suggest that IL-12 may be of benefit in immunization protocols, presumably as a result of its ability to potentiate IFN-γ production by NK cells and TH1 cells. Our laboratory has provided evidence regarding the beneficial effects of IL-12 in vivo. Mice treated with recombinant IL-12 were more resistant to *Listeria monocytogenes* infection, whereas mice treated with an anti-IL-12 antibody were more susceptible to the infection (93). This effect was seen after 1 day of infection, and the protective effect of IL-12 was largely abrogated by administration of an anti-IFN-γ MAb (93). Although these results are quite exciting and suggest that IL-12 contributes to host defense, we also noted signs of transient hepatic toxicity in IL-12–treated mice. This was reflected in increased levels of the liver-specific enzyme aspartate aminotransferase in plasma and by histopathologic evidence of small foci of hepatocyte necrosis (93). A similar effect of IL-12 on the liver in mice was reported previously by other investigators who administered IL-12 repeatedly over time (27). In addition, it has been reported that small amounts of recombinant IL-12 are beneficial and larger amounts are detrimental to resistance to experimental lymphocytic choriomeningitis virus infection in mice (67). Overall, these observations suggest that IL-12 likely plays an important role in host defense, but it may have other biological effects that must be better characterized if it is to be used in a clinical setting (15).

SUMMARY

The use of biological response modifiers may be entering a new era. Earlier goals of using microbe- or mammal-derived compounds as a radical new form of antimicrobial therapy have been replaced by a more conservative approach. Particular emphasis will be placed on evaluating the use of recombinant peptides (cytokines) as adjuncts to other forms of therapy. Hopefully, this will result in identification of new weapons in the battle against infectious diseases.

ACKNOWLEDGMENTS

Work done in my laboratory was supported by the U.S. Public Health Service (R01 AI21343) and the University of Wisconsin-Madison Graduate School. I am grateful to DNAX for supplying IL-10 and anti-IL-10 and to Genetics Institute (Cambridge, Mass.) and Hoffmann-LaRoche (Nutley, N.J.) for supplying IL-12 and anti-IL-12. I recognize the efforts of Mary Haak-Frendscho, Robin Kurtz, J. Roll, R. D. Wagner, J. F. Brown, N. Maroushek, and Y. Iizawa for work done in my laboratory that is described in this chapter. I also appreciate the assistance of the School of Veterinary Medicine Word Processing Center in preparing the manuscript.

REFERENCES

1. **Afonso, L. C. C., T. M. Scharton, L. Q. Vieira, M. Wysocka, G. Trinchieri, and P. Scott.** 1994. The adjuvant effect of interleukin-12 in a vaccine against *Leishmania major*. *Science* **263:**235–231.
2. **Agace, W. W., S. R. Hedges, M. Ceska, and C. Svanborg.** 1993. Interleukin-8 and the neutrophil response to mucosal gram negative infection. *J. Clin. Invest.* **92:**780–785.
3. **Antman, K. S., J. D. Griffin, A. Elias, M. A. Socinski, L. Ryan, S. A. Cannistra, D. Oette, M. Whitley, E. Frei III, and L. E. Schnipper.** 1988. Effect of recombinant human granulocyte-macrophage colony-stimulating factor on chemotherapy-induced myelosuppression. *N. Engl. J. Med.* **319:**593–598.
4. **Azuma, I.** 1992. Synthetic immunoadjuvants: application to non-specific host stimulation and potentiation of vaccine immunogenicity. *Vaccine* **10:**1000–1006.
5. **Babiuk, L. A., H. Bielfeldt-Ohmann, G. Gifford, C. W. Czarniecki, V. T. Scialli, and E. B. Hamilton.** 1985. Effect of bovine alpha interferon on bovine herpesvirus type-1 induced respiratory disease. *J. Gen. Virol.* **66:**2383–2394.
6. **Bean, A. G. D., R. A. Freiberg, S. Andrade, S. Menon, and A. Zlotnik.** 1993. Interleukin 10 protects mice against staphylococcal enterotoxin B-induced lethal shock. *Infect. Immun.* **61:** 4937–4939.
7. **Bermudez, L. E., and J. Champsi.** 1993. In-

fection with *Mycobacterium avium* induces production of interleukin-10 (IL-10), and administration of anti-IL-10 antibody is associated with enhancement of resistance to infection in mice. *Infect. Immun.* **61**:3093–3097.

8. **Biozzi, G., D. Mouton, C. Stiffel, and Y. Bouthiller.** 1984. A major role of the macrophage in quantitative genetic regulation of immunoresponsiveness and anti-infectious immunity. *Adv. Immunol.* **36**:189–234.

9. **Blanchard, D. K., J. Y. Djeu, T. W. Klein, H. Freidman, and W. E. Stewart III.** 1988. Protective effects of tumor necrosis factor in experimental *Legionella pneumophila* infections of mice via activation of PMN functions. *J. Leukocyte Biol.* **43**:429–435.

10. **Bloom, B. R., and C. J. L. Murray.** 1992. Tuberculosis: commentary on a reemergent killer. *Science* **257**:1055–1064.

11. **Bogdan, C., Y. Vodovotz, and C. Nathan.** 1991. Macrophage deactivation by interleukin 10. *J. Exp. Med.* **174**:1549–1555.

12. **Brown, W. C., W. C. Davis, A. E. Dobbelaere, and A. C. Rice-Ficht.** 1994. CD4$^+$ T-cell clones obtained from cattle chronically infected with *Fasciola hepatica* and specific for adult worm antigen express both unrestricted and Th2 cytokine profiles. *Infect. Immun.* **62**:818–827.

13. **Brunda, M. J.** 1994. Interleukin-12. *J. Leukocyte Biol.* **55**:280–288.

14. **Cassatella, M. A., L. Meda, S. Bonora, M. Ceska, and G. Constantin.** 1993. Interleukin 10 (IL-10) inhibits the release of proinflammatory cytokines from human polymorphonuclear leukocytes. Evidence for an autocrine role of tumor necrosis factor and IL-1β in mediating the production of IL-8 triggered by lipopolysaccharide. *J. Exp. Med.* **178**:2207–2211.

15. **Clerici, M., D. R. Lucey, J. A. Berzofsky, L. A. Pinto, T. A. Wynn, S. P. Blatt, M. J. Dolan, C. W. Hendrix, S. F. Wolf, and G. M. Shearer.** 1993. Restoration of HIV-specific cell-mediated immune responses by interleukin-12 *in vitro*. *Science* **262**:1721–1724.

16. **Czarniecki, C. W., and G. Sonnenfeld.** 1993. Interferon-gamma and resistance to bacterial infections. *APMIS* **101**:1–17.

17. **Daley, M., T. Williams, P. Coyle, G. Furda, R. Dougherty, and P. Hayes.** 1993. Prevention and treatment of *Staphylococcus aureus* infections with recombinant cytokines. *Cytokine* **5**:276–284.

18. **Denis, M., and E. Ghadirian.** 1993. IL-10 neutralization augments mouse resistance to systemic *Mycobacterium avium* infections. *J. Immunol.* **151**:5425–5430.

19. **Di-Macro, V., O. Lo-Iacono, M. Capra, S. Grutta, C. Ciaccio, C. Gerardi, A. Maggio, D. Renda, P. Almasio, and R. Pisa.** 1992. Alpha interferon treatment of chronic hepatitis C in young patients with homozygous beta-thalassemia. *Haematologica* **77**:502–506.

20. **Edwards, C. K., III, H. B. Hedegaard, A. Zlotnick, P. R. Gangadharan, R. B. Johnston, Jr., and M. J. Pabst.** 1986. Chronic infection due to *Mycobacterium intracellulare* in mice: association with macrophage release of prostaglandin E$_2$ and reversal by injection of indomethacin, muramyl dipeptide, or interferon-γ. *J. Immunol.* **136**:1820–1827.

21. **Ezekowitz, R. A. S. B., M. C. Dinauer, H. S. Jaffe, S. H. Orka, and R. E. Newburger.** 1988. Partial correction of the phagocyte defect in patients with X-linked chronic granulomatous disease by subcutaneous interferon gamma. *N. Engl. J. Med.* **319**:146–151.

22. **Farci, P., A. Mandas, A. Coiana, M. E. Lai, V. Desmet, P. Van Eyken, Y. Gibo, L. Caruso, S. Scaccabarozzi, D. Criscuolo, J. Ryff, and A. Balestrieri.** 1994. Treatment of chronic hepatitis D with interferon α-2a. *N. Engl. J. Med.* **330**:88–94.

23. **Flesch, I. E. A., and S. H. E. Kaufmann.** 1993. Role of macrophages and αβ T lymphocytes in early interleukin 10 production during *Listeria monocytogenes* infection. *Int. Immunol.* **6**:463–468.

24. **Frei, K., D. Nadal, H. Pfister, and A. Fontana.** 1993. *Listeria meningitis*: identification of a cerebrospinal fluid inhibitor of macrophage listericidal function as interleukin 10. *J. Exp. Med.* **178**:1255–1261.

25. **Frenck, R. W., G. Sarman, T. E. Harper, and E. S. Buescher.** 1990. The ability of recombinant murine granulocyte-macrophage colony-stimulating factor to protect neonatal rats from septic death due to *Staphylococcus aureus*. *J. Infect. Dis.* **162**:109–114.

26. **Fujiki, T., and A. Tanaka.** 1988. Antibacterial activity of recombinant murine beta interferon. *Infect. Immun.* **56**:548–551.

27. **Gately, M. K., R. R. Warrier, S. Honasoge, D. M. Carvajal, D. A. Faherty, S. E. Connaughton, T. D. Anderson, V. Sarmiento, B. R. Hubbard, and M. Murphy.** 1994. Administration of recombinant IL-12 to normal mice enhances cytolytic lymphocyte activity and induces production of IFN-γ *in vivo*. *Int. Immunol.* **6**:157–167.

28. **Gazzinelli, R. T., S. Hieny, T. A. Wynn, S. Wolf, and A. Sher.** 1993. Interleukin 12 is required for T-lymphocyte-independent induction of interferon γ by an intracellular parasite and induces resistance in T-cell-deficient hosts. *Proc. Natl. Acad. Sci. USA* **90**:6115–6119.

29. **Gazzinelli, R. T., I. P. Oswald, S. L. James, and A. Sher.** 1992. IL-10 inhibits parasite killing

and nitrogen oxide production by IFN-γ-activated macrophages. *J. Immunol.* **148**:1792–1796.

30. **Gérard, C., C. Bruyns, A. Marchant, D. Abramowicz, P. Vandenabeele, A. Delvaux, W. Fiers, M. Goldman, and T. Velu.** 1993. Interleukin 10 reduces the release of tumor necrosis factor and prevents lethality in experimental endotoxemia. *J. Exp. Med.* **177**:547–550.

31. **Ghalib, H. W., M. R. Piuvezam, Y. A. W. Skeiky, M. Siddig, F. A. Hashim, A. M. El-Hassan, D. M. Russo, and S. G. Reed.** 1993. Interleukin 10 production correlates with pathology in human *Leishmania donovani* infections. *J. Clin. Invest.* **92**:324–329.

32. **Gorin, N. C., B. Coiffier, M. Hayat, L. Fouillard, M. Kuentz, M. Flesch, P. Colombat, P. Boivin, S. Slavin, and T. Philip.** 1992. Recombinant human granulocyte-macrophage colony-stimulating factor after high-dose chemotherapy and autologous bone marrow transplantation with unpurged and purged marrow in non-Hodgkin's lymphoma: a double-blind placebo-controlled trial. *Blood* **80**:1149–1157.

33. **Gregory, S. H., E. J. Wing, D. J. Tweardy, R. K. Shadduck, and H. S. Lin.** 1992. Primary listerial infections are exacerbated in mice administered neutralizing antibody or macrophage colony-stimulating factors. *J. Immunol.* **149**:188–193.

34. **Groopman, J. E., J.-M. Molina, and D. T. Scadden.** 1989. Hematopoietic growth factors. *N. Engl. J. Med.* **321**:1449–1459.

35. **Grosh, W. W., and P. J. Quesenberry.** 1992. Recombinant human hematopoietic growth factors in the treatment of cytopenias. *Clin. Immunol. Immunopathol.* **62**:S25–S38.

36. **Gutterman, J. U.** 1994. Cytokine therapeutics: lessons from interferon α. *Proc. Natl. Acad. Sci. USA* **91**:1198–1205.

37. **Haak-Frendscho, M., J. F. Brown, Y. Iizawa, R. D. Wagner, and C. J. Czuprynski.** 1992. Administration of anti-IL-4 monoclonal antibody 11B11 increases the resistance of mice to *Listeria monocytogenes. J. Immunol.* **148**:3978–3985.

38. **Haak-Frendscho, M., and C. J. Czuprynski.** 1992. Use of recombinant interleukin-2 to enhance adoptive transfer of resistance to *Listeria monocytogenes* infection. *Infect. Immun.* **60**:1406–1414.

39. **Haak-Frendscho, M., R. S. Kurtz, and C. J. Czuprynski.** 1991. rIL-1 enhances adoptive transfer of resistance to *Listeria monocytogenes* infection. *Microb. Pathog.* **10**:385–392.

40. **Heinzel, F. P., M. D. Sadick, S. S. Mutha, and R. M. Locksley.** 1991. Production of interferon γ, interleukin 2, interleukin 4, and interleukin 10 by CD4+ lymphocytes *in vivo* during healing and progressive murine leishmaniasis. *Proc. Natl. Acad. Sci. USA* **88**:7011–7015.

41. **Heinzel, F. P., D. S. Schoenhaut, R. M. Rerko, L. E. Rosser, and M. K. Gately.** 1993. Recombinant interleukin 12 cures mice infected with *Leishmania major. J. Exp. Med.* **177**:1505–1509.

42. **Hirashima, K., Y. Yoshida, S. Asano, F. Takaku, M. Omine, S. Furusawa, T. Abe, H. Dohy, M. Tajiri, et al.** 1991. Clinical effect of recombinant human granulocyte colony-stimulating factor (rhG-CSF) on various types of neutropenia including cyclic neutropenia. *Biotherapy* **3**:297–307.

43. **Holaday, B. J., M. M. Pompeu, S. Jeronimo, M. J. Texeira, A. Sousa, A. W. Vasconcelos, R. D. Pearson, J. S. Abrams, and R. M. Locksley.** 1993. Potential role for interleukin-10 in the immunosuppression associated with kala azar. *J. Clin. Invest.* **92**:2626–2632.

44. **Holland, S. M., E. M. Eisenstein, D. B. Kuhns, M. L. Turner, T. A. Fleisher, W. Strober, and J. I. Gallin.** 1994. Treatment of refractory disseminated nontuberculous mycobacterial infection with interferon gamma. *N. Engl. J. Med.* **330**:1348–1355.

45. **Hshieh, C.-S., S. E. Macatonia, C. S. Tripp, S. F. Wolf, A. O'Garra, and K. M. Murphy.** 1993. Development of T$_{H1}$ CD4+ T cells through IL-12 produced by *Listeria*-induced macrophages. *Science* **260**:547–549.

46. **Hughes, H. P. A., M. Coupes, P. L. Godson, S. Van Drunen Little-vanden Hurk, L. McDougall, N. Rapin, T. Zamb, and L. A. Babuik.** 1991. Immunopotentiation of bovine herpes virus subunit vaccination by interleukin 2. *Immunology* **74**:461–466.

47. **Imashuku, S., M. Tsuchida, M. Sasaki, T. Shimokawa, H. Nakamura, T. Matsuyama, N. Taniguchi, M. Oda, S. Higuchi, K. Ishimoto, et al.** 1992. Recombinant human granulocyte-colony-stimulating factor in the treatment of patients with chronic benign granulocytopenia and congenial agranulocytosis. *Acta Paediatr.* **81**:133–136.

48. **Johnston, P. A., P. Coyle, J. Krisinski, W. Tao, R. Dougherty, and G. Furda.** 1992. Role of recombinant interleukin-1β in the prevention or therapy of *Staphylococcus aureus* mastitis, abstr. 88. *Conf. Res. Workers Anim. Dis.*

49. **Kaplan, G.** 1993. Recent advances in cytokine therapy in leprosy. *J. Infect. Dis.* **167**:S18–S22.

50. **Kaplan, G., A. Nusrat, E. N. Sarno, C. K. Job, J. McElrath, J. A. Porto, C. F. Nathan, and Z. A. Cohn.** 1987. Cellular responses to the intradermal injection of recombinant human gamma-interferon in lepromatous leprosy patients. *Am. J. Pathol.* **128**:345–353.

51. **Kasama, T., R. M. Strieter, N. W. Lukacs, M. D. Burdick, and S. L. Kunkel.** 1994. Reg-

ulation of neutrophil-derived chemokine expression by IL-10. *J. Immunol.* **152:**3559–3569.

52. **Khan, I. A., T. Matsuura, and L. H. Kasper.** 1994. Interleukin-12 enhances murine survival against acute toxoplasmosis. *Infect. Immun.* **62:** 1639–1642.

53. **Liu, Z., R. J. Simpson, and C. Cheers.** 1994. Role of IL-6 in activation of T cells for acquired cellular resistance to *Listeria monocytogenes. J. Immunol.* **152:**5375–5380.

54. **Locksley, R. M.** 1993. Interleukin 12 in host defense against microbial pathogens. *Proc. Natl. Acad. Sci. USA* **90:**5879–5880.

55. **Lord, B. I., M. H. Bronchud, S. Owens, J. Chang, A. Howell, L. Souza, and T. M. Dexter.** 1989. The kinetics of human granulopoiesis following treatment with granulocyte colony-stimulating factor *in vivo. Proc. Natl. Acad. Sci. USA* **86:**9499–9503.

56. **Magee, D. M., and E. J. Wing.** 1989. Secretion of colony-stimulation factors by T cell clones. Role in adoptive protection against *Listeria monocytogenes. J. Immunol.* **143:**2336–2341.

57. **Matsumoto, M., S. Matsubara, T. Matsuno, M. Tamura, K. Hattori, H. Nomura, M. Ono, and T. Yokota.** 1987. Protective effect of human granulocyte colony-stimulating factor on microbial infection in neutropenic mice. *Infect. Immun.* **55:**2715–2720.

58. **Mosmann, T. R., and R. L. Coffman.** 1987. Two types of mouse helper T-cell clones. *Immunol. Today* **8:**223–227.

59. **Mutis, T., E. M. Kraakman, Y. E. Cornelisse, J. B. Haanen, H. Spits, R. R. De Vries, and T. H. Ottenhoff.** 1993. Analysis of cytokine production by *Mycobacterium*-reactive T cells. Failure to explain *Mycobacterium leprae*-specific nonresponsiveness of peripheral blood T cells from lepromatous leprosy patients. *J. Immunol.* **150:**4641–4651.

60. **Nakamura, S., A. Minami, K. Fujimoto, and T. Kojma.** 1989. Combination effect of recombinant human interleukin-1 with antimicrobial agents. *Antimicrob. Agents Chemother.* **33:**1804–1810.

61. **Nash, A. D., S. A. Lofthouse, G. J. Barcham, H. J. Jacobs, K. Ashman, E. N. Meeusen, M. R. Brandon, and A. E. Andrews.** 1993. Recombinant cytokines as immunological adjuvants. *Immunol. Cell Biol.* **71:**367–379.

62. **Nemunaitis, J., K. Shannon-Dorcy, F. R. Appelbaum, J. Meyers, A. Owens, R. Day, D. Ando, C. O'Neill, D. Buckner, and J. Singer.** 1993. Long-term follow-up of patients with invasive fungal disease who received adjunctive therapy with recombinant human macrophage colony-stimulating factor. *Blood* **82:**1422–1427.

63. **Neu, H. C.** 1992. The crisis in antibiotic resistance. *Science* **357:**1064–1073.

64. **Nickerson, S. C., P. A. Baker, and P. Trinidad.** 1989. Local immunostimulation of the bovine mammary gland with interleukin-2. *J. Dairy Sci.* **72:**1764–1773.

65. **Obaid, K. A., S. Ahmad, H. M. Kahn, A. A. Mahdi, and R. Khanna.** 1989. Protective effect of *L. donovani* antigens using glucan as an adjuvant. *Int. J. Immunopharmacol.* **11:**229–235.

66. **Okamura, J., M. Yokoyama, I. Tsukimoto, A. Komiyama, M. Sakurai, S. Imashuku, S. Miyazaki, K. Ueda, Y. Hanawa, and F. Takaku.** 1992. Treatment of chemotherapy-induced neutropenia in children with subcutaneously administered recombinant human granulocyte colony-stimulating factor. *Pediatr. Hematol. Oncol.* **9:** 199–207.

67. **Orange, J. S., S. F. Wolf, and C. A. Biron.** 1993. Effects of IL-12 on the response and susceptibility to experimental viral infections. *J. Immunol.* **152:**1253–1264.

68. **Ozaki, Y., T. Ohashi, A. Minami, and S. Nakamura.** 1987. Enhanced resistance of mice to bacterial infection induced by recombinant human interleukin-1α. *Infect. Immun.* **55:**1436–1440.

69. **Pace, J. L., S. W. Russell, P. A. LeBlanc, and D. M. Muresko.** 1985. Comparative effects of various classes of mouse interferons on macrophage activation for tumor cell killing. *J. Immunol.* **134:**977–981.

70. **Powrie, F., S. Menon, and R. L. Coffman.** 1993. Interleukin-4 and interleukin-10 synergize to inhibit cell-mediated immunity *in vivo. Eur. J. Immunol.* **23:**2223–2229.

71. **Reddy, D. N., P. G. Reddy, H. C. Minocha, B. W. Fenwick, P. E. Baker, W. C. Davis, and F. Blecha.** 1990. Adjuvanticity of recombinant bovine interleukin-1 beta: influence on immunity, infection, and latency in bovine herpesvirus-1 infection. *Lymphokine Res.* **9:**295–307.

72. **Reddy, D. N., P. G. Reddy, W. Xue, H. C. Minocha, M. J. Daley, and F. Blecha.** 1992. Immunopotentiation of bovine respiratory disease virus vaccines by interleukin-1β and interleukin-2. *Vet. Immunol. Immunopathol.* **37:**25–38.

73. **Reddy, P. G., D. N. Reddy, S. E. Pruiett, M. J. Daley, J. E. Shirley, M. M. Chengappa, and F. Blecha.** 1992. Interleukin 2 treatment of *Staphylococcus aureus* mastitis. *Cytokine* **4:**227–231.

74. **Reed, S. G., and P. Scott.** 1993. T-cell and cytokine responses in leishmaniasis. *Curr. Opin. Immunol.* **5:**524–531.

75. **Reiner, S. L., S. Zheng, Z. Wang, L. Stowring, and R. M. Locksley.** 1994. *Leishmania* promastigotes evade interleukin 12 (IL-12) induction by macrophages and stimulate a broad range

of cytokines from CD4[+] T cells during initiation of infection. *J. Exp. Med.* **179**:447–456.

76. **Roll, J. T., K. M. Young, R. S. Kurtz, and C. J. Czuprynski.** 1990. Human rTNFα augments anti-bacterial resistance in mice: potentiation of its effects by recombinant IL-1α. *Immunology* **69**:316–322.

77. **Romani, L., P. Puccetti, A. Mencacci, E. Cenci, R. Spaccapelo, L. Tonnetti, U. Grohmann, and F. Bistoni.** 1994. Neutralization of IL-10 up-regulates nitric oxide production and protects susceptible mice from challenge with *Candida albicans. J. Immunol.* **152**:3514–3521.

78. **Roth, J. A.** 1987. Enhancement of nonspecific resistance to bacterial infection by biologic response modifiers, p. 329–342. *In* J. A. Roth (ed.), *Virulence Mechanisms of Bacterial Pathogens.* American Society for Microbiology, Washington, D.C.

79. **Rothermel, D. D., B. Y. Robin, and H. W. Murray.** 1983. γ Interferon is the factor in lymphokine that activates human macrophages to inhibit intracellular *Chlamydia psittaci* replication. *J. Immunol.* **131**:2542–2544.

80. **Seder, R. A., R. Gazzinelli, A. Sher, and W. E. Paul.** 1993. Interleukin 12 acts directly on CD4[+] T cells to enhance priming for interferon γ production and diminishes interleukin 4 inhibition of such priming. *Proc. Natl. Acad. Sci. USA* **90**:10188–10191.

81. **Sieling, P. A., J. S. Abrams, M. Yamamura, P. Salgame, B. R. Bloom, T. H. Rea, and R. L. Modlin.** 1993. Immunosuppressive roles for IL-10 and IL-4 in human infection. *J. Immunol.* **150**:5501–5510.

82. **Sironi, M., C. Muñoz, T. Pollicino, A. Siboni, F. L. Sciacca, S. Bernasconi, A. Vecchi, F. Colotta, and A. Mantovani.** 1993. Divergent effects of interleukin-10 on cytokine production by mononuclear phagocytes and endothelial cells. *Eur. J. Immunol.* **23**:2692–2695.

83. **Silva, J. S., P. J. Morrissey, K. H. Grabstein, K. M. Mohler, D. Anderson, and S. G. Reed.** 1992. Interleukin 10 and interferon γ regulation of experimental *Trypanosoma cruzi* infection. *J. Exp. Med.* **175**:169–174.

84. **Stein, D. S., J. G. Timpone, J. D. Gradon, J. M. Kagan, and S. M. Schnittman.** 1993. Immune-based therapeutics: scientific rationale and the promising approaches to the treatment of the human immunodeficiency virus-infected individual. *Clin. Infect. Dis.* **17**:749–771.

85. **Svetic, A., K. B. Madden, X. di Zhou, P. Lu, I. M. Kafona, F. D. Finkleman, J. F. Urban, Jr., and W. C. Gause.** 1993. A primary intestinal helminthic infection rapidly induces a gut-associated elevation of TH-2 associated cytokines and IL-3. *J. Immunol.* **150**:3434–3441.

86. **te Velde, A. A., R. Malefijt, R. J. F. Huijbens, J. E. de Vries, and C. G. Figdor.** 1992. IL-10 stimulates monocyte FcγR surface expression and cytotoxic activity. *J. Immunol.* **149**:4048–4052.

87. **Tripp, C. S., M. K. Gately, J. Hakimi, P. Ling, and E. R. Unanue.** 1994. Neutralization of IL-12 decreases resistance to *Listeria* in SCID and CB-17 mice. *J. Immunol.* **152**:1883–1887.

88. **Tripp, C. S., S. F. Wolf, and E. R. Unanue.** 1993. Interleukin 12 and tumor necrosis factor α are costimulators of interferon γ production by natural killer cells in severe combined immunodeficiency mice with listeriosis, and interleukin 10 is a physiologic antagonist. *Proc. Natl. Acad. Sci. USA* **90**:3725–3729.

89. **Vadas, M. A., A. F. Lopez, J. R. Gamble, and M. J. Elliot.** 1991. Role of colony-stimulating factors in leucocyte responses to inflammation and infection. *Curr. Opin. Immunol.* **3**:97–104.

90. **van Deuren, M., A. S. M. Dofferhoff, and J. W. M. van der Meer.** 1992. Cytokines and the response to infection. *J. Pathol.* **168**:349–356.

91. **Wagner, R. D., and C. J. Czuprynski.** 1993. Cytokine mRNA expression in livers of mice infected with *Listeria monocytogenes. J. Leukocyte Biol.* **53**:525–531.

92. **Wagner, R. D., N. M. Maroushek, J. Brown, and C. J. Czuprynski.** 1994. Treatment with anti-interleukin-10 monoclonal antibody enhances early resistance to but impairs complete clearance of *Listeria monocytogenes* infection in mice. *Infect. Immun.* **62**:2345–2353.

93. **Wagner, R. D., H. Steinberg, J. F. Brown, and C. J. Czuprynski.** 1994. Recombinant interleukin-12 enhances resistance to *Listeria monocytogenes* infection by induction of interferon-γ expression. *Microb. Pathog.* **17**:175–186.

94. **Wallis, R. S., and J. J. Ellner.** 1994. Cytokines and tubrculosis. *J. Leukocyte Biol.* **55**:676–681.

95. **Wang, Z., S. L. Reiner, S. Zheng, D. K. Dalton, and R. M. Locksley.** 1994. CD4[+] effector cells default to the Th2 pathway in interferon γ-deficient mice infected with *Leishmania major. J. Exp. Med.* **179**:1367–1371.

96. **Wogensen, L., X. Huang, and N. Sarvetnick.** 1993. Leukocyte extravasation into the pancreatic tissue in transgenic mice expressing interleukin 10 in the islets of Langerhans. *J. Exp. Med.* **178**:175–185.

97. **Yamamura, M., K. Uyenura, R. J. Dams, K. Weinberg, T. H. Rea, B. R. Bloom, and R. L. Modlin.** 1991. Defining protective responses to pathogenesis: cytokine profiles in leprosy. *Science* **254**:277–279.

98. **Zhang, M., M. K. Gately, E. Wang, J. Gong, S. F. Wolf, S. Lu, R. L. Modlin, and P. F. Barnes.** 1994. Interleukin 12 at the site of disease in tuberculosis. *J. Clin. Invest.* **93**:1733–1739.

VACCINATION STRATEGIES FOR MUCOSAL PATHOGENS

Suzanne M. Michalek, Noel K. Childers,
and Mark T. Dertzbaugh

17

It has been almost 200 years since Jenner showed that inoculation of a child with material scraped from a lesion on a dairymaid with cowpox provided protection from a lethal smallpox infection. It was from this work that the term vaccine (from the Latin word *vaccinus,* meaning from cows) and the concept of vaccination leading to immunity against infection originated. Jenner's concept gained acceptance almost 100 years later, when Pasteur proposed his germ theory of disease and demonstrated acquired immunity. In those studies, Pasteur and his colleagues reported that birds given feed supplemented with *Pasteurella avisepticum* were protected against chicken cholera. The idea of using the oral route for immunization to induce a protective response against diseases was pursued for many years at the Pasteur Institute (135). Then, in the early 1900s, Besredka (14) published his findings on oral vaccination of rabbits, which suggested the existence of a protective local immune system that

functions separately from the systemic immune system.

Despite these early studies of oral immunization and the evidence that oral immunization induces local or mucosal immune responses, the oral route has not been widely used for the development of vaccines for clinical use. Most vaccines have traditionally been given by the intramuscular or subcutaneous route, which induces systemic but not local or mucosal immune responses. This approach would be appropriate for inducing immune responses protective against diseases caused by infectious agents that penetrate the external or mucosal surfaces of the body. Most pathogens cause diseases by colonization of or invasion through the mucosa. Therefore, it is important to develop methods for inducing local immune responses that would protect mucosal surfaces against pathogens.

The primary immunoglobulin isotype found in external secretions such as saliva, milk, and gastrointestinal fluid is secretory immunoglobulin A (S-IgA), and numerous studies have shown that oral administration of antigen results in the induction of S-IgA antibodies (125, 131). Therefore, oral immunization should be the preferred route for eliciting host immune responses to pathogens at mucosal surfaces. Only recently has consideration been

Suzanne M. Michalek   Department of Microbiology, University of Alabama at Birmingham, 845 South 19th, BBRB 258, Birmingham, Alabama 35294-2170.   *Noel K. Childers*   Department of Community and Public Health Dentistry, School of Dentistry, University of Alabama at Birmingham, Birmingham, Alabama 35294-0007.   *Mark T. Dertzbaugh*   Applied Research Division, U.S. Army Medical Research Institute of Infectious Disease, Ft. Detrick, Maryland 21702-5011.

Virulence Mechanisms of Bacterial Pathogens, 2nd ed., Edited by J. A. Roth et al.
© 1995 American Society for Microbiology, Washington, D.C.

given to the development and clinical use of oral vaccines. This change is most likely due to our greater understanding of the relevance of the mucosal immune system and the increased awareness of society regarding infectious diseases such as AIDS. It is important to emphasize that oral vaccines have several advantages over parenteral types. They can be administered easily without the need of trained personnel and have, potentially, no side effects. The use of oral vaccines eliminates the need for needles and reduces the risk of contamination. Furthermore, the cost of preparing and administering oral vaccines would be less than the cost of parenteral vaccines. These are important considerations in determining the suitability of any vaccine for worldwide use. However, despite these advantages, oral vaccines also have a major limitation: their limited ability to survive the peristalsis and ciliary movement of epithelia, the mechanical (epithelial cell) and chemical (e.g., mucin) barriers, the acidic pH of the stomach, and the low pH and degradative enzymes present in the small intestine, which can eliminate or destroy the immunogens prior to their uptake and presentation to lymphoid cells in IgA inductive sites, e.g., the gut-associated lymphoid tissue (GALT).

Interestingly, studies have shown that particulate antigens, especially when presented as viable organisms, are more effective than soluble antigens in inducing local and generalized secretory and systemic immune responses (125, 131). At least three reasons can be given to explain this. First, the size and composition of particulate antigens may allow them to more effectively survive the environment of the gastrointestinal tract. Second, some particulates are more efficiently absorbed by the GALT, specifically, through the M (microfold) cells into the Peyer's patches. Third, soluble antigens cross the epithelial barrier of the gut in the form of low-molecular-weight peptides. The recognition of these peptides by lymphoid cells at sites other than in the Peyer's patches has been proposed as the stimulus that initiates systemic tolerance after antigen feeding (18) and

may provide a negative signal to the mucosal immune system. During the past decade, modern technology has facilitated the development of oral antigen delivery systems with immunopotentiating activity (125, 133, 139, 164).

In designing a vaccine, several factors need to be considered, such as its safety for use in animals, including humans; the convenience and cost of reproducibly generating the vaccine in large quantities; and its effectiveness in inducing the desired host response. Several strategies are currently being tested for use in oral vaccine development. These include use of genetically modified live bacterial and viral vectors, such as *Salmonella, Mycobacterium* (e.g., bacillus Calmette-Guérin [BCG]), adenovirus, poliovirus, and vaccinia virus, which express selected antigens from unrelated microorganisms. Nonreplicating antigen delivery systems include biodegradable microspheres and liposomes and oral adjuvants such as cholera toxin (CT) and its B subunit (CTB). Several of these vaccine strategies for mucosal pathogens are discussed in this chapter.

COMMON MUCOSAL IMMUNE SYSTEM

Most infectious agents cause disease either by colonization of or invasion through mucosal surfaces such as the gastrointestinal and respiratory tracts. The major humoral immune factor at these sites is locally produced S-IgA antibody (131). It has been estimated that 65 to 90% of immunoglobulin–producing cells in human secretory tissues synthesize IgA and that approximately 75% of the total immunoglobulin produced in humans is IgA. The induction, transport, and regulation of S-IgA involve mechanisms that are different from those involved in the regulation of systemic immune responses (21, 131, 134). Therefore, it is not surprising that current interest in vaccine development centers on consideration of mucosal routes for the induction of protective immune responses.

The predominance of IgA in external secretions is due to the selective transport of polymeric, J-chain-containing IgA through the epithelial cells into the secretions by the poly-Ig receptor, which becomes the secretory component of S-IgA. The presence of the secretory component renders the IgA more resistant to the proteolytic enzymes in secretions. S-IgA primarily occurs in a dimeric or tetrameric form and has a greater binding avidity than monomeric IgA; this avidity could contribute to its functions. Several functions for S-IgA at mucosal surfaces have been described (100, 102). These include neutralization of viruses (7, 57, 163) and of bacterial exotoxins and enzymes (114, 118, 193) and inhibition of organism adherence to surfaces (3, 87, 215) and of antigen absorption (213).

IgA is synthesized and secreted by plasma cells located in secretory tissues and glands. The antigen-specific S-IgA response can be induced by the direct application of antigen, which penetrates the tissue epithelium or infuses into the glands (79, 131, 152, 162), or by the injection of antigen into the glandular vicinity (128, 131). However, these methods for inducing S-IgA responses cannot explain the finding that secretions contain naturally occurring antibodies to organisms that are incapable of colonizing those mucosal tissues. For example, S-IgA antibodies to the oral bacterium *Streptococcus mutans* have been detected in human colostrum and milk (8) and in tears (4, 8). Since S-IgA in secretions is locally produced by plasma cells in the tissues or glands, it follows that the precursor IgA B cells home to these glands after prior stimulation at distant inductive sites. These findings suggest that mucosal immune responses can be induced through a unique system that has been termed the common mucosal immune system (CMIS) (131, 135). This system includes the inductive sites of the GALT and the bronchius-associated lymphoid tissues (BALT) and the effector sites, including the salivary, lacrimal, and mammary glands and the gastrointestinal, respiratory, and genital mucosae.

GALT and BALT are primary sites for induction of IgA responses to inhaled and injected antigens in mammals (16, 131, 132,

134). GALT consists primarily of Peyer's patches and the appendix. The Peyer's patches are discrete lymphoid follicles along the small intestine. The follicles can be single or aggregated, and their number and location can vary among species. The Peyer's patch consists of a dome region that contains T and B lymphocytes as well as dendritic cells and macrophages and an underlying B-cell follicle with germinal centers and parafollicular T-cell areas. Information on T-cell subsets involved in the CMIS has been reviewed elsewhere (126, 127). A unique epithelium consisting of cuboidal epithelial cells and specialized antigen-sampling M cells covers the dome region of the Peyer's patch (172). M cells have the important abilities to sample gut luminal antigens such as proteins (171), viruses (218), and bacteria (27) and to transport these antigens to the underlying lymphoid and antigen-presenting cells.

Once antigen is transported through the epithelium by M cells and presented to the lymphoid cells, antigen-stimulated IgA-committed B cells and T cells leave the Peyer's patches in the efferent lymphatics, pass through the mesenteric lymph nodes, and enter the systemic circulation via the thoracic duct. The circulating B cells eventually enter remote glandular and mucosal tissues such as mammary, lacrimal, and parotid glands and gastrointestinal mucosa, where they are preferentially retained. It is in these mucosal effector sites that the antigen-stimulated IgA-committed B cells clonally expand and differentiate into polymeric-IgA-producing cells. Polymeric IgA is then transported across the glandular or mucosal epithelial cells and released into the corresponding secretions as S-IgA antibodies with specificity to the antigen encountered in the Peyer's patch.

Most of the initial evidence for the existence of a CMIS came from studies of experimental animals. Craig and Cebra (43) showed that when lymphoid cells from Peyer's patches of rabbits are injected into irradiated allogeneic recipients, they repopulate the lamina propria of the intestine with IgA-producing plasma cells of the donor type. These findings were the first to suggest that Peyer's patches are enriched sources of precursor IgA B cells with the potential to populate mucosal tissues. Montgomery et al. (148) were the first to show that oral administration of a dinitrophenyl (DNP)-haptenated pneumococcal vaccine led to the appearance of S-IgA anti-DNP antibodies in the milk of immunized rabbits. These results have been confirmed and extended by the numerous studies of experimental animals that demonstrate that oral administration of any one of a variety of antigens results in an S-IgA response not only in the intestinal secretion but also in other secretions such as saliva and lacrimal, bronchial, and genital tract secretions (131).

The evidence for a CMIS in humans has by necessity been indirect (131, 134). It includes the findings that external secretions of glands distant from the site of antigen stimulation contain S-IgA antibodies that are specific for microbial and food antigens and that intragastric immunization with numerous infectious agents including viruses (13, 162, 212) and bacteria (51, 136) induces specific S-IgA antibodies in various external secretions but not in serum. Specific IgA antibody-secreting cells have been transiently detected in the peripheral blood of individuals orally immunized, and their presence preceded the appearance of S-IgA antibodies in saliva and tears (51, 53, 97). Collectively, these results indicate that the stimulation of GALT by oral immunization can be used to induce secretory immune responses to a variety of agents that infect mucosal surfaces distant from the gastrointestinal tract.

In addition to GALT and BALT, accumulating evidence suggests the existence of additional IgA inductive sites within the CMIS. Waldeyer's ring, consisting of palatine, lingual, and nasopharyngeal tonsils, is located at the top of the digestive and respiratory tracts and is continually exposed to inhaled or ingested antigens. This tissue appears to contribute IgA precursor cells to the upper respiratory and digestive tracts (19, 108). The unique architecture of the tonsils resembles that of lymph nodes (19) and GALT (93) in having antigen-presenting, T, B, and IgG- and IgA-containing plasma cells present in characteristic regions. Ogra (161) reported reduced nasopharyngeal antibody responses to perorally administered live poliovirus in tonsillectomized children and decreased nasopharyngeal resistance due to diminished S-IgA levels. The human nasal mucosa contains T cells (88, 217), HLA-DR-expressing dendritic and epithelial cells (189), and abundant IgA-secreting plasma cells that may originate in BALT or tonsillar tissue (20, 22). Intranasal immunization of humans against respiratory pathogens results in an antibody response in nasal secretion and serum (25, 41). The presence of antibodies in other mucosal effector sites has not been evaluated.

Part of the strategy to develop a successful vaccine is to induce an S-IgA immune response at the appropriate mucosal surface. For example, oral immunization may be most effective for inducing a response in the intestine, while intranasal immunization could be most effective for inducing responses in the oral cavity and the respiratory tract. Information suggesting compartmentalization within the CMIS and the route of immunization for optimal induction of S-IgA immune responses is currently accumulating.

LIPOSOMAL DELIVERY SYSTEM FOR MUCOSAL IMMUNIZATION

Characteristics of Liposomes

Liposomes are bilayered membrane vesicles composed of amphipathic molecules (polar and nonpolar portions) such as phospholipids (Fig. 1) that form multi- or unilayered (lamellar) vesicles, depending on the methods of preparation. Many of the phospholipids used to make liposomes are purified from food products (e.g., egg yolk); therefore, they are nontoxic and safe for human use. These vesicles form spontaneously when an aqueous solution is added to a dried film of the lipid components (10). The hydrophilic portion of the lipid molecule is oriented toward the inner or outer aqueous phase, and the hydrophobic portions are aligned inside the membranes. This method

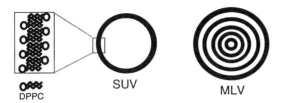

FIGURE 1 Schematic depiction of the two basic forms of liposomes. The small unilamellar vesicle (SUV) illustrates the bilayered membrane composed of dipalmitoyl phosphatidylcholine (DPPC) surrounding an aqueous core. The multilamellar vesicle (MLV) is characterized by several lipid bilayers separated by thin aqueous phases. (Reprinted from reference 139 with permission from Academic Press.)

results in a mixture of small (0.1-μm-diameter) to large (>1.0-μm-diameter) liposomes with mostly multilamellar liposomes. Methods used to produce more homogeneous and unilamellar liposomes consist of microemulsification (122), detergent dialysis (199), column chromatography (94), ultracentrifugation (11), and membrane extrusion (170). Electron microscopy has been used to determine the size and lamellar characteristics of liposomes (Fig. 2). Dynamic light scatter is another method of determining liposome size distribution. Although electron microscopy is the most definitive method of observing these small vesicles, liposome preparations can be analyzed by recording light scatter (214). If a fluorescence-activated cell sorter, which is available to most laboratories, is used, a practical method of qualitative analysis of liposome suspensions is recording laser light scatter by using flow cytometry (34).

Liposomes can be prepared with a net negative charge by using anionic phospholipids or by adding a negatively charged compound such as dicetylphosphate. Positively charged or neutral liposomes can also be made, depending on the components used in production. A common formulation for production of negatively charged liposomes consists of dipalmitoyl phosphatidylcholine and dicetylphosphate. Cholesterol is often added to increase membrane fluidity by interrupting the orderly interdigitation of the dipalmitoyl phosphatidylcholine molecules.

Initially, liposomes were studied for their ability to augment immune responses when given systemically. Their adjuvant quality was thought to be due to the uptake of liposomes (and antigen) by antigen-processing cells such as macrophages (83). Recent studies have shown that liposomes can also serve as an immunoadjuvant for mucosal immunization (32, 137, 139). Whether the adjuvant quality of liposomes is due to the more effective uptake by M cells of Peyer's patches and/or to their subsequent interaction with antigen-processing cells has not been shown. However, liposomes have been found in endocytic vesicles of M cells in close proximity to lymphocytes following intestinal administration (31) (Fig. 3). Because liposomes are bilayered membrane vesicles that mimic cell membranes, the uptake and processing of antigen may be enhanced by enclosing the antigen in the lipid vesicle. Another proposed mechanism for the adjuvant properties of liposomes is their ability to protect the antigen from acids and proteolytic enzymes of the gastrointestinal system. Although they are not completely resistant to lipases and bile salts found in the small intestines, cholesterol-containing liposomes can provide at least partial resistance (159, 182).

Liposomes and Dental Caries Vaccines

The use of liposomes containing antigen as oral vaccines was first investigated to determine their ability to augment protective salivary immune responses against the oral pathogen *S. mutans* in an experimental rat caries model. Initial investigations (141) compared liposome-*S. mutans* antigens to other vaccine preparations given orally. Liposome-*S. mutans* antigens enhanced salivary antibody activity and were more protective against *S. mutans*-induced dental caries than control whole cell or antigen only. However, the responses induced with the liposome-antigens were comparable to those seen in animals given *S. mutans* antigens with other adjuvants such as peptidoglycan, muramyl dipeptide, and water-in-oil emulsion. A later study (140) showed that adding muramyl dipeptide to liposome-antigen

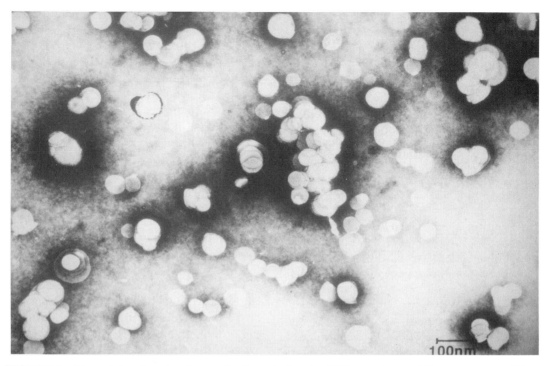

FIGURE 2 Transmission electron micrograph of negatively stained (2% uranyl acetate) liposomes prepared from dipalmitoyl phosphatidylcholine, cholesterol, and dicetylphosphate (molar ratio, 16:7:1, respectively). (Reprinted from reference 33 with permission from Plenum Press.)

further augmented the protective immune response to *S. mutans*. However, in another study, when the mucosal adjuvant CT was given in (or in addition to) the liposome-antigen preparation, no additional enhancement of the immune response was seen (unpublished studies). This may be partially due to the fact that incorporating CT into liposomes may interfere with the binding of CTB to GM_1 (see section below on CT).

Experimental data also indicate that the dose of *S. mutans* antigen necessary for producing a mucosal immune response following oral administration is much lower when antigen is incorporated into liposomes (84). This finding is most obvious when antigen associated with liposomes (i.e., antigen in the vesicles or noncovalently associated with or on the liposomal membrane) is isolated by removal of the unassociated antigen by means of gel chromatog-

raphy or ultracentrifugation (33, 36, 210). These studies and others (26, 138, 211) confirm the potential usefulness of a liposome-antigen delivery system for protection against mutans streptococcus–induced dental caries.

A challenge regarding the design of oral liposomal vaccines has been to determine the optimal characteristics of liposomes and the optimal form of liposome-antigen for induction of immune responses. In a recent study, various liposome preparations containing *S. mutans* glucosyltransferase (GTF) that differed in homogeneity and size of liposomes were evaluated for their abilities to induce protective immune responses (36). All liposome-GTF preparations tested induced immune responses that were more protective than that induced by GTF alone or by empty liposomes. Additionally, smaller, more homogeneous (unilamellar) liposomes were more effective in pro-

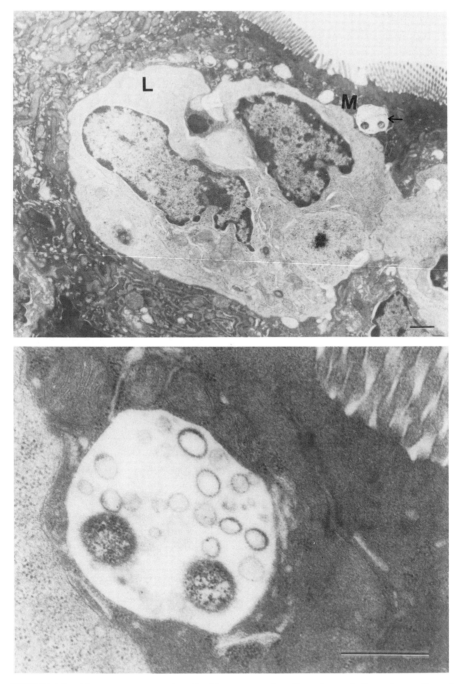

FIGURE 3 Electron micrographs of a thin section of rat Peyer's patch tissue from a ligated intestinal segment that was injected with small unilamellar liposomes. (Top) M cell (M) with an endocytic vesicle (arrow) and a closely associated lymphoid cell (L). Bar, 1.0 μm. (Bottom) Higher magnification of the M cell in the top panel, showing the endocytic vesicle containing numerous liposomes and two electron–dense particles. Bar, 0.2 μm. (Reprinted from reference 31 with permission from John Wiley & Sons, Inc.)

tecting rats against the development of dental caries.

Human studies utilizing oral liposome-*S. mutans* antigen vaccines have provided preliminary data indicating the safety and immunogenicity of these vaccines. Oral administration of liposome-*S. mutans* cell wall carbohydrate antigen in enteric coated capsules resulted in a transient salivary IgA anti-CHO response compared to preimmunization levels (35). A second immunization 6 to 9 months later resulted in a more rapid response of greater magnitude, which is characteristic of secondary responses. In that study, the liposome preparation was maintained in an aqueous phase, which is difficult to package. Therefore, in a subsequent human study, a dehydrated liposome-*S. mutans* GTF vaccine was tested (37). Volunteers swallowed capsules containing 500 µg of antigen for three consecutive days and were given a boost 4 weeks later. All subjects experienced increased salivary IgA anti-GTF responses, with a predominant salivary IgA2 anti-GTF response in more than half of the subjects. Since naturally occurring IgA anti-*S. mutans* protein responses are predominantly IgA1 (24), it is possible that the presentation of the antigen in liposomes accounted for the shift in the subclass of the IgA response. In recognition of the fact that IgA1 proteases are produced by some bacteria that colonize the oral cavity (and teeth; e.g., *Streptococcus sanguis*) (101), an IgA2 response could be more advantageous to the host because it would be resistant to degradation by this microbial protease.

Other Liposome Oral-Immunization Models Tested

The studies described thus far have concentrated on experimental data utilizing oral liposome vaccines containing antigens of *S. mutans* because of the number of investigations, the existence of an experimental animal model to study protection against *S. mutans*-induced dental caries, and the human studies that have been done. Collectively, these studies provide a strong basis for the use of this approach in the development of a caries vaccine for human use

(137). Nonetheless, other studies involving oral immunization with liposomes have been reported. Rhalem and coworkers (180) reported that oral immunization with antigens from *Nippostrongylus brasiliensis* were more protective against infection in mice with this helminth when the antigen was given in liposomes. Other studies have used various approaches to investigate the ability of liposomes to augment the mucosal responses to oral immunogens. The antigens used for these studies were obtained from pathogens such as *Clostridium tetanus* (89), *Vibrio cholerae* (28), *Porphyromonas gingivalis* (160), human immunodeficiency virus (205), and *Bordetella pertussis* (85). These studies showed augmentation of responses in the various systems employed; however, protective responses were not demonstrated owing to the lack of a useful model.

Clarke and Stokes (38) reported that mice fed liposome-antigens (CT or *Escherichia coli* cell wall extract) had significantly higher responses in serum but not in intestinal secretions than antigen-only controls. On the basis of these findings and other negative results (38), those authors concluded that liposomes are not a useful adjuvant for inducing mucosal responses to the antigens they tested. As mentioned above, our unpublished studies indicate that the use of CT in a liposome-antigen preparation does not result in an enhanced response to the antigen. Taken together, the results suggest the importance of evaluating each proposed liposome-antigen delivery system for its usefulness as an adjuvant.

Novel Approaches to Liposome Vaccination

Other strategies for improving mucosal vaccines that use liposomes have been proposed. One approach is to covalently link antigen or ligands to the surfaces of liposomes. The potential for specifically targeting vaccines to an antigen-processing cell or to M cells by using an antibody to a specific receptor or a ligand for the target cell surface may provide a more efficient system of immunization.

Mucosal immunization using liposomes is

not limited to oral administration. Abraham (1) tested intranasal application of large unilamellar liposomes containing bacterial polysaccharide antigens in BALB/c mice. That study reported enhanced immune responses in pulmonary secretions of mice immunized with the liposome-antigen compared to responses to antigen alone. The responses were much stronger than those following oral immunization (i.e., nasal immunization produced comparable antibody titers at 1/30 the oral dose). When liposome-*Pseudomonas aeruginosa* polysaccharide was administered intranasally, mortality of the animals after challenge was decreased. Abraham and Shah (2) also reported an improvement in immune responses when antigens were given in liposomes containing interleukin-2 (IL-2) (but not with liposomes containing IL-4). An 80-fold increase in pulmonary specific-antibody-producing plasma cells was found.

Another area of liposome research involves the packaging of RNA- and DNA-encoding antigens that could be inserted into antigen-processing or -presenting cells to provide a renewable source of antigen for continuous immunization (75). Whether insertion of nucleic acid-encoding antigen would be sufficient and what other genetic controls it would be necessary to include (e.g., genes for cytokines or promoters) in order to provide a long-lasting immunostimulating system are difficult questions that must be answered in order to bring this concept to reality.

BIODEGRADABLE MICROSPHERES AS ORAL VACCINES

The encapsulation of antigens into microspheres for the development of new and improved oral vaccines is currently receiving considerable attention. Over the past several years, those working with polymeric microspheres have made much progress toward the development of stable, single-dose, nontoxic vaccines that can be given orally (as well as via other mucosal routes, such as intranasally) and that effectively potentiate protective immune responses against infectious agents, including those that cause common childhood diseases.

Microspheres are now composed of nontoxic biodegradable material and are safe for use in animals and humans. The polymeric encapsulation of antigen prevents its acid and enzymatic degradation in the intestine. Furthermore, the polymer composition of microspheres can be changed, affecting the rate at which the antigen is released. Therefore, by combining microspheres of different compositions, it is possible to induce a primary antibody response after oral immunization as well as a secondary response weeks later. Other important considerations concerning the potential usefulness of microspheres as oral vaccines are their effectiveness in being taken up by IgA inductive tissue (e.g., Peyer's patches) and whether the desired immune responses are induced in the host.

Since the early findings of Strannegard and Yurchinson (200), it has been generally accepted that the particulate form of an antigen has an enhanced effectiveness as an oral immunogen. Currently, the only approved method of making adjuvant vaccines for human use is to adsorb protein antigens onto aluminum salts to create a particulate. These particulates potentiate the antibody response after systemic administration through creation of an antigen depot, induction of inflammation at the site of injection, and directed delivery of the particulate antigen into the draining lymph nodes. However, the sizes and properties of aluminum compounds are sensitive to slight changes in the conditions of their preparation and their age (150, 151), and they are not effective carriers for mucosal immunization. An alternative approach under investigation has been polymeric particulate adjuvants.

Microparticles

In 1965, Litwin and Singer (111) demonstrated a modest enhancement in the circulating-antibody response to human gamma globulin in rabbits immunized by the intravenous route with human gamma globulin adsorbed onto polystyrene latex particles. This adjuvant activity was dependent on adsorption of the antigen to the particles, was most apparent at limiting antigen doses, and was equivalent when particles of various diameters from 0.05 to 1.3 μm were tested. In contrast, Kreuter et al. (104, 105) reported that for subcutaneous injection of mice, adsorbing bovine serum albumin (BSA) to either poly(methyl methacrylate) or polystyrene particles potentiated the anti-BSA response to a greater extent than adsorption to aluminum hydroxide. However, it was Cox and Taubman (42) who proposed the use of polymeric particles coated with antigen for the induction of a mucosal antibody response. They reported that oral administration of DNP-bovine gamma globulin on glutaraldehyde-activated polyacrylamide beads (1 to 3 μm in diameter) was more effective at eliciting a salivary IgA anti-DNP response than was an equivalent dose of soluble DNP-bovine gamma globulin. However, it was impossible to determine from the data presented the degree to which the mucosal response was potentiated. More recently, O'Hagan et al. (167) tested poly(butyl-2-cyanoacrylate) particles with adsorbed ovalbumin (OVA) as carriers for mucosal immunization. Specific IgA antibodies in saliva were detected following oral administration to primed rats of either 0.1- or 3-μm diameter OVA-coated particles on four consecutive days. The 0.1-μm particles also induced a serum IgG response. In contrast, oral boosting with soluble OVA was without effect.

While it is clear that synthetic polymeric particles with antigen adsorbed to their surfaces can enhance systemic or mucosal immune responses, it is also clear that many antigens will not effectively adsorb to these surfaces. Furthermore, it is possible that the antigen exposed on the surfaces of the particles will be eluted or degraded. The incorporation of the vaccine antigen within the polymer can provide protection from degradation and control over the time of release for virtually any antigen. Several antigen-containing polymer systems have been investigated and are described below.

Poly(Methyl Methacrylate) Particles and Polymeric Implants

Poly(methyl methacrylate) particles potentiate responses to whole formalin-inactivated influenza virions (107) or split influenza vaccine (106). In those studies, methyl methacrylate was polymerized by gamma irradiation in the presence of antigen. Following subcutaneous injection of guinea pigs or intraperitoneal injection of mice with the vaccine, serum hemagglutination inhibition titers were higher than those induced by equivalent doses of the free vaccine or of vaccine with aluminum hydroxide as adjuvant. The adjuvant activity of the particle vaccine was dependent on particle size (104) and hydrophobicity (105), with smaller and more hydrophobic particles giving the best response.

Preis and Langer (176) used a solvent casting process to entrap BSA (100 μg) within ethylene-vinyl acetate copolymer pellets (0.3-mm diameter). Subcutaneous implantation of one pellet per mouse elicited a serum anti-BSA response. The response was equivalent to that stimulated by two injections of BSA (50 μg per dose) emulsified in complete Freund's adjuvant (CFA). It was proposed that the immunopotentiation afforded by this delivery system resulted from the prevention of antigen degradation and the sustained release of antigen by diffusion through the copolymer (176). In a subsequent study, BSA-containing pellets were prepared with the novel biodegradable polymer N-benzyloxycarbonyl-L-tyrosyl-L-tyrosine hexyl ester (CTTH)-iminocarbonate, which was formed by linking tyrosine dipeptide units together through hydrolytically labile bonds between the tyrosine side chain groups (103). This polymer was chosen because the degradation products include derivatives of L-tyrosine that exhibit adjuvant activity. Following implantation of the BSA-containing CTTH-iminocarbonate pellets, the antibody response induced was enhanced relative to the response obtained with the same dose of free BSA. Although the CTTH-iminocarbonate pellets did not potentiate humoral responses to the same extent as BSA-containing ethylene-vinyl acetate pellets (176), their in vivo biodegradation obviated the need for surgical removal.

POLYMERIC MICROSPHERES

The incorporation of vaccines within polymeric microspheres combines the advantages of a particulate adjuvant that protects and controls the release of the antigen with the ease of administration afforded by injection or mucosal application. Microencapsulation involves coating a bioactive agent, such as a vaccine antigen, with a protective wall material, which is generally polymeric. The antigen can be interspersed throughout the particles (microspheres), or each particle can consist of a core reservoir of antigen surrounded by an outer polymer shell (microcapsule). The microsphere product is a free-flowing powder of spherical particles that can range in size from <1 to >300 μm in diameter (Fig. 4). The systems that have been investigated for vaccine delivery include microspheres formed by cross-linking natural substances such as starch or albumin and synthetic polymers. Several antigen-containing polymeric microsphere systems have been investigated, and some are listed in Table 1 together with the conditions used to evaluate their effectiveness and the major experimental findings.

In an attempt to produce a biocompatible microsphere antigen delivery system from a natural substance, Artursson et al. (9) prepared microspheres (0.5 to 2.0 μm in diameter) containing human serum albumin (HSA) from acryl starch activated with ammonium peroxydisulfate and cross-linked with N,N,N',N'-tetramethylethylenediamine. The HSA-polyacryl starch microspheres elicited a serum anti-HSA antibody response in mice immunized by either the intramuscular or the intravenous route. The response peaked and remained for >200 days approximately 500-fold higher than the peak level induced with free

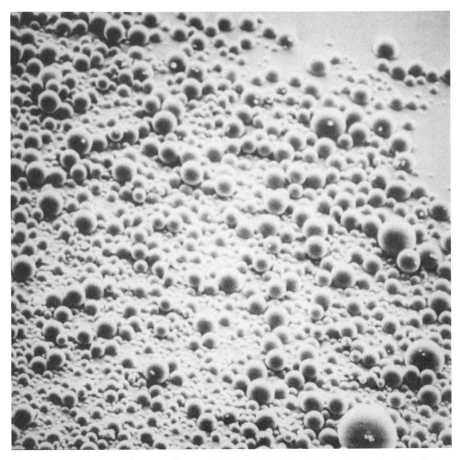

FIGURE 4 Scanning electron micrograph of DL–PLG microspheres. (Courtesy of Jay K. Staas, Southern Research Institute, Birmingham, Ala.)

TABLE 1 Selected immunization studies with antigens incorporated into polymeric microspheres[a]

Polymer (particle size)	Antigen	Host	Route of administration	Major findings	Reference(s)
Polyaryl-starch microspheres (0.5–2.0 μm)	HSA	Mice	i.m., i.v.	Enhanced serum response; antimatrix response induced	9
Polymerized RSA microspheres (100–200 μm)	Nodamura virus	Rabbits	i.m.	Enhanced response equivalent to that to FIA; antibody response to glutaraldehyde-polymerized RSA induced	119
Polyacrylamide microspheres (2.5 μm)	OVA	Rats	oral	Enhanced response; induction of salivary IgA in i.p.-primed host	168
DL-PLG microspheres (100 μm)	p72:TT	Mice	s.c., i.m.	Weak primary response, effective booster	6
1–10 μm	SEB toxoid	Mice	s.c., i.m., i.p., i.t., oral	Enhanced response; plasma IgG equivalent to CFA after i.m. administration; pulsed release; plasma, salivary, bronchial, gut IgA after oral or i.t. administration.	65–69
1–10 μm	Influenza virus	Mice	s.c., i.p., oral	Enhanced response; oral booster induced salivary IgA, enhanced serum HAI titer and protection	146, 147
1–10 μm	SIV	Rhesus macaques	i.m., oral, i.t.	Induction of serum and vaginal wash responses and protection	120
1–10 μm	OVA	Mice, rats	s.c., i.p.	Enhanced serum IgG response significantly greater than that with CFA in mice	165, 169
27 μm	CFA/I	Rabbits	i.g.	Enhanced serum IgG response; IgA coproantibody response	64
4.5 μm	CFA/II	Rabbits	i.d.	Induction of antigen-sensitive cells in Peyer's patches and splenic antibody-secreting cells	179
1–12 μm	AF/R1	Rabbits	i.d.	Protected against challenge	130
2.3–3.0 μm	OVA CTB	Mice	oral	Enhanced salivary IgA and serum IgG to OVA; splenic and MLN antibody-secreting cell response to CTB	166
	Tetanus toxoid	Mice	s.c.	Protein release influenced by polymer molecular weight and composition; serum IgG response equivalent to that with aluminum phosphate–absorbed toxoid	5

[a]Abbreviations: p72:TT, 28-amino-acid peptide from the surface glycoprotein of hepatitis B virus conjugated to tetanus toxoid; SIV, simian immunodeficiency virus; AF/R1, pilus adhesin of *E. coli* RDEC-1; s.c., subcutaneous; i.p., intraperitoneal; i.m., intramuscular; i.v., intravenous; i.t., intratracheal; i.d., intraduodenal; i.g., intragastric; MLN, mesenteric lymph nodes; HAI, hemagglutination inhibition; FIA, Freund's incomplete adjuvant.

HSA. However, an antibody response to the cross-linked starch was also demonstrated. In other studies, Martin et al. (119) investigated the use of a natural substance under conditions in which it is normally nonimmunogenic. Rabbits were injected by the intramuscular route with Nodamura virus encapsulated in microspheres (100 to 200 μm in diameter) formed by glutaraldehyde cross-linking autologous rabbit serum albumin (RSA). Serum antibody responses to the virus and to glutaraldehyde-cross-linked RSA were detected. Thus, while the number of studies examining cross-linked natural substances to formulate vaccine-containing microspheres has been limited, the induction of antibody responses to the microsphere matrix is a problem that appears to preclude their use in humans.

Antigen-containing synthetic polymer microspheres were examined by O'Hagan et al. (168) for their abilities to potentiate salivary and serum antibody responses. In that study, rats previously primed by intraperitoneal injection of OVA were subsequently given OVA (1 mg) entrapped in polyacrylamide microspheres orally on four consecutive days. The microspheres were macroporous particles (2.55 μm in diameter) with OVA partially exposed on the surface. A significant salivary IgA response in the absence of a rise in circulating antibodies was detected 65 days following oral boosting. Control animals that received soluble OVA as the oral booster did not respond. However, since polyacrylamide induces neurotoxic effects, the use of this formulation would be restricted to experimental applications.

DL-PLG Microspheres

More recently, a considerable effort has been made to evaluate microspheres formulated from poly(DL-lactide-co-glycolide) (DL-PLG) for the parenteral and oral delivery of vaccines because of the proven biocompatibility and biodegradability of this copolymer. DL-PLG is in the class of copolymers from which resorbable sutures, resorbable surgical clips, and controlled-release drug implants and microspheres are made (177). These polyesters are approved by the Food and Drug Administration and have a 30-year history of safe use in humans. DL-PLG induces only a minimal inflammatory response and biodegrades through hydolysis of ester linkages to yield the natural body constituents lactic and glycolic acids (206, 209) (Fig. 5). The polymer undergoes random non-enzymatic hydrolysis of its backbone ester linkages, and the rate at which DL-PLG biodegrades is a function of the ratio of lactide to glycolide in the copolymer (143). Thus, the rate of biodegradation of the polymer determines the time after administration when antigen release is initiated and the subsequent rate of release.

Altman and Dixon (6) examined the immune response induced by immunization with a 28-amino-acid peptide 72 (p72) from the surface glycoprotein of hepatitis B virus conjugated to tetanus toxoid (p72:TT). The p72:TT was incorporated into large DL-PLG microspheres (50 to 100 μm in diameter) in order to allow for the release of the antigen at a uniform rate over a period of 1 to 12 months. Subcu-

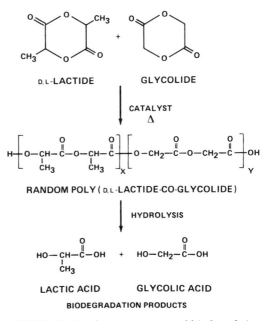

FIGURE 5 Synthesis, structure, and biodegradation products of DL-PLG. (Reprinted from reference 139 with permission from Academic Press.)

taneous injection of mice with microencapsulated p72:TT resulted in a very weak primary antibody response. However, the microencapsulated p72:TT retained antigenicity, as judged by its ability to prime for an anamnestic secondary response. The injection of microencapsulated p72:TT in incomplete Freund's adjuvant resulted in the induction of a high primary antibody response. These results suggest that the difference in responses between noninflammatory controlled-release DL-PLG microspheres and the previously described pellets (103, 176) may be related to inflammation induced by the implants.

In contrast to the results obtained with relatively large DL-PLG microspheres (6), systemic administration of staphylococcal enterotoxin B (SEB) (65–69), influenza vaccine (146, 147), simian immunodeficiency virus (120), and OVA (165, 169) in 1- to 10-μm DL-PLG microspheres results in strongly potentiated primary serum antibody responses. In one study, subcutaneous administration of SEB toxoid (50 μg) in 1- to 10-μm DL-PLG microspheres to mice resulted in a significantly higher (512-fold) plasma antitoxin neutralizing response than that induced by soluble toxoid (69). Furthermore, the magnitude and duration of the response were similar to those induced by the same dose of toxoid emulsified in CFA. Similar immunopotentiation was obtained in rhesus macaques immunized by intramuscular injection of microencapsulated whole formalin-inactivated simian immunodeficiency virus (120).

To determine the effect of the size of the microspheres on potentiating responses, 1- to 10- and 10- to 110-μm microspheres prepared from the same lot of DL-PLG and containing the same percentage of SEB toxoid by weight were used to immunize mice by the subcutaneous route (69). Mice immunized with the 1- to 10-μm microspheres had approximately 20-fold-greater immune enhancement than mice injected with the larger microspheres. The enhanced potentiation correlated with the phagocytosis and transportation of a large number of <10-μm microspheres to the lymph nodes draining the injection site. These data suggest that delivery of vaccine to antigen-presenting cells by <10-μm microspheres is a more effective approach to the enhancement of antibody responses than the depot release of free antigen that is provided by larger microspheres. This possibility can explain the poor adjuvancy observed in studies by Altman and Dixon (6). An accelerated rate of antigen release was also seen where the <10-μm microspheres were compared with the >10-μm microspheres after in vivo administration, presumably as a result of phagocytosis.

Mixtures of microspheres with varied antigen release characteristics based on size have been studied to determine their abilities to induce enhanced antibody responses. The single injection of a mixture of 1- to 10- and 20- to 50-μm DL-PLG microspheres containing SEB toxoid resulted in a biphasic antitoxin response (68). The first phase of antibody increase corresponded to the release of vaccine from the 1- to 10-μm component. The second phase corresponded to the release of vaccine from the 20- to 50-μm microspheres and stimulated an anamnestic secondary rise in antitoxin levels. The response was greater than that seen when microspheres of a single size were given alone. A similar enhancement of responses has been observed in mice given a single injection of a mixture of DL-PLG microspheres differing in their lactide-to-glycolide ratios (66a).

In order to establish the feasibility of using DL-PLG microspheres as a vehicle for oral immunization, it is important to show that these microspheres are taken up by Peyer's patches. Eldridge et al. (65, 66) used fluorochrome-containing microspheres to study the effect of size on uptake. Gut and other lymphoid tissues from mice given the fluorochrome-containing microspheres orally were analyzed by fluorescence microscopy. Shortly after oral administration of the microspheres, adsorption of the microspheres in the gut was restricted to the Peyer's patches (Fig. 6), and only microspheres <10 μm in diameter were adsorbed. Furthermore, the adsorbed microspheres were within Peyer's patch macrophages. With increasing

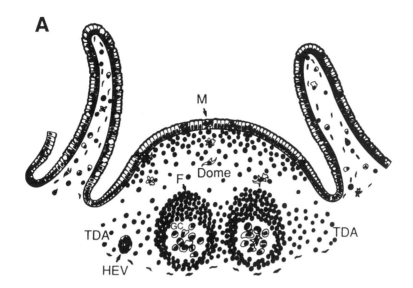

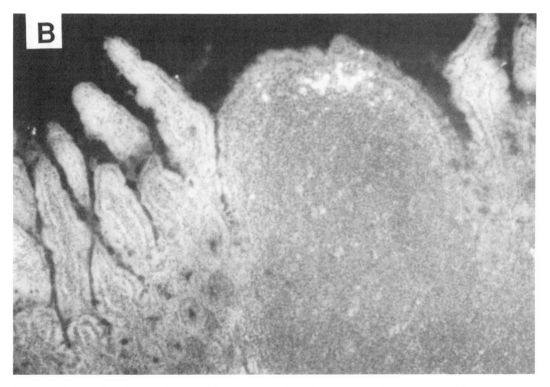

FIGURE 6 Uptake of coumarin-containing DL-PLG microspheres by Peyer's patches. (A) Cellular organization of a Peyer's patch. M, M cell; F, follicle; GC, germinal center; TDA, T-cell-dependent area; HEV, high endothelial venule; Dome, region containing macrophage and lymphocytes. (B) Presence of coumarin-containing DL-PLG microspheres in the specialized epithelium and dome region of the Peyer's patch revealed by fluorescence microscopy of a 6-μm-thick frozen section of duodenum obtained from a mouse 24 h after oral administration of a suspension of 1- to 10-μm microspheres. (Reprinted from reference 139 with permission from Academic Press.)

time after administration, the number of microspheres in the Peyer's patches fell, and the microspheres were subsequently observed in the mesenteric lymph nodes and spleen. The results of this study indicate that microspheres <5 μm in diameter pass through the Peyer's patches and move to the mesenteric lymph nodes and the spleen, while microspheres ranging in size from 5 to 10 μm in diameter remain in the Peyer's patches for up to 35 days. Mice given three oral administrations of SEB toxoid in 1- to 10-μm DL-PLG microspheres at 30-day intervals exhibited a steady rise in their plasma IgG antitoxin levels and significant levels of S-IgA antitoxin in saliva, gut fluids, and bronchoalveolar fluids 20 days following the third dose (65, 68). In contrast, mice orally immunized with nonencapsulated SEB toxoid mounted only a weak IgM response in plasma and none in mucosal secretions. In similar experiments, Moldoveanu et al. (146, 147) examined the immune response to influenza virus vaccine encapsulated in 1- to 10-μm DL-PLG microspheres in mice. They presented evidence that systemic administration of microencapsulated influenza virus potentiates the plasma hemagglutination inhibition titer, that encapsulation protects the influenza virus from acid degradation, and that oral boosting with encapsulated vaccine antigen is particularly effective in the induction of salivary IgA anti-influenza antibodies.

Recent studies have reported the use of microencapsulated colonization factor antigens of enterotoxigenic E. coli (64, 130, 179). Intragastric administration of the fimbrial adhesin CFA/I encapsulated in DL-PLG microspheres to rabbits resulted in systemic IgG anti-CFA/I responses (64). Little or no serum response was induced with unencapsulated CFA/I. S-IgA coproantibodies were detected in the stool of one of three rabbits immunized with microencapsulated CFA/I. In other studies, intraduodenal immunization of rabbits with microencapsulated CFA/II induced antigen-sensitized lymphocytes in the Peyer's patches and antibody-secreting cells in the spleen (179). Rabbits given microencapsulated AF/R1 by intraduodenal administration were protected from diarrhea induced by E. coli RDEC-1 (130).

To date, studies with DL-PLG microspheres containing virulence antigens from various infectious agents indicate that the microspheres are safe and effective in potentiating mucosal and systemic responses. Although additional studies are required, the current results serve to illustrate the potential of this approach for mucosal immunization with nonliving vaccine antigens. Current studies with DL-PLG microsphere vaccines are directed toward determining whether microsphere modifications will enhance the adsorption of the microspheres by Peyer's patches and other IgA inductive tissues, whether alternate routes of mucosal administration such as those involving the respiratory tract or the combination of different routes of immunization with these vaccines will enhance immune responses, and whether multirelease microsphere formulations are effective in inducing prolonged mucosal and systemic immune responses.

AVIRULENT *SALMONELLA* SPP. AS LIVE CARRIERS OF HETEROLOGOUS ANTIGENS

Live vaccines have the advantage over nonreplicating vaccines because they can persist and divide in the host, resulting in continuous exposure of the host to the vaccine antigen. Thus, it is possible that a single dose of live vaccine could represent an effective dose for the induction of a long-lasting immune response. The idea of using attenuated *Salmonella* organisms as an oral vaccine delivery system for antigens of heterologous pathogens began over a decade ago (45, 46, 48, 77, 86, 90, 181). This idea was based on separate but interrelated areas of research, including studies of the manner in which *Salmonella* spp. infect the host and of recombinant DNA technology. *Salmonella typhi* and *Salmonella typhimurium* colonize the intestines of humans and mice, respectively. These bacteria transverse the gut mucosa and can then be detected in systemic tissue, especially the spleen and liver. An initial site of *Sal-*

monella infection is the Peyer's patches (27). These bacteria are able to persist and replicate in the Peyer's patches and provide a continual supply of antigen to this IgA inductive site of the mucosal immune system. For use as a live vaccine, *Salmonella* strains had to be rendered avirulent while their ability to colonize and infect the host without causing disease was preserved. With the development of genetic techniques to incorporate recombinant DNA into *Salmonella* mutants, studies were designed to test avirulent *Salmonella* strains as potential live vectors for expression of cloned genes specifying colonization or virulence antigens of other pathogens (45, 46, 48, 86, 181). These live antigen delivery systems could be administered orally to induce mucosal IgA responses locally in the intestine and in other secretions as well as to induce systemic humoral and possibly cellular immune responses to the salmonellae and the cloned antigen.

The ability of oral *Salmonella* vaccines to express heterologous antigens and elicit immune responses is dependent on the viability of the carrier. Orally administered killed salmonellae are not effective in presenting heterologous antigens for the induction of immune responses (50, 74). A main concern, however, with regard to the use of a live vaccine is its safety. Therefore, for the development of an oral *Salmonella* vaccine, several factors must be addressed. The *Salmonella* mutants should be completely avirulent to the host. They should possess well-characterized mutations in specific genes and be stable. For safety of the vaccine strain, it should have two or more attenuating deletion mutations to preclude loss of the avirulence phenotype by reversion or gene transfer. The vaccine strain should be well tolerated and highly immunogenic. It should be effective in expressing the cloned genes that encode virulence factors of other pathogens for the induction of immune responses that render the host protected against infection and disease. It would be desirable if long-lasting immunity could be induced with a single dose.

The following sections discuss characteristics of several avirulent *Salmonella* mutants and

the effectiveness of recombinant avirulent *Salmonella* strains as bivalent-vaccine delivery systems for the induction of immune responses protective against various pathogens. The reader is referred to recent reviews (46, 181) for a more comprehensive coverage of this area.

Avirulent *Salmonella* Mutants

A number of avirulent mutants of the human pathogen *S. typhi* and the murine pathogens *S. typhimurium*, *Salmonella choleraesuis*, and *Salmonella dublin* have been derived and characterized for their potential usefulness as vaccine delivery systems. The *galE Salmonella* mutants are unable to synthesize the enzyme UDP-glucose-4-epimerase, which catalyzes the formation of UDP-galactose from glucose. These mutants are defective in lipopolysaccharide (LPS) biosynthesis (rough) when grown in the absence of galactose, but they synthesize a complete LPS (smooth) when grown in the presence of galactose. *S. typhimurium galE* mutants are avirulent in mice and retain their immunogenicity, as exemplified by their ability to induce responses protective against challenge with virulent *Salmonella* strains (80, 81, 92). The *S. typhi* Ty21a *galE* vaccine strain (82) produced by chemical mutagenesis is avirulent and immunogenic in humans. However, it is likely that attenuation is the result of mutations in addition to *galE*, since an *S. typhi galE* mutant generated by site-directed mutagenesis causes typhoid fever in human volunteers (91). *S. typhi* Ty21a has proved to be stable and is licensed for use in humans as an oral vaccine. Some *galE* mutants of *S. choleraesuis* retain virulence, probably because they require only small amounts of galactose for complete LPS synthesis (155).

The *aroA* mutants of *S. typhimurium* (90) are avirulent in mice, mainly because of their requirement for *p*-aminobenzoic acid (PABA), which is not provided by the animal host. PABA is synthesized from chorismate for the biosynthesis of a number of aromatic compounds, including amino acids. These auxotrophic mutants can colonize the gut, are detected in the spleen, and are immunogenic in mice. An *aroA* mutation of *S. choleraesuis* is also avirulent in mice (156). Subsequent studies with *S. typhimurium aroA* mutants have shown that the addition of a mutation in the gene coding for an enzyme in the purine pathway (i.e., *purE*) results in a less virulent strain (124, 158). Furthermore, the addition of the *purE* mutation to an *aroA S. typhi* strain leads to hyperattenuation with very low immunogenicity (109). *S. typhi* strains with *aroC* and *aroD* mutations are avirulent and immunogenic in humans, but they are nevertheless capable of causing fever and adverse reactions when given at high doses (203). Other studies have shown that *S. typhimurium* strains harboring mutations in two different *aro* genes are attenuated and induce responses in mice after a single dose (60).

Mutants defective in global gene regulation have also been evaluated for their influence on *Salmonella* colonization and virulence (46). This led to the discovery that *cya crp* mutants of *S. typhimurium,* which are unable to synthesize cyclic AMP and the cyclic AMP receptor protein (173), are highly attenuated and do not cause disease in mice. These mutants induce a protective response against challenge with a virulent strain when given orally to mice (47, 49, 50). *Salmonella cya crp* mutants grow slowly in vitro and in vivo, a property that may contribute to their attenuation (49). Mutations in these two genes affect numerous other genes, including those involved in carbohydrate and amino acid metabolism (50). An *S. typhi cya crp* mutant is both avirulent and immunogenic in humans; however, unlike the *aroC aroD* mutant, it causes symptoms in some volunteers (203). *cya crp Salmonella* mutants with a balanced lethal system are thought to be ideal vaccine strains for stable, high-level production of cloned gene products (50). This system is based on deletion of the chromosomal *asd* gene and an *asd*$^+$ gene on the vector (153). This combination eliminates the need for vector drug resistance markers, an important feature for live vaccines. The *asd* gene encodes aspartate ß-semialdehyde dehydrogenase, an enzyme common to the biosynthetic pathways of lysine, threonine, methionine, and diaminopimelic acid (DAP). DAP is an essential constituent of the peptidoglycan of the bacterial cell wall (185) but is not present in animal tissues. Therefore, an *asd* mutant without an *asd*$^+$ vector would not survive in an animal but would undergo DAP-less death and cell lysis.

The *ompR* gene product is part of a two-component regulatory system that affects the regulation of genes for a number of outer membrane proteins (*ompC* and *ompF*). It also influences other genes whose expression is dependent on changes in osmolarity. *S. typhimurium* mutants with mutations in the *ompR* gene are avirulent both orally and parenterally in mice (59). Furthermore, mice orally immunized with *ompR* mutants are protected against challenge with the virulent parent strain.

S. typhimurium phoP mutants were prepared in order to eliminate nonspecific acid phosphatases in strains to be used for transposon Tn*phoA* mutagenesis. These mutants are avirulent when administered to mice by either the oral or the intraperitoneal route (78), a property that may be due to their inability to reside intracellularly within macrophages (76). They are immunogenic and induce immune responses protective against challenge with virulent salmonellae (78). Mutants that constitutively express the *phoP* gene can positively or negatively affect the expression of a number of other *Salmonella* genes (144). *phoP* mutants that allow constitutive expression of *pagC* (*phoPc*) are totally avirulent and highly immunogenic for mice; however, these mutants revert at high frequency to a form that is virulent (144, 145). The *phoP* and *phoPc* mutations have been combined with *aroA* mutations in *S. typhimurium* for use as oral vaccines (145).

Bivalent *Salmonella* Vaccine Strains

A number of attenuated *Salmonella* mutant strains are oral vaccine candidates for use as live vectors for delivering heterologous antigens. These mutants can be detected in GALT following oral administration to the host but do not cause systemic disease. A variety of genes from bacteria, viruses, parasites, and even mammals have been cloned and are expressed in attenuated *Salmonella* spp. (Table 2). These recombinant strains have been used to immunize animals and in some cases humans. Studies have reported the induction of mucosal immune responses as well as serum antibody and cell-mediated immune responses specific for the salmonellae and the foreign antigen(s) expressed by the carrier strain. The effectiveness of the immune response in protection has also been assessed in several studies.

In a study by Poirier and coworkers (175), mice were orally immunized twice with *S. typhimurium* SL3261, which expresses in its cytoplasmic compartment the entire type 5 M protein (pepM5) of *Streptococcus pyogenes*. A serum anti-pepM5 response that peaked at 9 weeks was induced. A salivary IgA anti-pepM5 response was also induced. These orally immunized mice were protected against intraperitoneal and intranasal challenges with type 5 but not type 24 *S. pyogenes* M protein and against challenge with wild-type *S. typhimurium*. These results were among the first to demonstrate the induction of a protective mucosal immune response with an avirulent *Salmonella* spp. expressing a cloned virulence antigen of a heterologous bacterium.

Subsequent studies by Yang et al. (221) used *S. typhimurium* SL3261 expressing the gp63 protein of *Leishmania major* to establish its effectiveness in inducing protective immunity against *Leishmania* infection. Splenic T cells from orally immunized mice proliferated and produced gamma interferon and IL-2 but not IL-4 in the presence of *L. major* antigens. These results suggest the induction of a Th1 response. A serum anti-gp63 antibody but no delayed-type hypersensitivity response to the parasite antigen was seen in immunized animals. Orally immunized mice were able to control *Leishmania* infection. These results provided evidence for the protective role of Th1 responses to *Leishmania* proteins.

Other studies have used *Salmonella* mutants defective in global gene regulation in order to determine their effectiveness as oral vaccines in delivering heterologous antigens. Doggett et al. (58) evaluated a *cya crp asd* mutant of *S. typhimurium,* strain χ4072 for its ability to express antigenic determinants of SpaA, a virulence antigen of *Streptococcus sobrinus*. A single (pYA2901) and triple (pYA2905) repeat of an *spaA* gene fragment was inserted into an Asd⁺ vector and transformed into χ4072. Mice orally immunized with χ4072(pYA2905) had a demonstrable serum IgG and a slight salivary IgA anti-SpaA response. Neither a serum IgG nor a salivary IgA anti-SpaA response was detected in mice immunized with χ4072 (pYA2901). Interestingly, an intestinal IgA anti-SpaA response was detected in both groups of mice. Subsequent oral immunization of these mice resulted in an increased serum IgG but no anamnestic salivary IgA response. Studies from our laboratory also show that rats immunized with χ4072(pYA2905) elicit a salivary IgA and a serum IgG anti-SpaA response (178). We have also shown that rats eliciting a salivary IgA anti-SpaA response following oral immunization with χ4072(pYA2905) are protected against infection with *S. sobrinus* (unpublished findings).

The accumulating evidence from studies with recombinant *Salmonella* strains expressing a variety of heterologous antigens (Table 2) strongly supports the safety and effectiveness of this oral antigen delivery system for use in experimental animals and humans. However, a number of questions about this system remain unanswered. For example, it is still not fully known whether the amount, form, or location (e.g., cytoplasmic, periplasmic, or surface) of the heterologous virulence antigen(s) expressed by the *Salmonella* vector will influence the nature or magnitude of the immune response. In terms of the amount of cloned antigen, this will probably depend on the individual antigen and the type of response required. As discussed above, Doggett et al. (58) evaluated the responses of mice to different amounts of an antigenic fragment of SpaA expressed by a *cya crp asd* mutant of *S. typhimurium*. The single or triple repeat of the *spaA* gene fragment was in-

TABLE 2 Representative immunization studies with avirulent *Salmonella* spp. expressing heterologous antigen[a]

Salmonella mutant	Heterologous antigen	Source	Major finding			Reference(s)
			Ab	CMI	Protection	
S. typhimurium aroA	M protein (cytoplasmic)	*Streptococcus pyogenes*	+	ND	+	175
	gp63	*Leishmania major*	−	+	+	221
S. typhimurium aroA, aroA aroD, and *aroA aroC*	C fragment	*Clostridium tetani*	+	ND	+	30, 74, 202
S. typhimurium aroA and *galE*	K88	*Escherichia coli*	+	ND	+	62
S. typhimurium aroA S. enteridis aroA	LTB	*Escherichia coli*	+	ND	+	40, 61, 21
S. typhimurium aroA aroC aroD	p.69	*Bordetella pertussis*	−	ND	+	201
S. typhimurium aroA thyA asd	SpaA	*Streptococcus sobrinus*	+	+	ND	99
	GTF	*Streptococcus mutans*	+	ND	ND	99
S. typhimurium cya crp and *cya crp asd*	SpaA (cytoplasmic)	*Streptococcus sobrinus*	+	ND	ND	47, 58, 153, 178
	17-kDa protein	*Francisella tularensis*	+	+	+	191, 192
	LTB-HBsAg	Hepatitis B	−	+	ND	86
	HagB	*Porphyromonas gingivalis*	+	ND	ND	63
S. typhimurium cya crp S. choleraesuis cya crp cdt	31-kDa protein	*Brucella abortus*	+	−	ND	196–198
S. typhimurium Ty21a	O antigen	*Shigella flexneri*	+	ND	+	12
	O antigen	*Shigella sonnei*	+	ND	+	17

[a]Abbreviations: LTB, b subunit of heat-labile toxin; SpaA, surface protein antigen A; HBsAg, hepatitis B surface antigen; HagB, hemagglutinin; Ab, antibody responses (mucosal and/or systemic); CMI, cell-mediated immune response; ND not determined.

serted in the pYA292 vector and transformed into a *Salmonella* organism. Mice orally immunized with the mutant expressing the triple repeat of the SpaA fragment had higher serum IgG and salivary IgA antibody responses than mice immunized with the mutant expressing the single repeat of the fragment. These results suggest that the number of copies of an encoded determinant was important for the induction of a response. However, the actual amount of cloned protein produced by the two constructs was not determined.

Chatfield and coworkers (30) studied the effect of copy number, stability, and promoter on the immune response to *Salmonella*-produced fragment C of tetanus toxoid. In this study, the *tac* promoter from the plasmid encoding fragment C (pTETtac85) was replaced with the *nirB* promoter (pTETnir). Each plasmid was introduced into *aroA aroD S. typhimurium* BRD509 to produce BRD753 and BRD847, respectively. Fragment C was inducible in strain BRD847 under anaerobic conditions. Mice orally immunized once with this strain had a serum anti-fragment C response and were fully protected against tetanus toxin challenge. In BRD753, fragment C was constitutively produced, and mice immunized with this strain were only partially protected. On the basis of these results, it was suggested that the greater immunogenicity of BRD847 may be due to the more stable pTETnir plasmid (30).

The location and form of the heterologous antigen expressed by *Salmonella* spp. could affect the resulting response and probably depends on the protein. Several proteins, including ß-galactosidase (23), M protein (175), and SpaA (58, 178), are expressed cytoplasmically in *Salmonella* spp. These proteins induce systemic and secretory antibody responses as well as a cell-mediated immune response in the case of ß-galactosidase. Perhaps of importance for the induction of responses is that the cloned protein displays appropriate immunologic epitopes. In this regard, it is possible through immunologic and molecular biology techniques to define T- and B-cell epitopes of the antigens that are involved in the induction of mucosal responses and to incorporate the gene encoding these regions into the *Salmonella* vector. However, it will be important for the expressed protein to assume the proper configuration in order for the encoded epitopes to induce responses.

For proteins to be transported to the surface, correct folding as well as a secretion signal is important. If a natural secretion sequence for the protein cannot be used, then one could use the secretion signals from characterized proteins such as OmpA (157). Recently, OmpA has been used to express malarial antigens on the surfaces of *Salmonella* strains (188). However, stability in the intestinal tract and the immunogenicity of heterologous proteins on the surfaces of *Salmonella* strains require further investigation.

For optimal immunization, further information is also needed on the use of adjuvants in these delivery systems. Progress has already been made in the construction of vectors with a gene encoding a foreign antigen fused to the gene encoding CTB or *E. coli* enterotoxin (95, 154, 186). However, the ability of these constructs to enhance a specific immune response (adjuvant effect) has not been established. Further studies will be required to define the feasibility, safety, and effectiveness of these approaches for inducing the desired host response. It will also be important to determine whether *Salmonella* vaccine strains can induce tolerance and/or suppress the induction of a response based on host susceptibility to components of the vector, such as LPS, or to the expressed cloned antigen(s). The effect of the host immune response to the *Salmonella* vector and expressed cloned antigen on subsequent immunizations with the *Salmonella* strain and the same or different expressed cloned antigens or adjuvants will also require further investigation to define strategies for developing live-vaccine delivery systems for optimal induction of protective host responses to microbial pathogens.

CT AS AN ORAL IMMUNOADJUVANT

In recent years, several prospective oral immunoadjuvants have been identified (70), but none have consistently worked as well or shown as much promise as CT. The ability of CT to act as a mucosal vaccine adjuvant was discovered quite serendipitously. Unlike most proteins administered orally, CT was known to elicit very strong systemic and secretory antibody responses (174). In fact, CT and CTB are still two of the most potent mucosal immunogens identified. CT does not induce oral tolerance (71), a state of specific systemic unresponsiveness that develops upon repeated exposure of GALT to antigen (29, 149). CT also abrogates tolerance to the unrelated protein antigen keyhole limpet hemocyanin (KLH) when the two proteins are coadministered orally. Furthermore, an intestinal S-IgA response to KLH occurs only when KLH is coadministered orally with CT. These results indicate that CT could be used as an adjuvant for orally administered vaccines. Subsequent studies by a number of other investigators have confirmed the ability of CT to act as a mucosal adjuvant (54). Heat-labile enterotoxin of *E. coli* (LT), which shares 80% amino acid identity with CT, also possesses adjuvant activity (39). Although these two toxins are closely related, it is not known whether the sequence differences between CT and LT significantly affect any of their properties as adjuvants.

Properties and Biological Activities of CT

Like most bacterial toxins, CT is composed of two subunits. CTB is composed of five identical monomers, each 11.6 kDa in size, that mediate binding of the holotoxin to the surfaces of eukaryotic cells via high affinity for the monosialoganglioside GM_1 (44). This ganglioside is present on all nucleated cells, including intestinal epithelial cells, and is probably the mechanism by which CT is taken up from the intestinal lumen. CTA is cleaved posttranslationally into two peptides that are bound to each other by a disulfide bridge. The A1 peptide is the enzymatically active portion of the molecule and is responsible for ADP-ribosylation of the Gs protein that regulates the adenylate cyclase activity of the cell (195). The crystal structure of the closely related LT indicates that the A2 peptide mediates binding of A1 to CTB by forming a rigid finger that extends through the center and below the plane of the ring (190). This apparently allows CTA1 and CTB to function without interfering with one another.

Several factors may affect the ability of CT to act as a mucosal adjuvant. CT must be administered simultaneously with the antigen to the same mucosal surface (115). This finding suggests that CT may affect the regulatory balance within the local microenvironment of the mucosa-associated lymphoid tissue (MALT) so that it will respond positively to an antigen. The dose of CT reported for adjuvant activity has varied, suggesting that the dose may depend on the antigen being used. CT also elicits long-term immunologic memory not only to itself but also to other coadministered antigens (207), a desirable property for a vaccine adjuvant. The adjuvanticity of CT also varies with the genetic background of the host. By using congenic strains of mice, it was shown that high responders to CT are also high responders to KLH (72). Thus, the adjuvant and immunogenic properties of CT are probably closely related. These results also suggest that in an outbred population, CT may not work uniformly as an adjuvant. In fact, it may not work well as an adjuvant for all antigens. CT has been successfully used as an adjuvant for a variety of antigens ranging from proteins to polysaccharides to viruses; however, there are antigens for which CT does not augment S-IgA responses (216). Therefore, it may be necessary to demonstrate immunoadjuvant activity of CT for each antigen. CT is an effective adjuvant for inducing both secretory and systemic antibody responses to mucosally administered antigens. It may also be capable of stimulating cell-mediated immunity (123), but this property has not been well studied. Although CT is an effective mucosal adjuvant, there are some potential problems that could preclude its use in humans. Currently, it is not known whether CT is as effective an adjuvant in humans as it has been in mice. Furthermore, it is unlikely that CT will be administered to humans unless its toxicity can be separated from its adjuvanticity. Attempts to inactivate the A1 peptide by site-directed mutagenesis have failed (117). Another issue of great concern is the potential of CT to induce allergic reactions. In mice, the oral administration of antigen with CT followed by parenteral boosting with the antigen alone resulted in a rapid onset of histamine release, mast cell degranulation, elevated levels of serum IgE, and even death of the animals by anaphylaxis (194). This phenomenon occurred when CT but not CTB was used.

Antigen-CTB Conjugates

Because of some of the problems associated with using CT as an adjuvant, several investigators have been exploring the potential of CTB as a vaccine adjuvant and/or carrier. CTB is a potent mucosal immunogen, it is nontoxic, and it has an established record of safe use in humans as a component of an oral cholera vaccine (142). Unfortunately, the use of CTB as a mucosal adjuvant has met with mixed results. CTB has consistently shown more success as an adjuvant for intranasally administered vaccines than for orally administered ones (15, 54, 204, 220). This difference might be due to a difference in the environment of the nasopharyngeal region compared to the gastrointestinal tract that CTB and the antigen must traverse before reaching the local MALT.

Conjugation of antigen to CTB may be required to elicit an optimally enhanced antibody response. In an experiment using horseradish peroxidase (HRP), intraduodenal immunization with HRP mixed with CTB resulted in higher anti-HRP antibody responses in the serum and the intestinal secretion than were obtained when antigen was given alone (129). However, an even higher anti-HRP response was observed when HRP was chemically conjugated to CTB and used as the vaccine. Subsequent studies by several investigators (52, 98,

183) have confirmed the effectiveness of antigen-CTB conjugates in potentiating antibody responses to the antigen. However, other studies (110) with CTB-antigen conjugates did not result in enhanced responses. The poor response observed may be due to the method of conjugation. It has been reported that the immunogenicity of protein conjugates may vary depending on the coupling procedure used and the degree of cross-linking that occurs (208). None of the chemical conjugates that have been described in the literature have been examined for possible changes in the physical or biological properties of CTB. Such changes could affect the properties of CTB.

In an effort to circumvent these as well as other problems involved with making chemical conjugates, several investigators have genetically fused antigens to CTB (55, 184) or to the B subunit of LT (187). The immunogenicity of a streptococcal peptide given orally was significantly greater when it was fused to CTB then when it was fused to E. coli alkaline phosphatase (55). Although these results are promising, there are limitations in generating CTB-antigen fusion vaccines. Direct fusion of large peptides to CTB can significantly reduce the oral immunogenicity of CTB (56). The crystal structure of LT (190) indicates that it may now be possible to fuse fairly large antigens to CTB without affecting the structure and/or function of the subunit. In the genetic fusion of antigen to the N-terminal end of the A2 peptide, the natural rigidity of the peptide will act as a spacer arm, permitting the antigen and CTB to be conjugated to one another without affecting each other. The feasibility of this concept has already been demonstrated (96). However, the practical use of this system for mucosal immunization remains to be established.

Mechanisms of Action

The mechanisms by which CT and CTB are able to influence the outcome of the immune response to mucosally administered antigens have not been clearly defined. Several properties have been identified, including GM_1

binding (56), increased intestinal permeability (116), increased IL-1 production (113), and inhibition of suppressor cell function (73). The mechanism of action of CT has been discussed in detail in several recent reviews (54, 70). CT is consistently more potent as an adjuvant than CTB is and does not require conjugation to the antigen in order to be effective. This evidence indicates that CTA is an important contributor to this function. However, the mechanism involved has not been determined. While conjugation of antigens to CTB implies that CTB is simply delivering antigens to MALT, it must be stressed that CTB may be acting as more than just a protein carrier. CTB is a potent mucosal immunogen and has pharmacologic properties, including effects on B-cell isotype switching (112) and T-cell function (219). Since CTA is unable to exert any effect on lymphocyte function by itself, CTB must play an essential role in the adjuvant activity of CT. Both CT and CTB exert a multitude of effects on the immune system that are important in the response induced after antigen presentation to MALT.

Although CT induces protective immunity against mucosal pathogens (15, 204) and therefore shows great promise as an adjuvant for mucosally administered subunit vaccines, there is still much to be learned before it can be used in humans. Issues such as toxicity, when to use CT rather than CTB, what route of administration to use, and what type of antigens it can be used with will all have to be determined. We are just beginning to understand the properties essential for CT to act as a vaccine adjuvant. Once these properties have been identified, it may be possible to engineer mucosal vaccines to incorporate these characteristics.

ACKNOWLEDGMENTS

The studies from our laboratories that are summarized in this chapter were supported in part by U.S. Public Health Service grants DE 08182, DE 09081, DE 08228, DE 09846, DE 10607, and AI 33544 from the National Institutes of Health and by contract DAMD 117-90-C-0113 from the U.S. Army Medical Research Acquisition Activity.

We thank Philip B. Carter, Susan Jackson, and Janet Katz for their critical assessment of this review and Vickie Barron for her secretarial support.

REFERENCES

1. **Abraham, E.** 1992. Intranasal immunization with bacterial polysaccharide containing liposomes enhances antigen-specific pulmonary secretory antibody response. *Vaccine* **10**:461–468.
2. **Abraham, E., and S. Shah.** 1992. Intranasal immunization with liposomes containing IL-2 enhances bacterial polysaccharide antigen-specific pulmonary secretory antibody responses. *J. Immunol.* **149**:3719–3726.
3. **Abraham, S. N., and E. H. Beachey.** 1985. Host defenses against adhesion of bacteria to mucosal surfaces, p. 63–88. *In* J. F. Gallin and A. S. Fauci (ed.), *Advances in Host Defense Mechanisms*, vol. 4. Raven Press, Inc., New York.
4. **Allansmith, M. R., C. A. Burns, and R. R. Arnold.** 1982. Comparison of agglutinin titers to *Streptococcus mutans* in tears, saliva, and serum. *Infect. Immun.* **35**:202–205.
5. **Alonso, M. J., R. K. Gupta, C. Min, G. R. Siber, and R. Langer.** 1994. Biodegradable microspheres as controlled-release tetanus toxoid delivery systems. *Vaccine* **12**:299–304.
6. **Altman, A., and F. Dixon.** 1989. Immunomodifiers in vaccines. *Adv. Vet. Sci. Comp. Med.* **33**:301–343.
7. **Armstrong, S. J., and N. J. Dimmock.** 1992. Neutralization of influenza virus by low concentrations of hemagglutinin-specific polymeric immunoglobulin A inhibits viral fusion activity, but activation of the ribonucleoprotein is also inhibited. *J. Virol.* **66**:3822–3832.
8. **Arnold, R. R., J. Mestecky, and J. R. McGhee.** 1976. Naturally occurring secretory immunoglobulin A antibodies to *Streptococcus mutans* in human colostrum and saliva. *Infect. Immun.* **14:**355–362.
9. **Artursson, P., I.-L. Martensson, and I. Sjoholm.** 1986. Biodegradable microspheres. III. Some immunological properties of polyacryl starch microparticles. *J. Pharm. Sci.* **75:**697–701.
10. **Bangham, A. D., M. M. Standish, and J. C. Watkins.** 1965. Diffusion of univalent ions across the lamellae of swollen phospholipids. *J. Mol. Biol.* **13:**238–252.
11. **Barenholz, Y., D. Gibbes, B. J. Litman, J. Goll, T. E. Thompson, and R. D. Carlson.** 1977. A simple method for the preparation of homogeneous phospholipid vesicles. *Biochemistry* **16:**2806–2810.
12. **Baron, L. S., D. J. Kopecko, S. B. Formal, R. Seid, P. Guerry, and C. Powell.** 1987. Introduction of *Shigella flexneri* 2a type and group antigen genes into oral typhoid vaccine strain *Salmonella typhi* Ty21a. *Infect. Immun.* **55:**2797–2801.

13. **Bergmann, K.-C., R. H. Waldman, H. Tischner, and W.-D. Pohl.** 1986. Antibody in tears, saliva and nasal secretions following oral immunization of humans with inactivated influenza virus vaccine. *Int. Arch. Allergy Appl. Immunol.* **80:** 107–109.

14. **Besredka, A.** 1919. De la vaccination contre les etats typoides par voie buccale. *Ann. Inst. Pasteur* **33:**882–903.

15. **Bessen, D., and V. A. Fischetti.** 1990. Synthetic peptide vaccine against mucosal colonization by group A streptococci. *J. Immunol.* **145:** 1251–1256.

16. **Bienenstock, J., and A. D. Befus.** 1980. Mucosal immunology. *Immunology* **41:**249–270.

17. **Black, R. E., M. M. Levine, M. L. Clements, G. Losonsky, D. Herrington, S. Berman, and S. B. Formal.** 1987. Prevention of shigellosis by a *Salmonella typhi–Shigella sonnei* bivalent vaccine. *J. Infect. Dis.* **155:**1260–1265.

18. **Bland, P. W., and L. G. Warren.** 1986. Antigen presentation by epithelial cells of the rat small intestine. *Immunology* **58:**9–14.

19. **Brandtzaeg, P.** 1984. Immune functions of human nasal mucosa and tonsils in health and disease, p. 28–95. *In* J. Bienenstock (ed.), *Immunology of the Lung and Upper Respiratory Tract.* McGraw-Hill Book Co., New York.

20. **Brandtzaeg, P.** 1985. Cells producing immunoglobulins and other immune factors in human nasal mucosa. *Prot. Biol. Fluids* **32:**363–366.

21. **Brandtzaeg, P.** 1989. Overview of the mucosal immune system. *Curr. Top. Microbiol. Immunol.* **146:**13–25.

22. **Brandtzaeg, P., G. Karlsson, G. Hansson, B. Petruson, J. Bjorkander, and L. A. Hanson.** 1986. The clinical condition of IgA-deficient patients is related to the proportion of IgD- and IgM-producing cells in their nasal mucosa. *Clin. Exp. Immunol.* **67:**626–636.

23. **Brown, A., C. E. Hormaeche, R. DeMarco de Hormaeche, M. Winther, G. Dougan, D. J. Maskell, and B. A. D. Stocker.** 1987. An attenuated *aroA Salmonella typhimurium* vaccine elicits humoral and cellular immunity to cloned β-galactosidase in mice. *J. Infect. Dis.* **155:**86–92.

24. **Brown, T. A., and J. Mestecky.** 1985. Immunoglobulin A subclass distribution of naturally occurring salivary antibodies to microbial antigens. *Infect. Immun.* **49:**459–462.

25. **Brown, T. A., B. R. Murphy, J. Radl, J. J. Haaijman, and J. Mestecky.** 1985. Subclass distribution and molecular form of immunoglobulin A hemagglutinin antibodies in sera and nasal secretions after experimental secondary infection with influenza A virus in humans. *J. Clin. Microbiol.* **22:**259–264.

26. **Bruyere, T., D. Wachsmann, J. -P. Klein, M.** Scholler, and R. M. Frank. 1987. Local response in rat to liposome-associated *Streptococcus mutans* polysaccharide-protein conjugate. *Vaccine* **5:**39–42.

27. **Carter, P. B., and F. M. Collins.** 1974. The route of enteric infection in normal mice. *J. Exp. Med.* **139:**1189–1203.

28. **Chaicumpa, W., J. Pariaro, R. New, and E. Pongponratn.** 1990. Immunogenicity of liposome-associated oral cholera vaccine prepared from combined *Vibrio cholerae* antigens. *Asian Pac. J. Allergy Immunol.* **8:**87–94.

29. **Challacombe, S. J., and T. B. Tomasi.** 1980. Systemic tolerance and secretory immunity after oral immunization. *J. Exp. Med.* **152:**1459–1472.

30. **Chatfield, S. N., I. G. Charles, A. J. Makoff, M. D. Oxer, G. Dougan, D. Pickard, D. Slater, and N. F. Fairweather.** 1992. Use of the *nirB* promoter to direct the stable expression of heterologous antigens in *Salmonella* oral vaccine strains: development of a single-dose oral tetanus vaccine. *Bio/Technology* **10:**888–892.

31. **Childers, N. K., F. R. Denys, N. F. McGee, and S. M. Michalek.** 1990. Ultrastructural study of liposome uptake by M cells of rat Peyer's patch: an oral vaccine system for delivery of purified antigen. *Reg. Immunol.* **3:**8–16.

32. **Childers, N. K., and S. M. Michalek.** 1994. Liposomes, p. 241–254. *In* D. T. O'Hagan (ed.), *Novel Delivery Systems for Oral Vaccine Development.* CRC Press, Inc., Boca Raton, Fla.

33. **Childers, N. K., S. M. Michalek, F. R. Denys, and J. R. McGhee.** 1987. Characterization of liposomes for oral vaccines. *Adv. Exp. Med. Biol.* **216B:**1771–1780.

34. **Childers, N. K., S. M. Michalek, J. H. Eldridge, F. R. Denys, A. K. Berry, and J. R. McGhee.** 1989. Characterization of liposome suspensions by flow cytometry. *J. Immunol. Methods* **119:**135–143.

35. **Childers, N. K., S. M. Michalek, D. G. Pritchard, and J. R. McGhee.** 1991. Mucosal and systemic responses to an oral liposome-*Streptococcus mutans* carbohydrate vaccine in humans. *Reg. Immunol.* **3:**289–296.

36. **Childers, N. K., S. S. Zhang, E. Harokopakis, C. C. Harmon, and S. M. Michalek.** Properties of oral liposome-*Streptococcus mutans* glucosyltransferase (GTF) vaccine for practical use and for effective induction of protective immune responses. Submitted for publication.

37. **Childers, N. K., S. S. Zhang, and S. M. Michalek.** 1994. Oral immunization of humans with dehydrated liposomes containing *Streptococcus mutans* glucosyltransferase induces salivary immunoglobulin A2 antibody responses. *Oral Microbiol. Immunol.* **9:**146–153.

38. **Clarke, C., and C. Stokes.** 1992. The intestinal

and serum humoral immune response of mice to systemically and orally administered antigens in liposomes. I. The response to liposome-entrapped soluble proteins. *Vet. Immunol. Immunopathol.* **32:** 125–138.

39. **Clements, J. D., N. M. Hartzog, and F. L. Lyon.** 1988. Adjuvant activity of *Escherichia coli* heat-labile enterotoxin and effect on the induction of oral tolerance in mice to unrelated protein antigens. *Vaccine* **6:**269–273.

40. **Clements, J. D., F. L. Lyons, K. L. Lowe, A. L. Farrand, and S. El-Morshidy.** 1986. Oral immunization of mice with attenuated *Salmonella enteritidis* containing a recombinant plasmid which codes for production of the B subunit of heat-labile *Escherichia coli* enterotoxin. *Infect. Immun.* **53:** 685–692.

41. **Clements, M. L., and B. R. Murphy.** 1986. Development and persistence of local and systemic antibody responses in adults given live attenuated or inactivated influenza A virus vaccine. *J. Clin. Microbiol.* **23:**66–72.

42. **Cox, D. S., and M. A. Taubman.** 1984. Oral induction of the secretory antibody response by soluble and particulate antigens. *Int. Arch. Appl. Immunol.* **75:**126–131.

43. **Craig, S. W., and J. J. Cebra.** 1971. Peyer's patches: an enriched source of precursors for IgA-producing immunocytes in the rabbit. *J. Exp. Med.* **134:**188–200.

44. **Cuatrecasas, P.** 1973. Gangliosides and membrane receptors for cholera toxin. *Biochemistry* **12:** 3558–3566.

45. **Curtiss, R., III.** 1986. Genetic analysis of *Streptococcus mutans* virulence and prospects for an anticaries vaccine. *J. Dent. Res.* **65:**1034–1045.

46. **Curtiss, R., III.** 1990. Attenuated *Salmonella* strains as live vectors for the expression of foreign antigens, p. 161–188. *In* G. C. Woodrow and M. M. Levine (ed.), *New Generation Vaccines.* Marcel Dekker, Inc., New York.

47. **Curtiss, R., III, R. M. Goldschmidt, N. B. Fletchall, and S. M. Kelly.** 1988. Avirulent *Salmonella typhimurium Δcya Δcrp* oral vaccine strains expressing a streptococcal colonization and virulence antigen. *Vaccine* **6:**155–160.

48. **Curtiss, R., III, R. G. Holt, R. Barletta, J. P. Robeson, and S. Saito.** 1983. *Escherichia coli* strains producing *Streptococcus mutans* proteins responsible for colonization and virulence. *Ann. N.Y. Acad. Sci.* **409:**688–695.

49. **Curtiss, R., III, and S. M. Kelly.** 1987. *Salmonella typhimurium* deletion mutations lacking adenylate cyclase and cyclic AMP receptor protein are avirulent and immunogenic. *Infect. Immun.* **55:** 3035–3043.

50. **Curtiss, R., S. M. Kelly, P. A. Gulig, and K. Nakayama.** 1989. Selective delivery of antigens

by recombinant bacteria. *Curr. Top. Microbiol. Immunol.* **146:**35–49.

51. **Czerkinsky, C., S. J. Prince, S. M. Michalek, S. Jackson, M. W. Russell, Z. Moldoveanu, J. R. McGhee, and J. Mestecky.** 1987. IgA antibody producing cells in peripheral blood after antigen ingestion: evidence for a common mucosal immune system in humans. *Proc. Natl. Acad. Sci. USA* **84:**2449–2453.

52. **Czerkinsky, C., M. W. Russell, N. Lycke, M. Lindblad, and J. Holmgren.** 1989. Oral administration of a streptococcal antigen coupled to cholera toxin B subunit evokes strong antibody responses in salivary glands and extramucosal tissues. *Infect. Immun.* **57:**1072–1077.

53. **Czerkinsky, C., A.-M. Svennerholm, M. Quiding, R. Jonsson, and J. Holmgren.** 1991. Antibody-producing cells in peripheral blood and salivary glands after oral cholera vaccination of humans. *Infect. Immun.* **59:**996–1001.

54. **Dertzbaugh, M. T., and C. O. Elson.** 1991. Cholera toxin as a mucosal adjuvant, p. 119–131. *In* D. R. Spriggs and W. C. Koff (ed.), *Topics in Vaccine Adjuvant Research.* CRC Press, Inc., Boca Raton, Fla.

55. **Dertzbaugh, M. T., and C. O. Elson.** 1993. Comparative effectiveness of the cholera toxin B subunit and alkaline phosphatase as carriers for oral vaccines. *Infect. Immun.* **61:**48–55.

56. **Dertzbaugh, M. T., and C. O. Elson.** 1993. Reduction in the oral immunogenicity of cholera toxin B subunit by N-terminal peptide addition. *Infect. Immun.* **61:**384–390.

57. **Dimmock, N. J.** 1984. Mechanisms of neutralization of animal viruses. *J. Gen. Virol.* **65:**1015–1022.

58. **Doggett, T. A., E. K. Jagusztyn-Krynicka, and R. Curtiss III.** 1993. Immune responses to *Streptococcus sobrinus* surface protein antigen A expressed by recombinant *Salmonella typhimurium. Infect. Immun.* **61:**1859–1866.

59. **Dorman, C., S. Chatfield, C. Higgins, C. Hayward, and G. Dougan.** 1989. Characterization of porin and *ompR* mutants of a virulent strain of *Salmonella typhimurium: ompR* mutants are attenuated in vivo. *Infect. Immun.* **57:**2136–2140.

60. **Dougan, G., S. Chatfield, D. Pickard, J. Bester, D. O'Callaghan, and D. Maskell.** 1988. Construction and characterization of vaccine strains of *Salmonella* harbouring mutations in two different *aro* genes. *J. Infect. Dis.* **158:**1329–1335.

61. **Dougan, G., C. E. Hormaeche, and D. J. Maskell.** 1987. Live oral *Salmonella* vaccines: potential use of attenuated strains as carriers of heterologous antigens to the immune system. *Parasite Immunol.* **9:**151–160.

62. **Dougan, G., R. Sellwood, D. Maskell, K. Sweeney, F. Y. Liew, J. Beesley, and C. Hor-**

maeche. 1986. In vivo properties of a cloned K88 adherence antigen determinant. *Infect. Immun.* **52:**344–347.

63. **Dusek, D. M., A. Progulske-Fox, and T. A. Brown.** 1994. Systemic and mucosal immune responses in mice orally immunized with avirulent *Salmonella typhimurium* expressing a cloned *Porphyromonas gingivalis* hemagglutinin. *Infect. Immun.* **62:**1652–1657.

64. **Edelman, R., R. G. Russell, G. Losonsky, B. D. Tall, C. O. Tacket, M. M. Levine, and D. H. Lewis.** 1993. Immunization of rabbits with enterotoxigenic *E. coli* colonization factor antigen (CFA/I) encapsulated in biodegradable microspheres of poly(lactide-co-glycolide). *Vaccine* **1:**155–158.

65. **Eldridge, J. H., R. M. Gilley, J. K. Staas, Z. Moldoveanu, J. A. Meulbroek, and T. R. Tice.** 1989. Biodegradable microspheres: vaccine delivery system for oral immunization. *Curr. Top. Microbiol. Immunol.* **146:**59–66.

66. **Eldridge, J. H., C. J. Hammond, J. A. Meulbroek, J. K. Staas, R. M. Gilley, and T. R. Tice.** 1990. Controlled vaccine release in the gut-associated lymphoid tissues. I. Orally administered biodegradable microspheres target the Peyer's patches. *J. Controlled Release* **11:**205–214.

66a. **Eldridge, J. H., and S. M. Michalek.** Unpublished data.

67. **Eldridge, J. H., J. K. Staas, J. A. Meulbroek, J. R. McGhee, T. R. Tice, and R. M. Gilley.** 1990. Disseminated mucosal anti-toxin antibody responses induced through oral or intratracheal immunization with toxoid containing biodegradable microspheres, p. 375–378. *In* T. T. MacDonald, S. J. Challacombe, D. W. Bland, C. R. Stokes, R. V. Heatley, and A. M. Mowat (ed.), *Advances in Mucosal Immunology.* Kluwer Academic Publishers, London.

68. **Eldridge, J. H., J. K. Staas, J. A. Meulbroek, J. R. McGhee, T. R. Tice, and R. M. Gilley.** 1991. Biodegradable microspheres as a vaccine delivery system. *Mol. Immunol.* **28:**287–294.

69. **Eldridge, J. H., J. K. Staas, J. A. Meulbroek, J. R. McGhee, T. R. Tice, and R. M. Gilley.** 1991. Biodegradable and biocompatible poly(DL-lactide-co-glycolide) microspheres as an adjuvant for staphylococcal enterotoxin B toxoid which enhance the level of toxin-neutralizing antibodies. *Infect. Immun.* **59:**2978–2985.

70. **Elson, C. O., and M. T. Dertzbaugh.** 1994. Mucosal adjuvants, p. 391–402. *In* P. L. Ogra, J. Mestecky, M. E. Lamm, W. Strober, J. R. McGhee, and J. Bienenstock (ed.), *Handbook of Mucosal Immunology.* Academic Press, Inc., New York.

71. **Elson, C. O., and W. Ealding.** 1984. Cholera toxin feeding did not induce oral tolerance in mice and abrogated oral tolerance to an unrelated protein antigen. *J. Immunol.* **133:**2892–2897.

72. **Elson, C. O., and W. Ealding.** 1985. Genetic control of the murine immune response to cholera toxin. *J. Immunol.* **135:**930–932.

73. **Elson, C. O., and S. Solomon.** 1990. Activation of cholera toxin specific T cells in vitro. *Infect. Immun.* **58:**3711–3716.

74. **Fairweather, N. F., S. N. Chatfield, A. J. Makoff, R. A. Strugnell, J. Bester, D. J. Makell, and G. Dougan.** 1990. Oral vaccination of mice against tetanus by use of a live attenuated *Salmonella* carrier. *Infect. Immun.* **58:**1323–1326.

75. **Felgner, P. L., T. R. Gadek, M. Holm, R. Roman, H. W. Chan, M. Wenz, J. P. Northrop, G. M. Ringold, and M. Danielsen.** 1987. Lipofection: a highly efficient, lipid-mediated DNA-transfection procedure. *Proc. Natl. Acad. Sci. USA* **84:**7413–7417.

76. **Fields, P. I., R. V. Swanson, C. G. Haidaris, and F. Heffron.** 1986. Mutants of *Salmonella typhimurium* that cannot survive within the macrophage are avirulent. *Proc. Natl. Acad. Sci. USA* **83:**5189–5193.

77. **Formal, S. B., L. S. Baron, D. J. Kopecko, C. Powell, and C. A. Life.** 1981. Construction of a potential bivalent vaccine strain: introduction of *Shigella sonnei* form I antigen genes into the *galE Salmonella typhi* Ty21a typhoid vaccine strain. *Infect. Immun.* **34:**746–750.

78. **Galan, J., and R. Curtiss.** 1989. Virulence and vaccine potential of *phoP* mutants of *Salmonella typhimurium. Microb. Pathog.* **6:**433–443.

79. **Genco, R. J., and M. A. Taubman.** 1969. Secretory gA antibodies induced by local immunization. *Nature* (London) **221:**679–681.

80. **Germanier, R.** 1972. Immunity in experimental salmonellosis. III. Comparative immunization with viable and heat-inactivated cells of *Salmonella typhimurium. Infect. Immun.* **5:**792–797.

81. **Germanier, R., and E. Furer.** 1971. Immunity in experimental salmonellosis. II. Basis for the avirulence and protective capacity of *galE* mutants of *Salmonella typhimurium. Infect. Immun.* **4:**663–673.

82. **Germanier, R., and E. Furer.** 1975. Isolation and characterization of *galE* mutant Ty 21a of *Salmonella typhi:* a candidate strain for a live, oral typhoid vaccine. *J. Infect. Dis.* **131:**553–558.

83. **Gregoriadis, G.** 1992. Liposomes as immunological adjuvants: approaches to immunopotentiation including ligand-mediated targeting to macrophages. *Res. Immunol.* **143:**178–246.

84. **Gregory, R. L., S. M. Michalek, G. Richardson, C. Harmon, T. Hilton, and J. R. McGhee.** 1986. Characterization of immune response to oral administration of *Streptococcus sobrinus* ribosomal preparations in liposomes. *Infect. Immun.* **54:**780–786.

85. Guzman, C., G. Molinari, M. Fountain, M. Rohde, K. Timmis, and M. Walker. 1993. Antibody responses in the serum and respiratory tract of mice following oral vaccination with liposomes coated with filamentous hemagglutinin and pertussis toxoid. *Infect. Immun.* **61**:573–579.

86. Hackett, J. 1990. *Salmonella*-based vaccines. *Vaccine* **8**:5–11.

87. Hajishengallis, G., E. Nikolova, and M. W. Russell. 1992. Inhibition of *Streptococcus mutans* adherence to saliva-coated hydroxyapatite by human secretory immunoglobulin A antibodies to the cell surface protein antigen I/II: reversal of IgA1 protease cleavage. *Infect. Immun.* **60**:5057–5064.

88. Hameleers, D. M. H., A. E. Stoop, I. vander Ven, J. Bewenga, S. vander Baan, and T. Sminia. 1989. Intra-epithelial lymphocytes and non-lymphoid cells in the human nasal mucosa. *Int. Arch. Allergy Appl. Immunol.* **88**:317–322.

89. Hiraga, C., F. Ishii, and Y. Ichikawa. 1989. Oral immunization against tetanus, using liposome-entrapped tetanus toxoid. *Kansenshogaku Zasshi* **63**:1308–1312.

90. Hoiseth, S. K., and B. A. D. Stocker. 1981. Aromatic-dependent *Salmonella typhimurium* are non-virulent and effective as live vaccines. *Nature* (London) **291**:238–239.

91. Hone, D., S. Attridge, B. Forrest, R. Morona, D. Daniels, J. LaBrooy, R. Bartholomeusz, D. Shearman, and J. Hackett. 1988. A *galE via* (vi antigen-negative) mutant of *Salmonella typhi* Ty2 retains virulence in humans. *Infect. Immun.* **56**:1326–1333.

92. Hone, D., R. Marona, S. Attridge, and J. Hackett. 1987. Construction of defined *galE* mutants of *Salmonella* for use as vaccines. *J. Infect. Dis.* **156**:167–174.

93. Howie, A. J. 1980. Scanning and transmission electron microscopy on the epithelium of human palatine tonsils. *J. Pathol.* **130**:91–98.

94. Huang, C. 1969. Studies of phosphatidylcholine vesicles. Formation and physical characteristics. *Biochemistry* **8**:344–352.

95. Jagusztyn-Krynicka, E. K., J. E. Clark-Curtiss, and R. Curtiss. 1993. *Escherichia coli* heat-labile toxin subunit B fusions with *Streptococcus sobrinus* antigens expressed by *Salmonella typhimurium* oral vaccine strains: importance of the linker for antigenicity and biological activities of the hybrid proteins. *Infect. Immun.* **61**:1004–1015.

96. Jobling, M. G., and R. K. Holmes. 1992. Fusion proteins containing the A2 domain of cholera toxin assemble with B polypeptides of cholera toxin to form immunoreactive and functional holotoxin-like chimeras. *Infect. Immun.* **60**: 4915–4924.

97. Kantele, A., M. Aurilommi, and I. Jokinen. 1986. Specific immunoglobulin-secreting human blood cells after peroral vaccination against *Salmonella typhi*. *J. Infect. Dis.* **153**:1126–1131.

98. Katz, J., C. C. Harmon, G. P. Buckner, G. J. Richardson, M. W. Russell, and S. M. Michalek. 1993. Protective salivary immunoglobulin responses against *Streptococcus mutans* infection after intranasal immunization with *S. mutans* antigen I/II coupled to the B subunit of cholera toxin. *Infect. Immun.* **61**:1964–1971.

99. Katz, J., S. M. Michalek, R. Curtiss III, C. Harmon, G. Richardson, and J. Mestecky. 1987. Novel oral vaccines: the effectiveness of cloned gene products on inducing secretory immune responses. *Adv. Exp. Med. Biol.* **216B**: 1741–1747.

100. Kilian, M., J. Mestecky, and M. W. Russell. 1988. Defense mechanisms involving Fc-dependent functions of immunoglobulin A and their subversion by bacterial immunoglobulin A proteases. *Microbiol. Rev.* **52**:296–303.

101. Kilian, M., and J. Reinholdt. 1986. Interference with IgA defence mechanisms by extracellular bacterial enzymes, p. 173–208. *In* C. S. F. Easmon and J. Jeljaszewics (ed.), *Medical Microbiology*. Academic Press, Inc., London.

102. Kilian, M., and M. W. Russell. 1994. Function of mucosal immunoglobulin, p. 127–137. *In* P. L. Ogra, J. Mestecky, M. E. Lamm, W. Strober, J. McGhee, and J. Bienenstock (ed.), *Handbook of Mucosal Immunology*. Academic Press, Inc., San Diego, Calif.

103. Kohn, J., S. M. Niemi, E. C. Albert, J. C. Murphy, R. Langer, and J. G. Fox. 1986. Single-step immunization using a controlled release biodegradable polymer with sustained adjuvant activity. *J. Immunol. Methods* **95**:31–38.

104. Kreuter, J., U. Berg, E. Liehl, M. Soliva, and P. P. Speiser. 1986. Influence of particle size on the adjuvant effect of particulate polymeric adjuvants. *Vaccine* **4**:125–129.

105. Kreuter, J., E. Liehl, U. Berg, M. Soliva, and P. P. Speiser. 1988. Influence of hydrophobicity on the adjuvant effect of particulate polymeric adjuvants. *Vaccine* **6**:253–256.

106. Kreuter, J., R. Mauler, H. Gruschkau, and P. P. Speiser. 1976. The use of new polymethylmethacrylate adjuvants for split influenza vaccines. *Exp. Cell Biol.* **44**:12–19.

107. Kreuter, J., and P. P. Speiser. 1976. New adjuvants on a polymethylmethacrylate base. *Infect. Immun.* **13**:204–210.

108. Kuper, C. F., P. J. Koornstra, D. M. H. Hameleers, J. Biewenga, B. J. Spit, A. M. Duijvestijn, P. J. C. van Breda Vriesman, and T. Sminia. 1992. The role of nasopharyngeal lymphoid tissue. *Immunol. Today* **13**:219–224.

109. **Levine, M. M., D. Herrington, J. Murphy, J. Morris, G. Losonsky, B. Tall, A. Lindberg, S. Svenson, S. Baqar, M. Edwards, and B. Stocker.** 1987. Safety, infectivity, immunogenicity, and *in vivo* stability of two attenuated auxotrophic mutant strains of *Salmonella typhi*, 541Ty and 543Ty, as live oral vaccines in man. *J. Clin. Invest.* **79:**888–902.

110. **Liang, X. P., M. E. Lamm, and J. G. Nedrud.** 1988. Oral administration of cholera toxin-Sendai virus conjugate potentiates gut and respiratory immunity against Sendai virus. *J. Immunol.* **141:**1495–1501.

111. **Litwin, S. D., and J. M. Singer.** 1965. The adjuvant action of latex particulate carriers. *J. Immunol.* **95:**1147–1152.

112. **Lycke, N.** 1993. Cholera toxin promotes B cell isotype switching by two different mechanisms. *J. Immunol.* **11:**4810–4821.

113. **Lycke, N., A. K. Bromander, L. Ekman, U. Karlsson, and J. Holmgren.** 1989. Cellular basis of immunomodulation by cholera toxin *in vitro* with possible association to the adjuvant function *in vivo*. *J. Immunol.* **142:**20–27.

114. **Lycke, N., L. Eriksen, and J. Holmgren.** 1987. Protection against cholera toxin after oral immunization is thymus-dependent and associated with intestinal production of neutralizing IgA antitoxin. *Scand. J. Immunol.* **25:**413–419.

115. **Lycke, N., and J. Holmgren.** 1986. Strong adjuvant properties of cholera toxin on gut mucosal immune responses to orally presented antigens. *Immunology* **59:**301–308.

116. **Lycke, N., U. Karlsson, A. Sjolander, and K. E. Magnusson.** 1991. The adjuvant action of cholera toxin is associated with an increased intestinal permeability for luminal antigens. *Scand. J. Immunol.* **33:**691–698.

117. **Lycke, N., T. Tsuji, and J. Holmgren.** 1992. The adjuvant effect of *Vibrio cholerae* and *Escherichia coli* heat-labile enterotoxins is linked to their ADP-ribosyltransferase activity. *Eur. J. Immunol.* **22:**2277–2281.

118. **Majumdar, A. S., and A. C. Ghose.** 1981. Evaluation of the biological properties of different classes of human antibodies in relation to cholera. *Infect. Immun.* **32:**9–14.

119. **Martin, M. E. D., J. B. Dewar, and J. F. E. Newman.** 1988. Polymerized serum albumin beads possessing slow release properties for use in vaccines. *Vaccine* **6:**33–38.

120. **Marx, P. A., R. W. Compans, A. Gettie, J. K. Staas, R. M. Gilley, M. J. Mulligan, G. V. Yamshchikov, D. Chen, and J. H. Eldridge.** 1993. Protection against vaginal SIV transmission with microencapsulated vaccine. *Science* **260:**1323–1327.

121. **Maskell, D. J., K. J. Sweeney, D. O'Cal-**laghan, C. E. Hormaeche, F. Y. Liew, and G. Dougan. 1987. *Salmonella typhimurium aroA* mutants as carriers of the *Escherichia coli* heat-labile enterotoxin B subunit to the murine secretory and systemic immune system. *Microb. Pathog.* **2:**211–221.

122. **Mayhew, E., R. Lazo, W. J. Vail, J. King, and A. M. Green.** 1984. Characterization of liposomes prepared using a microemulsifier. *Biochim. Biophys. Acta* **775:**169–174.

123. **Mbawuike, I. N., and P. R. Wyde.** 1993. Induction of CD8$^+$ cytotoxic T cells by immunization with killed influenza virus and effect of cholera toxin B subunit. *Vaccine* **11:**1205–1209.

124. **McFarland, W. C., and B. A. D. Stocker.** 1987. Effect of different purine auxotrophic mutations on mouse virulence of a Vi-positive strain of *Salmonella dublin* and of two strains of *Salmonella typhimurium*. *Microb. Pathog.* **3:**129–141.

125. **McGhee, J. R., and J. Mestecky.** 1990. In defense of mucosal surfaces. Development of novel vaccines for IgA responses protective at the portal of entry of microbial pathogens. *Infect. Dis. Clin. N. Am.* **4:**315–341.

126. **McGhee, J. R., J. Mestecky, M. T. Dertzbaugh, J. H. Eldridge, M. Hirasawa, and H. Kiyono.** 1992. The mucosal immune system: from fundamental concepts to vaccine development. *Vaccine* **10:**75–88.

127. **McGhee, J. R., J. Mestecky, C. O. Elson, and H. Kiyono.** 1989. Regulation of IgA synthesis and immune response by T cells and interleukins. *J. Clin. Immunol.* **9:**175–199.

128. **McGhee, J. R., S. M. Michalek, J. Webb, J. M. Navia, A. F. R. Rahman, and D. W. Legler.** 1975. Effective immunity to dental caries: protection of gnotobiotic rats by local immunization with *Streptococcus mutans*. *J. Immunol.* **114:**300–305.

129. **McKenzie, S. J., and J. F. Halsey.** 1984. Cholera toxin B subunit as a carrier protein to stimulate a mucosal immune response. *J. Immunol.* **133:**1818–1824.

130. **McQueen, C. E., E. C. Boedeker, R. Reid, D. Jarboe, M. Wolf, M. Le, and W. R. Brown.** 1993. Pili in microspheres protect rabbits from diarrhea induced by *E. coli* strain RDEC-1. *Vaccine* **11:**201–206.

131. **Mestecky, J.** 1987. The common mucosal immune system and current strategies for induction of immune responses in external secretions. *J. Clin. Immunol.* **7:**265–276.

132. **Mestecky, J., R. Abraham, and P. L. Ogra.** 1994. Common mucosal immune system and strategies for the development of vaccines effective at the mucosal surface, p. 357–372. *In* P. L. Ogra, J. Mestecky, M. E. Lamm, W. Strober, J. R. McGhee, and J. Bienenstock (ed.), *Handbook*

of Mucosal Immunology. Academic Press, Inc., New York.

133. **Mestecky, J., and J. H. Eldridge.** 1991. Targeting and controlled release of antigens for the effective induction of secretory antibody responses. *Curr. Opin. Immunol.* **3:**492–495.

134. **Mestecky, J., and J. R. McGhee.** 1987. Immunoglobulin A (IgA): molecular and cellular interactions involved in IgA biosynthesis and immune response. *Adv. Immunol.* **40:**153–245.

135. **Mestecky, J., and J. R. McGhee.** 1989. Oral immunization: past and present. *Curr. Top. Microbiol. Immunol.* **146:**3–11.

136. **Mestecky, J., J. R. McGhee, R. R. Arnold, S. M. Michalek, S. J. Prince, and J. L. Babb.** 1978. Selective induction of an immune response in human external secretions by ingestion of bacterial antigen. *J. Clin. Invest.* **61:**731–737.

137. **Michalek, S. M., and N. K. Childers.** 1990. Development and outlook for a caries vaccine. *Crit. Rev. Oral Biol. Med.* **1:**37–54.

138. **Michalek, S. M., N. K. Childers, J. Katz, M. Dertzbaugh, S. Zhang, M. W. Russell, F. L. Macrina, S. Jackson, and J. Mestecky.** 1992. Liposomes and conjugate vaccines for antigen delivery and induction of mucosal immune responses. *Adv. Exp. Biol. Med.* **327:**191–198.

139. **Michalek, S. M., J. H. Eldridge, R. Curtiss III, and K. L. Rosenthal.** 1994. Antigen delivery systems: new approach to mucosal immunization, p. 373–386. *In* P. L. Ogra, J. Mestecky, M. E. Lamm, W. Strober, J. McGhee, and J. Bienenstock (ed.), *Handbook of Mucosal Immunology.* Academic Press, Inc., San Diego, Calif.

140. **Michalek, S. M., I. Morisaki, R. L. Gregory, S. Kimura, C. C. Harmon, S. Hamada, S. Kotani, and J. R. McGhee.** 1984. Oral adjuvants enhance salivary IgA responses to purified *Streptococcus mutans* antigens. *Prot. Biol. Fluids* **32:**47–52.

141. **Michalek, S. M., I. Morisaki, R. L. Gregory, H. Kiyono, S. Hamada, and J. R. McGhee.** 1983. Oral adjuvants enhance IgA responses to *Streptococcus mutans. Mol. Immunol.* **20:**1009–1018.

142. **Migasena, S., V. Desakorn, P. Suntharasamai, P. Pitisuttitham, B. Prayurahong, W. Supanaranond, and R. E. Black.** 1989. Immunogenicity and two formulations of oral cholera vaccine comprised of killed whole vibrios and the B subunit of cholera toxin. *Infect. Immun.* **57:**117–120.

143. **Miller, R. A., J. M. Brady, and D. E. Cutwright.** 1977. Degradation rates of resorbable implants (polylactates and polyglycolates): rate modification with changes in PLA/PGA copolymer ratios. *J. Biomed. Mater. Res.* **11:**711–719.

144. **Miller, S., and J. Mekalanos.** 1990. Consti-tutive expression of the PhoP regulon attenuates *Salmonella* virulence and survival within macrophages. *J. Bacteriol.* **172:**2485–2490.

145. **Miller, S. I., J. J. Mekalanos, and W. S. Pulkkinen.** 1990. *Salmonella* vaccines with mutations in the *phoP* virulence regulon. *Res. Microbiol.* **141:**817–821.

146. **Moldoveanu, Z., M. Novak, W.-Q. Huang, R. M. Gilley, J. K. Staas, D. Schafer, R. W. Compans, and J. Mestecky.** 1993. Oral immunization with influenza virus in biodegradable microspheres. *J. Infect. Dis.* **167:**84–90.

147. **Moldoveanu, Z., J. K. Staas, R. M. Gilley, R. Ray, R. W. Compans, J. H. Eldridge, T. R. Tice, and J. Mestecky.** 1989. Immune responses to influenza virus in orally and systemically immunized mice. *Curr. Top. Microbiol. Immunol.* **146:**91–99.

148. **Montgomery, P. C., J. Cohen, and E. T. Lally.** 1974. The induction and characterization of secretory IgA antibodies. *Adv. Exp. Med. Biol.* **45:**453–462.

149. **Mowat, A. M.** 1994. Oral tolerance and regulation of immunity to dietary antigens, p. 185–201. *In* P. L. Ogra, J. Mestecky, M. E. Lamm, W. Strober, J. McGhee, and J. Bienenstock (ed.), *Handbook of Mucosal Immunology.* Academic Press, Inc., San Diego, Calif.

150. **Nail, S. L., J. L. White, and S. L. Hem.** 1976. Structure of aluminum hydroxide gel. I. Initial precipitate. *J. Pharm. Sci.* **65:**1188–1191.

151. **Nail, S. L., J. L. White, and S. L. Hem.** 1976. Structure of aluminum hydroxide gel. II. Aging mechanism. *J. Pharm. Sci.* **65:**1192–1195.

152. **Nair, P. N. R., and H. E. Schroeder.** 1983. Local immune response to repeated topical antigen application in the simian labial mucosa. *Infect. Immun.* **41:**399–409.

153. **Nakayama, K., S. M. Kelly, and R. Curtiss.** 1988. Construction of an *asd* expression-cloning vector: stable maintenance and high level expression of cloned genes in a *Salmonella* vaccine strain. *Bio/Technology* **6:**693–697.

154. **Newton, S. M. C., C. O. Jacob, and B. A. D. Stocker.** 1989. Immune response to cholera toxin epitope inserted in *Salmonella* flagellin. *Science* **244:**70–72.

155. **Nnalue, N. A., and B. A. D. Stocker.** 1986. Some *galE* mutants of *Salmonella choleraesuis* retain virulence. *Infect. Immun.* **54:**635–640.

156. **Nnalue, N. A., and B. A. D. Stocker.** 1987. Test of the virulence and live-vaccine efficacy of auxotrophic and *galE* derivatives of *Salmonella choleraesuis. Infect. Immun.* **55:**955–962.

157. **O'Callaghan, D., A. Charbit, P. Martineau, C. Leclerc, S. van der Werf, C. Nauciel, and M. Hofnung.** 1990. Immunogenicity of foreign

peptide epitopes expressed in bacterial envelope proteins. *Res. Microbiol.* **141:**963–969.

158. **O'Callaghan, D., D. Maskell, F. Y. Liew, C. S. F. Easmon, and G. Dougan.** 1988. Characterization of aromatic- and purine-dependent *Salmonella typhimurium:* attenuation, persistence, and ability to induce protective immunity in BALB/c mice. *Infect. Immun.* **56:**419–423.

159. **O'Connor, C. J., R. G. Wallace, K. Iwamoto, T. Taguchi, and J. Sunamoto.** 1985. Bile salts damage of egg phosphatidylcholine liposomes. *Biochim. Biophys. Acta* **817:**95–102.

160. **Ogawa, T., H. Shimairchi, and S. Hamada.** 1989. Mucosal and systemic immune responses in BALB/c mice to *Bacteroides gingivalis* fimbriae administered orally. *Infect. Immun.* **57:**3466–3471.

161. **Ogra, P. L.** 1971. Effect of tonsillectomy and adenoidectomy on nasopharyngeal antibody response to poliovirus. *N. Engl. J. Med.* **284:**59–64.

162. **Ogra, P. L., and D. T. Karzon.** 1969. Distribution of poliovirus antibody in serum nasopharynx and alimentary tract following segmental immunization of lower alimentary tract with poliovaccine. *J. Immunol.* **102:**1423–1430.

163. **Ogra, P. L., E. E. Leibovitz, and G. Zhao-Ri.** 1989. Oral immunization and secretory immunity to viruses. *Curr. Top. Microbiol. Immunol.* **146:**73–81.

164. **O'Hagan, D. T. (ed.).** 1994. *Novel Delivery Systems for Oral Vaccines,* p. 1–268. CRC Press, Inc., Boca Raton, Fla.

165. **O'Hagan, D. T., H. Jeffery, M. J. J. Roberts, J. P. McGee, and S. S. Davis.** 1991. Controlled release microparticles for vaccine development. *Vaccine* **9:**768–771.

166. **O'Hagan, D. T., J. P. McGee, J. Holmgren, A. M. Mowat, A. M. Donachie, K. H. G. Mills, W. Gaisford, D. Rahman, and S. J. Challocombe.** 1993. Biodegradable microparticles for oral immunization. *Vaccine* **11:**149–154.

167. **O'Hagan, D. T., K. J. Palin, and S. S. Davis.** 1989. Poly(butyl-2-cyanoacrylate) particles as adjuvants for oral immunization. *Vaccine* **7:**213–216.

168. **O'Hagan, D. T., K. Palin, S. S. Davis, P. Artursson, and I. Sjoholm.** 1989. Microparticles as potentially orally active immunological adjuvants. *Vaccine* **7:**421–424.

169. **O'Hagan, D. T., D. Rahman, J. P. McGee, H. Jeffery, M. C. Davies, P. Williams, and S. S. Davis.** 1991. Biodegradable microparticles as controlled release antigen delivery systems. *Immunology* **73:**239–242.

170. **Olson, F., C. A. Hunt, F. C. Szoka, Jr., W. J. Vail, and D. Papahadjopoulos.** 1979. Preparation of liposomes of defined size distribution by extrusion through polycarbonate membranes. *Biochim. Biophys. Acta* **557:**9–23.

171. **Owen, R. L.** 1977. Sequential uptake of horseradish peroxidase by lymphoid follicle epithelium of Peyer's patches in the normal unobstructed mouse intestine: an ultrastructural study. *Gastroenterology* **72:**440–451.

172. **Owen, R. L., and A. L. Jones.** 1974. Epithelial cell specialization within human Peyer's patches: an ultrastructural study of intestinal lymphoid follicles. *Gastroenterology* **66:**189–203.

173. **Pastan, I., and S. Adhya.** 1976. Cyclic adenosine 5′-monophosphate in *Escherichia coli. Bacteriol. Rev.* **40:**527–551.

174. **Pierce, N. F., and J. L. Gowans.** 1975. Cellular kinetics of the intestinal immune response to cholera toxoid in rats. *J. Exp. Med.* **142:**1550–1563.

175. **Poirier, T. P., M. A. Kehoe, and E. H. Beachey.** 1988. Protective immunity evoked by oral administration of attenuated *aroA Salmonella typhimurium* expressing cloned streptococcal M protein. *J. Exp. Med.* **168:**25–32.

176. **Preis, I., and R. Langer.** 1979. A single-step immunization by sustained antigen release. *J. Immunol.* **28:**193–197.

177. **Redding, T. W., A. V. Schally, T. R. Tice, and W. E. Meyers.** 1984. Long acting delivery systems for peptides: inhibition of rat prostate tumors by controlled-release of D-Trp⁶-LH-RH from injectable microcapsules. *Proc. Natl. Acad. Sci. USA* **81:**5845–5851.

178. **Redman, T. K., C. C. Harmon, and S. M. Michalek.** 1994. Oral immunization with recombinant *Salmonella typhimurium* expressing surface protein antigen A of *Streptococcus sobrinus:* persistence and induction of humoral responses in rats. *Infect. Immun.* **62:**3162–3171.

179. **Reid, R. H., E. C. Boedeker, C. E. McQueen, D. Davis, L.-Y. Tseng, J. Kodak, K. Sau, C. L. Wilhelmsen, R. Nellore, P. Dalal, and H. R. Bhagat.** 1993. Preclinical evaluation of microencapsulated CFA/II oral vaccine against enterotoxigenic *E. coli. Vaccine* **11:**159–167.

180. **Rhalem, A., C. Bourdieu, G. Luffau, and P. Pery.** 1988. Vaccination of mice with liposome-entrapped adult antigens of *Nippostrongylus brasiliensis. Ann. Inst. Pasteur/Immunol.* **139:**157–166.

181. **Roberts, M., S. N. Chatfield, and G. Dougan.** 1994. *Salmonella* as carriers of heterologous antigens, p. 27–58. *In* D. T. O'Hagan (ed.), *Novel Delivery Systems for Oral Vaccines.* CRC Press, Inc., Boca Raton, Fla.

182. **Rowland, R. N., and J. F. Woodley.** 1980. The stability of liposomes *in vitro* to pH, bile salts

and pancreatic lipase. *Biochim. Biophys. Acta* **620:** 400–409.

183. **Russell, M. W., and H.-Y. Wu.** 1991. Distribution, persistence, and recall of serum and salivary antibody responses to peroral immunization with protein antigen I/II of *Streptococcus mutans* coupled to the cholera toxin B subunit. *Infect. Immun.* **59:**4061–4070.

184. **Sanchez, J., A. M. Svennerholm, and J. Holmgren.** 1988. Genetic fusion of a nontoxic heat-stable enterotoxin-related decapeptide-antigen to cholera toxin B subunit. *FEBS Lett.* **241:** 110–114.

185. **Schleifer, K. H., and O. Kandler.** 1972. Peptidoglycan types of bacterial cell walls and their taxonomic implications. *Bacteriol. Rev.* **36:**407–477.

186. **Schödel, F., G. Enders, M. C. Jung, and H. Will.** 1990. Recognition of a hepatitis B virus nucleocapsid T-cell epitope expressed as a fusion protein with the subunit B of *Escherichia coli* heat labile enterotoxin in attenuated salmonellae. *Vaccine* **8:**569–572.

187. **Schödel, F., and H. Will.** 1989. Construction of a plasmid for expression of foreign epitopes as fusion proteins with subunit B of *Escherichia coli* heat-labile enterotoxin. *Infect. Immun.* **57:**1347–1350.

188. **Schorr, J., B. Knapp, E. Hundt, H. A. Kupper, and E. Amann.** 1991. Surface expression of malarial antigens in *Salmonella typhimurium:* induction of serum antibody response upon oral vaccination of mice. *Vaccine* **9:**675–681.

189. **Sertl, K., T. Takenmura, E. Tschachler, V. J. Ferrans, M. A. Kaliner, and E. M. Shevach.** 1986. Dendritic cells with antigen-presenting capability reside in airway epithelium, lung parenchyma and visceral pleura. *J. Exp. Med.* **163:**436–451.

190. **Sixma, T. K., S. E. Pronk, K. H. Kalk, E. S. Wartna, B. A. M. van Zanten, B. Witholt, and W. G. J. Hol.** 1991. Crystal structure of a cholera toxin-related heat-labile enterotoxin from *E. coli. Nature* (London) **351:**371–377.

191. **Sjostedt, A., G. Sandstrom, and A. Tarnvik.** 1990. Immunization of mice with an attenuated *Salmonella typhimurium* strain expressing a membrane protein of *Francisella tularensis.* A model for identification of bacterial determinants relevant to the host defense against tularemia. *Res. Microbiol.* **141:**887–891.

192. **Sjostedt, A., G. Sandstrom, and A. Tarnvik.** 1992. Humoral and cell-mediated immunity in mice to a 17-kilodalton lipoprotein of *Francisella tularensis* expressed by *Salmonella typhimurium. Infect. Immun.* **60:**2855–2862.

193. **Smith, D. J., M. A. Taubman, and J. L. Ebersole.** 1985. Salivary IgA antibody to glu-

cosyltransferase in man. *Clin. Exp. Immunol.* **61:** 416–424.

194. **Snider, D. P., J. S. Marshall, M. H. Perdue, and H. Liang.** 1994. Cholera toxin (CTX) promotes IgE antibody and allergy during oral immunization. *FASEB J.* **8:**A282.

195. **Spangler, B. D.** 1992. Structure and function of cholera toxin and the related *Escherichia coli* heat-labile enterotoxin. *Microbiol. Rev.* **56:**622–647.

196. **Stabel, T. J., J. E. Mayfield, D. C. Morfitt, and M. J. Wannemuehler.** 1993. Oral immunization of mice and swine with an attenuated *Salmonella choleraesuis* [*cya-12(crp-cdt)19*] mutant containing a recombinant plasmid. *Infect. Immun.* **61:**610–618.

197. **Stabel, T. J., J. E. Mayfield, L. B. Tabatabai, and M. J. Wannemuehler.** 1990. Oral immunization of mice with attenuated *Salmonella typhimurium* containing a recombinant plasmid which codes for production of a 31 kilodalton protein of *Brucella abortus. Infect. Immun.* **58:** 2048–2055.

198. **Stabel, T. J., J. E. Mayfield, L. B. Tabatabai, and M. J. Wannemuehler.** 1991. Swine immunity to an attenuated *Salmonella typhimurium* mutant containing a recombinant plasmid which codes for production of a 31 kilodalton protein of *Brucella abortus. Infect. Immun.* **59:**2941–2947.

199. **Stahn, R., H. Schafer, M. Kunze, J. Malur, A. Ladhoff, and U. Lachmann.** 1992. Quantitative reconstitution of isolated influenza haemagglutinin into liposomes by the detergent method and the immunogenicity of haemagglutinin liposomes. *Acta Virol.* **36:**129–144.

200. **Strannegard, O., and A. Yurchison.** 1969. Formation of agglutinating and reaginic antibodies in rabbits following oral administration of soluble and particulate antigens. *Int. Arch. Allergy Appl. Immunol.* **35:**579–590.

201. **Strugnell, R., G. Dougan, S. Chatfield, I. Charles, N. Fairweather, J. Tite, J. L. Li, J. Beesley, and M. Roberts.** 1992. Characterization of a *Salmonella typhimurium aro* vaccine strain expressing the P.69 antigen of *Bordetella pertussis. Infect. Immun.* **60:**3994–4002.

202. **Strugnell, R. A., D. Maskell, N. Fairweather, D. Pickard, A. Cockayne, C. Penn, and G. Dougan.** 1990. Stable expression of foreign antigens from the chromosome of *Salmonella typhimurium* vaccine strains. *Gene* **88:**57–63.

203. **Tacket, C. O., D. M. Hone, R. Curtiss III, S. M. Kelly, G. Losonsky, L. Guers, A. M. Harris, R. Edelman, and M. M. Levine.** 1992. Comparison of the safety and immunogenicity of *aroC aro* and *cya crp Salmonella typhi*

strains in adult volunteers. *Infect. Immun.* **60:**536–541.

204. **Tamura, S., Y. Samegai, H. Kurata, T. Nagamine, C. Aizawa, and T. Kurata.** 1988. Protection against influenza virus infection by vaccine inoculated intranasally with cholera toxin B subunit. *Vaccine* **6:**409–413.

205. **Thibodeau, L., L. Constantineau, and C. Tremblay.** 1992. Oral priming followed by parenteral immunization with HIV-1 immunosomes induces HIV-specific salivary and circulatory IgA in mice and rabbits. *Vaccine Res.* **1:**233–240.

206. **Tice, T. R., and D. R. Cowsar.** 1984. Biodegradable controlled release parenteral systems. *Pharmacol. Technol. J.* **8:**26–31.

207. **Vajdy, M., and N. Y. Lycke.** 1992. Cholera toxin adjuvant promotes long-term immunological memory in the gut mucosa to unrelated immunogens after oral immunization. *Immunology* **75:**488–492.

208. **Verheul, A. F. M., A. A. Versteeg, M. J. DeReuver, M. Jansze, and H. Snippe.** 1989. Modulation of the immune response to pneumococcal type 14 capsular polysaccharide-protein conjugates by the adjuvant Quil A depends on the properties of the conjugates. *Infect. Immun.* **57:**1078–1083.

209. **Visscher, G. E., R. L. Robison, and G. I. Argentieri.** 1987. Tissue response to biodegradable injectable microcapsules. *J. Biomater. Appl.* **2:**118–131.

210. **Wachsmann, D., J.-P. Klein, M. Scholler, and R. M. Frank.** 1985. Local and systemic immune response to orally administered liposome-associated soluble *S. mutans* cell wall antigens. *Immunology* **54:**189–193.

211. **Wachsmann, D., J.-P. Klein, M. Scholler, J. Ogier, F. Ackermans, and R. M. Frank.** 1986. Serum and salivary antibody responses in rats orally immunized with *Streptococcus mutans* carbohydrate protein conjugate associated with liposomes. *Infect. Immun.* **52:**408–413.

212. **Waldman, R. H., J. Stone, K.-C. Bergmann, R. Khakoo, V. Lazzell, A. Jacknowitz, E. R. Waldman, and S. Howard.** 1986. Secretory antibody following oral influenza immunization. *Am. J. Med. Sci.* **292:**367–371.

213. **Walker, W. A., K. J. Isselbacher, and K. J. Bloch.** 1972. Intestinal uptake of macromolecules: effect of oral immunization. *Science* **177:**608–610.

214. **Wei, G. J., and V. A. Bloomfield.** 1980. Determination of size and charge distributions by combination of quasi-elastic light scattering and band transportation. *Anal. Biochem.* **101:**245–253.

215. **Williams, R. C., and R. J. Gibbons.** 1972. Inhibition of bacterial adherence by secretory immunoglobulin A: a mechanism of antigen disposal. *Science* **177:**697–699.

216. **Wilson, A. D., C. J. Clarke, and C. R. Stokes.** 1990. Whole cholera toxin and B subunit act synergistically as an adjuvant for the mucosal immune response of mice to keyhole limpet haemocyanin. *Scand. J. Immunol.* **31:**443–451.

217. **Winther, B., D. J. Innes, S. E. Mills, N. Mygind, D. Zito, and F. G. Hayden.** 1987. Lymphocyte subjects in normal airway mucosa of the human nose. *Arch. Otolaryngol. Head Neck Surg.* **113:**308–315.

218. **Wolf, J. L., R. S. Kauffman, R. Finberg, R. Dambrauskas, B. N. Fields, and J. S. Trier.** 1983. Determinants of reovirus interaction with the intestinal M cells and absorptive cells of murine intestine. *Gastroenterology* **85:**291–300.

219. **Woogen, S. D., W. Ealding, and C. O. Elson.** 1987. Inhibition of murine lymphocyte proliferation by the B subunit of cholera toxin. *J. Immunol.* **11:**3764–3770.

220. **Wu, H.-Y., and M. W. Russell.** 1992. Induction of mucosal immunity by intranasal application of a streptococcal surface protein antigen with cholera toxin B subunit. *Infect. Immun.* **61:**314–322.

221. **Yang, D. M., N. Fairweather, L. L. Button, W. R. McMaster, L. P. Kahl, and F. Y. Liew.** 1990. Oral *Salmonella typhimurium* (AroA$^-$) vaccine expressing a major leishmanial surface protein (gp63) preferentially induces T helper 1 cells and protective immunity against leishmaniasis. *J. Immunol.* **145:**2281–2285.

INDUCTION OF CYTOKINE
FORMATION BY BACTERIA AND
THEIR PRODUCTS

Anthony C. Allison and Elsie M. Eugui

18

Bacterial infections elicit in mammalian hosts local and systemic responses. Local responses include the recruitment of leukocytes and their activation for increased microbicidal activity (111). These responses can often eliminate the infections, so there is a transient local inflammatory reaction. Sometimes the bacteria cannot be eliminated, and then acute inflammation, e.g., septic arthritis, or chronic inflammation results. Well-known examples of the latter are the chronic granulomatous reactions associated with mycobacterial infections, e.g., tuberculosis, leprosy, and Johne's disease. An even commoner example is chronic inflammatory periodontal disease, which is caused by bacteria in dental plaque. *Borrelia burgdorferi* infections produce Lyme disease, one manifestation of which is a chronic inflammatory arthritis resembling rheumatoid arthritis (RA). Persistent *Pseudomonas aeruginosa* infections contribute to chronic bronchial inflammation in patients with cystic fibrosis. In all of these situations, cytokines such as tumor necrosis factor alpha (TNF-α) and interleukin-1β (IL-1β) play a role in pathogenesis.

Systemic responses occur when bacteria or their products enter the bloodstream. These responses can be moderate (fever, leukocytosis, influenza-like symptoms) or severe (the situation designated septic shock or the systemic inflammatory response syndrome). According to the Centers for Disease Control and Prevention, the number of sepsis cases linked to microbial infections tripled from 1979 to 1992 (118). Explanations include the increased number of elderly patients, including ill patients kept alive by improved medical treatments or placed at higher risk of infections by invasive medical procedures, and of AIDS patients. Both groups have weakened immune systems that predispose them to sepsis. Moreover, the proportion of patients with gram-negative bacterial infections has decreased to about 45% of the total (118), which limits the clinical utility of antibodies against lipopolysaccharide (LPS) endotoxins of gram-negative bacteria to prevent septic shock.

The clinical features of septic shock include refractory systemic hypotension, increased microvascular permeability, disseminated intravascular coagulation, acute respiratory failure, myocardial dysfunction, and ultimately, multiorgan failure. The attendant mortality is high, and there is no efficacious treatment. Evidence that several cytokines contribute to the pathogenesis of septic shock has accumulated. The

Anthony C. Allison DAWA, Inc., 2513 Hastings Drive, Belmont, California 94002. *Elsie M. Eugui* Syntex Research, Palo Alto, California 94304.

Virulence Mechanisms of Bacterial Pathogens, 2nd ed., Edited by J. A. Roth et al.
© 1995 American Society for Microbiology, Washington, D.C.

coagulation, complement, and kinin cascades are also activated, and cytokines interact with these. Probably many mediators are involved in the pathophysiology of shock (71, 97).

Circulating cytokines trigger the release of glucocorticoids that can suppress the formation of TNF-α and IL-1β. Potentially, this suppression, and cytokines such as IL-4 (55) and IL-10 (84), could provide feedback regulation of cytokine production, but it is not clear how effective this mechanism is in vivo. Moreover, glucocorticoids act synergistically with cytokines in some mechanisms that could aggravate shock, e.g., stimulation by IL-1 of the secretion of complement C3 and factor B and inhibition of the release of the control protein factor H by endothelial cells (23). This synergy may explain why glucocorticoids do not provide an effective treatment for septic shock, although they can be useful in local infections, e.g., meningitis.

During the past few years, attempts have been made to treat septic shock with antibodies against LPS or TNF-α and with recombinant IL-1 receptor antagonist. None of these treatments has been effective (118), which illustrates the complexity of the pathogenesis of shock. Since inhibition of the production or effects of single cytokines (TNF-α or IL-1β) is insufficient for treatment of shock, the desirability of identifying compounds that inhibit the production of several cytokines has become apparent. We have identified drugs that inhibit the transcription of cytokine genes (TNF-α, IL-1β, and IL-6) in a coordinate fashion (40), and drugs that selectively inhibit the translation of cytokine genes have also been described (77). During the next few years, it will be ascertained whether such drugs, perhaps used with an inhibitor of the complement and kinin cascades, can prevent the lethal effects of septic shock. At the local level, drugs can prevent cytokine-induced bone erosion in periodontal disease (5, 64).

This chapter begins with a profile of TNF-α and IL-1β, the major cytokines involved in pathophysiologic responses to bacteria, and some information on IL-6 and IL-8. Induction of cytokines by bacteria and their products is a broad subject that cannot be reviewed comprehensively here: obviously, this account is incomplete and reflects the interests of the authors.

PROINFLAMMATORY CYTOKINES

Proinflammatory, Catabolic, and Procoagulant Effects of TNF-α and IL-1β

In this chapter, we emphasize IL-1β as the major form of IL-1 released by activated cells of the monocyte-macrophage lineage. However, in some situations, IL-1α, which is derived mainly from other cell types, may predominate. This is true, for example, in periodontal disease, in which gingival epithelial cells produce IL-1α (88). In bioassays using human cells, human IL-1α and IL-1β have equipotent activities, although some differences are observed when murine cells are used (105). TNF-α and IL-1β exert proinflammatory effects by inducing the expression of adhesion molecules that recruit leukocytes, by priming leukocytes for oxidant production, and by inducing the production of prostaglandins and other mediators of inflammation (13). They exert catabolic effects by inducing the production of neutral metalloproteinases and plasminogen activators. They exert procoagulant effects by inducing endothelial cells to produce thromboplastin while downregulating the formation of plasminogen activators and thrombomodulin.

INDUCTION OF ADHESION MOLECULES

TNF-α and IL-1 induce the production and/or expression of adhesion molecules by endothelial cells, thereby increasing recruitment of leukocytes into sites of microbial infection or inflammation (100). Both TNF-α and IL-1 induce expression on endothelial cells of E selectin; ICAM-1, which by binding LFA-1 recruits neutrophils; and VCAM-1, which by binding VLA-4 recruits monocytes and lymphocytes (34). Antibody against TNF-α decreases recruitment of neutrophils following injection of immune complexes into the peritoneal cavity (132).

INDUCTION OF PROSTAGLANDIN SYNTHESIS

Prostaglandins are important mediators in the pathogenesis of inflammation and septic shock. Prostaglandin E_2 (PGE_2) from monocytes and fibroblasts and PGI_2 from endothelial cells are vasodilators, contributing to the heat, redness, and increased vascular permeability of inflammation. They are also comediators of pain. Vasodilatory prostaglandins are among the mediators of hypotension in septic shock. TNF-α and IL-1β induce the synthesis of prostaglandins in fibroblasts from rheumatoid synovial and periodontal tissues and of PGI_2 in endothelial cells (113). IL-1α and IL-1β increase the endogenous release of arachidonic acid in target cells as well as stimulating the synthesis of prostaglandins from exogenously added arachidonic acid (17). IL-1-mediated stimulation of phospholipase A_2 activity is correlated with induction of cytosolic phospholipase A_2 (81). However, cyclooxygenase undergoes irreversible self-inactivation, so that modulation of activity depends on continued synthesis of the enzyme (113).

Two genes encoding cyclooxygenases, termed *Cox-1* and *Cox-2,* have been cloned and expressed in functional form (58). The *Cox-1* gene is expressed ubiquitously in vivo and in vitro, whereas the *Cox-2* gene is expressed at very low levels in healthy tissues in vivo and in quiescent cells in culture. However, IL-1 rapidly induces expression of the *Cox-2* gene in cultured endothelial cells, with expression of the *Cox-2* isoenzyme and production of prostacyclin. In unstimulated monocytes, the constitutive enzyme is Cox-1. Stimulation by LPS has no effect on expression of the *Cox-1* gene but rapidly induces a high level of expression of the *Cox-2* gene (58). *Cox-1* and *Cox-2* transcripts have been found in synovia from patients with RA (24). Immunoprecipitation of in vitro-labeled proteins from freshly explanted rheumatoid synovial tissues shows that synthesis of Cox-2 protein is markedly increased by IL-1β. In the presence of dexamethasone, Cox-2 mRNA is not increased by IL-1β. Other reports of the stimulation by IL-1 of the expression of Cox-2 mRNA have been published (58, 96). Thus, one of the proinflammatory effects of IL-1β is induction of expression of the *Cox-2* gene, and one of the anti-inflammatory effects of glucocorticoids is suppression of the formation of Cox-2.

In some species, thromboxane produced by cells of the monocyte-macrophage lineage mediates pulmonary vasoconstriction and hypertension in septic shock (97). Thromboxane is one of the procoagulants responsible for disseminated intravascular coagulation in septic shock. The role of prostaglandins in bone erosion is discussed in the next section.

DEGRADATION OF CARTILAGE AND BONE

Inflammatory responses are associated with degradation of cartilage and bone. In septic arthritis, articular cartilage is eroded alongside the joint space. In RA, proliferating pannus grows into and erodes articular cartilage and bone in the vicinity, with progressive joint destruction; there is also variable osteoporosis in the adjacent bones. In periodontal disease, erosion of alveolar bone with consequent loss of teeth is a major clinical problem. It is generally accepted that TNF-α and IL-1β induce in connective tissue cells the production of neutral metalloproteinases (stromelysins and collagenases) that degrade both proteoglycan matrix and interstitial collagen in cartilage and bone (89). Addition of TNF-α, IL-1β, or IL-6 to organ cultures of rat fetal long bones augments catabolism, as shown by release of calcium as well as of fragments of proteoglycan and collagen into the medium (61, 83). When two or three of these cytokines are included in the culture medium, they have at least additive effects on bone resorption. Thus, all three cytokines are likely to contribute to bone erosion in RA and periodontal disease, and a drug inhibiting the production of all of them might have clinical utility.

Prostaglandins are comediators of bone catabolism induced by TNF-α, IL-1β, and IL-6. Cyclooxygenase inhibitors suppress IL-1-induced bone erosion in culture, although they vary markedly in potency in this assay; the most potent inhibitor of bone degradation so far identified is ketorolac (5), which raises the possibility of its use as a mouthwash to prevent alveolar bone erosion in periodontal disease.

The mechanisms by which cyclooxygenase inhibitors prevent IL-1-induced bone erosion are not fully understood. Factors that may be involved are the requirement for PGE_2 as a cofactor for IL-1-induced expression of plasminogen activators in cells of connective tissue type and the role of these activators in the activation of neutral metalloproteinases. Prostromelysins and procollagenase are secreted as inactive proenzymes that are converted into active enzymes by proteolytic cleavage or other processes. The relatively broad-spectrum serine proteinase plasmin can initiate a cascade of proteolytic events that result in metalloproteinase activation (89). The narrower-spectrum serine esterases, urokinase- and tissue-type plasminogen activators, convert plasminogen into plasmin. IL-1-induced expression of the genes for plasminogen activators requires PGE_2; it is inhibited by indomethacin and restored by adding PGE_2 again (80); PGE_2 alone does not induce production of plasminogen activators.

PROCOAGULANT EFFECTS OF TNF-α

As already discussed, TNF-α and IL-1 induce the production of prostaglandins. A prostaglandin released by activated cells of monocyte-macrophage lineage is thromboxane, which is a procoagulant. LPS induces in these cells the expression and activation of tissue factor, a potent procoagulant (102). TNF-α acts on endothelial cells to exert a net procoagulant effect, reviewed by Van der Poll et al. (127) and discussed below in the section on disseminated intravascular coagulation.

EFFECTS OF ADMINISTRATION OF TNF-α AND IL-1 IN HUMANS

It has been ethically justified to inject TNF-α and IL-1 into humans in order to define the effects and maximal tolerated doses of these cytokines. This has been done in the context of using TNF-α for cancer therapy and using IL-1 to accelerate recovery of hematopoiesis following depletion of bone marrow precursors by chemotherapy or radiotherapy of patients with cancer. The information obtained provides insight into the role of these cytokines in the pathogenesis of septic shock. Since the manifestations of this disorder vary in different experimental animal models, it is useful to have information on the major species for whom treatment of septic shock is intended.

Feinberg et al. (42) reported findings after 30-min intravenous infusions of escalating doses of TNF-α to humans with disseminated cancers. The principal toxicities were constitutional symptoms, including fever, chills, headache, and fatigue, that increased in severity with increasing dose. The maximum tolerated dose was 200 $\mu g/m^2$, the dose-limiting toxicities being constitutional symptoms and hypotension. After this dose, the level of TNF-α in the circulation was about 10 ng/ml. Counts of circulating granulocytes decreased, presumably owing to induction of adhesion molecules and margination. TNF-α infusion increased levels of circulating triglycerides and very-low-density lipoprotein. In this study, levels of circulating platelets decreased after TNF-α administration. Single injections of TNF-α (50 $\mu g/m^2$) into healthy humans activated the extrinsic pathway of blood coagulation but did not decrease platelet counts (127).

In another study, bolus intravenous injections of TNF-α (50 $\mu g/m^2$) into healthy men induced a transient increase in plasma levels of catecholamines, adrenocorticotropin, cortisol, and glucagon, which are associated with an early and sustained rise in plasma glucose concentrations (128). Glucose turnover was increased. Plasma concentrations of free fatty acid and glycerol increased transiently after TNF injection, as did free-fatty-acid turnover. The resting energy expenditure showed a transient rise after TNF injection. These metabolic changes are similar to those observed in septicemia.

Low doses (1 to 10 ng/kg of body weight) of IL-1 injected into humans increase the numbers of circulating neutrophils and platelets and of hematopoietic stem cells (123). These effects are attributed to costimulation of the proliferation of stem cells and augmented release of leukocytes from the bone marrow into the circulation. Higher doses of IL-1 decrease the numbers of circulating neutrophils, presumably because the cytokine increases the expression of adhesion molecules on endothelial cells, thereby augmenting the binding and margination of neutrophils. Moderate doses of IL-1 also induce fever, joint and muscle aches, increase sensitivity to pain, and lethargy. High doses of IL-1 ($\geq$100 ng/kg) reduce appetite, induce gastrointestinal disturbances, and produce hypotension that can reach dangerous levels when >300 ng/kg is injected (114).

IL-6

IL-6 is a 26-kDa pleiotropic cytokine produced by a variety of cells and acting on a wide range of tissues. For example, LPS and *Staphylococcus aureus* Cowan 1 induce the production of IL-6 by human monocytes (40). IL-6 stimulates the production by hepatocytes of C-reactive protein and other acute-phase reactants (56), levels of which are positively correlated with the severity of RA (33). The role of IL-6 in the pathogenesis of RA is less well defined: however, the cytokine is a cofactor for T-lymphocyte differentiation (120) and for the production of immunoglobulins by B lymphocytes in RA synovial tissue (93). Activated T lymphocytes (47a) and immune complexes (85) are thought to contribute to pathogenesis of RA and septic arthritis. Moreover, IL-6 can act synergistically with IL-1 in augmenting bone erosion (61). That could contribute to joint destruction in RA as well as to alveolar bone erosion in periodontal disease (103). Hence, a useful property of a drug could be inhibition of the production of IL-6 as well as of IL-1β and TNF-α. Levels of IL-6 in the circulation are high in humans with septic shock and are positively correlated with a fatal outcome (1). However, IL-6 appears to be a marker of the severity of septicemia rather than playing a role in the pathophysiology of shock. Indeed, by stimulating hepatocytes to synthe-size acute-phase proteins, including protease inhibitors, IL-6 could help counteract the effects of shock.

IL-8 and Related Chemokines

IL-8 is a low-M_r heat-stable cytokine (6 to 8 kDa) produced by several cell types, including monocytes, lymphocytes, fibroblasts, and endothelial cells (126). IL-8 has biological effects similar to those of other types of molecules chemotactic for neutrophils, including cell-derived leukotriene B$_4$ and platelet-activating factor, the plasma-derived anaphylotoxin C5a, and bacterial or synthetic formylmethionine peptides. LPS induces the production of IL-8 by monocytes, and IL-1 stimulates production of IL-8 by endothelial cells and fibroblasts (126). Release of IL-8 at sites of bacterial infection is one factor leading to the accumulation and activation of neutrophils, which themselves produce IL-8 when induced by TNF or IL-1. Following intravenous injection of *Escherichia coli* or LPS, IL-8 is found in the circulation with about the same time course as IL-6 (44). Indeed, the inducers leading to IL-6 production by various cell types are similar to those leading to IL-8 production. Common responsive elements leading to gene activation, e.g., NF-κB and NF–IL-6 motifs, may explain this similarity. Induction of IL-8 is also observed after IL-1 or TNF infusion (126).

IL-8 stimulates neutrophils to directed migration, as measured under agarose or in a Boyden chamber. It increases free Ca^{2+} levels in the cytoplasms of neutrophils and induces degranulation. IL-8 is a weak inducer of the formation of superoxide and hydrogen peroxide. Monocytes and eosinophils do not respond to IL-8. Injected into the skin, lungs, or peritoneal cavity, IL-8 produces neutrophil accumulation and neutrophil-dependent increased vascular permeability (126). Increased levels of IL-8 have been detected in bronchoalveolar lavage fluids from patients with adult respiratory distress syndrome and pulmonary fibrosis (20, 126). Injection of IL-8 intravenously leads to immediate neutropenia followed by neutrophil leukocytosis accompanied by the release of nonsegmented neutrophils from the marrow reservoir (126).

IL-8 belongs to a family of small inflammatory and growth-regulatory proteins that have four conserved cysteine residues. Depending on the arrangement of the two NH₂-terminal cysteines, the chemokine family is divided into two subfamilies. The C-X-C chemokines act on neutrophils and include, in addition to IL-8, the human platelet-derived factors platelet factor 4 and β-thromboglobulin. Forms of these products processed at the NH₂ terminus have activities similar to those of IL-8 but weaker (126).

The second subfamily comprises the C-C chemokines, which stimulate monocytes. Human monocyte chemotactic protein-1 (MCP-1) is coinduced with IL-8 in fibroblasts stimulated by IL-1 or TNF. Two structurally related human monocyte chemotactic proteins are MCP-2 and MCP-3, and other C-C chemokines include macrophage inflammatory protein-1 (MIP-1) and RANTES. MIP-1 has effects other than chemotaxis, including prostaglandin-independent pyrogenicity (27).

INDUCTION OF CYTOKINE FORMATION BY BACTERIA AND THEIR PRODUCTS

LPS of Gram-Negative Bacteria

Bacteria and their products are potent inducers of the production of cytokines by a wide range of cell types. LPS is the prototype inducer of cytokine production by cells of the monocyte-macrophage lineage, and this system has been used to define the molecular biology underlying cytokine production (40). LPS increases expression of the genes for TNF-α, IL-1β, IL-6, and IL-8 in these cells and induces steady-state levels of the corresponding mRNAs. In some cases, notably that of TNF-α, LPS also augments mRNA stability and increases the efficiency of translation (53). In monocytes and macrophages, inhibitors of protein synthesis do not affect LPS induction of cytokine gene transcription, suggesting that activation of preexisting transcription factors rather than formation of new transcription factors takes place.

The transcription factors involved are discussed below.

Staphylococcus aureus and Staphylococcal Toxins

Heat-killed *S. aureus* Cowan 1 (also known as Pansorbin) is frequently used as a polyclonal B-lymphocyte mitogen. The bacteria are also potent inducers of the production of TNF-α, IL-1β, and IL-6 by human peripheral blood mononuclear cells (PBMC), as shown in our laboratory (40) and elsewhere. Other investigators have documented that toxic shock syndrome-associated staphylococcal and streptococcal pyrogenic toxins are potent inducers of TNF production (41). Toxic shock syndrome toxin-1 induces production of TNF-α, TNF-β, and IFN-γ (67, 98).

Staphylococcal enterotoxin A (SEA) is a classic example of a bacterial superantigen, a polyclonal T-lymphocyte stimulator. This raises the question of whether production of proinflammatory cytokines by SEA is T cell dependent. We therefore compared induction of IL-1 synthesis and release by human PBMC when stimulated by SEA or LPS (50). Both bacterial products were found to be potent inducers of IL-1 synthesis, with SEA being more effective on a molar basis. Production after LPS induction was rapid: intracellular IL-1β was observed after 4 h, and maximal cell-free levels were attained by 12 h. Production after SEA stimulation was slower, with cell-free levels peaking between 24 and 48 h. The response of monocytes to LPS was T cell independent, whereas that to SEA required the presence of T cells. The CD4⁺ 45RO⁺ memory T-cell subset helped SEA induction of extracellular IL-1β more efficiently than did CD4⁺ 45RA⁺ naive T cells or CD8⁺ T cells. T-cell help for IL-1 induction could not be replaced by a panel of T-cell-derived recombinant lymphokines, including gamma interferon and TNF, indicating the participation of membrane-bound ligands or unidentified soluble mediators. In summary, superantigens, exemplified by SEA, are potent inducers of the production of IL-1 by human PBMC. This induction is T-cell dependent, whereas that by LPS is not. However, lymphokines such as gamma interferon can augment the production by monocytes of TNF-α and IL-1β.

P. aeruginosa

Although *P. aeruginosa* is only weakly pathogenic in healthy human hosts, it is an important cause of chronic progressive lung disease in patients with cystic fibrosis and of widespread disease and infection in immunocompromised individuals (59, 117). Several factors allow *P. aeruginosa* to colonize the lungs (59), but why the colonization persists and causes inflammatory damage remains uncertain. We found that exotoxin A of *P. aeruginosa* is immunosuppressive: the toxin inhibits proliferative responses of human lymphocytes and lymphokine production (117). That could facilitate colonization, particularly when immune responses are suboptimal.

Formalin-fixed, heat-killed *P. aeruginosa* bacteria are (117) potent inducers of TNF-α production by human PBMC. Measurable amounts of TNF-α are produced by 10^3 bacteria added to 2×10^5 PBMC. IL-1β is also induced by bacteria with the same time course (maximal production 24 h after stimulation). Polymyxin B does not inhibit TNF production by *P. aeruginosa,* suggesting that LPS is not required. Contrary to other claims, exotoxin A of *P. aeruginosa* does not induce TNF-α production. In summary, these findings suggest that immunosuppression by exotoxin A may facilitate persistent infection by *P. aeruginosa* and that induction by the bacteria of the production of TNF-α and IL-1β may contribute to chronic inflammation, e.g., in the lungs of cystic fibrosis patients.

Periodontal Disease

Periodontitis is a common disease in which the attachment tissues of the teeth and their alveolar bone housing are eroded, resulting in tooth loss. It is the major cause of tooth loss among adults living in industrial countries. It is widely accepted that gram-negative anaerobic bacteria in dental plaque, especially *Porphyromonas gingivalis, Actinobacillus actinomycetemcomitans,* and *Eikenella corrodens,* play a major role in the pathogenesis of periodontal disease through induction of cytokine production (88, 103). These bacteria, and the LPS derived from them, induce the production of TNF-α, IL-1α, IL-1β, and IL-6 by monocytes, macrophages, gingival fibroblasts, gingival epithelial cells, and endothelial cells. IL-1α and IL-1β are potent inducers of bone degradation; TNF-α and IL-6 have a synergistic action with IL-1 in this effect (61, 83).

Cytokine-induced bone degradation occurs by a prostaglandin-dependent mechanism, and cyclooxygenase inhibitors can prevent alveolar bone erosion in periodontal disease (5, 64). The high potency of some cyclooxygenase inhibitors in this effect (50% inhibitory concentration [IC_{50}] in the low nanogram range) (5) suggests that topical administration in the form of a twice-daily mouthwash may be sufficient to prevent tooth loss in periodontal disease.

MDP and Analogs

It has long been known that mycobacterial and other bacterial cell walls have adjuvant effects, increasing cell-mediated and humoral immune responses to unrelated antigens. The best-known example is the heat-killed mycobacteria in Freund's complete adjuvant. Ellouz et al. (36) showed that the minimal component of the mycobacterial cell wall with adjuvant activity is muramyl dipeptide (MDP; *N*-acetyl-muramyl-L-alanine-D-isoglutamine), which is a constituent of many bacterial cell walls. It is a potent inducer of the production of IL-1 by monocytes and macrophages (3); it is pyrogenic and can induce uveitis in experimental animals (130). Uveitis and arthritis are components of Reiter's syndrome, which complicates infections with some gram-negative bacteria, particularly in persons with the HLA-B27 haplotype (48). Cytokines are almost certainly involved in the pathogenesis of this type of arthritis and uveitis as well as in the pyrogenicity of MDP.

We and others synthesized more than 100 analogs of MDP in an effort to separate desired adjuvant activity from side effects such as pyrogenicity and capacity to induce uveitis. N-acetylmuramyl-L-threonyl-D-isoglutamine was formulated in a squalane, Pluronic copolymer microfluidized emulsion to produce Syntex adjuvant formulation (4). Used with recombinant and other subunit antigens, Syntex adjuvant formulation elicits cell-mediated immunity and antibodies of protective isotypes, e.g., immunoglobulin G2a (IgG2a) in the mouse (3). The adjuvant does not augment the formation of antibodies of the IgE isotype, which can produce hypersensitivity and are therefore undesirable. We have recently found that in lymph nodes draining sites of antigen injection in mice, Syntex adjuvant formulation augments the production of IFN-γ but not IL-4 (3). Gamma interferon stimulates the production of IgG2a, whereas IL-4 stimulates that of IgE. Thus, by preferentially stimulating cytokine production, MDP-type adjuvants can elicit selectively protective immune responses.

CYTOKINES AND OTHER MEDIATORS IN SEPTIC SHOCK

The pathogenesis of septic shock is complex and varies in different species of animals. For example, some animals are highly sensitive to LPS (humans, pigs, sheep, cows, goats, cats), whereas others are relatively resistant (baboons, dogs, rats, mice, rabbits) (30). Some animals (pigs, sheep, cows, goats, cats) have within the pulmonary vasculature many macrophages, which ingest a relatively high proportion of bacteria or other particles injected intravenously. In these animals, pulmonary vasoconstriction, hypertension, and edema are characteristic features of sepsis. Humans, baboons, dogs, rats, mice, and rabbits have few pulmonary intravascular macrophages (30). In humans, the majority of injected particles are cleared in the liver and spleen, with only a small fraction cleared in the lungs (18). Nevertheless, humans develop pulmonary hypertension with the adult respiratory distress syndrome (30). In all animals, peripheral vasodilatation and hypotension, which can be lethal, are features of septic shock.

Disseminated intravascular coagulation, with widespread deposition of fibrin in the microvasculature, is often observed in septic shock and is associated with the development of multiple organ failure (86). The mechanisms by which the clotting cascade is activated in septicemia are incompletely understood, but some are known. Bacterial products such as LPS act directly on monocytes to induce the expression of tissue factor, which activates the extrinsic pathway of coagulation (102) (Fig. 1). Contact-phase proteinases, including those in the intrinsic pathway of blood coagulation, are activated (Fig. 2). Bacterial products also induce the release from monocytes and macrophages of TNF-α, which acts on endothelial cells to exert a net procoagulant effect (see above). In patients with sepsis, plasma TNF-α levels are directly proportional to the extent of intravascular coagulation and fatal outcome (13, 49). Hence, one of the roles of TNF-α in septic shock is activation of disseminated intravascular coagulation.

Several questions arise regarding the role of cytokines in septic shock. First, what concentrations of cytokines are released into the circulation at different times? Since cytokines released early (TNF-α and IL-1β) can induce the production of those appearing later (IL-6 and IL-8), the next question is whether there is a cascade of cytokines (TNF-α inducing IL-1β, which in turn induces IL-6 and IL-8 [11, 16]). If that is the case, inhibiting the production or effects of TNF-α alone should provide effective prevention or early treatment of septic shock. However, if TNF-α is only amplifying LPS-induced IL-1β production (47) but there is parallel induction of cytokines and mediation of their effects, it will be necessary to inhibit the production of all cytokines that contribute significantly to the pathogenesis of septic shock. There is good evidence that TNF-α and IL-1β play a pathogenetic role in septic shock. As an academic exercise, it would be interesting to know the relative importance of late mediators of hemodynamic changes, disseminated intravascular coagulation, and metabolic disturbances (prostaglandins, leukotrienes, plate-

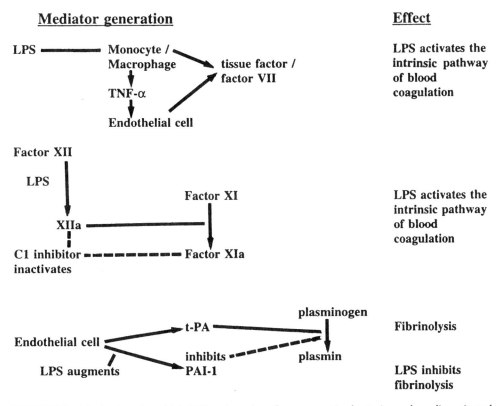

FIGURE 1 Mechanisms by which LPS endotoxins of gram-negative bacteria produce disseminated intravascular coagulation. t-PA, tissue-type plasminogen activator.

let-activating factor, activated complement components, kinins, histamine, nitric oxide, etc.). There are already indications that many late mediators are involved in septic shock, with their relative importance varying from species to species (97). In general, it seems likely that combinations of many inhibitors of the production and effects of late mediators will be required to treat septic shock, and that efficacy in experimental animals may not be predictive of efficacy in humans. Relying on inhibition of a single mediator, such as nitric oxide, may not be justified.

The sequence of cytokines observed in the circulation of nonhuman primates following injections of high doses of LPS or sublethal doses of *E. coli* is instructive (44, 47) (Fig. 3). A large amount of TNF-α is released first, peaking at 2 h and returning to baseline by 4

h. This is followed by a lower peak of IL-1β, which is maximal at 3 to 4 h and decreases more slowly than that of TNF-α. Still later, IL-6 and IL-8 are found in the circulation, where they remain much longer than TNF-α and IL-1β. In humans injected with small doses of LPS, similar but lower peaks of TNF-α and IL-6 have been observed, but IL-1β was usually undetectable (13, 90); in one study, IL-1β was observed at a single time point after LPS infusion (19).

Administration of TNF-α to otherwise healthy nonhuman primates induces cardiovascular collapse, multi-organ system failure, and death similar to that seen in lethal septicemia and clinical septic shock. An early peak of circulating IL-1β and a later peak of IL-6 are observed, both attenuated by giving the animals antibody against TNF-α (47). Neutrali-

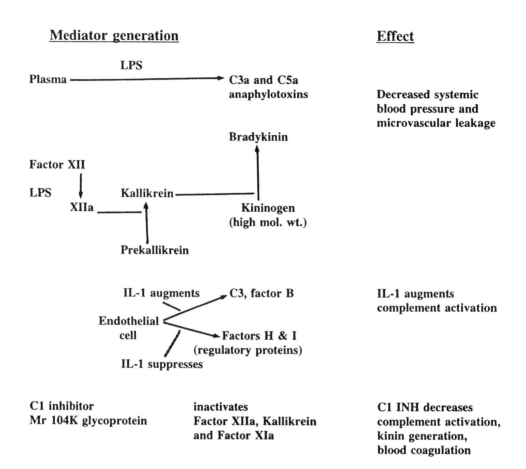

Mediator generation

Effect

Decreased systemic blood pressure and microvascular leakage

IL-1 augments complement activation

C1 INH decreases complement activation, kinin generation, blood coagulation

FIGURE 2 Activation of contact-phase proteinases by LPS in the complement and kinin cascades.

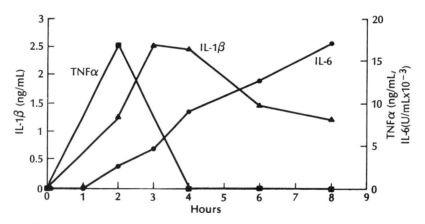

FIGURE 3 Sequence of cytokines in the circulation of the baboon *Papio anubis* following injection of *E. coli* (J. Kenney).

zation of TNF-α also reduces hypoglycemia and hypertriglyceridemia but does not prevent a fatal outcome in animals given a high dose of LPS. These findings implicate TNF-α as a mediator of septic shock that acts in part by amplifying the formation of IL-1β in response to LPS. However, they make it unlikely that TNF-α is the only important mediator of septic shock. Administration of an antibody against TNF-α also fails to prevent lethality after cecal ligation and puncture in mice (39). However, neutralization of TNF-α increases the survival of pigs injected with *E. coli* (65).

Injection of IL-1β into baboons produces hypotension and other disorders similar to those produced by TNF-α but less severe (43). These effects suggest that IL-1β could be another major mediator of septic shock. Consistent with that interpretation is the finding that administration of recombinant IL-1ra to baboons during lethal *E. coli* septic shock attenuates the decrease in mean arterial blood pressure and cardiac output and improves survival (44). However, following an encouraging phase II trial of IL-1ra administration in patients with septic shock (45), the results of a phase III trial were disappointing (118). This suggests that TNF-α (levels of which are unaffected by IL-1ra administration) and other mediators exert effects through mechanisms independent of IL-1; an example would be procoagulant effects (see above). TNF-α is more toxic than IL-1 administered to baboons (43, 47), presumably because it is not only inducing IL-1 production.

In humans with septic shock, high levels of TNF-α, IL-1, and IL-6 in the circulation are positively correlated with severe morbidity and a lethal outcome (1, 129). IL-6 is a major mediator of acute-phase responses, inducing the production by hepatocytes of the proteins involved (1, 56). Measurement of IL-6 levels in body fluids could provide a useful indication of the severity of the microbial challenge in sepsis, but it is not clear that IL-6 plays a role in mediating the hemodynamic and metabolic disturbances in shock. Starnes et al. (116) reported that treatment of rats with a monoclonal

antibody against murine IL-6 improved survival in patients with *E. coli* shock. However, administration of large doses of IL-6 to mice (44) or dogs (101) had no hemodynamic or hematologic effects. The general conclusion is that it is desirable to inhibit the production and/or effects of both TNF-α and IL-1β to prevent or treat septic shock. Additional activities, such as inhibiting the production of thromboplastin as well as the complement, kinin, and intrinsic coagulation cascades, would probably also be useful.

Disseminated Intravascular Coagulation

A serious complication of septic shock is disseminated intravascular coagulation. Bacterial products activate the coagulation cascade directly and through mediators such as TNF-α (Fig. 1). LPS and other bacterial products induce in monocytes the production of tissue factor, which in the correct phospholipid microenvironment activates factor VII and the intrinsic pathway of coagulation (102). By acting directly on the contact phase, LPS activates factor XII to factor XIIa (Hageman factor), which through factor XI triggers the intrinsic pathway of coagulation (68) (Fig. 2). Both the extrinsic and intrinsic pathways lead to the activation of thrombin, the control enzyme of the clotting cascade. TNF-α acts on endothelial cells to exert a net procoagulant effect (127). TNF-α also increases the expression on endothelial cells of tissue factor and concurrently downregulates thrombomodulin and thereby impairs activation of the regulatory protein C.

The fibrinolytic and thrombolytic activity of blood is determined by a balance of plasminogen activators and inhibitors of these enzymes (PAIs). The most important activator in circulating plasma is tissue-type plasminogen activator, which is derived from vascular endothelial cells; TNF-α suppresses the release of this activator from these cells (37). The predominant inhibitor in circulating plasma is PAI-1. LPS induces the synthesis of PAI-1 by endothelial and smooth muscle cells; however, it is not known whether these are the principal sources of the PAI-1 that is increased in circulating plasma of humans and rats after endotoxin injection (37). Dexamethasone does not decrease levels of circulating PAI-1 following LPS injection and actually increases the baseline PAI-1 level (37). TNF-α and IL-1 can induce PAI-1 in rats, whereas IL-6 does not; however, neither TNF-α nor IL-1 seems to be significantly involved in the induction of PAI-1 by LPS in rats (37). Infusion of dogs with high doses of TNF-α results in microvascular thrombosis (125). Recombinant human TNF-α, administered as an intravenous bolus injection to healthy men, induces an early and short-lived rise in circulating levels of the activation peptide of factor X, followed by a prolonged increase in the plasma concentration of the prothrombin fragment F_{1+2} (127). These findings show activation of factor X and prothrombin. There were no signs that the intrinsic pathway of blood coagulation was activated. Thus, a single injection of TNF-α elicits a rapid and sustained activation of the common pathway of coagulation, probably through the extrinsic route. This effect of TNF-α could play an important part in the pathogenesis of disseminated intravascular coagulation in septicemia.

The Complement and Kinin Cascades

LPS added to plasma activates the complement cascade, with the formation of the anaphylotoxins C3a and C5a (68) (Fig. 2). These decrease systemic blood pressure and augment capillary leakage and are believed to contribute to the pathogenesis of septic shock. Endothelial cells secrete alternative complement pathway proteins C3 and factor B and the regulatory proteins factor H and factor I. IL-1 modulates the secretion of these proteins in a direction that would favor complement activation in the vicinity of the endothelium: C3 and factor B secretions are significantly stimulated, whereas factor H secretion is decreased in the presence of IL-1 (23). Glucocorticoids act synergistically with IL-1 to increase C3 and factor B secretion by endothelial cells (23).

LPS activates factor XII to factor XIIa (Hageman factor), which leads to the generation of kallikrein from prekallikrein; in turn, kallikrein converts high-molecular-weight kininogen to bradykinin, causing hypotension and capillary leakage (68). Evidence that activation of the contact-phase proteases contributes to the pathogenesis of septic shock comes from the use of selective inhibitors or antagonists (32).

An inhibitor of the contact-phase proteases and the classic complement pathway is the C1 esterase inhibitor (C1 INH). It is a glycoprotein that contains 35% carbohydrate and has an M_r of 104,000. C1 INH inhibits complement and inactivates factor XIIa, kallikrein, and factor XIa. After binding to these proteases, C1 INH is cleaved and forms a stable covalent 1:1 bimolecular enzyme inhibitor complex with its target protease, thereby acting as a suicide substrate (32). Administration of C1 INH has beneficial effects in LPS-induced sepsis in the rat (32), and the possible use of C1 INH with an inhibitor of cytokine production for treatment of shock is discussed below.

INHIBITORS OF CYTOKINE PRODUCTION

Agents Raising cAMP Levels Suppress Production of TNF-α but Not IL-1β

Activated monocytes and macrophages produce prostaglandins, notably PGE_2, which might provide feedback inhibition of cytokine production in vivo. E-type prostaglandins bind to EP_1 receptors, which activate adenylate cyclase, and cyclic AMP (cAMP) mediates many of their effects. Adrenalin binds to and activates β-adrenergic receptors, which also activate adenylate cyclase. Adrenalin circulating in patients with septic shock might likewise act as a feedback inhibitor of cytokine production. cAMP phosphodiesterase (PDE) inhibitors, in particular those active against the PDE-IV isozyme that is present in human monocytes (82), would be expected to have similar effects on cytokine production. Reference PDE-IV inhibitors are rolipram and nitroquazone (82, 92). Evidence for a differential effect of these agents on the production of TNF-α and IL-1β is accumulating.

Several groups of investigators have found that elevation of cAMP levels in human monocytes by activation of adenylate cyclase or inhibition of PDE suppresses LPS-induced TNF-α production with little effect on IL-1β production (14, 38, 92, 104, 108) (Figs. 4 and 5). A typical experiment is shown in Fig. 4; rolipram and nitroquazone inhibit TNF-α synthesis by LPS-activated monocytes, whereas inhibitors of other cAMP PDE isozymes have no significant effect (92). The nonselective PDE inhibitors pentoxifylline and isobutylmethyl xanthine reduce steady-state levels of TNF-α mRNA and suppress TNF-α production by LPS-stimulated monocytes and macrophages (104). Adding dibutyryl cAMP to LPS-stimulated murine macrophages lowers the rate of transcription of the TNF-α gene, decreases steady-state mRNA levels, and inhibits production of the cytokine (121). In the murine macrophage cell line RAW264.7, Han et al. (53) found that pentoxifylline blocked LPS-induced TNF-α mRNA accumulation

but had no effect on reporter mRNA translation. In contrast, dexamethasone had little effect on mRNA accumulation but strongly impeded translational derepression.

β-Adrenergic agonists also raise cAMP levels in THP-1 cells and suppress the production of TNF-α (109). Inhibition of TNF-α production is abolished by a β-adrenergic antagonist but not by an α-adrenergic antagonist, and the β-adrenergic agonist isoproterenol inhibits TNF-α production.

The effects on the production of IL-1β of elevating the level of cAMP are less pronounced and consistent, depending on the cells used and the experimental conditions. It was initially reported that PGE_2 inhibits the production of IL-1 by LPS-activated human monocytes (73). More recently, evidence that increased cAMP levels to some extent augment IL-1α and IL-1β gene expression and cytokine production in monocytes and macrophages has been obtained (60, 69). Others have found no significant effect on IL-1β production of elevating cAMP levels by using PDE-IV inhibitors (92) (Fig. 5). However, increasing cAMP

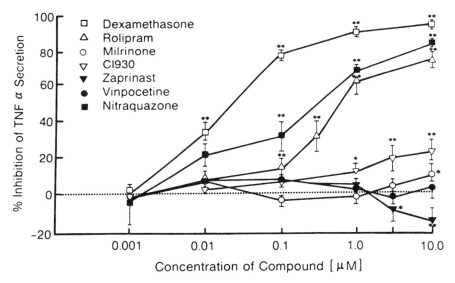

FIGURE 4 Suppression by dexamethasone and cAMP PDE-IV inhibitors (rolipram, nitroquasone) of the secretion of TNF-α by LPS-activated human monocytes. Inhibitors of other cAMP PDE isozymes have no effect on TNF-α secretion, whereas the inhibitors of cGMP PDE zaprinast augment secretion to some extent (58).

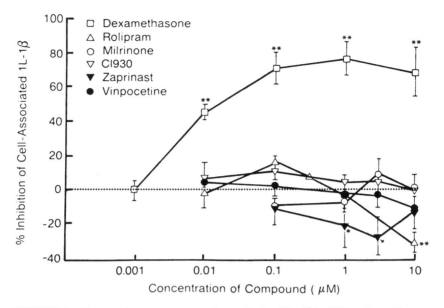

FIGURE 5 Dexamethasone suppresses the synthesis of IL–1β by LPS–activated human monocytes, whereas inhibitors of cAMP PDEs (including PDE-IV isozyme expressed in monocytes) have no effect. The inhibitor of the cGMP PDE zaprinast augments production of IL–1β to some extent. These observations are of cell-associated IL–1β; the effects of drugs on secreted IL–1β were comparable (58).

levels in monocytes can have some suppressive effect on secretion of IL–1β (60, 92).

Elevating cGMP levels by activating guanylate cyclase with sodium nitroprusside, adding exogenous cyclic nucleotides, or using zaprinast to inhibit cGMP-selective PDE-V somewhat augments the production of TNF-α in murine macrophages (92); this contrasts with the stronger suppressive effect on TNF-α production of raising cAMP levels (Fig. 4). Exogenous dibutyryl cGMP has no effect on IL–1β production by human monocytes (60). One group of investigators reported that raising cGMP levels with SIN-1 inhibits IL–1β production by human mononuclear cells (38), whereas other investigators found that the PDE-V-selective inhibitor zaprinast somewhat augments IL–1β secretion (92) (Fig. 5).

In summary, agents elevating cAMP levels in cells of monocyte-macrophage lineage, including E-type prostaglandins, β-adrenergic agonists, and cAMP PDE-IV inhibitors, sup-

press the production of TNF-α. They do not suppress the production of IL–1β but may have some inhibitory effect on its secretion.

Effects of Glucocorticoids on IL–1β Production

Although glucocorticoids are among the most potent and widely used anti-inflammatory drugs, their mode of action is incompletely understood. The most popular theory has been that glucocorticoids induce the formation of a group of proteins, collectively termed lipocortins, that inhibit phospholipase A_2 activity, thereby decreasing the production of proinflammatory prostaglandins and leukotrienes. However, when members of the lipocortin family were cloned, it became clear that they are present in rather high concentrations in many cell types, are not steroid inducible, and are phospholipid-binding proteins rather than phospholipase inhibitors. Thus, the role of lipocortins as mediators of the anti-inflammatory effects of steroids is in doubt. An alternative explanation is that glucocorticoids inhibit the formation of TNF-α, IL-1β, and other proinflammatory mediators (6).

It is generally accepted that glucocorticoids inhibit the production of IL-1 by LPS-induced murine macrophages (115) and human monocytes (7, 70) as well as by U937 human promonocytic cells (72, 79). A representative experiment, showing suppression by dexamethasone of the production of IL-1α and IL-1β in LPS-stimulated human monocytes (7), is illustrated in Fig. 6. There is a consensus that dexamethasone decreases the steady-state level of IL-1β mRNA in LPS-activated human monocytes. Since many effects of glucocorticoids are exerted at the level of transcription, it was surprising to find that dexamethasone had no detectable effect on LPS-augmented transcription of the IL-1β gene in monocytes (7, 70). However, dexamethasone markedly decreased the stability of IL-1β mRNA, as shown both by steady-state measurements and pulse-labeling. Dexamethasone-induced instability of IL-1β mRNA required protein synthesis; a glucocorticoid antagonist blocked its effects.

In U937 cells primed with phorbol myris-

tate acetate and stimulated with LPS, dexamethasone had some inhibitory effect on the transcription of the IL-1β gene (79). However, in these cells, the transcription factors required for LPS-induced expression of the IL-1β gene are not already present, and protein synthesis is required for their induction: dexamethasone may have been inhibiting formation of the transcription factors rather than the transcription process itself. In U937 cells, dexamethasone also decreases IL-1β mRNA stability (79).

To examine whether glucocorticoid treatment would accelerate degradation of all mRNAs containing an A + U-rich consensus sequence in the 3′ untranslated region, Fos-mRNA was also investigated. In U937 cells, glucocorticoids had no effect on Fos-mRNA stability under conditions when IL-1β mRNA showed a marked decrease in stability (79). Thus, an A + U-rich sequence does not appear to be the common recognition target required for dexamethasone-induced mRNA degradation.

Since the glucocorticoid-induced lowering of IL-1β mRNA levels in LPS-treated monocytes parallels the decrease in intracellular and secreted cytokine levels measured by immunoassay, we conclude that glucocorticoids do not have a major effect on translational efficiency or on secretion (7). Knudsen et al. (73) suggested that glucocorticoids decrease to some extent the efficiency of translation of the IL-1β message, and Kern et al. (70) postulated some inhibition of IL-1β secretion in human monocytes. The dose-response curves for dexamethasone-induced inhibition of intracellular and secreted IL-1β in LPS-induced monocytes reported by Molnar-Kimber et al. (92) (Fig. 5) are also very similar, suggesting that there is no major effect on secretion. In an assay that discriminates between effects on IL-1β synthesis and secretion, dexamethasone inhibited the production of IL-1β with an IC_{50} of 0.2 μM but had little effect on secretion (22).

In summary, the principal mechanism by which glucocorticoids inhibit the formation of

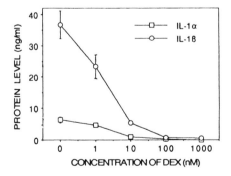

FIGURE 6 Dose-dependent inhibition by dexamethasone (DEX) on the production of IL-1α and IL-1β by LPS-activated human monocytes. Cells were incubated with LPS for 24 h, and cell-associated cytokines were measured by ELISA (73).

IL-1β in monocytes is by decreasing the stability of their mRNAs.

Effects of Glucocorticoids on TNF-α Production

Dexamethasone in a dose-dependent manner decreases LPS-induced production of TNF-α in mouse peritoneal macrophages, the RAW264.7 murine macrophage cell line (53), and human monocytes (92) (Fig. 4). Macrophage activation induces a rise in the rate of transcription of the TNF-α gene and a concurrent increase in the efficiency with which TNF-α mRNA is translated (53); glucocorticoids partially inhibit TNF-α mRNA accumulation in LPS-activated RAW264.7 cells, but their major effect is to depress the translation of the TNF-α mRNA that is produced. RAW264.7 cells were transfected with two constructs designed to assess posttranscriptional activation of TNF-synthesis. In each construct, chloramphenicol acetyltransferase (CAT) transcription was driven by the simian virus 40 late-gene promoter. In construct A, the CAT-encoding sequence was followed by most of the human TNF-α 3′ untranslated region. In control construct B, no TNF-α sequence was present. The presence of the TNF-α 3′ untranslated region permitted translational derepression to occur in cells containing construct A; dexamethasone inhibited this CAT expression, whereas pentoxiphylline did not. Cells transfected with construct A were not induced by LPS or suppressed by dexamethasone.

In summary, dexamethasone only weakly inhibits the accumulation of TNF-α mRNA but strongly inhibits translational derepression. In LPS-activated human monocytes, the great majority of TNF-α protein is extracellular, only about 5% being intracellular; nevertheless, dose-response curves of dexamethasone inhibition of TNF-α in both compartments are comparable (92), so there is no obvious effect on secretion of the cytokine. Inhibition of cytokine production by a novel steroid, mometasone furoate, has also been reported (10).

Bicyclic Imidazoles as Inhibitors of IL-1β and TNF-α Formation

A group at SmithKline Beecham Pharmaceutical Laboratories synthesized a novel group of anti-inflammatory drugs by combining the structurally related pharmacophores of the cyclooxygenase inhibitors flumizole and tiflamizole with the immunomodulatory agent levamisole (51, 52, 54, 87). This led to the class of fused bicyclic 2,3-dihydroimidazo [2,1-b] thiazolines (77). SK&F 81114 inhibits adjuvant arthritis in the rat. Replacement of a substituted phenyl ring in this compound with a 4-pyridyl ring gives SK&F 86002, with improved absorption and in vivo pharmacology. SK&F 86002 inhibits the synthesis of IL-1 in human monocytes (78); this compound is a more potent inhibitor of IL-1 production than are the products of its oxidative metabolism, the sulfoxide (SK&F 86096) and the sulfone (SK&F 104343).

A second series of bicyclic imidazoles was synthesized by replacing the cyclic sulfur atom of the imidazothiazole with a methylene (CH_2) group to produce 6,7-dihydro-[5H]-pyrollo-[1,2-a]imidazoles (79). Of these, SK&F 105561 and its sulfoxide, SK&F 105809, have been most thoroughly investigated. The inactive prodrug SK&F 105809 is reductively metabolized to the sulfoxide SK&F 105561, which is a potent inhibitor of cytokine production in vitro and is also active in vivo. The active sulfide is in turn converted in vivo by oxidative metabolism to the inactive sulfone SK&F 105942. The lead imidazothiazoline SK&F 86002 and the lead imidazopyrrole SK&F 105561 are potent inhibitors of the production of TNF-α and IL-1β in LPS-stimulated human monocytes.

SK&F 86002 and 105561 also inhibit the production of IL-6 and IL-8 by human monocytes (77). When low concentrations of LPS are used to stimulate the cells, the IC_{50}s are comparable to those required for inhibition of TNF-α and IL-1β synthesis; when higher concentrations of LPS are used, more drug is required to inhibit IL-6 and IL-8 formation. However, the same drugs have no effect on the formation of the IL-1 receptor antagonist, granulocyte-macrophage colony stimulating factor, or IFN-α (79). Thus, the effect of the bicyclic imidazoles on induced cytokine formation in monocytes is selective.

SK&F 86002 also inhibits TNF-α production by LPS-stimulated, oil-elicited murine peritoneal macrophages (77). However, the IC$_{50}$ for mouse macrophages (5 to 8 μM) is higher than that for human monocytes (0.5 μM); when the drug is used in nontoxic concentrations, cytokine production is completely inhibited in human cells but only 50% inhibited in mouse cells. The effects of the drug in the mouse macrophage cell line RAW264.7 are similar to those in murine peritoneal macrophages. In summary, the bicyclic imidazoles inhibit TNF-α production by both murine and human monocyte cultures. However, their potency as inhibitors is 5 to 10 times greater in human monocytes than in mouse primary macrophages or macrophage cell lines.

Inhibition by Bicyclic Imidazoles of IL-1β and TNF-α Message Translation

Lee et al. (77) analyzed the mechanism by which bicyclic imidazoles inhibit the synthesis of TNF-α and IL-1β in human monocytes and in THP-1 cells stimulated with LPS (0.5 to 1 μg/ml). Northern (RNA) blots show no effect of SK&F 86002 (5 μM) on IL-1β mRNA but a twofold reduction in the level of TNF-α mRNA. In the same cells, Western blotting (immunoblotting) shows a 10-fold decrease in TNF-α protein and a 2-fold decrease in IL-1β protein. The transcript size and the ratio of intracellular to secreted proteins are unaffected by the drug. SK&F 86002 inhibits the rate of translation of IL-1β message but does not alter the half-life of IL-1β as measured by [^{35}S]-methionine pulse-chase.

In THP-1 cells, SK&F 86002 (1 to 5 μM) has no effect on TNF-α mRNA levels but decreases TNF-α protein levels by almost twofold (77). These findings suggest that SK&F 86002 inhibits IL-1β and TNF-α synthesis at the translational level. To confirm that hypothesis, Lee et al. (77) compared the kinetics

of action of SK&F 86002 with those of fast-acting inhibitors of transcription (actinomycin D) and translation (anisomycin). Parallel cultures of THP-1 cells were simultaneously activated with LPS, and individual cultures were treated with several inhibitors at various times after stimulation. All cultures were harvested 2.5 h after LPS activation, and secreted TNF-α levels were measured by an enzyme-linked immunosorbent assay (ELISA). The results showed that the kinetics of action of SK&F 86002 coincide with those of anisomycin rather than those of actinomycin D, suggesting that SK&F 86002 inhibits a step close to the onset of TNF-α mRNA translation. Glucocorticoids also inhibit TNF-α production mainly at the level of translation (53).

To examine directly the effect of the compounds on the efficiency of TNF-α mRNA translation, cytosolic extracts of human monocytes were fractionated on sucrose gradients. Fractions were analyzed by RNA PCR for TNF-α and cyclophilin. Quiescent monocytes contained a substantial amount of TNF-α mRNA that was primarily associated with 43S preribosomal complexes. Upon activation, the upregulation of TNF-α mRNA was correlated with a proportional increase in TNF-α-specific message associated with polyribosomes. Treatment with active but not inactive analogs of the bicyclic imidazoles resulted in a marked accumulation of TNF-α mRNA in the 43S complex-containing fractions and a concomitant reduction in the polyribosome pool. Neither activation nor drug alone affected the distribution of cyclophilin mRNA in the same fractions. These results suggest that the initiation of TNF-α mRNA translation induced by activation is inhibited by the bicyclic imidazoles (77).

Chin and Kostura (22) found that human blood monocytes stimulated with low doses of LPS synthesize IL-1β but do not secrete it. Secretion can be induced by higher doses of LPS or heat-killed staphylococci. This assay can be used to discriminate between inhibitors of IL-1β synthesis and secretion. SK&F 86002 is a potent inhibitor of IL-1β secretion, suggesting that the drug affects events other than translation.

Effects of Bicyclic Imidazoles in a Model of Shock

A murine model of endotoxic shock in which the animals were injected with LPS in combination with D-galactosamine was studied. Oral administration of SK&F 86002 or SK&F 105809 at 30 or 60 min before LPS injection produced a dose-dependent suppression of serum TNF-α levels (9). This treatment increased the survival of mice in a dose-related manner that was correlated with suppression of levels of circulating TNF-α.

In summary, the bicyclic imidazoles represent a novel class of anti-inflammatory compounds with potent inhibitory effects on the production of the proinflammatory cytokines TNF-α and IL-1β in vitro and in vivo. These effects are exerted mainly at the level of translation of the specific mRNAs. In addition, the bicyclic imidazoles are potent inhibitors of cyclooxygenase and less potent inhibitors of 5-lipoxygenase. No immunosuppressive effects have been demonstrated. The bicyclic imidazoles are active in an acute model of septic shock and in models of acute and chronic inflammation in vivo. The clinical utility of the lead bicyclic imidazoles has been limited by toxicity, but analogs with better safety profiles may be found.

Inhibition by Antioxidants of Transcription of Genes for Proinflammatory Cytokines

Another strategy for identifying small molecules able to inhibit the production of TNF-α, IL-1β, and IL-6 is using antioxidants that prevent the activation of transcription factors required for induced expression of cytokine genes. Several reports on the inhibitory effects of antioxidants on the production of individual cytokines by LPS-activated mouse peritoneal macrophages, human monocytes, THP-1 cells, and human whole blood have been published (Table 1). Eugui et al. (40) investigated the phenomenon systematically, comparing effects of different types of antioxidants on the production of TNF-α, IL-1β, and IL-6 in cultured PBMC activated by LPS and in other ways. Some antioxidants, but not others, were found to be potent inhibitors of the production of all three cytokines. The molecular basis of antioxidant-mediated inhibition was analyzed, and results showed that antioxidants can inhibit the activation of transcription factors NF-κB and AP-1 and the transcription of cytokine genes. The in vitro observations were then confirmed by in vivo experiments in mice.

Initially, a series of compounds was tested for their capacities to inhibit the production of IL-1β in LPS-stimulated human PBMC cultures. Tetrahydropapaveroline (THP), a tetrahydroisoquinoline derivative, was found to be a potent inhibitor of IL-1β production (IC$_{50}$, about 1.5 μM) (Fig. 7). Because they are structurally related to THP, 10,11-dihydroxyaporphine (apomorphine) and norapomorphine were tested and found to inhibit IL-1 production efficiently (Table 2). The $R(+)$ and $R(-)$ stereoisomers of dihydroxyaporphine were equipotent in this assay, showing separation from dopamine against activity.

Antioxidants with widely different structures were then tested for their capacities to inhibit the production of IL-1β in LPS-activated human PBMC. IL-1β released into culture supernatants and cell-associated protein (cell lysates) were evaluated by using a two-site ELISA. Several moderately lipophilic antioxidants, including butylated hydroxyanisole (BHA) and nordihydroguaiauretic acid (NDGA), were found to be potent inhibitors of IL-1β production (IC$_{50}$, 4 μM or lower) (Table 2). NDGA acid is an inhibitor of 5-lipoxygenase as well as of lipid peroxidation. However, another redox 5-lipoxygenase inhibitor, zileuton, did not affect IL-1β formation, suggesting that 5-lipoxygenase products are not involved in the signal transduction system leading to cytokine production in LPS-activated monocytes.

The more hydrophilic antioxidants tested, ascorbic acid and trolox, had no effect on IL-1β production at concentrations up to 200 μM.

TABLE 1 Inhibition by antioxidants of production of TNF-α and IL-1 in monocytes and macrophages

Cells used	Compound	Cytokine tested	Reference
Murine macrophages	BHA	TNF	21
Murine macrophages	Probucol	IL-1	74
THP-1	Probucol, α-Tocopherol	IL-1β	2
Monocytes	N-Acetylcysteine	TNF	99
Human whole blood	Dimethyl sulfoxide, mannitol	IL-8	29

Mannitol, a hydroxyl radical scavenger, was inactive at 100 mM concentration. The same was true of the physiologic lipophilic antioxidant α-tocopherol as well as some classic antioxidants, i.e., butylated hydroxytoluene, quercetin, and N,N'-diphenyl-p-phenylene diamine. N-Acetylcysteine had some inhibitory effect (IC$_{50}$, 42 mM), but this effect was much lower than those of several lipophilic antioxidants (Table 2).

The question arose whether the inhibitory effect of THP is selective for the signaling pathway initiated by LPS or applies also to other effective inducers, such as silica and *S. aureus* Cowan I (Pansorbin). THP proved to be as potent in the inhibition of IL-1β production with all of the other inducers tested. The drug also inhibited IL-1β production when a weaker inducer, zymosan, was used. These observations show that THP can inhibit the production of IL-1β by human PBMC stimulated with several inducers. Similar observations were made with other antioxidants.

THP was also used to analyze the reversibility of the inhibitory effect. Adherent PBMC were cultured with LPS and THP for 2 h and repeatedly washed. Fresh medium containing the same amount of LPS was added, and the cultures were incubated overnight. Duplicate cultures were maintained with LPS and THP for the same period without washing. Removal of the drug did not reverse its inhibitory effect, so the inhibition of IL-1β production by THP appears to be irreversible, at least until protein synthesis replaces inactivated molecules.

Tested in parallel with inhibition of cytokine synthesis, THP, 10,11-dihydroxyapor-

TABLE 2 Antioxidant variations in potency as inhibitors of cytokine formation (40)

Activity and antioxidant	IC$_{50}$ (μM)
High	
BHA	2.9
THP	1.0
Apomorphine	2.6
Norapomorphine	1.6
NDGA	1.3
Mepacrine	3.0
Low[a]	
Ascorbic acid	—
α-Tocopherol	—
Mannitol	—
Trolox	—
Butylated hydroxytoluene	—
Quercetin	—
N,N'-diphenyl-p-phenylene diamine	—
Zileuton (5-lipoxygenase inhibitor)	—

[a]—, insignificant inhibition at 50 to 200 μM.

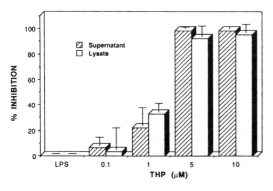

FIGURE 7 Dose-dependent inhibition by THP of the production of IL-1β in human PBMC activated by LPS. Means and standard errors of intracellular (lysate) and extracellular cytokines are shown (40).

phine, and BHA did not inhibit protein synthesis, as shown by incorporation of labeled leucine. Emetine, which is structurally related to THP, strongly inhibited protein synthesis in parallel experiments using monocytes.

BHA, THP, 10,11-dihydroxyaporphine, and NDGA were further tested for their effects on production of TNF-α and IL-6 by LPS-stimulated human PBMC. The three compounds are approximately equipotent as inhibitors of the production of the three cytokines (Table 2).

To ascertain whether this effect is exerted in all cell types, human dermal fibroblasts and fibroblast-type cells from synovial tissues of patients with RA were studied. In these cells, IL-1 induces the production of IL-6. THP, BHA, 10,11-dihydroxyaporphine, and norapomorphine at concentrations higher than those required to totally inhibit the production of IL-6 in PBMC had no demonstrable effect on IL-6 production in fibroblasts. Thus, antioxidant-mediated inhibition of cytokine production does not occur in all cell types.

Some of the compounds in Table 1 were selected for analysis of in vivo effects on cytokine production. TNF-α and IL-1β were measured in serum following a lethal challenge with LPS (200 μg per mouse). Levels of TNF-α in circulating blood peak at about 1.5 h post-injection, and those of IL-1β peak at 4 h after LPS injection. Subcutaneous administration of a single dose of 10,11-dihydroxyaporphine (100 mg/kg) given 30 min before challenge inhibits TNF-α production by 95% (Fig. 8). Before IL-1β reached its peak, mice were given a second dose of dihydroxyaporphine (50 mg/kg) 2 h after LPS injection and 2 h before bleeding. This treatment reduced the circulating levels of IL-1β by 88% (Fig. 8). Thus, in vivo cytokine production is strongly inhibited by THP and by dihydroxyaporphine.

Molecular Biological Effects of Antioxidants

To analyze the mechanism by which antioxidants suppress cytokine formation, levels of cytokine mRNAs were determined in LPS-stim-

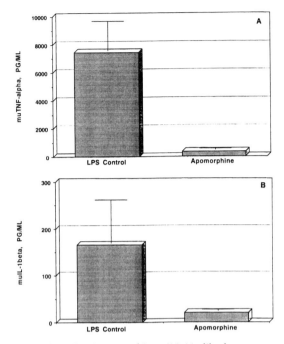

FIGURE 8 Apomorphine (10,11-dihydroxyaporphine) treatment markedly decreases levels of TNF-α and IL-1β in the circulation of mice challenged with LPS (40).

ulated PBMC (40). Previous studies had demonstrated that LPS increases oxidant production (66) and IL-1β mRNA. LPS also increases levels of TNF-α, IL-6, and IL-8 mRNAs in PBMC. The antioxidants tested decreased TNF-α, IL-1β, and IL-6 mRNA levels to baseline expression but had much less effect on IL-8 mRNA. Some of these effects are shown in Fig. 9. To ascertain whether these effects are due to changes in transcription or changes in mRNA stability, nuclear transcription assays were performed. LPS markedly stimulated transcription of the IL-1β gene, and THP antagonized the stimulatory effect of LPS on transcription. In view of observations that antioxidants decrease AP-1 activity, one of the reference messages studied was c-fos. However, no effect of THP on transcription of the c-fos gene could be detected.

The experiments just described suggest that antioxidants inhibit LPS-stimulated transcription of some cytokine genes. To ascertain

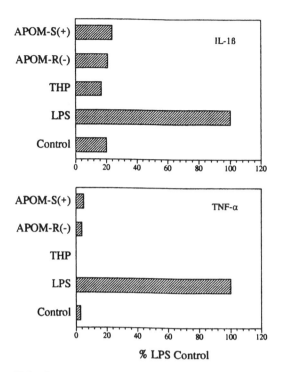

FIGURE 9 Decreased steady-state levels of IL-1β and TNF-α mRNAs in human PBMC treated with stereoisomers of apomorphine [APOM-S(+) and APOM-R(−)] and THP (40).

whether this inhibition is correlated with inhibited activation of transcription factors, electrophoretic mobility shift assays were performed. NF-κB and AP-1 were analyzed first because of reports that they are subject to redox regulation (62, 107). Nuclear extracts of unstimulated PBMC showed NF-κB activity, which increased following LPS stimulation. Specificity of binding was shown by competition with unlabeled oligonucleotides. When PBMC were treated with THP in the presence or absence of LPS, nuclear NF-κB activity was markedly decreased or eliminated altogether; similar observations were made with AP-1 (40). Thus, THP inhibited activation of NF-κB and AP-1 in intact cells but had no effect on the binding of these transcription complexes to cognate DNA recognition sequences. In cells treated with THP, no effect on several other transcription factors (SP-1, CRE, CTF/

NF-1, OCT) could be demonstrated. Inhibition of the activation of transcription factors, including NF-κB (63), AP-1 (25, 31), and others (26, 63, 94), by antioxidants could explain the antioxidant suppression of cytokine gene transcription.

It is generally accepted that a c-*fos* serum-responsive element and NF-κB play major roles in promoting expression of the IL-6 gene (56). At least one member of the NF–IL-6 transcription factor complex also associates with the p50 subunit of NF-κB (76). Evidence implicating κB sequences in LPS-induced expression of the TNF-α gene in macrophages has also been presented (110, 119). Less is known about regulation of the expression of the IL-1β gene, which is complicated by the fact that DNA sequences both proximal and distal to the transcription start site are involved (112). A phorbol myristate-responsive IL-1β enhancer element contains a DNA motif similar to that of the AP-1-binding site of the collagenase gene (112). The importance of NF–IL-6 transcription factors in the activation of IL-1 has also been reported (112). The presence of a functional NF-κB site in the human IL-1β promoter has recently been described (57). Mutation of the NF-κB site in the context of the IL-1β promoter decreases the responsiveness of the promoter to LPS in human monocytic lineage cells. Those authors concluded that the IL-1β gene may be considered an important additional member of the family of cytokine genes regulated in part by the NF-κB/*rel* family of transcription factors (57). Thus, inhibition of NF-κB and AP-1 activation may contribute to the observed effect of antioxidants on transcription of the IL-1β gene, although specific internal deletions of sequences binding transcription factors will be required to establish their role.

Comments on Suppression of Cytokine Synthesis by Antioxidants

Several antioxidants are potent inhibitors of the production of TNF-α and IL-1β in LPS-stimulated human PBMC. The same concentrations of antioxidants are equally effective in the inhibition of cytokine production in PBMC irrespective of the stimulus (LPS, staphylococci, silica, and zymosan). Effects of antioxidants are not due to overall inhibition of protein synthesis and are gene selective. The antioxidants do not affect transcription of the β-actin and c-*jun* genes, and they actually increase production of the IL-1 receptor antagonist as well as of lysozyme and lysosomal enzymes in PBMC (131).

Remarkably, the effects of the antioxidants are also cell type selective. Antioxidants that are potent inhibitors of IL-6 production in PBMC do not inhibit production of the same cytokine in dermal or synovial fibroblasts. The mechanisms stimulating IL-6 production are diverse (56), and the transcription factors used in fibroblasts are presumably less susceptible to redox regulation than those used in PBMC. Other cell types should be studied, but the findings raise the possibility that selected antioxidants may be able to suppress the formation of proinflammatory cytokines in cells of the monocyte-macrophage lineage without causing major side effects.

Antioxidants show a wide variation in potency as inhibitors of cytokine production. One of the most active compounds is THP, a tetrahydroisoquinoline. Two compounds with some structural relationship to THP, 10,11-dihydroxyaporphine (apomorphine) and norapomorphine, are also potent inhibitors. Since the two stereoisomers of dihydroxyaporphine are equipotent as inhibitors of cytokine production, there is separation of this activity from dopamine agonist activity.

The classic antioxidants BHA and NDGA are strong inhibitors of cytokine production. NDGA is an inhibitor of 5-lipoxygenase and of lipid peroxidation. However, zileuton, used in concentrations that inhibit 5-lipoxygenase activity, has no effect on cytokine production. This suggests that the antioxidant activity of NDGA rather than its 5-lipoxygenase inhibitory activity explains its effect on cytokine production. The common feature of the structurally diverse compounds with high activity

listed in Table 2 is their capacity to function as antioxidants and moderate lipophilicity.

The water-soluble antioxidants ascorbic acid and Trolox are inactive, suggesting that the antioxidants have to function in a lipid environment. However, this must be a specialized microenvironment within target cells. The physiologic membrane antioxidant α-tocopherol is ineffective as an inhibitor of cytokine production, as are several classic antioxidants listed in Table 2 (butylated hydroxytoluene, quercetin, N,N'-diphenyl-*p*-phenylene diamine). Further work is necessary to define structure-activity relationships of inhibitors of cytokine production, the intracellular compartment in which they are active, and the mechanism by which they block the activation of transcription factors.

Antioxidants Augment Expression of IL-1ra

Coordinate inhibition by a drug of the production of TNF-α, IL-1β, and IL-6 in cells of the monocyte-macrophage lineage could obviously exert useful anti-inflammatory effects. If the drug also augments production of IL-1ra, that could be an additional advantage. Waters et al. (131) found that some antioxidants known to be inhibitors of TNF-α, IL-1β, and IL-6 production increase the production of IL-1ra in cultures of human monocytes. The increase may not seem spectacular, but the amounts of IL-1ra produced are more than 100 times those of IL-1β, and the ratio of production of IL-1β and IL-1ra is substantially changed by the drugs.

It has been postulated that a balance between the production of proinflammatory cytokines (such as TNF-α and IL-1β) and IL-1ra plays an important role in determining the course of RA (8). The same may be true of the arthritis associated with Lyme disease, which is caused by the bacterium *B. burgdorferi*. Patients with high concentrations of IL-1ra and low concentrations of IL-1β in synovial fluid have rapid resolution of attacks of arthritis, whereas patients with the reverse pattern of cytokine concentrations have long intervals to recovery

(91). These observations are consistent with the interpretation that induction of IL-1 release by products of *B. burgdorferi* contributes to the arthritis of Lyme disease. Using drugs to manipulate the balance of production of cytokines and IL-1ra in such a way as to produce antiinflammatory effects is a novel therapeutic strategy.

Effects of Antioxidants in Experimental Animal Models of Septic Shock

Chaudhri and Clark (21) made mice sensitive to LPS by low-level infection with malaria. Pretreatment with the antioxidant BHA or with the iron chelator desferal decreased TNF levels in the circulation following LPS challenge. BHA is a potent inhibitor of the production of both TNF-α and IL-1β by human monocytes (Table 2). Nonaka et al. (95) found that after intraperitoneal injection of LPS, mice have increased lipid peroxide levels (measured by peroxidase activity of hemoglobin) in their circulation. An antioxidant, 2-octadecylascorbic acid, decreased levels of circulating lipid peroxides and significantly increased the survival of mice injected with LPS. Calves are sensitive to shock induced by LPS infusion. The lazaroid tirilazad mesylate, a 21-aminosteroid, was shown by Rose and Semrad (106) to mitigate clinical signs of endotoxemia in calves, attenuating hypoglycemia and preventing lactic acidosis. Tirilazad mesylate is an antioxidant preventing iron-dependent lipid peroxidation (15). The protective effects of antioxidants in models of septic shock are likely to be due, at least in part, to inhibition of cytokine production.

POSSIBLE THERAPIES FOR SEPTIC SHOCK

During the past few years, anecdotal reports of inhibition of TNF-α and IL-1 synthesis by drugs have been replaced by systematic studies of natural and synthetic regulators of cytokine production. These studies have illuminated the mechanisms by which feedback regulation of cytokine production occurs both locally and

systemically. Circulating monocytes and tissue macrophages in healthy animals can be described as being in a neutral state. They can be activated into an inflammatory state by LPS, MDP, other microbial products, immune complexes, or activated complement components to become proinflammatory and procoagulant. LPS and other stimulants rapidly induce the formation of TNF-α by activating transcription and increasing the efficiency of translation of TNF-α mRNA. Soon after TNF-α induction, synthesis of IL-1β is induced, and then IL-6 and IL-8 are induced. The synthesis of TNF-α is turned off first, the synthesis of IL-1β continues for a while, and that of IL-6 and IL-8 continues longer. A local feedback regulator of TNF-α formation is PGE$_2$, which increases cAMP levels and thereby inhibits the synthesis of TNF-α but not that of IL-1β. A second local feedback regulator is IL-10, which is released by LPS-activated monocytes later than the aforementioned cytokines. IL-10 suppresses the synthesis of TNF-α and IL-1β while stimulating the formation of IL-1ra (84).

Excess TNF-α and IL-1β that pass into the circulation, as in septic shock, trigger the release of systemic feedback inhibitors of their production. They induce release from the pituitary of corticotropin-releasing factor, which in turn induces the release into the circulation of adrenocorticotropin and glucocorticoids (12). Glucocorticoids inhibit the production of TNF-α and IL-1β with little (if any) effect on transcription. Glucocorticoids decrease the stability of IL-1β mRNA and the efficiency of TNF-α mRNA translation. In patients with septic shock, levels of circulating adrenalin also increase. Acting through β-adrenergic receptors, adrenalin increases cAMP levels in monocytes and macrophages and inhibits the formation of TNF-α.

For the prevention and treatment of septic shock and for the treatment of diseases with inflammatory pathogenesis, it is desirable to inhibit the formation of both TNF-α and IL-1β. In periodontal disease and RA, an additional effect, i.e., inhibiting the synthesis of IL-6, a comediator of bone erosion, is desirable. Further useful properties of a drug would be augmentation of the formation of the IL-1 receptor antagonist and of IL-10. This list of requirements excludes most known drugs. For example, pentoxyfylline and cAMP PDE-IV inhibitors suppress the formation of TNF-α but not IL-1β. An inhibitor of the enzyme that activates pro-IL-1β (124) would have no effect on TNF-α, IL-1α, or IL-6.

Two groups of compounds (bicyclic imidazoles and some antioxidants) show that it is possible to selectively inhibit the formation of proinflammatory cytokines, with one group exerting effects at the level of transcription and the other exerting effects at the level of translation. Stimulating cells of the monocyte-macrophage lineage into an inflammatory state involves activation of transcription factors and coordinate induction of the expression of genes for proinflammatory cytokines. Some antioxidants prevent the activation of transcription factors required for the coordinate expression of TNF-α, IL-1β, and IL-6 genes in response to stimulation of monocytes and macrophages by LPS and in other ways. These inhibitory effects are gene selective: the same compounds can augment the formation of IL-1ra and of lysosomal enzymes, while the expression of most genes is unaffected. The effects of the antioxidants are also cell type selective, inhibiting IL-6 production in monocytes but not in fibroblasts. A role for oxidant damage in inflammation and septic shock, and in particular in respiratory distress syndromes, has been postulated (29, 97). Antioxidants could inhibit this type of tissue damage as well as the induction of TNF-α and IL-1β gene expression.

Bicyclic imidazoles inhibit selectively the translation of TNF-α and IL-1β mRNAs. Doubtless other classes of drugs coordinately inhibiting cytokine production will be identified. In principle, they could provide the first successful treatment of septic shock as well as novel therapies for inflammatory diseases free from the limiting side effects of glucocorticoids and cyclooxygenase inhibitors (35, 75). Since the pathogenesis of shock is complex, it may be necessary to administer a drug that suppresses cytokine production together with the C1 INH, which blocks the contact-phase protease pathways. However, the drug suppressing cytokine production might provide useful prophylaxis against shock in susceptible patients, e.g., those in intensive care. Clinical studies to be carried out during the next few years should establish whether the promise of drugs suppressing the production of proinflammatory cytokines is fulfilled.

ACKNOWLEDGMENTS

We are indebted to John Kenney, who collaborated in several studies of cytokine production by bacterial products, and to Simon Lee, Barbara De Lustro, Sussan Rouhafza, and Mariola Ilnicka, who collaborated in our studies of inhibitors of cytokine production. K. Molnar-Kimber kindly provided Figs. 4 and 5, and Linda Miencier's help in the preparation of the manuscript is gratefully acknowledged.

REFERENCES

1. Aarden, L. A. 1989. Hybridoma growth factor. *Ann. N.Y. Acad. Sci.* **557**:192–199.

2. Akeson, A. L., C. W. Woods, L. B. Mosher, C. E. Thomas, and R. L. Jackson. 1991. Inhibition of IL-1β expression in THP-1 cells by probucol and tocopherol. *Atherosclerosis* **86**:261–270.

3. Allison, A. C. 1994. Adjuvants and immune enhancement. *Int. J. Technol. Assessment Health Care* **10**:107–120.

4. Allison, A. C., and N. E. Byars. 1986. An adjuvant formulation that selectively elicits the formation of antibodies of protective isotypes and of cell-mediated immunity. *J. Immunol. Methods* **95**:157–168.

5. Allison, A. C., C. Chin-R, and Y. Cheng. 1993. Cyclooxygenase inhibitors vary widely in potency for preventing cytokine-induced bone resorption. *Ann. N.Y. Acad. Sci.* **696**:149–170.

6. Allison, A. C., and S. W. Lee. 1989. The mode of action of anti-rheumatic drugs. I. Anti-inflammatory and immunosuppressive effects of glucocorticoids. *Prog. Drug Res.* **33**:63–81.

7. Amano, Y., S. W. Lee, and A. C. Allison. 1993. Inhibition by glucocorticoids of the formation of interleukin-1α, interleukin-1β and interleukin-6: mediation by decreased mRNA stability. *Mol. Pharmacol.* **43**:176–182.

8. Arend, W. P., and J. M. Dayer. 1990. Cytokines and cytokine inhibitors or antagonists in rheumatoid arthritis. *Arthritis Rheum.* **33**:305–315.

9. Badger, A. M., D. L. Olivera, J. E. Talmadge, and N. Hanna. 1989. Protective effect

of SK&F 86002, a novel dual inhibitor of arachidonic acid metabolism, in murine models of endotoxic shock: inhibition of tumor necrosis factor as a possible mechanism of action. *Circ. Shock* **27:** 51–61.

10. **Barton, B. E., J. P. Jakway, S. R. Smith, and M. I. Siegel.** 1991. Cytokine inhibition by a novel steroid, mometasone furoate. *Immunopharmacol. Immunotoxicol.* **13:**251–261.

11. **Bertin, P. B., J. S. Kenney, M. R. Welch, H. B. Lindsley, R. Treves, and A. C. Allison.** 1992. IL-1 inhibitors block IL-6 and IL-8 secretion by fragments of human rheumatoid arthritis synovium. *Arthritis Rheum.* **35:**C189.33.

12. **Besedovsky, H., A. del Rey, E. Sorkin, and C. A. Dinarello.** 1986. Immunoregulatory feedback between interleukin-1 and glucocorticoid hormones. *Science* **233:**652–654.

13. **Beutler, B.** 1992. *Tumor Necrosis Factors: the Molecules and Their Emerging Role in Medicine,* p 574. Raven Press, New York.

14. **Brandwein, S. R.** 1986. Regulation of interleukin 1 production by mouse peritoneal macrophages. *J. Biol. Chem.* **261:**8624–8632.

15. **Braughler, J. M., J. F. Pregenzer, J. L. Chase, L. A. Duncan, E. J. Jacobsen, and J. M. McCall.** 1987. Novel 21-aminosteroids as potent inhibitors of iron-dependent lipid peroxidation. *J. Biol. Chem.* **262:**10438–10440.

16. **Brennan, F. M., D. Chantry, A. Jackson, R. Maini, and M. Feldmann.** 1989. Inhibitory effect of TNF-α antibodies on synovial cell interleukin-1 production in rheumatoid arthritis. *Lancet* **ii:**244.

17. **Breviario, F., P. Proserpio, F. Bertocchi, M. G. Lampugnani, A. Mantovani, and E. Dejana.** 1990. Interleukin-1 stimulates prostacyclin production by cultured human endothelial cells by increasing arachidonic acid mobilization and conversion. *Arteriosclerosis* **10:**129–134.

18. **Buchanan, J. W., and H. N. Wagner.** 1985. Regional phagocytosis in man, p. 147–170. *In* S. M. Reichard and J. P. Filkins (ed.), *The Reticuloendothelial System—a Comprehensive Treatise.*

19. **Cannon, J. G., R. G. Tompkins, J. A. Gelfand, H. R. Michie, G. G. Stanford, Jos W. M. van der Meer, S. Endres, G. Lonnemann, J. Corsetti, B. Chernow, D. W. Wilmore, S. M. Wolff, J. F. Burke, and C. A. Dinarello.** 1990. Circulating interleukin-1 and tumor necrosis factor in septic shock and experimental endotoxin fever. *J. Infect. Dis.* **161:**79–84.

20. **Carre, P. C., R. L. Martenson, T. E. King, P. W. Noble, C. L. Sable, and P. W. H. Rickes.** 1991. Increased expression of the interleukin-8 gene by alveolar macrophages in idiopathic pulmonary fibrosis. A potential mechanism for the recruitment and activation of neutrophils in lung fibrosis. *J. Clin. Invest.* **88:**1802–1810.

21. **Chaudhri, G., and I. A. Clark.** 1989. Reactive oxygen species facilitate the *in vitro* and *in vivo* lipopolysaccharide-induced release of tumor necrosis factor. *J. Immunol.* **143:**1290–1294.

22. **Chin, G., and M. G. Kostura.** 1993. Dissociation of IL-1β synthesis and secretion in human blood monocytes stimulated with bacterial cell wall products. *J. Immunol.* **151:**5574–5585.

23. **Coulpier, M., A. Andrew, H. Dauchel, O. Costa, O. Lees, M. Fontaine, and J. Ripoche.** Glucocorticoids enhance the IL-1 augmentation of complement C3 and factor B production by human endothelial cells. *Eur. J. Cell. Biol.,* in press.

24. **Crofford, L. J., R. L. Wilder, A. P. Ristimäki, H. Sano, E. F. Remmers, H. R. Epps, and T. Hla.** 1994. Cyclooxygenase-1 and -2 expression in rheumatoid synovial tissues: effects of interleukin-1β, phorbol ester and corticosteroids. *J. Clin. Invest.* **93:**1095–1101.

25. **Datta, R., D. E. Hallahan, S. M. Kharbanda, E. Rubin, M. L. Sherman, E. Huberman, R. R. Weichselbaum, and D. W. Kufe.** 1992. Involvement of reactive oxygen intermediates in the induction of c-jun gene transcription by ionizing radiation. *Biochemistry* **31:**8300–8306.

26. **Datta, R., N. Taneja, V. P. Sukhante, S. A. Querishi, R. Weichselbaum, and D. W. Kufe.** 1993. Reactive oxygen intermediates target CC (A/T) 6 GG sequences to mediate activation of the early growth response 1 transcription factor gene by ionizing radiation. *Proc. Natl. Acad. Sci. USA* **90:**2419–2422.

27. **Davatelis, G., S. D. Wolpe, B. Sherry, J.-M. Dayer, R. Chicheportiche, and A. Cerami.** 1988. Macrophage inflammatory protein-1: a prostaglandin-independent endogenous pyrogen. *Science* **243:**1066–1068.

28. **Debets, J. M. H., R. Kampmeijer, M. P. M. H. Van der Linden, W. A. Buurman, and C. J. Van der Linden.** 1989. Plasma tumor necrosis factor and mortality in critically ill septic patients. *Crit. Care Med.* **17:**489–494.

29. **De Forge, L. E., J. C. Fantone, J. S. Kenney, and D. G. Remick.** 1992. Oxygen radical scavengers selectively inhibit interleukin 8 production in human whole blood. *J. Clin. Invest.* **90:**2123–2129.

30. **Dehring, D. J.** 1993. Sheep and pigs as animal models of bacteremia, p. 1060–1075. *In* G. Schlag and H. Redl (ed.), *The Pathophysiology of Shock, Sepsis and Organ Failure.* Springer Verlag, Berlin.

31. **Devary, Y., R. A. Gottlieb, L. F. Lau, and M. Karin.** 1991. Rapid and preferential activation of the c-jun gene during the mammalian UV response. *Mol. Cell. Biol.* **11:**2804–2811.

32. **Dickneite, G.** 1993. Influence of C1 inhibitor

on inflammation, edema and shock. *Behring Inst. Mitt.* **93:**299–305.

33. Dixon, J. S., H. A. Bird, N. G. Sitton, M. E. Pickup, and V. Wright. 1984. C-reactive protein in the serial assessment of disease activity in rheumatoid arthritis. *Scand. J. Immunol.* **13:**39–44.

34. Elices, M., L. Osborn, Y. Takada, C. Crause, S. Luhowskyj, M. Hemler, and R. R. Lobb. 1990. VCAM-1 on activated endothelium interacts with the leukocyte integrin VLA-4 at a site distinct from the fibronectin-binding site. *Cell* **60:**577–584.

35. Elliott, M. J., R. N. Maini, M. Feldmann, R. O. Williams, F. M. Brennan, and C. Q. Chu. 1993. Treatment of rheumatoid arthritis with chimeric monoclonal antibodies to TNF-α: safety, clinical efficacy and control of the acute-phase response. *J. Cell. Biochem.* **145:**17B .

36. Ellouz, F., A. Adam, R. Ciorbaru, and E. Lederer. 1974. Minimal structural requirements for adjuvant activity of bacterial peptidoglycans. *Biochem. Biophys. Res. Commun.* **59:**1317–1324.

37. Emeis, J. J., and G. M. van den Hoogen. 1992. Pharmacological modulation of the endotoxin-induced increase in plasminogen activator inhibitor activity in rats. *Blood Coagulation Fibrinolysis* **3:**575–581.

38. Endres, S., H. J. Fuelle, B. Sinha, et al. 1991. Cyclic nucleotides differentially regulate the synthesis of tumor necrosis factor α and interleukin 1 beta by human mononuclear cells. *Immunology* **72:**56–60.

39. Eskandari, M. K., G. Bolgos, C. Miller, D. T. Nguyen, L. E. De Forge, and D. G. Remick. 1992. Anti-tumor necrosis factor antibody therapy fails to prevent lethality after cecal ligation and puncture or endotoxemia. *J. Immunol.* **148:**2724–2730.

40. Eugui, E. M., B. DeLustro, S. Rouhafza, M. Ilnicka, S. W. Lee, R. Wilhelm, and A. C. Allison. 1994. Some antioxidants inhibit, in a coordinate fashion, the production of TNF-α, IL-1β and IL-6 by human peripheral blood mononuclear cells. *Int. Immunol.* **6:**409–422.

41. Fast, D. J., P. M. Schlievert, and R. D. Nelson. 1989. Toxic shock syndrome-associated staphylococcal and streptococcal pyrogenic toxins are potent inducers of tumor necrosis factor production. *Infect. Immun.* **57:**291–294.

42. Feinberg, B., R. Kurzrock, M. Talpaz, M. Blick, S. Saks, and J. U. Gutterman. 1988. A phase 1 clinical trial of intravenously administered recombinant tumor necrosis alpha in cancer patients. *J. Clin. Oncol.* **6:**1328–1334.

43. Fischer, E., M. A. Marano, A. Barber, A. Hudson, K. Lee, C. S. Rock, A. S. Hawes, R. C. Thompson, T. J. Hayes, T. D. Anderson, W. R. Benjamin, S. F. Lowry, and L. L.

Moldawer. 1991. A comparison between the effects of interleukin 1α administration and sublethal endotoxemia in primates. *Am. J. Physiol.* **261:**R442–R452.

44. Fischer, E., M. A. Marano, K. van Zee, C. S. Rock, A. S. Hawes, W. A. Thompson, L. de Forge, J. S. Kenney, D. G. Remick, D. C. Blaedow, R. C. Thompson, S. F. Lowry, and L. L. Moldawer. 1992. Interleukin-1 receptor blockade improves survival and hemodynamic performance in *Escherichia coli* septic shock, but fails to alter host responses to sublethal endotoxemia. *J. Clin. Invest.* **89:**1551–1557.

45. Fisher, C. J., Jr., G. J. Slotman, S. Opal, J. Pribble, D. Stiles, and M. Catalano. 1994. Initial evaluation of human recombinant interleukin-1 receptor antagonist in the treatment of sepsis syndrome: a randomized open label, placebo-controlled multi-center trial. *Crit. Care Med.* **22:**12–21.

46. Fong, Y., M. A. Marano, L. L. Moldawer, H. Wei, S. E. Calvano, J. S. Kenny, A. C. Allison, A. Cerami, G. T. Shires, and S. F. Lowry. 1990. The acute splanchnic and peripheral tissue metabolic response to endotoxin in man. *J. Clin. Invest.* **85:**1896–1904.

47. Fong, Y., K. J. Tracey, L. L. Moldawer, D. G. Hesse, K. B. Manogue, J. S. Kenny, A. T. Lee, G. C. Kuo, A. C. Allison, S. F. Lowry, and A. Cerami. 1989. Antibodies to cachectin/tumor necrosis factor reduce interleukin-1β and interleukin-6 appearance during lethal bacteremia. *J. Exp. Med.* **170:**1627–1633.

47a. Gaston, J. S. H., S. Strober, J. J. Solvera, D. Gandour, N. Lane, D. Schurman, R. T. Hoppe, R. C. Chin, E. M. Eugui, J. H. Vaughn, and A. C. Allison. 1988. Dissection of the mechanisms of immune injury in rheumatoid arthritis using total lymphoid irradiation. *Arthritis Rheum.* **31:**21–30.

48. Geczy, A. F., K. Alexander, H. V. Bashir, J. P. Edmonds, L. Upfold, and J. Sullivan. 1983. HLA-B27, *Klebsiella* and ankylosing spondylitis: biological and chemical studies. *Immunol. Rev.* **70:**23–50.

49. Girardin, E., G. E. Giran, J.-M. Dayer, P. Roux-Lombard, and P.-H. Lambert. 1988. Tumor necrosis factor and interleukin-1 in the serum of children with severe infectious purpura. *N. Engl. J. Med.* **319:**397–400.

50. Gjörloff, A., H. Fischer, G. Hedlund, J. Hansson, J. S. Kenney, A. C. Allison, H.-O. Sjogren, and M. Dohlsten. 1991. Induction of interleukin-1 in human monocytes by the superantigen staphylococcal enterotoxin A requires the participation of T cells. *Cell. Immunol.* **137:**61–71.

51. Griswold, D. E., P. C. Hillegas, P. C. Meunier, M. J. Di Martino, and N. Hanna. 1988.

Effect of inhibitors of eicosanoid metabolism in murine collagen-induced arthritis. *Arthritis Rheum.* **31**:1406–1412.

52. **Griswold, D. E., S. Hoffstein, P. J. Marshall, E. F. Webb, P. E. Bender, and N. Hanna.** 1989. Inhibition of inflammatory cell infiltration by bicyclic imidazoles SK&F 86002 and SK&F 104493. *Inflammation* **13**:727–739.

53. **Han, J., P. Thompson, and B. Beutler.** 1990. Dexamethasone and pentoxyfylline inhibit endotoxin-induced cachectin/tumor necrosis factor synthesis at separate points in the signaling pathway. *J. Exp. Med.* **172**:391–394.

54. **Hanna, N., P. J. Marshal, and J. Newton, Jr., et al.** 1990. Pharmacological profile of SK&F 105809, a dual inhibitor of arachidonic acid metabolism. *Drugs Exp. Clin. Res.* **16**:137–147.

55. **Hart, P. H., G. F. Vitti, D. R. Burgess, G. A. Whitty, D. S. Piccoli, and J. A. Hamilton.** 1989. Potential anti-inflammatory effects of interleukin 4: suppression of human monocyte tumor necrosis factor-α, interleukin-1 and prostaglandin E$_2$. *Proc. Natl. Acad. Sci. USA* **86**:3803–3807.

56. **Hirano, T., S. Akira, T. Taga, and T. Kishimoto.** 1990. Biological and clinical aspects of interleukin 6. *Immunol. Today* **11**:443–449.

57. **Hiscott, J., J. Merois, J. Garoufalis, M. S'Addario, A. Roulston, I. Kwan, N. Pepin, J. Lacoste, H. Nguyen, G. Bensi, and M. Fenton.** 1993. Characterization of a functional NF-κB site in the human interleukin-1β promoter: evidence for a positive autoregulatory loop. *Mol. Cell. Biol.* **13**:6231–6240.

58. **Hla, T., A. Ristimäki, S. Appleby, and J. G. Barriocanal.** 1993. Cyclooxygenase gene expression in inflammation and angiogenesis. *Ann. N.Y. Acad. Sci.* **696**:197–204.

59. **Hoiby, N., and C. Koch.** 1990. *Pseudomonas aeruginosa* infection in cystic fibrosis and its management. *Thorax* **45**:881–884.

60. **Hurme, M.** 1990. Modulation of interleukin-1β production by cyclic AMP in human monocytes. *FEBS Lett.* **263**:35–37.

61. **Ishimi, Y., C. Miyaura, C. H. Jin, T. Akatsu, E. Abe, Y. Nakamura, A. Yamaguchi, S. Yoshiki, T. Matsuda, T. Hirano, T. Kishimoto, and T. Suda.** 1990. IL-6 is produced by osteoblasts and induces bone resorption. *J. Immunol.* **145**:3297–3303.

62. **Israel, N., M.-A. Gougerat-Pocidalo, A. Aillet, and J.-L. Virelizier.** 1992. Redox status influences constitutive or induced NF-κB translocation and HIV long terminal repeat activity in human T and monocytic cell lines. *J. Immunol.* **149**:3386–3393.

63. **Ivanov, V., M. Merkenschlager, and R. Ceredig.** 1993. Antioxidant treatment of thymic organ cultures decreases NF-κB and TCF1(α) tran-

scription factor activities and inhibits αβ T cell development. *J. Immunol.* **151**:4694–4704.

64. **Jeffcoat, M. K., M. S. Reddy, L. W. Moreland, and W. J. Koopman.** 1993. Effects of nonsteroidal antiinflammatory drugs bone loss in chronic inflammatory disease. *Ann. N.Y. Acad. Sci.* **696**:292–302.

65. **Jesmok, G., C. Lindsey, M. Duerr, M. Fournel, and T. Emerson, Jr.** 1992. Efficacy of monoclonal antibody against human tumor necrosis factor in *E. coli*-challenged swine. *Am. J. Pathol.* **141**:1197–1207.

66. **Johnston, R. B., Jr.** 1981. Enhancement of phagocytosis-associated oxidative metabolism as a manifestation of macrophage activation, p. 33–52. *In* E. Pick (ed.), *Lymphokines,* vol. 3. Academic Press, Inc., New York.

67. **Jupin, C., S. Anderson, C. Damais, J. E. Alouf, and M. Parant.** 1988. Toxic shock syndrome toxin 1 as an inducer of tumor necrosis factors and gamma interferon. *J. Exp. Med.* **167**: 752–761.

68. **Kaplan, A. P., and M. Silverberg.** 1987. The coagulation-kinin pathway of human plasma. *Blood* **70**:1–15.

69. **Kassis, S., J. C. Lee, and N. Hanna.** 1989. Effects of prostaglandins and cAMP levels on monocyte IL-1 production. *Agents Actions* **27**:274–276.

70. **Kern, J. A., R. J. Lamb, J. C. Reed, R. P. Daniele, and P. C. Nowell.** 1988. Dexamethasone inhibition of interleukin-1β production by human monocytes. *J. Clin. Invest.* **81**:237–244.

71. **Killbourn, R. G., S. S. Gross, A. Jubran, J. Adams, R. Levi, O. W. Griffith, and R. F. Lodocto.** 1990. NG-methyl-L-arginine attenuates TNF-induced hypotension: implications for the involvement of nitric oxide. *Proc. Natl. Acad. Sci. USA* **87**:3629–3632.

72. **Knudsen, P. J., C. A. Dinarello, and T. B. Strom.** 1986. Prostaglandins post-transcriptionally inhibit monocyte expression of interleukin 1 activity by increasing intracellular cyclic adenosine monophosphate. *J. Immunol.* **137**:3189–3194.

73. **Knudsen, P. J., C. A. Dinarello, and T. B. Strom.** 1987. Glucocorticoids inhibit transcriptional and post-transcriptional expression of interleukin 1 in U937 cells. *J. Immunol.* **139**:4129–4134.

74. **Ku, G., N. S. Doherty, J. A. Wolos, and R. L. Jackson.** 1988. Inhibition by probucol of interleukin 1 secretion and its implication in atherosclerosis. *Am. J. Cardiol.* **62**:77B–81B.

75. **Lebsack, M. E., C. C. Paul, D. C. Bloedow, F. X. Burch, M. A. Sade, W. Chase, and M. A. Catalano.** 1991. Subcutaneous IL-1 receptor antagonist in patients with rheumatoid arthritis. *Arthritis Rheum.* **34**:545.

76. **Le Clair, K. P., M. A. Blanar, and P. A.**

Sharp. 1992. The p50 subunit of NF-κB associates with the NF-IL-6 transcription factor. *Proc. Natl. Acad. Sci. USA* **89:**8145–8149.

77. **Lee, J. C., A. M. Badger, D. E. Griswold, D. Dunnington, A. Truneh, B. Votta, J. R. White, P. R. Young, and P. E. Bender.** 1993. Bicyclic imidazoles as a novel class of cytokine biosynthesis inhibitors. *Ann. N.Y. Acad. Sci.* **696:** 149–170.

78. **Lee, J. C., D. E. Griswold, B. Votta, and N. Hanna.** 1988. Inhibition of monocyte IL-1 production by the anti-inflammatory compound, SK&F 86002. *Int. J. Immunopharmacol.* **10:** 835–843.

79. **Lee, S. W., A. P. Tsou, and H. Chan.** 1988. Glucocorticoids selectively inhibit the transcription of the IL-1β gene and decrease the stability of IL-1β mRNA. *Proc. Natl. Acad. Sci. USA* **85:** 1204–1208.

80. **Leizer, T., B. J. Clarris, P. E. Ash, J. VanDamme, J. Saklatvala, and J. Hamilton.** 1987. Interleukin-1-beta and interleukin-1-alpha stimulate the plasminogen activator activity and prostaglandin E$_2$ levels of human synovial cells. *Arthritis Rheum.* **30:**562–566.

81. **Lin, L.-L., A. Y. Lin, and D. L. DeWitt.** 1992. Interleukin-1α induces the accumulation of cytosolic phospholipase A$_2$ and the release of prostaglandin E$_2$ in human fibroblasts. *J. Biol. Chem.* **267:**23451–23454.

82. **Livi, G. P., P. Kmetz, M. M. McHale, L. B. Cieslinski, G. M. Sathe, D. P. Taylor, R. L. Davis, T. J. Torphy, and J. M. Balcarek.** 1990. Cloning and expression of cDNA for a human low-K_m rolipram-sensitive cyclic AMP phosphodiesterase. *Mol. Cell. Biol.* **10:**2678–2686.

83. **MacDonald, B. R., and M. Gowen.** 1992. Cytokines and bone. *Br. J. Rheumatol.* **31:**149–155.

84. **Malefyt, R. de W., J. Abrams, B. Bennett, C. G. Figdor, and J. E. de Vries.** 1991. Interleukin 10 (IL-10) inhibits cytokine synthesis by human monocytes: an autoregulatory role of IL-10 produced by monocytes. *J. Exp. Med.* **174:**1209–1220.

85. **Mannik, M., and F. A. Nardella.** 1983. Self-associating IgG rheumatoid factors, p. 124–130. *In* Y. Shiokawa, T. Abe, and Y. Yamauchi (ed.), *New Horizons in Rheumatoid Arthritis.* Excerpta Medica, Amsterdam.

86. **Marder, V. J., S. E. Martin, C. W. Francis, and R. W. Colman.** 1987. Consumptive thrombo-hemorrhagic disorders, p. 975–1015. *In* R. W. Colman, J. Hirsh, V. S. Marder, and E. W. Salzman (ed.), *Hemostasis and Thrombosis: Basic Principles and Clinical Practice,* 2nd ed. J.B. Lippincott, Philadelphia.

87. **Marshall, P. J., D. E. Griswold, J. Breton, E. F. Webb, L. M. Hillegass, H. M. Sarau, J.**

Newton, Jr., J. C. Lee, P. E. Bender, and N. Hanna. 1991. Pharmacology of the pyrroloimidazole, SK&F 105809. I. Inhibition of inflammatory cytokine production and of 5-lipoxygenase- and cyclooxygenase-mediated metabolism of arachidonic acid. *Biochem. Pharmacol.* **42:**813–824.

88. **Masada, M. P., R. Persson, J. S. Kenney, S. W. Lee, R. C. Page, and A. C. Allison.** 1990. Measurement of interleukin-1α and -β in gingival crevicular fluid: implications for the pathogenesis of periodontal disease. *J. Periodontal Res.* **25:**156–163.

89. **Matrisian, L. M.** 1990. Metalloproteinases and their inhibitors in matrix remodeling. *Trends Genet.* **6:**121–125.

90. **Michie, H. R., K. R. Manogue, D. R. Spriggs, A. Revhaug, S. O'Dwyer, C. A. Dinarello, A. Cerami, S. M. Wolff, and D. W. Wilmore.** 1988. Detection of circulating tumor necrosis factor after endotoxin administration. *N. Engl. J. Med.* **318:**1481–1486.

91. **Miller, L. C., E. A. Lynch, S. Isa, J. W. Logan, C. A. Dinarello, and A. Steere.** 1993. Balance of synovial fluid IL-1β and IL-1 receptor antagonist and recovery from Lyme arthritis. *Lancet* **341:**146–148.

92. **Molnar-Kimber, K. L., L. Yonno, R. J. Heaslip, and B. M. Weichman.** 1992. Differential regulation of TNF-α and IL-1β production from endotoxin stimulated human monocytes by phosphodiesterase inhibitors. *Mediators Inflammation* **1:**411–417.

93. **Nawata, Y., E. M. Eugui, S. W. Lee, and A. C. Allison.** 1989. IL-6 is the principal factor produced by synovia of patients with rheumatoid arthritis that induces B-lymphocytes to secrete immunoglobulin. *Ann. N.Y. Acad. Sci.* **557:** 230–239.

94. **Nguyen, T., and C. B. Pickett.** 1992. Regulation of rat glutathione 5-transferase Ya subunit gene expression. DNA-protein interaction at the antioxidant responsive element. *J. Biol. Chem.* **267:**13535–13539.

95. **Nonaka, A., T. Manabe, and T. Tobe.** 1990. Effect of a new synthetic free radical scavenger, 2-octadecyl ascorbic acid, on the mortality in mouse endotoxemia. *Life Sci.* **47:**1933–1939.

96. **O'Banion, M. K., V. D. Winn, and D. A. Young.** 1992. cDNA cloning and functional activity of a glycocorticoid-regulated inflammatory cyclooxygenase. *Proc. Natl. Acad. Sci. USA.* **89:** 4888–4892.

97. **Olson, N. C., J. R. Dodam, and K. T. Kruse-Elliott.** 1992. Endotoxemia and gram-negative bacteremia in swine: chemical mediators and therapeutic considerations. *J. Am. Vet. Med. Assoc.* **200:**1884–1893.

98. **Parsonnet, J., and Z. A. Gillis.** 1988. Production of tumor necrosis factor by human monocytes in response to toxic-shock syndrome toxin-1. *J. Infect. Dis.* **158:**1026–1033.

99. **Peristeris, P., B. D. Clark, S. Gatti, R. Faggiani, A. Mantovani, M. Mengozzi, S. Orencole, M. Sironi, and P. Ghezzi.** 1992. N-acetylcysteine and glutathione as inhibitors of tumor necrosis factor production. *Cell. Immunol.* **140:**390–399.

100. **Pober, J. S., M. P. Bevilacqua, D. L. Neudrido, L. A. Lapiere, W. Fiers, and M. A. Gimbrone.** 1986. Two distinct monokines, interleukin-1 and tumor necrosis factor, each independently induce biosynthesis and transient expression of the same antigen on the surface of cultured human vascular endothelial cells. *J. Immunol.* **136:**1680–1687.

101. **Preiser, J. C., D. Schwarz, J. van der Linden, J. Content, P. V. Bussche, W. Buurman, W. Sebald, E. Dupont, M. R. Pinsky, and J.-L. Vincent.** 1991. Interleukin 6 administration has no acute hemodynamic or hematologic effect in the dog. *Cytokine* **3:**1–4.

102. **Prydz, H., and A. C. Allison.** 1978. Tissue thromboplastin activity of isolated human monocytes. *Thrombosis Hematosis* **39:**582–591.

103. **Reinhardt, R. A., M. P. Masada, W. B. Kaldahl, L. M. DuBois, K. L. Kornman, and A. C. Allison.** 1993. Gingival fluid IL-1 and IL-6 levels in refractory periodontitis. *J. Clin. Periodontol.* **20:**225–231.

104. **Renz, H., J. H. Gong, A. Schmidt, M. Nain, and D. Gemsa.** 1988. Release of tumor necrosis factor α from macrophages; enhancement and suppression are dose-dependently regulated by prostaglandin E_2 and cyclic nucleotides. *J. Immunol.* **141:**2388–2393.

105. **Rodeke, H. H., M. Martin, N. Topley, V. Kaemer, and K. Resch.** 1991. Differential biological activities of human interleukin-1α and interleukin-1β. *Eur. Cytokine Net.* **2:**51–59.

106. **Rose, M. L., and S. D. Semrad.** 1992. Clinical efficacy of tirilazad mesylate for treatment of endotoxemia in neonatal calves. *Am. J. Vet. Res.* **53:**2305–2310.

107. **Schreck, R., P. Rieber, and P. A. Bauerle.** 1991. Reactive oxygen intermediates as apparently widely used messengers in the activation of NFκB transcription factor and HIV-1. *EMBO J.* **10:**2247–2258.

108. **Semmler, J., H. Wachtel, and S. Endres.** 1993. The specific type IV phosphodiesterase inhibitor rolipram suppresses tumor necrosis factor production by human mononuclear cells. *Int. J. Immunopharmacol.* **15:**409–413.

109. **Severn, A., N. T. Rapson, C. A. Hunter, and F. Y. Liew.** 1992. Regulation of tumor necrosis factor production by adrenaline and beta-adrenergic agonists. *J. Immunol.* **148:**3441–3445.

110. **Shakhov, A. N., M. A. Collart, P. Vassali, S. A. Nedospasov, and C. V. Jongeneel.** 1990. Kappa B-type enhancers are involved in lipopolysaccharide-mediated transcriptional activation of the tumor necrosis factor alpha gene in primary macrophages. *J. Exp. Med.* **171:**35–47.

111. **Shalaby, M. R., B. B. Aggarwal, E. Rinderknecht, L. P. Svedersky, and M. A. Palladino.** 1985. Activation of human PMN functions by interferon and tumor necrosis factor. *J. Immunol.* **135:**2069–2073.

112. **Shirakawa, F., K. Saito, C. A. Bonagura, D. L. Galson, M. J. Fenton, A. C. Webb, and P. E. Auron.** 1993. The human pro-interleukin β gene requires DNA sequences both proximal and distal to the transcription start site for tissue-specific induction. *Mol. Cell Biol.* **13:**1332–1344.

113. **Smith, W. L.** 1992. Prostanoid biosynthesis and mechanism of action. *Am J. Physiol.* **263:**F181–F191.

114. **Smith, J. W., II., W. J. Urba, R. G. Steis, et al.** 1990. *Am. Soc. Clin. Oncol.* **9:**717.

115. **Snyder, D. S., and E. R. Unanue.** 1982. Corticosteroids inhibit murine macrophage Ia expression and interleukin-1 production. *J. Immunol.* **129:**1803–1805.

116. **Starnes, H. F., Jr., M. K. Pearce, J. H. Tewari, J. H. Yim, J. C. You, and J. S. Abrams.** 1990. Anti-IL-6 monoclonal antibodies protect against lethal *Escherichia coli* infection and lethal tumor necrosis factor-α challenge in mice. *J. Immunol.* **145:**4185–4191.

117. **Staugas, R. E., D. P. Harvey, A. Ferrante, M. Nandoskar, and A. C. Allison.** 1992. Induction of tumor necrosis factor (TNF) and interleukin-1 (IL-1) by *Pseudomonas aeruginosa* and exotoxin A-induced suppression of lymphoproliferation and TNF, lymphotoxin, gamma interferon and IL-1 production in human leukocytes. *Infect. Immun.* **60:**3162–3168.

118. **Stone, R.** 1994. Search for sepsis drugs goes on despite past failures. *Science* **264:**365–367.

119. **Sung, S. J., J. A. Walters, J. Hudson, and J. M. Gimble.** 1991. Tumor necrosis factor-α mRNA accumulation in human myelomonocytic cell lines. *J. Immunol.* **147:**2047–2054.

120. **Takai, Y., G. G. Wong, S. C. Clark, S. J. Burkoff, and S. H. Herrmann.** 1988. B cell stimulatory factor-2 is involved in the differentiation of cytotoxic T lymphocytes. *J. Immunol.* **140:**508.

121. **Tannenbaum, C. S., and T. A. Hamilton.** 1989. Lipopolysaccharide induced gene expression in murine peritoneal macrophages is selec-

tively suppressed by agents that elevate cAMP. *J. Immunol.* **142:**1274–1280.

122. **te Velde, A. A., R. J. F. Huijbens, K. Heije, J. E. de Vries, and C. G. Figdor.** 1990. Interleukin 4 (IL-4) inhibits secretion of IL-1β, tumor necrosis factor-α and IL-6 by human monocytes. *Blood* **76:**1392–1397.

123. **Tewari, A., W. C. Buhles, Jr., and H. F. Starnes, Jr.** 1990. Preliminary report: effects of interleukin-1 on platelet counts. *Lancet* **336:**712–714.

124. **Thornberry, N. A., H. G. Bull, J. R. Calaycay, K. T. Chapman, A. D. Howard, M. J. Kostura, D. K. Miller, S. M. Molineaux, J. R. Weidner, J. Aunins, K. O. Elliston, J. M. Ayala, F. J. Casano, J. Chin, G. J.-F. Ding, L. A. Egger, E. P. Gaffney, G. Limjuco, O. C. Palyha, S. M. Raju, A. M. Rolando, J. P. Salley, T. T. Yamin, T. D. Lee, J. E. Shively, M. MacCross, R. A. Mumford, J. A. Schmidt, and M. J. Tocci.** 1992. A novel heterodimeric cysteine protease is required for interleukin-1β processing in monocytes. *Nature* (London) **356:**768–774.

125. **Tracey, K. J., S. F. Lowry, T. J. Fahey III, J. D. Albert, Y. Fong, D. Hesse, B. Beutler, K. R. Manogue, S. Calvano, H. Wei, A. Cerami, and G. T. Shires.** 1987. Cachectin/tumor necrosis factor induces lethal shock and stress hormone responses in the dog. *Surg. Gynecol. Obstet.* **164:**415–422.

126. **van Damme, J.** 1994. Interleukin-8 and related chemokines, p. 185–208. *In* A. W. Thompson (ed.), *The Cytokine Handbook,* 2nd ed. Academic Press, London.

127. **Van der Poll, T., H. R. Büller, H. ten Cate, C. H. Wortel, K. A. Bauer, Sander J. H. van Deuenter, C. E. Hack, H. P. Sauerwine, R. D. Rosenberg, and J. W. ten Cate.** 1990. Activation of coagulation after administration of tumor necrosis factor to normal subjects. *N. Engl. J. Med.* **322:**1622–1629.

128. **van der Poll, T., J. A. Romijn, E. Endert, J. J. Borm, H. Büller, and H. P. Sauerwein.** Tumor necrosis factor mimics the metabolic response to acute infection in healthy humans, *J. Clin. Invest,* in press.

129. **Waage, A. P., P. Brandzaeg, A. Haltenoten, P. Kierulf, and T. Espevik.** 1989. The complex pattern of cytokines in serum from patients with meningococcal septic shock; association between interleukin-6, interleukin-1 and fatal outcome. *J. Exp. Med.* **169:**333–338.

130. **Waters, R. V., T. G. Terrell, and G. H. Jones.** 1986. Uveitis induction in the rabbit by muramyl dipeptides. *Infect. Immun.* **51:**816–825.

131. **Waters, R. V., D. Webster, and A. C. Allison.** 1994. Mycophenolic acid and some antioxidants induce differentiation of monocytic lineage cells and augment production of the IL-1 receptor antagonist. *Ann. N.Y. Acad. Sci.* **696:**185–196.

132. **Zhang, Y., B. F. Rames, and B. A. Jakschik.** 1992. Neutrophil recruitment by tumor necrosis factor from mast cells in immune complex peritonitis. *Science* **258:**1957–1959.

PAST, PRESENT, AND FUTURE STUDIES

VI

THE STATE AND FUTURE OF STUDIES
ON BACTERIAL PATHOGENICITY

H. Smith

19

In June 1987, I assessed the state and future of studies on bacterial pathogenicity at the end of a symposium, "Virulence Mechanisms of Bacterial Pathogens," held in Ames, Iowa (104). Seven years later, again in Ames, I repeated the exercise in a second symposium of the same title. Much has happened since the first symposium, and more is likely to occur in the future. The reason for this is that bacterial pathogenicity continues to be one of the most popular and exciting areas of microbiology.

The background for my method of assessment is as follows. Pathogenicity (virulence) is a multifactorial property with five cardinal requirements: infection of mucous surfaces, entry into host tissues through these surfaces, multiplication in the environment in vivo, interference with host defense mechanisms, and damage to the host. The molecular bases of these biological requirements are the determinants of pathogenicity (112). Our research objectives are to recognize each determinant, identify it, and relate its structure to function. The logical steps to achieve the final goal are as follows.

1. Obtain a relevant animal model for virulence determinations and investigations.

2. Select or generate strains of high and low virulence.

3. Recognize determinants by comparing the strains in biological tests related to the five requirements for pathogenicity.

4. Identify the determinant and prove that it causes the selected biological property.

5. Show that the determinant is produced in vivo and contributes to disease (virulence).

6. Obtain the chemical structure of the determinant.

7. Relate the structure to biological action.

The methods and difficulties of taking each of these steps, including the enormous benefit derived from use of genetic manipulation and the importance of examining bacteria grown in vivo for hitherto unknown virulence determinants, have been described elsewhere (118, 119, 122).

To assess the present state of the subject, each of the five requirements for pathogenicity is examined to see how far research has moved up the seven steps of the research ladder toward the goal of identifying the determinants and relating their structure to biological function. A few special topics, such as opportunistic pathogens, are treated in a similar manner. This should reveal popular areas, neglected areas, and areas receiving some attention but needing

H. Smith The Medical School, University of Birmingham, Edgebaston, Birmingham B15 2TT, United Kingdom.

Virulence Mechanisms of Bacterial Pathogens, 2nd ed., Edited by J. A. Roth et al.
© 1995 American Society for Microbiology, Washington, D.C.

more in the future. Examples are taken from the symposium and elsewhere. At the end, I deal with regulation of virulence determinant production, a subject only just emerging at the time of the previous symposium but now one of the most popular in bacterial pathogenicity.

PRESENT STATUS OF RESEARCH ON BACTERIAL PATHOGENICITY

The overall picture concerning present research on bacterial pathogenicity is summarized in Table 1. Each topic is discussed here in turn.

Infection of Mucous Surfaces

There are four aspects of successful infection of mucous surfaces: effective competition with the commensals that abound on some surfaces,

TABLE 1 Present status of research on the five requirements for bacterial pathogenicity and some special areas

Topic	Research step[a]
Infection of mucous surfaces	
Competition with commensals	3
Moving through mucus to epithelial surfaces	4 and 5
Adherence to epithelial cells	6 and 7
Entry into host	
Invasion of epithelial cells	4 and 5
Spread into deeper tissues	3
Multiplication in vivo	3[b]
Interference with host defense mechanisms	
Nonspecific humoral bactericidins and phagocytes	6 and 7
Immune responses: long-term persistence	3
Damage to host	
Toxins	7
Induction of cytokines and inflammation	4 and 5
Immunopathologic reactions	3
Special areas	
Opportunistic pathogens	4 and 5[c]
Change from carrier to invasive state	3
Host and tissue specificities	4 and 5[c]
Mixed infections	4 and 5[c]

[a]The research steps are described in the text. In summary: 3, relate biological observations to virulence; 4, identify determinants concerned; 5, show that determinants operate in vivo; 6, obtain structure; 7, relate structure to function.
[b]Except for usage of iron, which is steps 6 and 7.
[c]Steps 4 and 5 have been attained but in only a few cases.

movement to the surface through the overlying mucus, adherence to the epithelial cells, and interference with host defenses. The first three are considered here. The fourth is dealt with later, but note that bacterial interactions with humoral and cellular defense systems on mucous surfaces do not receive the attention they deserve in view of their importance in initial invasion and in evoking mucosal immunity (see chapters 1, 3, and 17 of this volume).

COMPETITION WITH COMMENSALS

The normal microbiota of the lower bowel, the upper respiratory tract, and the female lower urogenital tract have a strong protective action. The mechanisms involved are nutrient deprivation, occupation of mucosal surface space, and production of inhibitors (114, 137, 141; see chapter 15 of this volume). This protection is overcome by relatively small numbers of shigellae, streptococci, *Escherichia coli,* and other mucosal pathogens. The mechanisms they use are known in general terms. In the bowel, for example, the following appear to be important: successful competition for carbon and energy sources under anaerobic conditions, good adherence to mucosal sites not occupied by the indigenous microorganisms, and a short lag phase (see chapter 15 of this volume). However, the specific bacterial determinants that are responsible for overcoming commensal protection by a particular pathogen, e.g., shigellae in the colon, remain a mystery. Research on the subject is at step 3. Participation of microbial physiologists and development of new experimental approaches are needed to fill this major gap in our knowledge of pathogens.

The microbiota of the mouse gut have been held sufficiently stable in anaerobic continuous culture in vitro for studies of the behavior of an inoculum of streptomycin-resistant *E. coli* (see chapter 15 of this volume). One approach, therefore, is to combine observations of a specific pathogen in an appropriate animal model with those conducted in vitro in such cultures of the relevant microbiota. In addition to using antibiotic-resistant strains of the pathogen to facilitate isolation from mixed cultures, techniques (possibly immunologic) for separating the pathogen directly from such cultures are needed, and animal models for biological and chemical examinations are also needed.

MOVING THROUGH MUCUS TO EPITHELIAL SURFACES

Experiments in vivo and in vitro with *Vibrio cholerae,* salmonellae, and *Campylobacter* spp. indicate that mobility (i.e., possession of flagellae) and chemotaxis are important in mucous penetration (29, 35; see chapter 3 of this volume). Nonchemotactic mutants of *Campylobacter jejuni,* for example, are cleared from the intestines of suckling mice more quickly than the motile wild type (130). Flagellar proteins are virulence determinants, and antibodies to them have been detected in patients in some cases, e.g., in cases of *C. jejuni* infection (139). Specific host nutrients that cause chemotaxis of particular pathogens are known in only isolated cases, e.g., 1-fucose and 1-serine for *C. jejuni* (35, 139). Recently, two additional factors that affect penetration of mucus have been recognized. First, there may be a blocking action either by host receptors for the adhesins of the invading bacteria in the mucus (121) or by agglutinating antibodies (85). Second, ability to grow in mucus may aid penetration. Experiments with *E. coli* (121), *Salmonella typhimurium* (82), and *Yersinia enterocolitica* (79) support this idea; receptors or other blocking agents in the mucus are overwhelmed. Mucinases that have been recognized for *Shigella flexneri* (100), *E. coli* (121), and *Y. enterocolitica* (79) may aid penetration of mucus more by providing nutrients for growth (100) than by reducing the viscosity of mucus. Overall, this subject is entering the determinant stage (steps 4 and 5).

ADHERENCE TO EPITHELIAL CELLS

Convenient biological tests have made the topic of adherence popular for more than twenty years. Most bacterial pathogens have been studied, and for many (e.g., *E. coli, Neisseria gonorrhoeae, Pseudomonas aeruginosa,* and *Bordetella pertussis*), the determinants of adherence (adhesins) have been identified and shown to operate in vivo (6, 9, 35, 39, 47, 113, 124; see chapters 1 and 3 of this volume). Most adhesins are proteins, and some, but not all, are present as fimbriae (pili). Their host cell receptors, mostly carbohydrate but some proteins, have been identified in many cases. Studies are at the structure and function levels (steps 6 and 7).

Adhesins operating in the alimentary and urogenital tracts have received the most attention. The position is well illustrated by the fimbrial adhesins of *E. coli* (35, 47, 56, 124). The most prominent fimbriae are those involved in adherence to gut epithelium (K88, K99, CFA/I and CFA/II) and types 1, P, and S, which appear to be significant in urinary tract infections. They are composed of protein units arranged in helices. Individual structures vary. In general, however, major units form a scaffolding for minor units that are either localized at the tips of the fimbriae or inserted along their length. The regions that determine receptor specificity are on the minor units of type 1, P, and S fimbriae and the major protein units of fimbriae involved in gut adhesion. Amino acid sequences have been determined for both types of units. Also, some host cell receptors are known, e.g., β-D-Gal groups for K88 fimbriae; Neu-5-Glcα(2-3)-Lac structures for K99, D-mannosides for type 1, α-D-GaL-(1-4)-β-D-Gal groups for type P, and α-sialyl-(2-3)-β-Gal residues for type S. The amino acid domains that react with the host receptors have been identified by site-directed mutagenesis. Lysine and arginine at positions 132 and 136, respectively, appear to be important for the K99 protein unit (47, 59), and lysine and arginine at positions 116 and 118, respectively, are important for the Sfa minor unit of S fimbriae (47, 90). Some structure function studies have been conducted on adhesins of respiratory pathogens (see chapter 1 of this volume). Those on *B. pertussis* are perhaps the best examples. Filamentous hemagglutinin and pertussis toxin are the major adhesins that determine attachment of *B. pertussis* to ciliated cells of the human respiratory tract. Their lectin domains (amino acid residues 1141 to 1279 for the former and residues 40 to 54 of subunit S2 of the latter) bind to the same host cell receptor unit, a galactose-*N*-acetyl galactosamine group, and show significant similarity in some amino acid regions (see chapter 1 of this volume).

Chapters 1 and 3 of this volume indicate a recent trend in studies of adhesion to epithelial cells. Some attention is now being directed to biological and biochemical changes occurring in the host cells when they interact with bacteria such as pneumococci and enteropathogenic strains of *E. coli*.

Entry Into Host

Entry into the host has two aspects. First, the pathogen enters epithelial cells, grows within them, and spreads from one cell to another. Second, having either been released from infected epithelial cells or passed between them, the pathogen penetrates beyond the lamina propria to cause invasive disease.

INVASION OF EPITHELIAL CELLS

At the last symposium, the subject of invasion was just entering the determinant stage (steps 4 and 5), spearheaded by studies of *Shigella flexneri*. Since then, there has been an explosion of research (103; see chapters 2 and 3 of this volume). Most studies have combined genetic manipulation with observations on invasion of cell lines. Some, but not all, of the putative determinants identified by these studies (step 4) have been shown to be present in vivo and to be relevant to infection (step 5).

Shigella flexneri infects the surface layers of the colon. A large plasmid is required for virulence in vivo and invasion of cell lines in vitro (43, 105). The latter occurs by active endocytosis involving actin filaments. Three gene products trigger the entry process: IpaB (invasion plasmid antigen B) (62 kDa), IpaC (42 kDa), and IpaD (38 kDa). IpaB, a hemolysin, is the most important. The antigens are surface proteins produced in vivo: patients have antibodies to them. Soon after *Shigella flexneri* enters tissue culture cells, IpaB lyses the engulfment vacuoles. Outside the vacuoles, the bacteria grow rapidly within the cytoplasm. The shigellae move within cells and spread from one to another. This movement is due to the formation of a "tail" of polymerized host actin under the influence of the intracellular spread gene *icsA,* which codes for a 120-kDa protein. Transfer from one cell to another occurs by shigellae entering protrusions in the surface of one cell that are inserted into invaginations in an adjacent cell. The portions containing bacteria are cut off, and the dual membranes are lysed by the product of another plasmid gene, *icsB*. The liberated bacteria engage in further cycles of growth, movement, and transfer. These mechanisms of invasion and the determinants concerned were elucidated by studies of HeLa cells. These cells do not have a brush border as do colonic enterocytes in vivo and a confluent growth of CaCo-2 cells in vitro. Shigellae cannot invade through a brush border. Experiments with CaCo-2 cells showed that invasion could take place only if the sides and bases of the cells were exposed. Experimental pathology of infected primates and ligated intestinal loops in rabbits together with interactions of shigellae with macrophage cell lines in vitro (146) suggests that in vivo epithelial cells are exposed for invasion as follows. Shigellae enter the colonic mucosa through lymphoid follicles corresponding to the Peyer's patches of the small bowel. In the lamina propria, they infect macrophages and cause programmed cell death (apoptosis). Cytokines are liberated and induce inflammation. The influx of phagocytes destroys the basal membrane and disrupts the epithelium. The cells are not now covered by a brush border and hence are vulnerable to plasmid–induced invasion as described above. The determinant of apoptotic death of macrophages is the IpaB protein (146). This work on *Shigella flexneri* underlines the point made about good models in chapter 3 of this volume: it is a paradigm of how cell culture work should be integrated with observations in vivo to reveal what actually happens during infection of mucous surfaces.

The human pathogen *Listeria monocytogenes*

attacks the lining of the gut and produces invasive disease. Its mechanisms of invasion of cultured epithelial cells are similar to those of *Shigella flexneri:* ingestion, escape from the vacuole, rapid growth, polymerization of actin, movement, and transfer to adjacent cells through protrusions. Experiments with CaCo-2 cells (27, 40, 41, 67, 99) indicate the following putative determinants. Internalin, an 800-amino-acid outer membrane protein (OMP) is responsible for entry into cells. Listerolysin O, an excreted hemolysin, causes escape from vacuoles. A 610-amino-acid protein, the product of the *actA* gene, evokes polymerization of actin. These studies in vitro have not been followed by observations in vivo at the same depth as for *Shigella flexneri.*

Y. enterocolitica is a common human pathogen, and *Yersinia pseudotuberculosis,* a rodent pathogen, causes occasional human disease. Both invade through intestinal surfaces, possibly via M cells of Peyer's patches in the ileum (35, 88). Hence, interaction with epithelial cells may not be as important in pathogenesis as countering nonspecific defense mechanisms (17). Nevertheless, invasion of cell lines (HEp-2 and CHO cells) has received much attention (17, 35, 57, 88; see chapter 2 of this volume). Active endocytosis occurs but not escape from the ingestion vacuoles. Growth is therefore slow. The chromosomal genes responsible for invasion, *inv* (invasion) and *ail* (accessory invasion locus), code for two OMPs of 92 and 17 kDa, respectively. The former, called invasin, binds to proteins (integrins) on cell surfaces. Both *Y. enterocolitica* and *Y. pseudotuberculosis* produce invasin, but only the former produces the 17-kDa protein. The possession of functional *inv* and *ail* loci by strains of *Y. enterocolitica* correlates with causation of clinical disease. Also, *inv* mutants of *Y. pseudotuberculosis* infect mice more slowly than wild types do. On the other hand, less invasin is produced at 37°C than at 30°C, and its production by different strains of *Y. enterocolitica* in HEp-2 and CHO cells does not correlate with recovery of strains from mesenteric lymph glands of intragastrically inoculated mice. Hence, the situation with these putative virulence determinants is still uncertain (17, 35). Recently, a plasmid-derived invasion factor of *Y. pseudotuberculosis,* the product of the *yadA* locus, has been demonstrated (143).

Salmonellae may also penetrate the surface layers of the gut via M cells of the Peyer's patches (16, 35; see chapter 3 of this volume). However, observations of intestinal infection of guinea pigs (131) indicate direct invasion of epithelial cells. The invasion begins with degeneration of the brush border near the bacteria followed by engulfment of bacteria into a membrane-bound vacuole and restoration of the brush border. Also, some bacteria pass through intercellular junctions into the lamina propria. Salmonellae invade HeLa, HEp-2, or Henle 407 cells by active endocytosis (16, 34, 35; see chapters 2 and 3 of this volume). They remain in vacuoles, grow slowly, move through the cell, and transcytose through the opposite surface. Ability to invade these cells correlates with virulence in mice. Unfortunately, the genetic basis for the invasion process and the determinants concerned are not yet clear (16, 34; see chapter 2 of this volume). Many genes are involved, and although some may have been characterized (4, 42), their relative importance still has to be resolved.

As for adhesion to epithelial cells, changes in cell biology that accompany bacterial invasion are coming under scrutiny (e.g., the changes in Ca^{2+} concentrations and enzymes resulting from invasion by salmonellae) (see chapters 2 and 3 of this volume).

SPREAD INTO DEEPER TISSUES

Passage of pathogens such as salmonellae between epithelial cells and spread beyond the lamina propria to cause systemic disease has not yet been studied intensively at the determinant level (Table 1). Overcoming the action of macrophages in the lamina propria, regional lymph glands, spleen, and liver must be an important aspect of the process. Indeed, survival within migrating macrophages could be an effective method of spread. It should be possible to derive from fully invasive strains mutants that are checked at the lamina propria or regional lymph glands when examined in vivo. These mutants could be compared with the wild types on sodium dodecyl sulfate-polyacrylamide gels to indicate the determinants concerned. In a study of this type, production of aerobactin by a strain of *E. coli* was associated with deeper penetration into the tissues of gnotobiotic lambs (20). Another aspect of spread is penetration of blood-borne pathogens through the endothelia of blood vessels into tissues such as the brain; recent use of an endothelial cell culture system indicated that lipopolysaccharide (LPS) may be important in such penetration by meningococci (see chapter 3 of this volume).

Multiplication In Vivo

Bacterial multiplication in vivo is essential for pathogenicity, a fact shown by the low virulence of *aro* and *pur* auxotrophs of *S. typhimurium* (16; see chapters 7 and 17 of this volume). Rapid multiplication is an advantage in acute disease: it enables the pathogen to overwhelm the initial defenses and cause damage before the immune system becomes effective. Slow multiplication or no growth at all may be an advantage in chronic disease, because the host may be less aware of the presence of persisting bacteria (98). It is surprising, therefore, in view of its importance, that no universally applicable methods for measuring growth rate in tissues are available. Also, with a few exceptions, the nutrients and environmental factors that determine growth rates for particular pathogens in tissues are a mystery (120). The present position is summarized as follows.

Early attempts to measure doubling times in vivo relied on a genetic marker distributing to only one of the two daughter cells in each generation (81). The method indicated long doubling times (2 to 5 h) for salmonellae and *E. coli* in mice. Later, the increase in ratios of wild-type organisms (which multiply in vivo) to temperature-sensitive mutants (which should not grow in vivo) showed that growth rates of *E. coli* and *P. aeruginosa* were relatively fast (doubling time, 20 to 33 min) (53, 125). More

recently, a temperature-sensitive plasmid, pHSG422, of salmonellae, which in vivo distributes to only one of the two progeny at each division, has been used (46). Unfortunately, these promising studies have not yet been extended. In research on nutrients and metabolism affecting bacterial growth in vivo, only methods of overcoming iron restriction have received adequate attention at the molecular level. Here, the amount of work on many different pathogens is prodigious; the subject progresses constantly and is reviewed frequently (45, 73; see chapter 6 of this volume). At one time, the relation of bacterial nutrition to tissue localization was investigated. For example, urea and erythritol, growth stimulants for *Proteus mirabilis* and brucellae, respectively, contribute to localization in the human kidney in the first case and in the placentas and fetal membranes of domestic animals in the second (120). These promising leads have not been followed up in other cases of tissue localization.

Hence, apart from the continuing work on overcoming iron restriction, multiplication in vivo remains at step 3 of the research ladder as one of the most understudied areas of bacterial pathogenicity. How can the situation be improved? First, microbial physiologists should be encouraged to enter the field of pathogenicity. Second, the use of temperature-sensitive mutants described above should be extended to measure growth rates in vivo of a wide range of pathogens. Third, the effect of potentially important nutrients on pathogens in vivo, e.g., $PO_4^{3/4-}$, Zn^{2+}, Mg^{2+}, amino acids, and other nutrients, could be studied by methods used for iron restriction. Finally, a new approach would be to obtain information on the concentrations of potential nutrients in particular tissues by infecting them with bacteria that contain *lacZ* gene fusions that respond to nutritional parameters (34; see chapter 3 of this volume). Then, the growth rate of the pathogen in vitro could be determined in the presence of the relevant concentration of nutrients to see if it was similar to that in vivo. If necessary, molecular studies could follow.

The possible importance of zero growth rate

in chronic disease and carrier states has been mentioned. Recently, interest in the stationary phase of the bacterial growth cycle has been mounting. Depletion of essential nutrients can cause vegetative *E. coli* to enter a remarkably resistant stationary state able to survive prolonged periods of starvation. Global changes in gene expression occur. These changes and the nutritional and metabolic mechanisms involved are being investigated (12). Similar studies are progressing for *P. aeruginosa* (see chapter 6 of this volume). When more information is available from these in vitro studies, it may be possible to investigate whether pathogens enter the stationary phase in vivo and, if they do, which metabolic mechanisms are involved.

Interference with Host Defense Mechanisms

On mucous surfaces, during the primary lodgment phase of infection (112) and as bacteria spread through the lymph nodes, spleen, and liver, pathogenic bacteria must offset destruction by the nonspecific action of humoral bactericidins and phagocytes. To persist in the host and cause chronic disease or carrier states, the pathogen must either prevent the occurrence of the immune response or (more often) circumvent its action.

INTERFERENCE WITH NONSPECIFIC HUMORAL BACTERICIDINS AND PHAGOCYTES

The determinants of interference with nonspecific mechanisms have received much attention since the term "aggressin" was coined for them by Bail at the turn of the century (28, 32, 35, 66, 94, 104, 115, 140). The number of papers on the subject in this volume (see chapters 1, 4, 7, 8, and 9) shows that it is still a popular area. Numerous determinants have been identified and proved to be relevant in vivo (steps 4 and 5). Many are polysaccharides present in capsules around bacteria, e.g., *Neisseria meningitidis, Klebsiella pneumoniae, Haemophilus influenzae,* and K types of *E. coli.* Some are LPSs, such as those of *E. coli,* salmonellae, *N. gonor-*

rhoeae, and *N. meningitidis.* Others are proteins. The M protein of group A streptococci, the listerolysin O of *L. monocytogenes,* and the macrophage-induced stress proteins of *S. typhimurium* are examples. Still others, such as the capsular polyglutamic acid of *Bacillus anthracis* (112) and the lipoarabinomannan of *Mycobacterium leprae,* are unusual compounds (see chapter 10 of this volume). These determinants use many different strategies to interfere with nonspecific defense mechanisms. The main methods of counteracting the action of complement are as follows. (i) Stop initiation of the cascade by masking bacterial surface components that activate it. (ii) Prevent binding of early complement components such as C3. (iii) Interrupt the cascade by binding to essential intermediates such as C4bp. (iv) Inhibit access of either the opsonic C3b or the lytic C5b-9 complexes to their target sites. Turning to inhibition of phagocytes, some determinants kill them either directly or by inducing programmed (apoptotic) death. Others are less drastic in action. They inhibit chemotaxis of phagocytes, prevent contact with bacteria, or stop ingestion by interfering with opsonization. Intracellular survival is of much current interest (see chapters 7, 8, and 10 of this volume). The relevant pathogens can stop the oxidative burst or withstand its effect, prevent phagolysosome fusion or resist the released enzymes, reduce production of cytokines or resist their effects, or lyse the membrane of phagolysosomes so that the pathogen can escape into the less bactericidal cytoplasm. In some cases, the determinants are known, e.g., listerolysin O for *L. monocytogenes* and the lipoarabinomannan of *M. leprae,* but in other cases, e.g., for *S. typhimurium, Francisella tularensis,* and *Mycobacterium tuberculosis,* they are not clear (see chapters 7, 8, 10, and 14 of this volume).

Many studies are now at the structure-function level (steps 6 and 7). Progress has not, however, been as rapid as that on toxins for two reasons. First, many of the determinants are polysaccharides or LPSs, for which our knowledge of chemical structure and our ability to manipulate are more limited than for proteins (127). Second, knowledge of the chemistry of the biological systems that are counteracted by the determinants, i.e., the complement cascade and ingestion and killing by phagocytes, is not at the same level as that for biological systems affected by toxins, e.g., protein synthesis and membrane structure (2).

Protein determinants have been investigated with little difficulty. For example, a terminal region of the M protein of group A streptococci contains a concentration of acidic amino acids that occupy external positions on the coiled structure. This concentration may prevent phagocytosis by creating electrostatic repulsion between streptococci and phagocytes (36). Studies on carbohydrates and LPSs have been confined to recognition of side chain sugars that are involved in biological activity. A long O side chain and the presence of certain sugars seem important in the LPSs of *E. coli* and salmonellae (see chapters 4 and 11 of this volume). Sialyl groups on capsular polysaccharides of K1 *E. coli*, b and c types of *N. meningitidis*, and group B streptococci may prevent the activation and action of the complement cascade (35, 61, 62).

Also, terminal Galβl-4GlcNAc residues on the side chain of a conserved 4.5-kDa LPS component in gonococci and meningococci are sialylated. This confers resistance to serum killing, resistance to ingestion by phagocytes, and other biological properties related to pathogenicity (31, 48, 78, 123). Structural studies and genetic manipulation of the LPS structure are making progress (52, 64, 91, 101, 136; see chapter 11 of this volume). When this work is related to the results of serum killing and phagocytosis tests, structure-function relationships might emerge.

INTERFERENCE WITH IMMUNE RESPONSES: LONG-TERM PERSISTENCE

One of the most serious problems in human and veterinary medicine is long-term persistence of pathogens in the host. It occurs in chronic diseases, which cause distress to so many people, and in carrier states, which have epidemiologic impact as continuing foci of infection. Suggesting reasons why the usually effective immune defenses fail to eliminate the pathogens is not hard (15, 98, 116). Antigen-presenting cells, B lymphocytes, or T lymphocytes could be inhibited either directly or indirectly by stimulating suppressor cells. Subversion of the immune response could occur by poor antigenicity of virulence determinants, antigenic variation, proteolysis of antibody, sequestration in nonprofessional phagocytes, survival within long-lived macrophages by impairment of their killing mechanisms, and sanctuary in immunologically deficient sites.

There is no doubt that some pathogens can accomplish one or more of these processes. Staphylococcal toxins are potent suppressors of the antibody response to T-cell-dependent antigens (3). The capsular polysaccharides of serogroup B meningococci and of K1, K4, and K5 *E. coli* are poor antigens, probably because of structural similarities to host components (60, 61). The pili of *N. gonorrhoeae* and *E. coli* undergo antigenic variation (103, 124). *N. gonorrhoeae*, *N. meningitidis*, *H. influenzae*, and *Streptococcus pneumoniae* form immunoglobulin A proteases (35). Enteric pathogens invade and grow in epithelial cells, i.e., nonprofessional phagocytes (35, 106). Many bacteria can survive and grow within long-lived macrophages, and the determinants are sometimes known, e.g., the listerolysin O of *L. monocytogenes* (see chapters 7, 8, 10, and 13 of this volume). *Proteus mirabilis* and *E. coli* localize in the kidney, where hypertonicity and ammonium ions inhibit host defenses (49, 117). At the molecular level, antigenic variation has captured much attention recently. For example, the pilin antigen of *N. gonorrhoeae* changes randomly by two pathways, intragenomic recombinational exchange to activate silent genes and transformation by chromosal DNA released from lysed gonococcal cells, and the surface antigens of *Mycoplasma hyorhinis* vary by a process similar to the first pathway (103; see chapter 4 of this volume).

Unfortunately, most of the molecular studies have been conducted on bacterial pathogens that are not usually implicated in serious chronic diseases and carrier states (98). The troublesome pathogens such as *M. tuberculosis* and *Treponema pallidum* get far less attention. Bacteria that cause persistent diseases in domestic animals and can be transmitted to humans (the zoonoses), e.g., *Brucella* spp., *Salmonella* spp., and *Campylobacter* spp., are similarly neglected in this respect. There are, however, exceptions. Antigenic variation of *Borrelia hermsii*, which causes relapsing fever, occurs by a mechanism similar to one used by *N. gonorrhoeae*. Different immunodominant lipoproteins are encoded by at least 27 variable genes carried on linear plasmids. The majority of them are promoterless silent copies that are expressed only after transposition to a single expression site on another linear plasmid (8, 103). Also, the lipoarabinomannan of *M. leprae* appears to be an important determinant of intracellular survival within macrophages (see chapter 10 of this volume).

To sum up, although there is speculation on how important persistent pathogens (see above) deal with the immune response, proof that the suggested mechanisms operate in patients is usually lacking, and there are hardly any investigations at the molecular level. Some examples are as follows. Suppression of the immune response may occur in patients with syphilis but is not proven (98). In patients with leprosy, the cellular immune response is known to be weak but the cellular and molecular mechanisms are not clear (98). The immune response in patients with chronic mycobacterial infections is inefficient, but the reasons are unknown (144). Lipid-rich surface components on mycobacteria, *T. pallidum,* and other spirochetes may be poorly antigenic, but more evidence is needed (98). Antigenic variation of *Borrelia hermsii* (see above) plays a role in persistence in relapsing fever (8), but there is no evidence for this in syphilis and Lyme disease (98). Persistence in epithelial cells of the gallbladder and urinary tract with continuous shedding of the infected cells may be responsible for the carrier state caused by *Salmonella typhi.* Such persistence may also occur in long-

term kidney infections with *Proteus mirabilis* and *Leptospira* spp. Unfortunately, the molecular mechanisms of invasion of the relevant epithelial cells are not known for these organisms. Long-term survival of *M. tuberculosis, Brucella* spp., and salmonellae in macrophages probably occurs in chronic disease (98, 116), but the determinants concerned are unknown.

This subject has a low research rating (Table 1) because insufficient attention is being given to relevant pathogens. The recent upsurge in tuberculosis appears to be focusing more effort in this area, particularly on survival and growth within phagocytes.

Damage to the Host

Ability to damage the host distinguishes pathogenicity from infectivity. Bacteria damage the host in three ways: production of toxins, direct stimulation of cytokines, and induction of immunopathologic reactions.

TOXINS

Studies of bacterial toxins are at the top of the research ladder (step 7) and constitute one of the most exciting areas of all molecular biology. This is apparent from chapters 11, 12, 13, and 14 of this volume. It is impossible in the small space available here to do justice to the advances made since the last symposium. The reader is referred to recent reviews (2, 38, 142) and the relevant chapters in this volume. A few examples are provided to illustrate the position. Cholera toxin and the heat-labile toxin of *E. coli* are two-unit toxins (142). A ring of five identical noncovalently associated B subunits forms the binding moiety that interacts with ganglioside GM_1 on cell surfaces. This binding requires a tryptophan group at position 88 on one B subunit and a glycine group at position 33 on an adjacent B subunit (142). The toxic moiety is the A subunit, which is split proteolytically into parts A1 and A2 when the toxin binds to the cell membrane. The A1 part enters the cell and catalyzes the ADP-ribosylation of arginine 201 on the G50 regulatory component of the adenylate cyclase system. This promotes elevated intracellular levels of cyclic

AMP, which is thought to be responsible for water and electrolyte loss in disease (142). The toxin of *B. pertussis* is another two-unit toxin that catalyzes the ADP-ribosylation of a cysteine residue in the α subunit of a trimeric GTP-binding protein, thereby disrupting the latter's reaction with its receptor, which uncouples signal transduction (see chapter 14 of this volume). Active sites of toxins are usually clefts in the polypeptide chains. Those of diphtheria toxin, *P. aeruginosa* exotoxin A, *E. coli* LT-Ip, *B. pertussis* toxin, and Shiga-like toxin I involve glutamic acid residues at positions 148, 553, 112, 129, and 167 on the polypeptide chains, respectively (2, 26, 134, 142; see chapter 14 of this volume). The two neurotoxins tetanus toxin and botulinum B toxin are zinc-dependent endopeptidases that selectively cleave synaptobrevin, a protein that is present in the membranes of synaptic vesicles and is involved in neurotransmitter release (55). Many toxins cause pore formation in the membranes of erythrocytes and other cells (see chapters 12 and 13 of this volume). The α toxin of *Staphylococcus aureus* binds to the membranes and then diffuses laterally to form hexameric oligomers that produce the pores (1 to 1.5 nm in diameter) by mechanisms that are not completely understood (see chapter 13 of this volume).

INDUCTION OF CYTOKINES AND INFLAMMATION

The subject of cytokines and inflammation has become popular since the last symposium because of increasing knowledge of the cytokines (interleukins [IL] 1, 2, 3, 6, and 8; tumor necrosis factors alpha and beta [TNF-α and TNF-β]; gamma interferon; and colony stimulating factor) and other mediators of the acute-phase response and chronic inflammation (2, 21, 22, 50, 133) (see chapters 9, 11, 16, and 18 of this volume). Interest has been encouraged by the ready availability of methods for measuring these compounds. Their mechanisms of action and interaction are too complex to describe here. One should remember that induction of cytokines increases resistance to infectious dis-

ease, a point stressed in chapters 8 and 16 of this volume. However, this section is concerned with the damage that occurs through overproduction of cytokines. In this respect, it is relevant to record that induction of these mediators can result from direct action of bacterial products on phagocytes, T and B lymphocytes, and other cells as well as from the immune response in classic immunopathologic reactions. Furthermore, the bacterial products concerned, i.e., the determinants, are being recognized, and many have been shown to be produced in vivo (steps 4 and 5).

Some of the products are toxins. The LPS (endotoxin) of gram-negative bacteria was the first to be investigated (21, 22). It is now under intensive reexamination in this respect because of recent advances in knowledge of LPS structure and of the genes and enzymes involved in biosynthesis. LPS-binding proteins on macrophages and the pathways by which the latter are activated to produce cytokines and other mediators of endotoxic shock have been recognized (see chapter 11 of this volume). More recently, superantigen toxins, e.g., staphylococcal enterotoxin and toxic shock syndrome toxin, streptolysin O, listeriolysin O, *P. aeruginosa* hemolysins, and *E. coli* alpha-hemolysin, have been described. These superantigens have profound T-cell mutagenic properties and liberate many cytokines from host cells (3, 50, 68, 108). Mycoplasmas can also produce superantigens with similar properties (see chapter 9 of this volume).

Perhaps more important than the toxins are bacterial cell wall components that stimulate cytokines in opportunistic infections and chronic disease. Schwab (109, 110) considers that chronic inflammatory processes such as arthritis and possibly enteritis can arise from long-term persistence of peptidoglycan-polysaccharide complexes from many bacteria. He used a rat arthritis model and peptidoglycan-polysaccharide complexes from group A streptococci to demonstrate these effects. Other examples are as follows. *Staphylococcus aureus, Staphylococcus epidermidis,* and purified staphylococcal peptidoglycans stimulate TNF production by human monocytes (132) and peptidoglycan from *S. epidermidis* induced arthritis in a mouse model (96). Staphylococcal and streptococcal lipoteichoic acids induce IL-8 production by human monocytes (127). Two outer surface lipoproteins from *Borrelia burgdorferi* stimulate B-cell mitogenic activity, induce cytokine production, and produce arthritis in a mouse model (74): they may cause arthritis in Lyme disease. The cell wall lipoarabinomannan of *M. leprae* is another example; it induces TNF and arthritis (see chapter 10 of this volume). Clearly, this subject is at steps 4 and 5, and work on it is likely to increase in the future because of strong clinical implications that we may be able to prevent either the production or the action of cytokines (see chapters 8, 11, and 18 of this volume).

IMMUNOPATHOLOGIC REACTIONS

Hardly any attention has been given to the subject of immunopathologic reactions since the last symposium. In type II (cytotoxic) immunopathologic reactions, antibody reacts with antigens on host cells, thus priming them for lysis by complement or killing by phagocytes. If bacterial antigens are similar to those on host cells, antibodies to them can be destructive. Heart damage and other rheumatoid sequelae to streptococcal infections may be caused in this way (97, 145). Type III or Arthus-type reactions occur when antigen-antibody complexes are deposited on tissues. Complement is fixed, phagocytes are attracted, and enzymes released from them damage the tissues. These reactions occur in chronic lung infections with *P. aeruginosa* lepromatous leprosy and post-streptococcal glomerular nephritis (51, 97). Type IV or delayed hypersensitivity reactions are harmful manifestations of cell-mediated immunity (138). Macrophages and other cytotoxic cells mobilized to the site of persistent infection release enzymes and cytokines that damage tissue. These reactions are found in patients with tuberculosis and other diseases caused by intracellular bacteria (19, 97). Some immunopathologic reactions can be enhanced by superantigens (see chapter 9 of this volume).

Little progress has been made at the determinant level. The fundamental difficulty is that the reactions are probably caused by several antigens, none of which stand out from the others. Most studies have been of either mixed antibodies arising in clinical cases or animal models inoculated with whole bacteria rather than individual antigens (19, 51, 97, 145). Only a few specific determinants have been investigated. The M protein of streptococci, which evokes type II reactions, shares common epitopes with antigens of cardiac sarcolemma membranes (107) and articular cartilage (7). Ankylosing spondylitis may be caused by antibodies to capsular antigens of *Klebsiella* spp. reacting with the antigens of human HLA B27 lymphocytes (30, 145). Mycobacteria produce many antigens (65, 83) that evoke antibodies and T cells. At present, the antigens that cause lung damage and arthritis are not clear. The position may change, however, in the future, because many genes of *M. tuberculosis* are being cloned. This will allow production of appreciable quantities of pure antigens for biological tests (65, 83). For example, the 65-kDa antigen of *M. tuberculosis* is a heat shock protein similar to those produced by other prokaryotic and eukaryotic cells. In mice, it is immunodominant among other mycobacterial antigens, and in rats, it evokes antibody and T-cell responses and produces arthritis (65). This type of work could raise the subject from step 3 to the determinant stage (steps 4 and 5) of research.

Special Areas

OPPORTUNISTIC PATHOGENS

Bacteria that are either nonpathogenic or weakly pathogenic for healthy people are causing increasing trouble in hospitalized patients who have been debilitated by immunosuppressive drugs (in transplantation surgery and treatment of cancer); broad-spectrum antibiotics (which remove protective commensals); insertion of catheters, heart valves, and other prostheses; and AIDS (11, 25, 58). The most common opportunistic pathogens, e.g., *E. coli*, *Staphylococcus epidermidis*, *Klebsiella* spp., *Enter-*

obacter spp., *Serratia* spp., and *Pseudomonas* spp., are members of the normal human flora that are introduced into new anatomical sites by medical procedures. They flourish because host defenses have been virtually eliminated. Indeed, opportunistic infections may be a fertile field for application of some of the methods of enhancing host resistance to bacteria discussed in chapter 16 of this volume.

The general pattern of pathogenicity studies (see above) cannot be applied to opportunistic pathogens. Animal models that reflect the human situation are not easily developed. Generation and comparison of virulent and less virulent strains, perhaps the most productive tools in studies of pathogenicity, are not possible because of the low inherent pathogenicity of clinical isolates. Research should be concentrated on only two of the five facets of pathogenicity: multiplication in vivo and causation of harm to the host. Survival and penetration of mucous surfaces are not relevant, because the organisms are introduced directly into the tissues. Interference with host defense has a low priority, because defenses are already drastically reduced by the condition of the patient.

As for normal pathogens (see above), multiplication in vivo is not being studied at present. Causation of harm to the host is receiving some attention. Stimulation of cytokines and inflammatory mediators by nontoxic bacterial products may be more important than toxin production. Heat-killed *Staphylococcus epidermidis* induced TNF and IL-1 in rabbits and a shocklike state and tissue injury without endotoxemia (135), and IL-1 antagonists reduced the effects (1). The determinant of cytokine production and damage appeared to be peptidoglycan (96, 132).

Some opportunistic pathogens, e.g., *Staphylococcus epidermidis,* adhere to catheters and other prostheses. Then they produce glycocalices (polysaccharide layers) that inhibit access of antibiotics and also of host defense factors if they are present (18). This aspect of opportunistic infection is open to the approaches used to study normal pathogens in regard to adherence to mucous membranes and formation of

polysaccharide capsules. When adherence tests with catheters were done, the formation of a capsular polysaccharide adhesin by *Staphylococcus epidermidis* was widespread among clinical isolates (93). Transposon-induced mutants deficient in capsular polysaccharide adhesin showed less catheter adhesion than wild types (92).

The studies of *Staphylococcus epidermidis* are at steps 4 and 5 and show what could be accomplished for other opportunistic bacterial pathogens.

CHANGE FROM CARRIER TO INVASIVE STATE

The change from a carrier state to an invasive state is epitomized by meningococcal infection. Many people carry meningococci in the upper respiratory tract. Occasionally, and usually in relatively few people, these meningococci cause disease, sometimes with fatal consequences. Is the conversion from a carrier to an invasive state due to a change in the host or in the pathogen or in both? The phenomenon is clinically important but rare. It is virtually impossible to create a change from a carrier to an invasive condition sufficiently regularly in an animal model for meaningful investigations. The only recourse is to investigate natural outbreaks in depth, which is not easy in emergency situations. It is not surprising that research is at step 3. What can be done in the future?

In regard to change of host, the strength of defense mechanisms (humoral bactericidins, phagocytes, antibodies, cell-mediated immunity) of patients and carriers should be compared. Also, the body fluids (serum, spinal cord fluid) of patients may stimulate growth of the pathogen more than those of carriers do. Turning to changes in the bacteria, organisms from patients and carriers obtained either directly or after minimal subculture should be compared for virulence determinants. A recent study of meningococci sets a pattern. The LPS components of some meningococcal immunotypes, including L3-7-9, possess terminal Galβ1-4Glc NAc- groups that can be sialylated by cytidine 5'-monophospho-N-acetylneuraminic acid (CMP-NANA) either endogenously (serogroups B and C) or exogenously (serogroup A) (78). This sialylation contributes to resistance of meningococci to serum killing and to ingestion by phagocytes (31, 48). The LPS components of other immunotypes, including L1-8-10, lack Galβ1-4Glc NAc-groups and are not sialylated. In an epidemic caused by group B meningococci, 97% of case isolates but only 24% of carrier isolates expressed the L3-7-9 immunotype. In contrast, 70% of carrier isolates but only 13% of case isolates expressed the L1-8-10 immunotype (63). Furthermore, in an infant mouse model, LPS immunotype L3-7-9 was associated with invasive disease, but L1-8-10 was not (75). This work should stimulate molecular studies of changes from carrier to invasive state in outbreaks of disease caused by other pathogens such as staphylococci and pneumococci.

HOST AND TISSUE SPECIFICITIES

Why do some bacteria cause disease in one animal species and not in others? For example, why is gonorrhoea specifically a human disease? Also, why in a single host is there a tendency to infect and damage one tissue rather than another, e.g., for shigellae, the colon but not the small intestine? Host and tissue specificities entail variations in the host, not the bacterial pathogen. Nevertheless, two points should be made in relation to this review. The host variations that determine specificity of infection relate to the five bacterial requirements for pathogenicity. Only in a few cases are the determinants known.

Examples are as follows. Receptors for mucosal adherence and invasion can vary from host to host and tissue to tissue. *N. gonorrhoea* adheres to and penetrates the surfaces of cells from organ cultures of human oviduct but not of rabbit, pig, or cow oviduct (62). *B. pertussis* targets ciliated cells of the human respiratory tract and not the same cells of other species that do not acquire pertussis infection (see chapter 1 of this volume). Nutrients that promote growth in vivo can be present in some tissues

and some hosts but not in others. A growth stimulant for brucellae, erythritol, is present in the fetal placentas and fluids but not in maternal tissues of cattle, sheep, and pigs. This explains the intense localization of brucellae in the fetal tissues of these species and the consequent abortions. Furthermore, lack of erythritol in corresponding fetal tissues of humans, mice, and guinea pigs corresponds to the absence of contagious abortion in brucellosis in these species (112, 120). Host defense mechanisms can differ in different hosts and tissues. Inbred strains of mice differ in resistance or susceptibility to *S. typhimurium,* and this difference is reflected in the abilities of their macrophages to allow intracellular growth (72). One factor contributing to the susceptibility of kidney tissue to attack by pathogens may be the reduction in efficiency of phagocytes due to hypertonicity (49). Finally, hosts and tissues can vary in susceptibility to toxins. Diphtheria toxin is highly toxic for humans but not for rats (128). The toxin of *Pasteurella multocida* has a predilection for the bones of pig nasal turbinates and causes the twisted snout of atrophic rhinitis (33, 37). Although in a few instances this subject is at steps 4 and 5, it needs more attention.

MIXED INFECTIONS

Mixed infections are of growing importance. A few have been investigated at the determinant level. Exacerbation of bacterial infections by viruses, the most serious problem, cannot be discussed here. Bacterial interactions occur in periodontal disease, abdominal abscesses, and opportunistic infections (58, 84, 114).

The following point should be stressed in relation to investigations of mixed infections at the determinant level. Two or more bacterial types that are individually unable to fulfil the five requirements for pathogenicity may complement each other to provide a full armory of determinants. This was first demonstrated in heel abscess in sheep (102), which is caused by a mixture of *Fusobacterium necrophorus* and *Corynebacterium pyogenes.* The low ability of *Fusobacterium necrophorus* to grow in the tissues of the host is enhanced by a growth stimulant derived from *Corynebacterium pyogenes*. The relative inability of the latter to inhibit phagocytic defenses is compensated for by a product from the former that is toxic to phagocytes. Similar examples of compensation between different bacterial types have been demonstrated in animal models related to periodontal disease (114) and abdominal abscesses (13, 95, 105). Hence, the determinants of mixed infection can be investigated successfully. Lack of interest, not lack of opportunity, is the problem.

AREAS REQUIRING ATTENTION IN THE FUTURE

The reason for assessing the present state of research in bacterial pathogenicity is to identify areas that need more attention in the future. In Table 2, the topics that have been discussed are listed under three categories. The popular topics will continue to be attractive, and many of them will be taken higher up the research ladder. The virulence determinants involved are the foci of attention in the design of vaccines (see chapter 17 of this volume). The neglected areas require urgent attention at the determinant level (steps 4 and 5), not least because

TABLE 2 Categorization of areas in relation to need for future attention

Popular areas
    Adherence to epithelial cells
    Invasion of epithelial cells
    Interference with nonspecific humoral bactericidins
      and phagocytes
    Toxins
    Induction of cytokines and inflammation

Neglected areas
    Competition with commensals
    Spread into deeper tissues
    Multiplication in vivo
    Immunopathologic reactions
    Change from carrier to invasive state

Areas needing more attention
    Moving through mucus to epithelial surfaces
    Interference with immune responses: long-term
      persistence
    Opportunistic infections
    Host and tissue specificities
    Mixed infections

most of them have important implications in public health and epidemiology. Applying this criterion to the topics in the third category, those requiring more attention than at present, bacterial persistence in the face of the immune response and opportunistic pathogens are perhaps the most important areas.

REGULATION OF PRODUCTION OF VIRULENCE DETERMINANTS: EFFECT OF THE ENVIRONMENT IN VIVO

Studies of the regulation of virulence determinant production in vitro and the influence on it of environmental factors are extremely popular at present (54; see chapter 5 of this volume). This interest is an offshoot of demonstrations in other areas of bacteriology that organisms have global regulatory networks by which a given environmental signal (temperature, osmotic pressure, oxygen status, pH, or nutrient availability) causes coordinate induction or repression of diverse and unlinked genes (44). These studies have encouraged speculation on environmental control of production of virulence determinants in vivo. The current studies in vitro are reviewed, and their influence on the situation in vivo is discussed.

Regulation of Virulence Determinant Production In Vitro

Investigations of pathogens soon showed that production of different virulence determinants is often regulated by a common mechanism. The ToxR gene system of *V. cholerae*, for instance, regulates formation of A and B enterotoxin components, adhesins, and some OMPs. It is modulated by osmolarity, pH, temperature, oxygen status, and availability of amino acids (87).

Global regulation of gene expression involves a hierarchical network of regulons, i.e., groups of genes under the control of a common regulator (24). Regulators are usually specific DNA-binding proteins that recognize the control regions of their subservient genes. "Ground-level" regulons are controlled by other regulons higher in the network and so on, with DNA supercoiling and DNA-associated proteins influencing some prominent members of the hierarchy. The regulators of many regulons contain two components. The environmental parameter is sensed by one protein, and the other protein has the regulatory function. In the histidine protein kinase response regulator family, information from the first protein is passed to the second by a phosphorylation reaction. Cross talking between regulons can occur at the phosphorylation stage; i.e., sensor molecules from one signal transduction pathway modify response regulons from another. Members of this family, e.g., the EnvZ-OmpR and PhoP-PhoQ systems that respond to osmotic and phosphate signals, respectively, regulate many cellular processes of both pathogens and nonpathogens (24). The cyclic AMP receptor protein system and the various heat shock regulons also control many genes with diverse functions.

In addition to these regulons with wide functions, pathogens sometimes contain other regulons that are concerned primarily with producing virulence determinants. The ToxR system is a regulatory cascade in which some genes are activated by ToxR, a 32-kDa protein, while others need participation of an intermediate regulator, ToxT (23). The *bvg* system is a virulence regulon formed by *B. pertussis*. It regulates production of adenylate cyclase, filamentous hemagglutinin, toxin, and hemolysin. Their formation is reduced by lowering the temperature from 37°C and increasing the concentrations of $MgSO_4$ and nicotinamide (86). *Yersinia* spp. have two regulatory networks (17) that work independently. One responds to temperature and regulates nearly all the virulence determinants of yersiniae (enterotoxin, invasins, adhesins, LPS, and *Yersinia* OMPs [Yops]). The other regulon responds to Ca^{2+} and regulates only Yop production (129). *Yersinia* spp. grow well at 37°C provided Ca^{2+} is present, but Yop production is low. Removal of Ca^{2+} restricts growth but encourages Yop production. *Shigella flexneri* has a *virR* regulon that responds to temperature and causes production at 37°C of the virulence

determinants concerned with epithelial-cell invasion (80).

Work on regulatory networks is at the highest level of molecular biology and therefore is exciting and attractive to many people. It appears to me, however, that a note of caution may be needed in relation to studies of bacterial pathogenicity. It is not surprising that virulence is affected by regulons that control general cell processes (e.g., mutations in *S. typhimurium* that turn off the EnvZ-OmpR or PhoP-PhoQ system) and reduce virulence for mice (24). This situation and the fascination of the field may tempt researchers to engage in prolonged investigations of complex systems that control general bacterial metabolism. Perhaps effort on bacterial pathogenicity should be concentrated on those regulons that are most immediately involved with virulence determinants and on the environmental factors in vivo that affect them.

Effect of the Environment In Vivo on Production of Virulence Determinants

GENERAL CONSIDERATIONS

Almost all investigations of pathogenicity are conducted with bacteria grown in vitro. This environment is different from that in the tissues of diseased animals, where bacteria produce their whole armory of virulence determinants. Not only is the in vivo environment complex, but it also changes as infection proceeds (111, 120). When pathogens are moved from animals to laboratory cultures and vice versa, phenotypic change and selection can take place (120). These changes have two implications for studies of pathogenicity. First, putative determinants of pathogenicity indicated by experiments in vitro may not be formed in vivo. Second, one or more of the full armory of virulence determinants that are necessarily produced in vivo for virulence to be manifested may not be formed under arbitrarily chosen conditions of growth in vitro. These implications have been heeded in studies of pathogenicity over the past few years.

Observations of bacteria obtained directly

from patients or animal models have confirmed the presence of putative virulence determinants indicated by in vitro studies and have revealed hitherto unknown determinants (120). In these investigations, the multiple influences of the undefined environmental conditions in vivo have been accepted as a whole. Only in two instances have particular host factors been identified and shown to cause the production in vivo of defined virulence determinants. Iron restriction in vivo results in many pathogenic species forming new OMPs that enable them to gain enough iron for growth (45; see chapter 6 of this volume). Host CMP-NANA is used as a substrate by gonococci in vivo to sialylate a Galβ1-4Glc NAc- group on the side chain of a 4.5-kDa LPS component that is present in >90% of gonococci that have been examined. This sialylation confers resistance to both serum killing and opsonophagocytosis as well as affecting other properties connected with pathogenicity (123). The future goal in studies of bacterial pathogenicity should be to identify specific host factors and their corresponding bacterial determinants at different stages of infection and in different anatomical sites.

IMPACT OF CURRENT STUDIES ON VIRULENCE DETERMINANT REGULATION

The environmental influences investigated in studies of regulons in vitro (temperature, osmolarity, oxygen status, pH, ion concentrations, and nutrient availability) apply in vivo and probably differ according to the stage of the infection and its anatomical site. The difficulty is in proving this to be so for a particular influence, a named pathogen and during infection of a relevant host. Apart from iron restriction and CMP-NANA (see above), only the effect of temperature on virulence determinant production is known to operate in vivo, e.g., for *B. pertussis* and *Shigella flexneri*. Even with temperature, there are anomalies. Cholera toxin is formed in patients, yet production in vitro is lower at 37°C than at 30°C (10). Clearly, interaction of environmental influences with regulons in vivo is more complex than that in defined conditions in vitro. Another example is Ca^{2+} regulation of Yop production by yersiniae, which has been demonstrated in vitro but may not occur in vivo. Yops are produced in infected animals, where yersiniae spread and multiply largely extracellularly in high Ca^{2+} concentrations that in vitro are nonpermissive for Yop production (17, 129). To sum up, the current interest in the effect of environmental factors on regulation of virulence determinant production has promoted speculation on what happens in vivo but has not provided much information. How can the gap be filled?

First, information is needed on the existence in vivo of those environmental factors that appear from in vitro studies to be important. Much information on temperature, osmolarity, oxygen status, pH, cations, anions, amino acids, sugars, and other compounds in blood, mucous surface secretions, body fluids, and tissue extracts is already available, for example in the *Geigy Scientific Tables* (69–71). If necessary, measurements of host factors by established methods can be made at potential sites of infection. Intracellular concentrations of Na, Mg, P, S, Cl, K, and Ca were measured by X-ray microanalysis (89) in mouse intestinal cells of control and rotavirus-infected animals (126). Fluorescent dyes that react to environmental parameters can be introduced into cells on carriers, and the reactions can be measured by quantitative fluorescence microscopy, e.g., for measurements of pH in phagolysosomes of macrophages (5). Also, *lacZ* gene fusions (reporter genes) that respond to environmental and nutritional influences (Ca^{2+}, pH, Fe, Mg, O_2, glucose, and mannose) have been used in studies with *Y. pestis* and salmonellae (34; see chapter 3 of this volume). In summary, information on environmental parameters in vivo either is already available or can easily be gained.

The second requirement is to ascertain what virulence genes are expressed in vivo that may not be expressed in vitro. As described above, this has been done in the past either by examining organisms obtained directly from infected animals for virulence determinants or by looking for antibodies to them in convalescent-phase sera. However, the in vivo expression technology (IVET) has been described elsewhere (77). *S. typhimurium* was used in the initial work. Segments of DNA from a wild-type *S. typhimurium* were linked to a synthetic operon containing a promoterless *purA* gene and a promoterless *lacZ* operon in positions where the segments could provide promoters for *purA* and *lacZ* expression. The mixed genes were then inserted into a *purA* auxotroph of *S. typhimurium* that, unless complemented, does not grow well in vivo. When injected into mice, some bacteria grew in the spleen, indicating *purA* production under the influence of promoters in genes provided by the segments of *Salmonella* DNA. Bacteria with *Salmonella* genes not expressed in vivo were virtually eliminated. The recovered bacterial population contained members that had genes expressed in vivo and in vitro and others that had genes expressed only in vivo. The difference was revealed by plating on MacConkey lactose agar. The majority of the colonies were red (Lac$^+$ and Pur$^+$), indicating gene expression in vitro as well as in vivo. However, 5% were white (Lac$^-$ and Pur$^-$), indicating that they contained genes expressed only in vivo. Genetic and sequence analysis of 15 of the Lac$^-$ strains identified two previously uncharacterized genes. Mutations in these *ivi* genes reduced virulence for mice, indicating that their products were virulence determinants that have yet to be identified.

Two extensions have widened the scope of the IVET system. In the first, chloramphenicol resistance is substituted for *purA* complementation as the selection system (76). The promoterless *purA* gene is replaced by a promoterless chloramphenicol acetyltransferase (*cat*) gene, and the mice used for selection are dosed with chloramphenicol. The remainder of the procedure is the same. The second extension uses genetic recombination to report gene expression (14). A resolvase is produced from a promoterless copy of the *tnpR* gene of the transposable element γδ under the influence of genes expressed by the pathogen (*V. cholerae*) in vivo. The resolvase excises a tetracycline reporter gene, and hence the bacteria containing the genes that are expressed in vivo become tetracycline susceptible. This susceptibility is detected by appropriate tests of bacteria recovered from the infected animal.

Finally, the mixed environmental parameters identified in vivo should be simulated in vitro, and the production of virulence determinants controlled by various regulons should be studied. Such studies, using some of the mixed parameters found in vivo, may explain anomalies like the production of cholera toxin and Yops in patients (see above). Also, environmental influences possibly acting in vivo, e.g., the effect of contact between *Yersinia* spp. and phagocytes, can be investigated (129). Interesting results can emerge from such studies. The IVET Lac-*cat* system selected clones of *V. cholerae* that expressed fusion genes in the intestines of orally infected mice. However, none of these clones formed the products of ToxR-gene controlled genes when grown under conditions in vitro that activate expression of those genes (76). The genetic recombination system showed that the *irgA* gene of *V. cholerae* (which codes for Fe-regulated OMPS) was expressed in iron-limiting media and the peritoneal cavities of mice. Surprisingly however, expression did not occur when *V. cholerae* grew in the disease-relevant sites, i.e., mouse intestine and rabbit ileal loops (14).

To sum up, current interest in the regulation of production of virulence determinants and the influence of environmental factors on this regulation has been beneficial in evoking speculation on the role such factors play in vivo. However, it has not yet provided much information on the host factors specifically involved in particular bacterial infections. More effort should be directed to identifying the genes expressed in vivo and the host factors concerned by using the new technology that is now emerging.

CONCLUDING REMARKS

The remarks I made at the last symposium still apply. Our subject has made great progress and is in good heart. There is much to do in the future. What more could we wish?

REFERENCES

1. **Aiuro, K., J. A. Gelfand, J. F. Burke, R. C. Thompson, and C. A. Dinarello.** 1993. Interleukin-1 (IL-1) receptor antagonist prevents *Staphylococcus epidermidis*-induced hypertension and reduces circulating levels of tumor necrosis factor and IL-1β in rabbits. *Infect. Immun.* **61:** 3342–3350.

2. **Alouf, J. E., and J. H. Freer.** 1991. *Sourcebook of Bacterial Toxins.* Academic Press, London.

3. **Alouf, J. E., H. Knoll, and V. Kohler.** 1991. The family of mitogenic shock-inducing and superantigenic toxins from staphylococci and streptococci, p. 367–414. *In* J. E. Alouf and J. H. Freer (ed.), *Sourcebook of Bacterial Toxins.* Academic Press, London.

4. **Altmeyer, R. M., J. K. McNern, J. C. Bossio, I. Rosenshine, B. B. Finlay, and J. E. Galan.** 1993. Cloning and molecular characterization of a gene involved in *Salmonella* adherence and invasion of cultured epithelial cells. *Mol. Microbiol.* **7:**89–98.

5. **Aranda, C. M. A., J. A. Swanson, W. P. Loomis, and S. I. Miller.** 1992. *Salmonella typhimurium* activates virulence gene transcription within acidified macrophage phagosomes. *Proc. Natl. Acad. Sci. USA* **89:**10079–10083.

6. **Arp, L. H.** 1988. Bacterial infection of mucosal surfaces: an overview of cellular and molecular mechanisms, p. 3–27. *In* J. A. Roth (ed.), *Virulence Mechanisms of Bacterial Pathogens.* American Society for Microbiology, Washington, D.C.

7. **Baird, R. W., M. S. Bronz, W. Kraus, H. R. Hill, L. G. Veasey, and J. B. Dale.** 1991. Epitopes of group A streptococcal M protein shared with antigens of articular cartilage and synovium. *J. Immunol.* **146:**3132–3137.

8. **Balour, A. G.** 1990. Antigenic variation of a relapsing fever *Borrelia* species. *Annu. Rev. Microbiol.* **44:**155–171.

9. **Beachy, E. H.** 1981. Bacterial adherence: adhesin-receptor interactions mediating the attachment of bacteria to mucosal surfaces. *J. Infect. Dis.* **143:**325–345.

10. **Betley, M. J., V. L. Miller, and J. J. Mekalanos.** 1986. Genetics of bacterial enterotoxins. *Annu. Rev. Microbiol.* **40:**577–605.

11. **Bisno, A. L., and F. A. Waldvogel.** 1989. *Infections Associated with Indwelling Medical Devices.* American Society for Microbiology. Washington, D.C.

12. **Bohannon, D. E., N. Connell, J. Keener, A. Tormo, M. Espinosa-Urgel, M. M. Zambrano, and R. Kolter.** 1991. Stationary-phase-inducible "Gearbox" promoters: differential effects of KatF mutations and role of σ⁷⁰. *J. Bacteriol.* **173:**4482–4492.

13. **Brook, I.** 1986. Encapsulated anaerobic bacteria in synergistic infections. *Microbiol. Rev.* **50:**452–457.

14. **Camilli, A., D. T. Beattie, and J. J. Mekalanos.** 1994. Use of genetic recombination as a reporter of gene expression. *Proc. Natl. Acad. Sci. USA* **91:**2634–2638.

15. **Campa, M.** 1984. Bacterial interference with cell mediated immunity, p. 135–157. *In* G. Falcone et al. (ed.), *Bacterial and Viral Inhibition and Modulation of Host Defense.* Academic Press, London.

16. **Chatfield, S., J. L. Li, M. Sydenham, G. Douce, and G. Dougan.** 1992. Salmonella genetics and vaccine development. *Symp. Soc. Gen. Microbiol.* **49:**299–312.

17. **Cornelis, G. R.** 1992. *Yersiniae,* finely tuned pathogens. *Symp. Soc. Gen. Microbiol.* **49:**231–265.

18. **Costerton, J. W., R. T. Irvin, and K. J. Cheng.** 1981. The bacterial glycocalyyx in nature and disease. *Annu. Rev. Microbiol.* **35:**299–324.

19. **Dannenberg, A. M., Jr.** 1989. Immune mechanisms in pathogenesis of pulmonary tuberculosis. *Rev. Infect. Dis.* **11:**S369–S378.

20. **Der Vartanian M., B. Jeffeux, M. Contrepois, M. Chvarot, J. P. Girandeau, Y. Bertin, and C. Martin.** 1992. Role of aerobactin in systemic spread of an opportunistic strain of *Escherichia coli* from the intestinal tract of gnotobiotic lambs. *Infect. Immun.* **60:**2800–2807.

21. **Dinarello, C. A.** 1989. Interleukin I and its biologically related cytokines. *Adv. Immunol.* **44:** 153–205.

22. **Dinarello, C. A., J. G. Cannon, and S. M. Wolff.** 1988. New concepts on the pathogenesis of fever. *Rev. Infect. Dis.* 10:168–189.

23. **DiRita, V. J., C. Parsot, G. Jander, and J. J. Mekalanos.** 1991. Regulatory cascade controls virulence in *Vibrio cholerae. Proc. Natl. Acad. Sci. USA* **88:**5403–5407.

24. **Dormaan, C. J., and N. N. Bhriain.** 1992. Global regulation of gene expression during environmental adaptation: implications for bacterial pathogens. *Symp. Soc. Gen. Microbiol.* **49:**193–230.

25. **Dougherty, S. H.** 1988. Pathology of infection in prosthetic devices. *Rev. Infect. Dis.* **10:**1102–1117.

26. **Douglas, C. M., and R. J. Collier.** 1987. Exotoxin A of *Pseudomonas aeruginosa.* Substitution of

glutamic acid 553 with aspartic acid drastically reduces toxicity and enzyme activity. *J. Bacteriol.* **169:**4967–4971.

27. **Draamsi, S., C. C. Knocks, C. Forestier, and P. Cossart.** Internalin-mediated invasion of epithelial cells by *Listeria monocytogenes* is regulated by the bacterial growth state, temperature and the pleotropic activator *prfA. Mol. Microbiol.* **9:**931–941.

28. **Dubos, R. J., and J. C. Hirsch.** 1965. *Bacterial and Mycotic Infections of Man,* 4th ed. J. B. Lippincott, Philadelphia.

29. **Easton, K. A., D. R. Morgan, and S. Krakowka.** 1989. *Campylobacter pylori:* virulence factors in gnotobiotic pigs. *Infect. Immun.* **57:**1119–1125.

30. **Ebringer, R. W.** 1980. HLA-B27 and the link with rheumatic diseases: recent developments. *Clin. Sci.* **59:**405–410.

31. **Estabrook, M. M., N. C. Christopher, J. M. Griffiss, C. J. Baker, and R. E. Mandrell.** 1992. Sialylation and human neutrophil killing of group C *Neisseria meningitidis. J. Infect. Dis.* **166:**1079–1088.

32. **Falcone, G., M. Campa, H. Smith, and G. M. Scott.** 1984. *Bacterial and Viral Inhibition and Modulation of Host Defense.* Academic Press, London.

33. **Felix, R., H. Fleisch, and P. C. Frandlsen.** 1992. Effect of *Pasteurella multocida* toxin on bone resorption in vitro. *Infect. Immun.* **60:**4984–4988.

34. **Finlay, B. B.** 1992. Molecular genetic approaches to understanding bacterial pathogenesis. *Symp. Soc. Gen. Microbiol.* **49:**33–45.

35. **Finlay, B. B., and S. Falkow.** 1989. Common themes in microbial pathogenicity. *Microbiol. Rev.* **53:**210–230.

36. **Fischetti, V. A.** 1989. Streptococcal M protein: molecular and biological behavior. *Clin. Microbiol. Rev.* 2:285–314.

37. **Foged, M. T.** 1992. *Pasteurella multocida* toxin. The characterization of the toxin and its significance in diagnosis and prevention of progressive atrophic rhinitis in pigs. *APMIS* **100**(25):1–56.

38. **Freer, J., R. Aitken, J. E. Alouf, G. Boulmois, P. Falmage, F. Febrenbach, G. Montecucco, Y. Piemont, R. Rappuoli, T. Wadstrom, and B. Witholt.** 1994. *Bacterial Protein Toxins.* Gustav Fischer Verlag, Stuttgart.

39. **Freter, R., and G. W. Jones.** 1983. Models for studying the role of bacterial attachment in virulence and pathogenesis. *Rev. Infect. Dis.* **5:**S647-S658.

40. **Gaillard, J. L., P. Berche, C. Fremhel, E. Gouin, and P. Cossart.** 1991. Entry of *L. monocytogenes* into cells is mediated by interlanin, a repeat protein reminiscent of surface antigens from Gram positive cocci. *Cell* **65:**1127–1141.

41. **Gaillard, J. L., P. Berche, J. Mounier, S. Richard, and P. J. Sansonetti.** In vitro model of penetration and intracellular growth of *Listeria monocytogenes* in the human enterocyte-like cell line CaCo-2. *Infect. Immun.* **55:**2822–2829.

42. **Ginocchio, C., J. Pace, and J. E. Galan.** 1992. Identification and molecular characterization of *Salmonella typhimurium* gene involved in triggering internalization of *Salmonella* into cultured epithelial cells. *Proc. Natl. Acad. Sci. USA* **89:**5976–5980.

43. **Goldberg, M. B., and P. J. Sansonetti.** 1993. *Shigella* subversion of the cellular cytoskeleton: a strategy for epithelial colonization. *Infect. Immun.* **61:**4941–4946.

44. **Gottesman, S.** 1984. Bacterial regulation: global regulatory networks. *Annu. Rev. Genet.* **18:**415–441.

45. **Griffiths, E.** 1991. Iron and bacterial virulence—a brief overview. *Biol. Metals* **4:**7–13.

46. **Gulig, P. A., and T. J. Doyle.** 1993. The *Salmonella typhimurium* virulence plasmid increases the growth rate of salmonellae in mice. *Infect. Immun.* **61:**504–511.

47. **Hacker, J.** 1992. Role of fimbrial adhesins in the pathogenesis of *Escherichia coli* infections. *Can. J. Microbiol.* **38:**720–727.

48. **Hammerschmidt, S., O. Ebeling, C. Birkholtz, V. Zahringer, B. D. Robertson, J. Van Putten, and M. Frosch.** 1994. Contribution of genes from the capsule gene complex (cps) to lipooligosaccharide biosynthesis and serum resistance in *Neisseria meningitidis. Mol. Microbiol.* **11:**885–896.

49. **Hampton, M. B., S. T. Chambers, M. O. M. Vissers, and C. C. Winterbaum.** 1994. Bacterial killing by neutrophils in hypertonic environments. *J. Infect. Dis.* **169:**839–846.

50. **Hewitt, C. R. A., J. D. Hayball, J. R. Lamb, and R. E. O'Hehir.** 1992. The superantigenic activity of bacterial toxins. *Symp. Soc. Gen. Microbiol.* **49:**149–172.

51. **Hoiby, M., G. Doring, and P. O. Schiotz.** 1986. The role of immune complexes in pathogenesis of bacterial infections. *Annu. Rev. Microbiol.* **40:**29–53.

52. **Holst, O., and H. Brade.** 1992. Chemical structure of the core region of lipopolysaccharides, p. 135–170. *In* D. C. Morrison and J. L. Ryan (ed.), *Bacterial Endotoxic Lipopolysaccharides,* vol 1., *Molecular Biochemistry and Cellular Biology.* CRC Press, Inc., Boca Raton, Fla..

53. **Hooke, A. M., D. O. Sordelli, M. C. Cerquetti, and A. J. Vogt.** 1985. Quantitative determination of bacterial replication in vivo. *Infect. Immun.* **49:**424–427.

54. **Hormaeche, C. E., C. W. Penn, and C. J. Smyth.** 1992. *Molecular Biology of Bacterial Infection.* Cambridge University Press, Cambridge.

55. **Huttner, W. B.** 1993. Snappy exocytoxins. *Nature* (London) **365:**104–105.

56. **Isaacson, R. E.** 1988. Molecular and genetic basis of adherence for enteric *Escherichia coli* in animals, p. 28–44. *In* J. A. Roth (ed.), *Virulence Mechanisms of Bacterial Pathogens.* American Society for Microbiology, Washington, D.C.

57. **Isberg, R. R., D. L. Voorhis, and S. Falkow.** 1987. Identification of invasin: a protein that allows enteric bacteria to penetrate cultured mammalian cells. *Cell* **50:**769–778.

58. **Isenberg, H. D.** 1988. Pathogenicity and virulence: another view. *Clin. Microbiol. Rev.* **1:** 40–53.

59. **Jacobs, A. A. C., B. H. Simons, and F. K. De Graaf.** The role of lysine-132 and arginine-136 in the receptor binding domain of the K99 fibrillar subunit. *EMBO J.* **6:**1805–1808.

60. **Jann, K., and B. Jann.** 1990. Bacterial capsules. *Curr. Top. Microbiol.* **150.**

61. **Jann, K., and B. Jann.** 1992. Capsules of *Escherichia coli*: expression and biological significance. *Can. J. Microbiol.* **38:**705–710.

62. **Johnson, A. P., D. Taylor-Robinson, and Z. A. McGee.** Species specificity of attachment and damage of oviduct mucosa by *Neisseria gonorrhoeae*. *Infect. Immun.* **18:**833–839.

63. **Jones, D. M., R. Borrow, A. J. Fox, S. Gray, K. A. Cartwright, and J. T. Poolman.** 1992. The lipooligosaccharide immunotype as a virulence determinant in *Neisseria meningitidis. Microb. Pathog.* **13:**219–224.

64. **Kastowsky, M., T. Gutberlet, and H. Bradaczek.** 1992. Molecular modeling of the three-dimensional structure and conformational flexibility of bacterial lipopolysaccharide. *J. Bacteriol.* **174:**4798–4806.

65. **Kaufmann, S. H. E.** 1993. Immunity to intracellular bacteria. *Annu. Rev. Immun.* **11:**129–163.

66. **Kaufmann, S. H. E., and I. E. A. Flesch.** 1992. Life within phagocytic cells. *Symp. Soc. Gen. Microbiol.* **49:**97–106.

67. **Knocks, C., E. Gouin, M. Tabouret, P. Berche, H. Ohayon, and P. Cossart.** 1992. *L. monocytogenes*-induced actin assembly requires the *actA* gene product, a surface protein. *Cell* **68:**521–531.

68. **Konig, W., J. Scheffer, J. Knoller, W. Schonfeld, J. Brom, and M. Koller.** 1991. Effects of bacterial toxins on activity and release of immunomodulators, p. 461–490. *In* J. E. Alouf and J. H. Freer (ed.), *Sourcebook of Bacterial Toxins.* Academic Press, London.

69. **Lentner, C.** 1984. *Geigy Scientific Tables,* vol. 3. *Physical Chemistry, Composition of Blood, Hematology, Sonamtometric Data.* Ciba Geigy Ltd., Basel.

70. **Lentner, C.** 1990. *Geigy Scientific Tables,* vol. 5. *Heart and Circulation.* Ciba Geigy Ltd., Basel.

71. **Lentner, C.** 1991. *Geigy Scientific Tables,* vol. 1. *Units of Measurement, Body Fluids, Composition of the Body, Nutrition.* Ciba Geigy Ltd., Basel.

72. **Lissner, C. R., D. L. Weinstein, and A. D. O'Brien.** 1985. Mouse chromosome 1. *ity* locus regulates microbicidal activity of isolated peritoneal macrophages against a diverse group of intracellular and extracellular bacteria. *J. Immun.* **135:** 544–559.

73. **Litwin, C. M., and S. B. Calderwood.** 1993. Role of iron in regulation of virulence genes. *Chem. Microbiol. Rev.* **6:**137–149.

74. **Ma, Y., and J. J. Weis.** 1993. *Borrelia burgdorferi* outer surface lipoproteins OspA and OspB possess B-cell mitogenic and cytokine stimulating properties. *Infect. Immun.* **61:**3843–3853.

75. **Mackinnon, F. G., R. Borrow, A. R. Gorringe, A. J. Fox, D. M. Jones, and A. Robinson.** 1993. Demonstration of lipooligosaccharide immunotype and capsule as virulence factors for *Neisseria meningitidis* using an infant mouse intranasal infection model. *Microb. Pathog.* **15:**359–366.

76. **Mahan, M. J., J. M. Slauch, P. C. Hanna, A. Camilli, J. W. Tobias, M. K. Waldor, and J. J. Mekalanos.** 1994. Selection for bacterial genes that are specifically induced in host tissues: the hunt for virulence factors. *Infect. Agent. Dis.* **2:** 263–268.

77. **Mahan, M. J., J. M. Slauch, and J. J. Mekalanos.** 1993. Selection of bacterial virulence genes that are specifically induced in host tissues. *Science* **259:**686–688.

78. **Mandrell, R. E., J. M. Griffiss, H. Smith, and J. A. Cole.** 1993. Distribution of a lipooligosaccharide-specific sialyl transferase in pathogenic and nonpathogenic *Neisseria. Microb. Pathog.* **14:**315–327.

79. **Mantle, M., and C. Rombough.** 1993. Growth in and breakdown of purified rabbit small intestinal mucus by *Yersinia enterocolitica. Infect. Immun.* **61:**4131–4138.

80. **Maurelli, A. T., and P. J. Sansonetti.** 1988. Identification of a chromosomal gene controlling temperature-regulated expression of *Shigella* virulence. *Proc. Natl. Acad. Sci. USA* **85:**2820–2824.

81. **Maw, J., and G. G. Meynell.** 1968. The true division and death rates of *Salmonella typhimurium* in the mouse spleen determined with superinfecting phage P_{22}. *Br. J. Exp. Pathol.* **49:**597–613.

82. **McCormick, B. A., B. A. D. Stocker, D. C. Laux, and P. S. Cohen.** 1988. Roles of motility, chemotaxis and penetration through and growth in intestinal mucus in the ability of an avirulent strain of *Salmonella typhimurium* to colonize the large intestine of streptomycin-treated mice. *Infect. Immun.* **56:**2209–2217.

83. **McFadden, J.** 1990. *Molecular Biology of the Mycobacteria.* Surrey University Press, Guildford, United Kingdom.

84. **McGowen, J. F.** 1985. Changing etiology of nosocomial bacteremia and fungemia and other hospital acquired infections. *Rev. Infect. Dis.* **7:**S357-S370.

85. **McSweegan, E., D. H. Burr, and R. I. Walker.** 1987. Intestinal mucus gel and secretory antibody are barriers to *Campylobacter jejuni* adherence to INT407 cells. *Infect. Immun.* **55:**1431-1435.

86. **Melton, A. R., and A. A. Weiss.** 1989. Environmental regulation of expression of virulence determinants in *Bordetella pertussis. J. Bacteriol.* **171:**6206-6212.

87. **Miller, J. F., J. J. Mekalanos, and S. Falkow.** 1989. Coordinate regulation and sensory transduction in the control of bacterial virulence. *Science* **243:**916-922.

88. **Miller, V. L., B. B. Finlay, and S. Falkow.** 1988. Factors essential for the penetration of mammalian cells by *Yersinia. Curr. Top. Microbiol. Immunol.* **138:**15-39.

89. **Morgan, A. J.** 1985. *X-Ray Micro-Analysis: Electron Microscopy for Biologists.* Oxford University Press, Oxford.

90. **Morschhauser, I., H. Hoschutzky, K. Jann, and J. Hacker.** 1990. Functional analysis of the sialic acid-binding adhesin SfaS of pathogenic *Escherichia coli* by site-specific mutagenesis. *Infect. Immun.* **58:**2133-2138.

91. **Moxon, E. R., and D. Maskell.** 1992. *Haemophilus influenzae* lipopolysaccharide: the biochemistry and biology of a virulence factor. *Symp. Soc. Gen. Microbiol.* **49:**75-96.

92. **Muller, E., T. Hubner, N. Gutterrez, S. Takeda, D. A. Goldmann, and G. B. Pie.** 1993. Isolation and characterization of transposon mutants of *Staphylococcus epidermidis* deficient in capsular polysaccharide/adhesin and slime. *Infect. Immun.* **61:**551-558.

93. **Muller, E., S. Takeda, H. Shiro, D. Goldmann, and G. B. Pier.** 1993. Occurrence of capsular polysaccharide/adhesin amongst clinical isolates of coagulase negative staphylococci. *J. Infect. Dis.* **168:**1211-1218.

94. **O'Grady, F., and H. Smith.** 1981. *Microbial Perturbation of Host Defenses.* Academic Press, London.

95. **Onderdork, A. B., R. C. Cisneros, R. Finburg, J. H. Crabb, and D. L. Kasper.** 1990. Animal model system for studying virulence of and host response to *Bacteriodes fragilis. Rev. Infect. Dis.* **12:**Sl69-Sl77.

96. **Onta, J., M. Sashida, N. Fujii, S. Sugawara, H. Rihiishi, and K. Kumagai.** 1993. Induction of acute arthritis in mice by peptidoglycan derived from gram-positive bacteria and its possible role in cytokine production. *Microbiol. Immunol.* **37:**573-582.

97. **Parish, W. E.** 1972. Host damage resulting from hypersensitivity to bacteria. *Symp. Soc. Gen. Microbiol.* **22:**157-192.

98. **Penn, C. W.** 1992. Chronic infections, latency and the carrier state. *Symp. Soc. Gen. Microbiol.* **49:**107-125.

99. **Portnoy, D. A., P. S. Jacks, and D. J. Hinricks.** 1988. Role of haemolysin for intracellular growth of *Listeria monocytogenes. J. Exp. Med.* **167:**1459-1471.

100. **Prizont, R.** 1982. Degradation of intestinal glycoprotein by pathogenic *Shigella flexneri. Infect. Immun.* **36:**615-620.

101. **Rietschel, E. T., U. Seydel, U. Zahringen, U. F. Schade, L. Brade, H. Loppnow, V. Feist, W. H. Wang, A. J. Ulmer, H. D. Flad, K. Bradenburg, T. Kirikae, D. Grimmecke, O. Hoist, and H. Brade.** 1991. Bacterial endotoxin: molecular relationships between structure and activity. *Infect. Dis. Clin. N. Am.* **5:**753-779.

102. **Roberts, D. S.** 1969. Synergic mechanisms in certain mixed infections. *J. Infect. Dis.* **120:**720-724. (Editorial.)

103. **Robertson, B. D., and T. F. Meyer.** 1992. Antigenic variation in bacterial pathogens. *Symp. Soc. Gen. Microbiol.* **49:**61-73.

104. **Roth, J. A. (ed.).** 1988. *Virulence Mechanisms of Bacterial Pathogens.* American Society for Microbiology, Washington, D.C.

105. **Rotstein, O. D., T. L. Pruett, and R. I. Simmons.** 1985. Mechanisms of microbial synergy in polymicrobial surgical infections. *Rev. Infect. Dis.* **7:**151-170.

106. **Sansonetti, P. J.** 1992. Molecular and cellular biology of epithelial invasion by *Shigella flexneri* and other enteroinvasive pathogens. *Symp. Soc. Gen. Microbiol.* **49:**47-60.

107. **Sargent, S. J., E. H. Beachey, C. E. Corbett, and J. B. Dale.** 1987. Sequence of protective epitopes of streptococcal M proteins shared with sarcolemmal membranes. *J. Immunol.* **139:**1285-1290.

108. **Schlievert, P. M.** 1993. Role of super-antigens in human disease. *J. Infect. Dis.* **167:**997-1002.

109. **Schwabe, J. H.** 1983. Bacterial interference with immunospecific defenses. *Phil. Trans. R. Soc. London Ser. B.* **303:**123-135.

110. **Schwabe, J. H.** 1993. Phlogistic properties of peptidoglycan-polysaccharide polymers from cell walls of pathogenic and normal floral bacteria which colonize humans. *Infect. Immun.* **61:**4535-4539.

111. **Smith, H.** 1958. The use of bacteria grown in

vivo for studies on the basis of their pathogenicity. *Annu. Rev. Microbiol.* **12:**77–102.

112. **Smith, H.** 1968. Biochemical challenge of microbial pathogenicity. *Bacteriol. Rev.* **32:**164–184.

113. **Smith, H.** 1977. Microbial surfaces in relation to pathogenicity. *Bacteriol. Rev.* **41:**475–500.

114. **Smith, H.** 1982. The role of microbial interactions in infectious disease. *Phil. Trans. R. Soc. London Ser. B.* **297:**551–561.

115. **Smith, H.** 1983. The elusive determinant of bacterial interference with non-specific host defenses. *Phil. Trans. R. Soc. London Ser. B.* **303:**99–113.

116. **Smith, H.** 1984. Bacterial subversion rather than suppression of immune defenses, p. 171-190. *In* G. Falcone et al. (ed.), *Bacterial and Viral Inhibition and Modulation of Host Defenses.* Academic Press, London.

117. **Smith, H.** 1984. The biochemical challenge of microbial pathogenicity. *J. Appl. Bacteriol.* **57:**395–404.

118. **Smith, H.** 1988. The state and future of studies on bacterial pathogenicity, p. 365–382. *In* J. A. Roth (ed.), *Virulence Mechanisms of Bacterial Pathogens.* American Society for Microbiology, Washington, D.C.

119. **Smith, H.** 1989. The mounting interest in bacterial and viral pathogenicity. *Annu. Rev. Microbiol.* **43:**1–22.

120. **Smith, H.** 1990. Pathogenicity and the microbe *in vivo. J. Gen. Microbiol.* **136:**377–393.

121. **Smith, H.** 1992. Virulence determinants of *Escherichia coli:* present knowledge and questions. *Can. J. Microbiol.* **38:**747–752.

122. **Smith, H.** The revival of interest in mechanisms of bacterial pathogenicity. *Biol. Rev.,* in press.

123. **Smith, H., J. A. Cole, and N. J. Parsons.** 1992. Sialylation of gonococcal lipopolysaccharide by host factors: a major impact on pathogenicity. *FEMS Microbiol. Lett.* **100:**287–292.

124. **Smyth, C. J., and S. G. J. Smith.** 1992. Bacterial fimbriae: variation and regulatory mechanisms. *Symp. Soc. Gen. Microbiol.* **49:**267–298.

125. **Sordelli, D. O., M. C. Cerquetti, and A. K. Hooke.** 1985. Replication rate of *Pseudomonas aeruginosa* in the murine lung. *Infect. Immun.* **50:**388–391.

126. **Spencer, A. J., M. P. Osborne, S. J. Haddon, J. Collins, W. G. Starkey, D. C. A. Candy, and J. Stephen.** 1990. X-ray microanalysis of a rotavirus infected mouse intestine: a new concept of diarrhoeal secretion. *J. Pediatr. Gastroenterol. Nutr.* **10:**516–529.

127. **Standiford, T. J., D. A. Arenburg, J. M. Danforth, S. L. Kunkel, G. M. Van Otteren, and R. M. Strieter.** 1994. Lipoteichoic acid induces secretion of interleukin 8 from human blood monocytes; a cellular and molecular analysis. *Infect. Immun.* **62:**119–125.

128. **Stephen, J., and R. A. Pietrowski.** 1986. *Bacterial Toxins,* 2nd ed. Thomas Nelson and Sons, Ltd., Walton-on-Thames, United Kingdom.

129. **Straley, S. C., G. V. Piano, E. Skrzpek, P. L. Haddix, and K. A. Fields.** 1993. Regulation by Ca^{2+} in Yersinia low-Ca^{2+} response. *Mol. Microbiol.* **8:**1005–1010.

130. **Takata, T., S. Fujimoto, and K. Amako.** 1992. Isolation of non-chemotactic mutants of *Campylobacter jejuni* and their colonization of the mouse intestinal tract. *Infect. Immun.* **60:**3596–3600.

131. **Takeuchi, A.** 1967. Electron microscope studies of experimental Salmonella infection. I. Penetration into the intestinal epithelium by *Salmonella typhimurium. Am. J. Pathol.* **50:**109–136.

132. **Timmerman, C. P., E. Mattsson, L. Martinez-Martinez, L. de Graaf, J. A. G. Van Strijp, H. A. Verbrugh, J. Verhoef, and A. Fleer.** 1993. Induction of release of tumor necrosis factor from human monocytes by staphylococci and staphylococcal peptidoglycans. *Infect. Immun.* **61:**4167–4172.

133. **Tunkel, A. R., and W. M. Scheld.** 1993. Pathogenesis and pathophysiology of bacterial meningitis. *Clin. Microbiol. Rev.* **6:**118–136.

134. **Tweten, R. K., J. T. Barbieri, and R. J. Collier.** 1985. Diphtheria toxin. Effect of substituting aspartic acid for glutamic acid on ATP-ribosyltransferase activity. *J. Biol. Chem.* **260:**10392–10394.

135. **Vakabayashi, G., J. A. Gelfand, W. K. Jung, R. J. Connolly, J. F. Burke, and C. A. Dinarello.** 1991. *Staphylococcus epidermidis* induces complement activation, tumor necrosis factor and interleukin 1, a shock-like state and tissue injury in rabbits without endotoxaemia. *J. Clin. Invest.* **87:**1925–1935.

136. **Valvano, M. A.** 1992. Pathogenicity and molecular genetics of O-specific side chain lipopolysaccharides of *Escherichia coli. Can. J. Microbiol.* **38:**711–719.

137. **Van Der Waaji, D.** 1992. Mechanisms involved in the development of the intestinal microflora in relation to the host organism: consequences for colonization resistance. *Symp. Soc. Gen. Microbiol.* **49:**1–12.

138. **Virella, G., J. M. Goust, and H. H. Fudenberg.** 1990. *Introduction to Medical Immunology,* 2nd ed. Marcel Dekker, Inc., New York.

139. **Walker, R. I., M. B. Caldwell, E. C. Lee, P. Guerry, T. J. Trust, and G. M. Rutz-Palacios.** 1986. Pathophysiology of *Campylobacter enteritis. Microbiol. Rev.* **50:**81–94.

140. **Wilson, G. S., and A. A. Miles.** 1946. *In* W.

W. C. Topley and G. S. Wilson (ed.), *Principles of Bacteriology and Immunity,* 3rd ed. Edward Arnold Ltd., London.

141. **Wilson, K. H., and F. Perini.** 1988. Role of competition for nutrients in suppression of *Clostridium difficile* by colonic microflora. *Infect. Immun.* **56:**2610–2614.

142. **Wren, B. W.** 1992. Bacterial enterotoxin interactions. *Symp. Soc. Gen. Microbiol.* **49:**127–147.

143. **Yang, Y., and R. R. Isberg.** 1993. Cellular internalization in the absence of invasin expression is promoted by *Yersinia pseudotuberculosis yadA* product. *Infect. Immun.* **61:**3907–3913.

144. **Young, D., T. Garbe, R. Lothigra, and C. Abou-Zeid.** 1990. Protein antigens: structure, function and regulation, p. 1–35. *In* I. McFadden (ed.), *Molecular Biology of the Mycobacteria.* Surrey University Press, Guildford, United Kingdom.

145. **Zabriskie, J. B.** 1983. Immunopathological mechanisms in bacterial-host interactions. *Phil. Trans. R. Soc. London Ser. B.* **303:**177–187.

146. **Zychlinsky, A., B. Kenney, R. Menard, M. C. Prevost, I. B. Holland, and P. J. Sansonetti.** 1994. IpaB mediates macrophage apoptosis induced by *Shigella flexneri. Mol. Microbiol.* **11:**619–628.

INDEX